Evolutionary Biology: Concepts and Theories

Evolutionary Biology: Concepts and Theories

Editor: Richard Arber

R CALLISTO REFERENCE

www.callistoreference.com

Callisto Reference,
118-35 Queens Blvd., Suite 400,
Forest Hills, NY 11375, USA

Visit us on the World Wide Web at:
www.callistoreference.com

ISBN: 978-1-63239-924-3 (Hardback)

Cataloging-in-Publication Data

Evolutionary biology : concepts and theories / edited by Richard Arber.
    p. cm.
Includes bibliographical references and index.
ISBN 978-1-63239-924-3
1. Evolution (Biology). 2. Evolutionary developmental biology. I. Arber, Richard.
QH366.2 .E96 2018
576.8--dc23

Table of Contents

Preface

Modern evolutionary biology studies evolution by using technological advancements from fields like computer science and molecular genetics. Comprehensive insights in this field are applied in related areas of interest such as population dynamics, evolutionary developmental biology and genomic phylostratigraphy. Topics in this book comprehensively cover the processes, patterns and mechanisms of evolution. Different approaches, evaluations, methodologies and advanced studies on the dynamics of evolutionary principles have been included in this book. It aims to serve as a resource guide for students and experts alike and contribute to the growth of the discipline.

This book is the end result of constructive efforts and intensive research done by experts in this field. The aim of this book is to enlighten the readers with recent information in this area of research. The information provided in this profound book would serve as a valuable reference to students and researchers in this field.

At the end, I would like to thank all the authors for devoting their precious time and providing their valuable contribution to this book. I would also like to express my gratitude to my fellow colleagues who encouraged me throughout the process.

Editor

Distinct and Diverse: Range-Wide Phylogeography Reveals Ancient Lineages and High Genetic Variation in the Endangered Okapi (*Okapia johnstoni*)

David W. G. Stanton[1]*, John Hart[2], Peter Galbusera[3], Philippe Helsen[3], Jill Shephard[3¤], Noëlle F. Kümpel[4], Jinliang Wang[5], John G. Ewen[5], Michael W. Bruford[1]

1 School of Biosciences, Cardiff University, Cardiff, United Kingdom, 2 Lukuru Foundation, Projet Tshuapa-Lomami-Lualaba (TL2), Kinshasa, Democratic Republic of Congo, 3 Centre for Research and Conservation, Royal Zoological Society of Antwerp, Antwerp, Belgium, 4 Conservation Programmes, Zoological Society of London, London, United Kingdom, 5 Institute of Zoology, Zoological Society of London, London, United Kingdom

Abstract

The okapi is an endangered, evolutionarily distinctive even-toed ungulate classified within the giraffidae family that is endemic to the Democratic Republic of Congo. The okapi is currently under major anthropogenic threat, yet to date nothing is known about its genetic structure and evolutionary history, information important for conservation management given the species' current plight. The distribution of the okapi, being confined to the Congo Basin and yet spanning the Congo River, also makes it an important species for testing general biogeographic hypotheses for Congo Basin fauna, a currently understudied area of research. Here we describe the evolutionary history and genetic structure of okapi, in the context of other African ungulates including the giraffe, and use this information to shed light on the biogeographic history of Congo Basin fauna in general. Using nuclear and mitochondrial DNA sequence analysis of mainly non-invasively collected samples, we show that the okapi is both highly genetically distinct and highly genetically diverse, an unusual combination of genetic traits for an endangered species, and feature a complex evolutionary history. Genetic data are consistent with repeated climatic cycles leading to multiple Plio-Pleistocene refugia in isolated forests in the Congo catchment but also imply historic gene flow across the Congo River.

Editor: Valerio Ketmaier, Institute of Biochemistry and Biology, Germany

Funding: This project was jointly funded by the UK Natural Environment Research Council (NERC) and ZSL (ZSL as NERC CASE industry partner and via an Erasmus Darwin Barlow Expedition grant). The authors also acknowledge the financial support provided by GIC (the Okapi Conservation Project), the US Fish and Wildlife Service, the UK's Darwin Initiative and the Mohamed bin Zayed Species Conservation Fund. The funders had no role in study design, data collection and analysis, decision to publish, or preparation of the manuscript.

Competing Interests: The authors have declared that no competing interests exist.

* Email: stantondw@cardiff.ac.uk

¤ Current address: School of Veterinary and Life Sciences, Murdoch University, Perth, Australia

Introduction

The okapi (*Okapia johnstoni*) is an evolutionarily distinctive even-toed ungulate endemic to the Democratic Republic of Congo (DRC) that has recently been reclassified as 'Endangered' by the IUCN [1]. The okapi also holds iconic status among the Congolese people, appearing on bank notes and as the icon of the Congolese conservation agency (the ICCN; Institut Congolais pour la Conservation de la Nature), and is thus a potentially important conservation flagship and umbrella species for the region. However, the species is under major on-going threat from habitat fragmentation, human encroachment, regional armed conflict and poaching [2]. The okapi was recognised as a member of the Giraffidae family in 1901 [3] and to date has only been the subject of one long-term *in situ* ecological study [4]. No photograph of a live, free-ranging, wild okapi was believed to be in existence until the release of a camera-trap image in 2008 [5]. The enigmatic nature of this species is due to its elusive behaviour, affinity for dense rainforest, and the on-going political instability in the regions of the DRC where it occurs, severely limiting scientific study. One important component in conservation management of endangered species is an understanding of the genetic structure of species and populations. This includes an understanding of the causes of any observed genetic differentiation, such as major geographic and demographic barriers in the ancient and recent past [6]. Virtually nothing is known of the diversity or details of the evolutionary history of the *Okapia* genus, which has almost no fossil record, a likely consequence of the okapi's adaptation to closed-canopy forest where the conditions for fossilisation are poor [7]. Although there is a paucity of phylogeographic studies within the Congo Basin, several studies have been carried out on related taxa, across a broader geographic region within Africa [8–14]. Including these other realated taxa in a comparative phylogeographic approach can help contextualise the history and diversity of each of the taxa, and help asses the wider implications of the findings.

The historic range of the okapi is thought to have included large sections of the central/eastern Congo Basin, although it is likely that they are currently confined to a small fraction of their former distribution [15]. This relatively wide historic range potentially makes them an important model for investigating historical processes governing the biogeography of the fauna of this region, a subject that remains under considerable debate [16–18]. A phylogeographic approach can give insights into gene-flow, divergence times and effective population sizes, which has been done a number of times with widely distributed African species [9,11,12,19–21], but comprehensive investigations within the Congo Basin have been much less common [22,23]. This may be particularly useful in the absence of fossils. The Congo appears to have a profound effect in partitioning faunal diversity. For example, the river is implicated in maintaining one million years of evolutionary divergence between chimpanzees and bonobos [23–26], and is thought to be the most important feature for structuring species diversity of *Praomys* spp. (family: Muridae) in the Congo Basin [27]. Many questions regarding central African biogeography cannot, however, be resolved currently due to a paucity of studies. In particular, there are very few studies investigating the role of the Congo River on *within* species genetic diversity [27]. Okapi are a potential model large mammal to help test competing biogeographic theories, and investigate the role of the Congo River on within species genetic diversity due to the okapis close association with closed-canopy rainforest and relatively wide historic distribution (compared to other studied taxa) across the Congo Basin, including both sides of the Congo River (Fig. 1).

Here we used a comparative phylogeographic approach, utilising mitochondrial and nuclear DNA sequences to provide the first molecular-informed description of the evolution of the okapi, and to investigate biogeographic hypotheses in the Congo Basin.

Methods

Study area and sampling

This study analysed 69 okapi samples, including feces (n = 37), museum specimens (n = 19 preserved skin samples; sampled with permission from the museum of Central Africa, Tervuren, Belgium; museum sample numbers: 12604, 8305, 14235, 14906, 12517-a, 14454, 11043, 1193, 8011, 9726, 9727, 13991, 13242, 14236, 14234, 909, 13336, 15298, 15299) and clippings of dried skin (n = 13) from artefacts found in villages in the DRC (Fig. 1). The sampling methods used in the present study were therefore all non-invasive. Permission for sampling was provided by the Institut Congolais pour la Conservation de la Nature (ICCN; permit numbers: 0996/ICCN/DG/ADG/MG/KBY/2011 and 090/ICCN/ADG/DG/KV/2012). Fecal samples were collected either by, a) walking randomly placed transects through forest sites and collecting any feces observed, or b) by identifying okapi sign and searching the surrounding area for feces. Sampling methodology a) was used in areas of high okapi density (the Okapi Faunal Reserve [RFO; Fig. 1]), and sampling methodology b) was used in areas of low okapi density (everywhere else in the range that fecal samples were found). Skin samples were in the form of clippings taken from skins owned by individuals living in villages in, or near field sites. Museum samples (skin and bone) were sampled with permission from the Royal Museum of Central Africa, Tervuren. These samples were collected between September 1911 and May 1939 and their locations were obtained from information accompanying the samples, usually the name of a town/village, likely representing the closest habitation to where the individuals were hunted. Samples were grouped into one of four broad sampling 'regions' (see Fig. 1) for later analysis and as a descriptive reference.

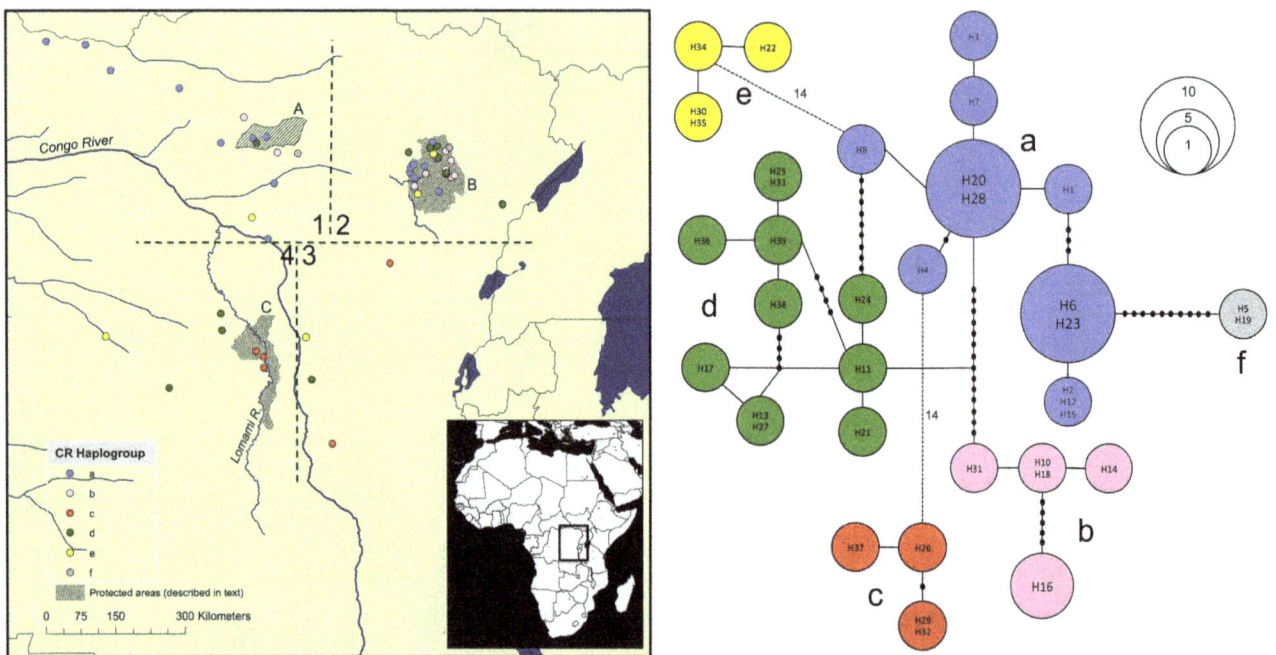

Figure 1. Okapi samples used in the present study, with the colour relating to the adjacent network [30]**, based on 833 bp of mitochondrial DNA.** For the network, TCS connected alleles with a 95% confidence limit, those that did not fall within that limit are connected with dotted lines (with numbers corresponding to the number of mutations). Haplotypes are grouped into haplogroups (number of mutations within always less than between a haplogroup) by colour. Some haplotypes contain more than one label due to different programs using missing data in different ways. Sampling locations are arbitrarily labelled 1–4 for reference in the text. Key protected areas are labelled A (Rubi-Tele Hunting Reserve), B (Okapi Faunal Reserve, RFO), C (Lomami National Park), D (Lomami River).

Molecular methodology

Five pairs of mitochondrial DNA primers were designed using an available okapi sequence (Genbank accession number: NC_020730.1; OJ1-F [15162–15180]: ATGAATCGGAGGA-CAACCA, OJ1-R [15359–15380]: GGCCTCTTCTTTGAGT-CTTAGG, 217 bp; OJ2-F [15359–15380]: CCTAAGACTCAA-AGAAGAGGCC, OJ2-R [15525–15542]: TGCTGCGTTAAG-GCTGTG, 184 bp; OJ3-F [15495–15515]: CCCACAACAAC-CAACACAAAC, OJ3-R [15741–15761]: CGGGATACGCAT-GTTGACRAG, 247 bp; OJ4-F [15645–15665]: ATATGCCC-CATGCATATAAGC, OJ4-R [15885–15905]: CCCTGAAGA-AAGAACCAGATG, 263 bp; OJ5-F: CTACCATGAGGACA-AATATCATT, OJ5-R: CATTCAGGTTTGATATGAGG). Mitochondrial DNA PCR was carried out with a total volume of 25 µl with 4 µg BSA (New England Biolabs, Ipswich, MA, USA), 1× PCR buffer (Invitrogen, Merelbeke, Belgium), 2.5 mM $MgCl_2$, 0.2 mM dNTPs, 0.5 µM each primer, 1 unit of GoTaq (Invitrogen) and 2 µl of DNA. PCR conditions were as follows: 94°C for 3 mins; 60× cycles of 94°C for 30 secs, 58°C for 35 secs, 72°C for 45 secs; and a final extension of 72°C for 5 mins. PCR products were visualised on a 3% agarose gel and sequenced by Macrogen Europe. Of the 69 samples, a subset of 28 were used to produce nuclear intron EPIC [28] sequences. Twelve pairs of nuclear DNA primers were designed (Table S1) from forty-eight primer pairs (selected from the 'best 96' loci from Aitken et al. [28]), tested on DNA extracted from blood samples of two captive individuals (White Oaks Conservation Center, Forida; Studbook numbers: 486 and 578; individuals not used for subsequent analysis), preferentially choosing loci that were reported to amplify a single band in *Bos taurus*. PCR conditions followed Aitken et al. [28]. PCR products were visualised on a 3% agarose gel, and primers that produced a single band (n = 20) were tested in four dried skin samples from wild okapi (two from region one; one from region three; and one from region four). PCR mix and conditions were the same as Aitken et al. [28], except annealing times were increased to 1 min and the number of cycles on the second step of the touchdown PCR was increased to 40. PCR products for all four samples were sequenced (Eurofins MWG Operon, Ebersberg, Germany) in the forward and reverse direction. Primers were redesigned to amplify shorter fragments (~100 bp) for use with non-invasive samples in fragments that contained at least one SNP in the four samples that they were tested in (n = 14). All 14 primer pairs were then tested in 6 non-invasively collected samples (feces and dried skins), and the 12 most consistent primer pairs were selected. These primer pairs were then used for the full set of 28 samples in this study.

Sequence analysis

Sequences were aligned in Sequencher 4.9 [29] and four mitochondrial DNA contig alignments were created. These consisted of 370 bp of the Cyt *b* gene, the tRNA-Pro (66 bp) and tRNA-Thr (69 bp) genes and 328 bp of the CR, as well as a concatenation of all four genes (833 bp). Contigs consisted of shorter fragments than the original PCR product amplified in order to minimize missing data. To visualise the sequence data, a network of the complete mtDNA fragment was drawn in TCS 1.21 [30]. Pairwise and average nucleotide diversities were calculated in DAMBE v4.2.13 [31], as were amino acid translations for Cyt *b* sequences, and haplotype diversities were calculated in DNAsp v5 [32]. The presence of nuclear inserts of mitochondrial DNA (*Numt* [33][33]) was assessed by, i) the presence of a single band on an agarose gel, ii) comparison with known mitochondrial DNA sequence on GenBank, iii) for Cyt *b* sequences: the lack of stop codons in the translated amino acid

sequence and the lack of any markedly distinct amino acid substitutions. In all cases, SNPs were scored manually in Sequencher 4.9 [29] by sequencing individual PCR amplicons in the forward and reverse direction. To characterise population genetic diversity for this set of 12 EPIC loci, we calculated summary statistics in Arlequin v3.5 [34], using the full set of 28 samples.

Okapi mitochondrial CR nucleotide diversity (π) was compared to CR sequence diversity of a number of other African ungulates. The taxa compared were hartebeest (*Alcelaphus buselaphus spp*; six subspecies), bontebok (*Damaliscus pygargus*), giraffe (*Giraffa camelo-pardalis spp*; six subspecies), roan antelope (*Hippotragus equinus spp*; five subspecies), African buffalo (*Syncerus caffer*), common eland (*Taurotragus oryx*) and bushbuck (*Tragelaphus scriptus spp*; 21 subspecies), chosen based on the availability of CR sequences in Genbank (studies and GenBank IDs are given in Table S2). The sequences from all taxa were aligned, including any flanking regions, and the start position of the CR for this complete contig was identified, based on the annotation from GenBank. This was to ensure that a homologous section was being compared between taxa. A ubiquitous section of the complete contig was then separated out into the taxon groups shown in Table 1 and re-aligned. This re-alignment consisted of between 268–275 bp for each taxon, including indels. Indels were included in all π calculations. π and SD for Position 31–306 of the CR was calculated for each of the eight taxon groups described in Table 1, in DAMBE v4.2.13 [31]. The contigs for each of these taxon groups are given in Files S1, S2, S3, S4, S5, S6, S7, S8.

Partitioning of genetic diversity

To partition relative contributions of genetic diversity, AMOVA statistics were calculated using Arlequin v3.5 [34] between the four sampling regions defined in Fig. 1. These regions were designed to compare the extent of genetic structure between individuals at spatial extremes of the okapis range on the same, versus opposite sides of the Congo, in order to investigate the influence of this River on the genetic structure of fauna in the Congo Basin. The sampling regions comprised the Rubi-Tele Hunting Reserve and surrounding areas (region one), the RFO and surrounding areas (region two), the Aruwimi/Lindi/Tshopo (ALT) Rivers and Maiko National Park (region three) and the Tsuapa/Lomami/Lualaba (TL2) Rivers (region four). In order to asses the importance of how these sampling regions were delineated, the AMOVA analyses were repeated, with the line that separates sampling region one and two moved 200 km East, and 200 km West (from the mid-point between the two closest samples between each group). The lines delineating sampling region three could not be moved due to low sample number from this region. The lines delineating sample region four could not be changed as this sampling region is naturally delineated by the Congo River. All 69 mtDNA sequences were used for this AMOVA analysis.

A total of 28 individuals from the four sampling regions (region one, n = 5; region two, n = 14; region three, n = 4; region four, n = 5) were used with the EPIC loci. Of the twelve EPIC loci investigated, four contained greater than one SNP. All SNPs within one sequence were presumed to be linked. Therefore, for analyses using only SNPs (i.e. not the intron sequences), one SNP with high polymorphism was chosen from each of those three intron sequences. AMOVA statistics and F-statistics were calculated on the SNP data using Arlequin v3.5 [34] with the same "groups" and "populations" used for the mitochondrial DNA.

Table 1. Nucleotide diversity in 268–275 bp of homologous CR sequences of African ungulates, sorted on pi.

Taxon	pi	SD	Number of haplotypes
Tragelaphus scriptus spp	0.151295	0.072901	197
Tragelaphus stricptus barkeri	0.092222	0.044905	5
Alcelaphus buselaphus spp.	0.085421	0.041796	92
Tragelaphus stricptus roualeyni	0.075681	0.037075	13
Tragelaphus stricptus massaicus	0.069069	0.033945	18
Alcelaphus buselaphus lichtensteini	0.058252	0.028857	56
Tragelaphus stricptus scriptus	0.056282	0.027889	28
Tragelaphus stricptus sylvaticus	0.053813	0.02672	11
Giraffe camelopardalis spp.	0.051564	0.025654	29
Giraffa camelopardalis rothschildi	0.050059	0.024946	4
Hippotragus equinus spp.	0.046574	0.023565	43
Alcelaphus buselaphus swaynei	0.04647	0.02328	4
Syncerus caffer caffer	0.045189	0.022629	60
OKAPIA JOHNSTONI	0.045015	0.022561	26
Alcelaphus buselaphus lelwel	0.043154	0.021669	5
Alcelaphus buselaphus major	0.041606	0.020936	6
Giraffa camelopardalis giraffa	0.041507	0.020894	4
Tragelaphus stricptus dianae	0.038925	0.019665	4
Giraffa camelopardalis reticulata	0.03833	0.019387	8
Taurotragus oryx	0.037956	0.019322	50
Tragelaphus stricptus knutsoni	0.037463	0.018972	23
Tragelaphus stricptus fasciatus	0.037455	0.018967	5
Tragelaphus strepsiceros strepsiceros	0.035514	0.01807	24
Tragelaphus stricptus heterochrous	0.034188	0.017418	3
Tragelaphus stricptus meruensis	0.033939	0.0173	3
Alcelaphus buselaphus caama	0.033708	0.017214	10
Alcelaphus buselaphus cokei	0.030588	0.015737	11
Giraffa camelopardalis tippelskirchi	0.029038	0.014982	11
Tragelaphus stricptus sassae	0.024374	0.012757	8
Hippotragus equinus cottoni	0.021818	0.011541	3
Tragelaphus stricptus pictus	0.021818	0.011541	3
Tragelaphus stricptus decula	0.01705	0.009268	8
Tragelaphus stricptus meridionalis	0.014545	0.00807	3
Tragelaphus stricptus bor	0.013857	0.00774	5
Tragelaphus stricptus meneliki	0.012165	0.006927	9
Giraffa camelopardalis angolensis	0.011679	0.006697	5
Tragelaphus stricptus cottoni	0.007273	0.004552	7
Tragelaphus stricptus dama	0.007273	0.004552	3
Tragelaphus stricptus johannae	0.007273	0.004552	5
Tragelaphus stricptus phaleratus	0.007273	0.004552	4
Damaliscus pygargus	0.004902	0.003371	3
Giraffa camelopardalis antiquorum	0.004866	0.003356	3

Key taxa are shown in bold. These often correspond to the combined calculations for several species or subspecies.

Population and sequence divergences

To investigate inter- and intra-specific okapi mitochondrial lineage divergences, a time-calibrated phylogeny was created for okapi and giraffe, using 505 bp of homologous mtDNA from GenBank (accession numbers in Table S2 and S3), using BEAST v1.7.5 (Bayesian Evolutionary Analysis by Sampling Trees; [35]),

with red deer (*Cervus elaphus*) as an outgroup (due to the phylogenetic proximity, but distinctiveness of Cervidae to Giraffidae; [36]). An HKY +Gamma model was used, selected by jModelTest v2.1.1 [37,38]. The okapi and giraffe tree was constructed using lognormal relaxed and strict clocks (using a mutation rate of 0.1 s/s/MY, used previously for in the Giraffidae family [10]), and with Yule speciation, coalescent constant, coalescent expansion growth and speciation birth death tree models. The most appropriate model was then selected by comparing the 2*ln Bayes factors for these trees, calculated using TRACER v1.5 with 1000 bootstrap replicates. A value of greater than ten was taken as strong evidence for supporting a model, following Kass and Raftery [39]. The MCMC chain was set at 20,000,000 iterations, with three repeats combined to create the final tree. TRACER v1.5 [40] was used to asses the MCMC output of all BEAST runs. Divergence times (and corresponding standard deviations) between okapi and giraffe, and Giraffidae and Cervidae were taken from Hassanin et al. [36]. Hassanin et al. [36] used an extensive dataset which included complete mitochondrial genome alignments for 210 taxa, utilising six fossil calibration points. To contextualise the results of the Giraffidae phylogeny, a phylogeny was also constructed including okapi, giraffe, duiker and bushbuck jointly, using the same approach as for the Giraffidae phylogeny. Pig (*Sus scrofa*) and collared peccary (*Pecari tajacu*) were selected as outgroups, based on the mtDNA phylogeny of Hassanin et al. [36] giving good support for these species occurring within Cetartiodactyla, but outside Ruminantia. The okapi, giraffe, duiker and bushbuck tree used the 274 bp of Cyt *b* sequence that was overlapping between the present study, and the sequences from Genbank (Table S3), and used the node divergence estimates from Hassanin et al. ([36]; Cetartiodactyla mean 86.8, SD 11.5; Giraffidae mean 15.7, SD 3.3; Pecora mean 28.1, SD 4.5; Bovidae mean 20.0, SD 1.9). The short Cyt *b* sequence for this analysis is therefore due to the limited amount of homologous sequences available for comparison.

PopABC [41] was used to model divergence times between present-day okapi populations, as well as to infer present-day and ancestral effective population sizes, using the EPIC and mitochondrial sequences. For the EPIC loci, haplotypes were reconstructed using Phase v2.1.1 [42]. Pairwise analyses were carried out between sampling regions one versus two, two versus four and one versus four (Fig. 1). The number of iterations used was $1e^{6}$, with the rejection step set at $1e^{-5}$. In order to determine prior ranges, preliminary runs were carried out, starting with very wide priors and altering them until all the posterior distributions were distinct from the priors (priors are given in Table S4).

Results

Primer design

Following Bonferroni correction, none of the nuclear loci were found to be in linkage disequilibrium or to be out of Hardy-Weinberg Equilibrium (HWE). For HWE testing, sampling regions one to four (Fig. 1) were analysed separately. A summary of nuclear SNP variation is given in Table S1.

Sequence analysis

A network (TCS) of mtDNA is shown in Fig. 1. Six distinct haplogroups were recovered (number of pairwise differences was always higher between haplogroups than between any two haplotypes within a haplogroup), and while some were geographically restricted, most were not. Haplogroups a, b and f were restricted to the northeast side of the Congo River, and haplogroup d was much more common southwest of the Congo River (42.9% of all samples from this side) than northeast (4.3% of all samples from this side). Haplogroup c was found in 42.9% and 4.3% of all samples from the southwest and northeast sides of the Congo River respectively. Haplogroup e was found in 14.3% and 9.7% of all samples from the southwest and northeast sides of the Congo River respectively. The general pattern therefore was one

of widespread haplogroups throughout the okapis range, with some spatial patterning relating to the Congo River (Fig. 1).

Haplotype and nucleotide diversities, and number of polymorphic sites for the mtDNA genes investigated in the present study are given in Table S5. A list of sample ID's, GPS coordinates and haplotype information for the mtDNA dataset is also given in Table S6. CR nucleotide diversity was compared to that of a number of other African species (see Table 1). Based on a common 275 bp of CR sequence, the combined bushbuck (*Tragelaphus scriptus* spp.) dataset showed the highest haplotype diversity (0.151), with bushbuck ecotypes showing highly variable nucleotide diversity estimates (0.007–0.092; Table 1). Nucleotide diversity in okapi (0.045) was slightly lower than the combined giraffe dataset (0.052), and very similar to the African buffalo (0.045), and higher than the eland antelope (0.038).

Partitioning of genetic diversity

Partitioning of mitochondrial and nuclear DNA sequence variation was investigated across the geographic range, between all sampling regions, and between sampling regions on the northeast side of the Congo River versus the southwest side. For mtDNA, 15.34% of the molecular variation was explained between sampling regions (p<0.001) and 11.41% of the molecular variation was explained by grouping the samples either side of the Congo River (p = 0.257). For the SNPs, 4.82% of the molecular variation was explained between sampling regions (p = 0.041) and 4.07% of the molecular variation was explained by grouping the samples either side of the Congo River (p = 0.249). F_{ST} values were high and significant for all pairwise comparisons between sampling regions one, two and four for mtDNA, except for the comparison between sampling region one and two, which was low, but significant (Table 2). F_{ST} values were non-significant for all pairwise comparisons for the SNP dataset (data not shown).

To investigate the consequence of the choice of delineation of the sampling regions, we repeated the AMOVA analysis with the line that separates sampling regions one and two moved 200 km East and 200 km West. The results of these changes are given in File S9 and did not notably influence the results.

Population and sequence divergences

To investigate phylogenetic relationships between haplotypes, BEAST [35] was used to clarify phylogenetic relationships and to infer divergence times of the lineages. In all model comparisons, TRACER identified that a relaxed clock was more appropriate than a strict clock (all 2*ln Bayes factors >10). The Bayes factors for pairwise comparisons of the different relaxed clock models were all low, however in every pairwise comparison, the 2*ln Bayes factor for the Yule speciation model was the highest (min 1.0, max 2.0), and was therefore used. A phylogeny was constructed for okapi and giraffe, using 505 bp of homologous mitochondrial DNA (Fig. 2). The phylogeny identified several deep lineages within okapi, including one ancient divergence that divides okapi mtDNA into two groups. BEAST analysis estimated the most ancestral okapi divergence as occurring at 1.7–12.8 (Fig. 2; 95% HPD; mean, 6.83; posterior probability of 0.96) mya. Six of the ten other okapi divergence events were also estimated at greater than one million years old. The giraffe section of the phylogeny (Fig. 2) showed divergence events of a similar magnitude, with the most ancestral divergence estimated at 2.0–12.6 (95% HPD; mean, 6.3; posterior probability of 0.99) mya.

In order to further understand the okapi phylogeny, trees for okapi, giraffe, duiker and bushbuck, were also reconstructed jointly (Fig. 3). This was done in an attempt to address some of the discrepancies that can be encountered when using dated

Table 2. Pairwise F_{ST} values for mitochondrial DNA.

	Sampling region 1	Sampling region 2	Sampling region 3	Sampling region 4
Sampling region 1	0			
Sampling region 2	0.072**	0		
Sampling region 3	0.247**	0.186**	0	
Sampling region 4	0.310***	0.192**	0.021^{NS}	0

***$p<0.001$,
**$p<0.01$,
^{NS} = Not Significant.

phylogenies, such as faulty calibration points [43], rate heterogeneity among lineages [44], and time dependent of rates of evolution [45,46]. The comparative approach addresses these issues by simply providing relative divergence estimates using a single methodology, rather than trying to estimate absolute dates using different methodologies. Despite using only 274 bp of sequence data, the alignment for this phylogeny included 136 variable sites. The phylogeny also contained a large number of nodes supported with a posterior probability of greater than 0.95 (Figure 3). This phylogeny (relaxed lognormal) gave estimates of TMRCA (Time to Most Recent Common Ancestor) for okapi of 2.0–7.9 mya and for giraffe of 2.7–9.3 mya. The topology of the section of the tree containing bushbuck and duiker species was broadly concordant with phylogenies of these species created in previous studies (Moodley and Bruford [11]; Johnston and Anthony [47] respectively). The 95% confidence intervals of the

divergence times of the duiker species in the phylogeny from the present study all overlap with the intervals in Johnston and Anthony [47]. However, the inferred dates of the coalescent events of the Cyt *b* lineages for bushbuck, and the *T. scriptus* and *T. sylvaticus* lineages in this study were considerably higher than Moodley and Bruford [11] (Table 3). This difference is likely due to a combination of, 1) the larger mtDNA fragment investigated in that study, 2) the more comparative approach in the present study, utilising a more inclusive taxon set for the phylogeny, 3) the use of different programs for constructing phylogenies between the present study and that one. Based on this joint phylogeny, the divergence of the two most divergent okapi lineages predates the divergence of several major duiker species, including *C. jentinki* from *C. dorsalis* (Fig. 3, node 16); *C. rufilatus* (node 17); *C. nigrifrons* from *C. harveyi* (node 17); *C. natalensis* (node 17); and *C. spadix* from *C. silvicultor* (node 19). These duiker lineages (nodes 16, 17 and 19)

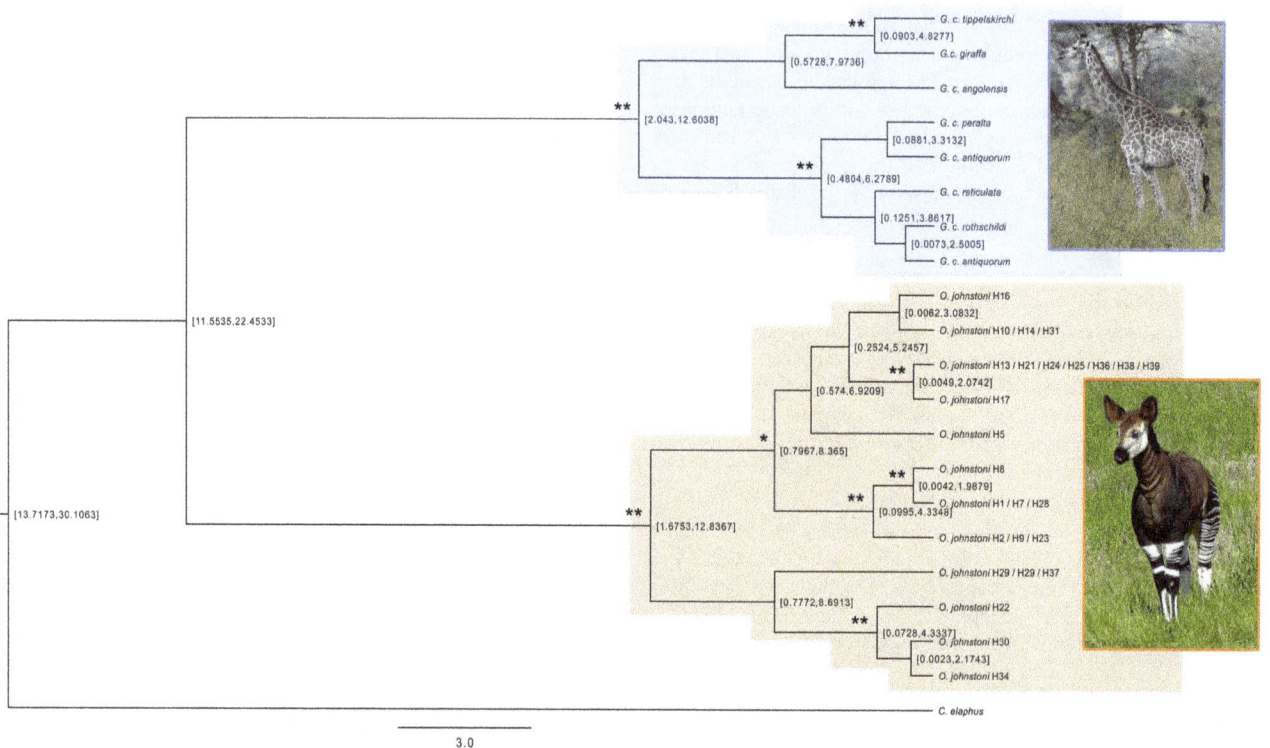

Figure 2. Giraffidae phylogeny drawn in BEAST v1.7.5 [35], with red deer (*Cervus elaphus*) as an outgroup, using 505 bp of mtDNA. Posterior probabilities of >0.8 are highlighted with a single asterisk and posterior probabilities of >0.95 are highlighted with a double-asterisk. Haplotype labels refer to the haplotypes in Fig. 1.

have previously been estimated to have diverged between 1.74–3.54, 1.18–2.38 to 0.80–1.91 mya respectively [47]. The divergence of the okapi lineages also appears to pre-date the emergence of many of today's described bushbuck subspecies, for example *T. sylvaticus sylvaticus* from *T. sylvaticus meneliki* and *T. sylvaticus powelli*, and approximately twice as old as the emergence of both *T. scriptus decula* and *T. sylvaticus ornatus*. The TMRCA for the okapi is similar to that of all the giraffe subspecies.

PopABC [41] was used to infer divergence times, migration rates, and present-day and historic effective population sizes of pairwise combinations of samples from sampling regions one, two and four, using both mitochondrial DNA and nuclear loci (Table 4; for posterior distributions see Files. S10, S11, S12 [A–G]; samples from sampling region three were excluded due to low sample number). Migration rate was inferred to be consistently lower when comparing populations northeast verses southwest of the Congo River compared to the same side, and three of the four inferred migration rates across the Congo were an order of magnitude lower than the two migration rates on the same side. In every instance ancestral effective population size (NeA1) was considerably higher than any of the inferred present-day effective population sizes (Ne1 and Ne2), implying a reduction in population size since these populations became separated. Time since divergence of all the populations was inferred at approximately 200 kya, and interestingly, was the same for all population comparisons. Inferred mutation rates, however, varied substantially among pairwise comparisons, as did the effective population size for region two.

Discussion

Okapi genetic diversity and evolutionary history

Paleontological records of *Okapia spp.* are virtually non-existent, with no known fossils predating the Pleistocene, except *Okapia stillei* (Dietrich [48] in Van der Made and Morales [49]), which has since been reclassified as *Giraffa* [7]. Giraffidae are first known

from the late early Miocene in Africa, and by the Late Miocene giraffids were very widespread and diverse. During the Early Pliocene they became rare in Eurasia, but remained diverse in Africa [7,50–52]. Okapi and giraffe are thought to share a common ancestor approximately 16 mya [36,53]. Based on the okapi and giraffe phylogeny, the present study estimates the most ancient divergence within okapi mitochondrial lineages to be minimally 1.7 mya (divergence of haplotypes H22, H29, H30, H34 and H37 from the remaining haplotypes [Fig. 2]; with maximum sequence divergences of 7.10% and 3.49% for CR and Cyt *b* sequences respectively). This result implies that okapi mitochondrial DNA haplotype divergence dates to at least the early Pleistocene. Sequence divergences of this magnitude are more consistent with divergence dates detected between African species or subspecies (e.g. divergence of *Elephas* and *Loxodonta* elephant genera [54]; the *Phacochoerus africanus massaicus*, *P. a. sundevallii* and *P. a. africanus* warthog subspecies divergences [21]; spotted hyena divergences [55]; and the Scriptus and Sylvaticus bushbuck species divergences [11]), yet there is no suggestion that okapis comprise more than one taxon. This estimate of intraspecific divergence time for okapi is also at the upper limit for what has previously been estimated for the emergence of the extant giraffe subspecies (0.54–1.62 mya [10]).

The present study constructed a phylogeny (Fig. 3), and calculated genetic diversities (Table 1) in a comparative manner, that included multiple ungulate taxa. Based on the combined phylogeny, the divergence of the two most ancestral okapi mitochondrial lineages (divergence that splits haplotypes H22, H29, H30, H34 and H37 from the remaining haplotypes) predates the divergence of several major duiker lineages, which have previously been estimated to have diverged between 0.80–3.54 mya. This gives further support to a divergence of at least 1.7 mya for the most ancestral okapi mtDNA lineage. The divergence of okapi is again estimated to be similar to that of all the giraffe subspecies, as well as the emergence of many of the bushbuck subspecies (e.g. *T. scriptus decula* and *T. s. ornatus*). This is a

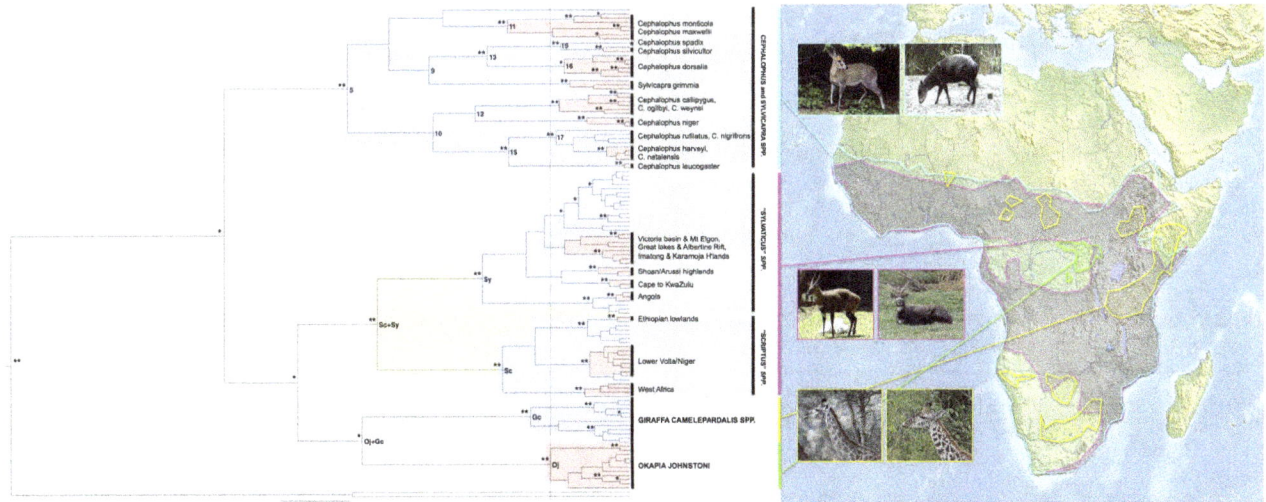

Figure 3. Okapi (*Okapia johnstoni*), giraffe (*Giraffa camelopardalis*), bushbuck (*Tragelaphus scriptus spp.*) and duiker (Cephalophinae *spp.*) tree drawn in BEAST v1.7.5. Posterior probabilities of >0.8 are highlighted with a single asterisk and posterior probabilities of >0.95 are highlighted with a double-asterisk. Dotted line indicates the most ancestral divergence within okapi. The shading on the tree shows when taxonomic units can be monophyletically grouped, with the different colours corresponding to different levels of inclusiveness for these groupings. For example, for bushbuck, Victoria Basin & Mt Elgon, Great Lakes & Albertine Rift and Imatong & Karamoja Highlands ecoregions could be grouped monophyletically, and are shaded red. The next monophyletic taxonomic grouping are the "scriptus" species (shaded blue), and then all bushbuck (shaded yellow).

Table 3. 95% HPD intervals for dates of divergences (mya) for Fig. 3 of the present study, and from the original studies (nodes 5–17, Johnston and Anthony [48]; nodes Sc/Sy, Moodley and Bruford [14]).

Node	Dates (Previous study; mya)	Dates (present study; mya)
5	6.27–11.43	10.47–20.91
9	4.16–7.78	6.60–16.44
10	3.58–6.69	6.50–16.20
11	2.68–5.31	3.71–10.93
12	2.52–4.97	4.53–13.67
13	2.53–4.93	4.38–12.52
15	2.13–4.27	3.16–11.75
16	1.74–3.54	1.51–6.62
17	1.18–2.38	1.87–7.30
19	0.80–1.91	1.35–7.28
Sc	2.0–3.0	4.01–10.97
Sy	2.0–3.0	4.65–12.65
Sc+Sy	3.9–6.5	9.46–19.01

surprising result, particularly when one considers the morphological and geographic variation that is contained within these giraffe, duiker and bushbuck taxa [10,11,47]. The results of the CR nucleotide diversity comparison showed similar results. Okapi nucleotide diversity was similar to the combined giraffe subspecies and African Cape buffalo, and higher than the eland antelope. This comparative methodology provides a much more useful and meaningful means of comparing interspecific genetic diversity than simply stating genetic diversities out of context. Table 1 shows okapi to be one of the more genetically diverse of the ungulate species investigated in this study, implying a rich and diverse evolutionary history.

Evolutionary biogeography of the Congo Basin

The most ancestral mitochondrial DNA divergence in okapi is dated at greater than 1.7 mya (Fig. 2). The Congo River is a likely candidate for the cause of the split of the most ancestral mtDNA sequence lineages in okapi, however, it is not possible to prove this definitively due to the possibility of retention of ancestral polymorphism in populations either side of the Congo River. The mitochondrial DNA network shows six distinct lineages

(Fig. 1), and divergences of several of the other major okapi mtDNA lineages from the BEAST phylogeny are also dated at greater than one million years ago (the divergence of both of the monophyletic clades is *at least* 0.8 mya, Fig. 2). These dates may be explained by the Congo Basin fragmenting into refugia at various stages throughout the Pleistocene. This is consistent with a hypothesis of increases in African climate variability [56] and aridity [57] at approximately 2.8–2.5, 1.9–1.7 and 1.1–0.9 mya. Okapi are however known to be highly selective folivores and currently occupy a disjunct distribution within the Congo Basin. Refugia may therefore have provided isolated regions of suitable forest type, rather than simply comprising patches of forest separated by savannah. Cowling et al. [16] simulated the paleovegetation of Central Africa and LGM simulations indicate that although tropical broadleaf forest may not have been severely displaced by expanding grassland in central Africa, the structure of the forests may have been very different from today (with forests characterized by lower leaf area indices, lower tree heights and lower carbon content).

The inferred Approximate Bayesian relative divergence times and migration rates between the okapi sampling regions were relatively consistent. As would be expected, migration rates of sampling regions one versus two (same side of river) were consistently higher than regions one versus four and two versus four (opposite sides of river). Interestingly however, divergence times between all population comparisons were the same (~200 kya). When taken together, these results imply that although populations on the same side of the Congo River maintained much higher gene-flow since the time of divergence, they nonetheless diverged around the same time to those on opposite sides of the river. This would suggest that initial population divergence between okapi populations either side of the Congo River was primarily linked to the same biogeographic process that separated those populations on the same side of the river. A possible explanation for this could again be forest fragmentation, linked to repeated glaciation events during the Pleistocene [57–59]. This estimate of population divergence is considerably more recent than the estimates of sequence divergence discussed earlier (>1.7 mya). This implies greater than one population fragmentation event, again suggestive of repeated cooling events. These results add to a growing body of evidence that tropical forest refugia can play an important role in driving evolutionary diversification, but that this role has been much more prominent in tropical Africa [47,60–64], than in the Amazon [65–69].

These results are also in accordance with the distribution of mtDNA haplotypes. Deep genetic divergences indicate historic population isolation, but the presence of these haplotypes on each

Table 4. Values with the highest posterior probabilities for the parameters investigated for the popABC analysis, with comparisons between regions one, two and four (R1, R2 and R4).

Parameter	R1vR4	R1vR2	R2vR4
AvMutS	$5e^{-4}$	$<1e^{-5}$	$1.5e^{-2}$
mig1	0.2	4	1
mig2	0.2	2	0.1
NeA1	6000	4500	14000
t1	$2e^{5}$	$2e^{5}$	$2e^{5}$

Parameters investigated are average mutation of the sequence (AvMutS), migration into each population (mig1 and mig2), effective population size of the ancestral population (NeA1) and time of population splitting (t1). Effective population size of each population (Ne1 and Ne2) were omitted from the table due to lack of convergence.

side of the Congo River suggests relatively recent gene-flow. Bonobos and chimpanzees provide a particularly interesting comparison to okapi, as their combined range spans the Congo River. Bonobos and chimpanzees are estimated to have diverged ~1 mya [23–26], with chimpanzees restricted to the northeast side of the Congo River, whereas bonobos are restricted to the southwest side. The diversification of chimpanzee sub-species, and bonobo haplogroups are explained by fluctuations in climate during the Pleistocene and the associated changes in forest cover [23,70]. Taken together, these studies suggest periods of Pleistocene forest expansion that genetically differentiated southern, eastern and western populations in large numbers of savannah taxa. This would imply that okapi have at some point in the past been able to either cross or go around the Congo River, allowing admixture between mitochondrial lineages, whereas chimpanzees and bonobos have not. It is currently unknown if okapi are able to cross large rivers, and in the case of the Congo it may intuitively sound unlikely due to the River's considerable size, and as it is likely to have existed roughly in the same formation for tens or even hundreds of millions of years [71]. However, geomorphic mechanisms do exist that may make this possible. Neck cutoff and oxbow lake formation could theoretically allow populations of organisms to move to the opposite side of a river without actually crossing it. A second possible explanation, anastomosis, is a common mechanism where the path of a river is broken into islands with channels of much smaller size. This process could have led to each of the individual channels being surmountable when the entire width of the river is not.

Partitioning of present-day genetic diversity

We show that deeply divergent mitochondrial haplotypes are ubiquitous across the okapi's range. This suggests that historic biogeographic processes have shaped the structure of genetic diversity in this species, and these processes pre-date the present-day distribution of okapi. Nonetheless, present-day geography also contributes to the structuring of genetic diversity in okapi. AMOVA consistently showed a very high percentage of genetic differentiation between sampling regions. This, in combination with high F_{ST} values for mitochondrial DNA sequence data between populations on the same, and opposite sides of the Congo River, particularly between sampling regions one and four, highlights the importance of the Congo River in structuring present-day genetic diversity in okapi. In comparison, mtDNA F_{ST} values between sampling regions one and two were much lower. The level of present-day population genetic differentiation seen in okapi is within the range of what is seen among chimpanzee populations [70]. No known morphological or behavioral differences separate these chimpanzee populations, however they are regarded as separate sub-species [72,73], again emphasizing the remarkable genetic diversity seen within okapi.

The findings presented here therefore add to the evidence that a combination of the Congo River [23–26], and Pleistocene forest refugia [47,60–64] are the most important factors structuring contemporary genetic diversity of large mammals in the Congo Basin. However, interestingly, the Lomami River (a tributary running parallel southwards with the upper stretches of the Congo River) is the feature that delineates the range of the okapi population on the southwest side of the Congo River. This river has recently been shown to limit the range of a recently described primate, the "lesula" (*Circopithecus lomamiensis*, [74]) and has also recently been shown to be the only river to be a strong barrier to gene-flow in bonobos [23]. A future avenue for research could therefore involve a multi-taxon analysis of the combined role of the Congo and Lomami Rivers in structuring species and genetic diversity in this area.

The future viability of okapi is under considerable doubt [1]. The present study identifies a rich and diverse evolutionary history for this emblematic and elusive species. This is likely a result of a dynamic historical biogeography in the Congo Basin, leading to expansion and contraction of Pleistocene refugia. Contemporary okapi populations contain high levels of genetic diversity, with mtDNA haplotypes widespread. The pinpointing of evolutionarily significant units on the basis of molecular data alone is therefore complex for this species and perhaps not justifiable given the equivocal nature of the data and the overall threats to the species across its range. It is certainly noteworthy, however, that the okapis remaining southwest of the Congo River have divergent allele frequencies and are at low density and we therefore suggest that they be treated as a separate management unit (*sensu* Moritz [75], for example) to the population north of the river, re-emphasising the biogeographic importance of this region (referred to as "TL2" [74]). However, in general conservation efforts should aim to protect as large a proportion of the okapis range as possible.

Supporting Information

Table S1 EPIC primers designed and tested in this study. Observed and expected heterozygosities were generated using the same 28 individuals as the AMOVA and F-statistic analysis in the present study. Multiple SNPs occurring on a single sequence have been notated with a suffixed letter.

Table S2 Study names and genbank IDs of sequences used in the comparative analysis of CR nucleotide diversity. CR section refers to the DNA fragment available for use from genbank, with numbers referring to the position of the fragment, relative to the start of the CR (0) based on the genbank annotations.

Table S3 Study names and genbank IDs of sequences used for the 274 bp phylogeny of bushbuck, duiker, giraffe and okapi, including pig and collared peccary outgroups. Original* and final* haplotypes refer to number of haplotypes in the study in which those haplotypes were originally sequenced, and haplotypes based on 274 bp sequences in the present study respectively.

Table S4 Table of prior values for PopABC analyses. All priors (Ne, effective population size; NeA, ancestral effective population size; t, divergence time; mig, migration rate) used a uniform distribution, except mutAvS (average sequence mutation rate), which PopABC only gives the option of a normal or lognormal distribution. Priors were determined by carrying out preliminary runs, and altering the prior value until all posterior distributions were distinct from the prior distributions.

Table S5 Nucleotide and haplotype diversities, and number of polymorphic sites for the mtDNA genes used in the present study.

Table S6 Table containing individual sample ID's, sequence data for the complete 833 bp fragment used in the present study, and GenBank accession numbers for the corresponding submissions to GenBank (sub-

mitted separately for the CR haplotypes, and a fragment that contains the Cyt b, tRNA-Thr and tRNA-Pro genes).

File S1 Contig of position 31–306 of the 92 CR haplotypes for *Alcelaphus buselaphus*. Nucleotide diversity for these haplotypes is described in Table 1.

File S2 Contig of position 31–306 of the three CR haplotypes for *Damaliscus pygargus*. Nucleotide diversity for these haplotypes is described in Table 1.

File S3 Contig of position 31–306 of the 29 CR haplotypes for *Giraffe camelopardalis*. Nucleotide diversity for these haplotypes is described in Table 1.

File S4 Contig of position 31–306 of the 43 CR haplotypes for *Hippotragus equinus*. Nucleotide diversity for these haplotypes is described in Table 1.

File S5 Contig of position 31–306 of the 26 CR haplotypes for *Okapia johnstoni*. Nucleotide diversity for these haplotypes is described in Table 1.

File S6 Contig of position 31–306 of the 60 CR haplotypes for *Syncerus caffer*. Nucleotide diversity for these haplotypes is described in Table 1.

File S7 Contig of position 31–306 of the 50 CR haplotypes for *Taurotragus oryx*. Nucleotide diversity for these haplotypes is described in Table 1.

File S8 Contig of position 31–306 of the 197 CR haplotypes for *Tragelaphus scriptus*. Nucleotide diversity for these haplotypes is described in Table 1.

File S9 Description of AMOVA analysis, repeated after sampling region delineations had been changed.

File S10 Prior (black) and posterior (blue) distribution for PopABC [41] analysis of region one versus region two (Fig. 1). Parameters investigated are mutation rate (mut rate [A]), migration into sampling regions one and two (labelled mig1 [B] and mig2 [C] respectively), effective population size of sampling regions one and two (labelled Ne1 [D] and Ne2 [E] respectively), effective population size of the ancestral population (NeA [F]) and time since divergence of sampling regions one and two (labelled t1 [G]).

File S11 Prior (black) and posterior (blue) distribution for PopABC analysis of sampling region one versus sampling region four. Parameters investigated are mutation rate (mut rate [A]), migration into sampling regions one and four (labelled mig1 [B] and mig2 [C] respectively), effective population size of sampling regions one and four (labelled Ne1 [D] and Ne2 [E] respectively), effective population size of the ancestral population (labelled NeA [F]) and time since divergence of sampling regions one and four (labelled t1 [G]).

File S12 Prior (black) and posterior (blue) distribution for PopABC analysis of region two versus region four. Parameters investigated are mutation rate (mut rate [A]), migration into sampling regions two and four (labelled mig1 [B] and mig2 [C] respectively), effective population size of sampling regions two and four (labelled Ne1 [D] and Ne2 [E] respectively), effective population size of the ancestral population (labelled NeA [F]) and time since divergence of sampling regions two and four (labelled t1 [G]).

Acknowledgments

We gratefully acknowledge the assistance of the *Institut Congolais pour la Conservation de la Nature* (ICCN), in particular Director J J Mapilanga for allowing the field collection and export of samples. We also thank all the museum curators that sent historic samples or allowed us to collect them. In particular, from the Chicago Field Museum: Bill Stanley and Lawrence Heaney (with the help of Keith Dobney, Aberdeen Univ. and Greger Larson, Durham Univ); from Copenhagen NHM: Hans J. Baagøe, Kristian Gregersen and Mogens Andersen; from Paris NHM: Joséphine Lesur-Gebremariam; from the Royal Museum for Central Africa, Tervuren, Belgium: Wim Wendelen (with the help of Floris van der Sman). We thank Gilman International Conservation, the Okapi Conservation Project, the Frankfurt Zoological Society, the Wildlife Conservation Society, the Lukuru Foundation/TL2 project, the Zoological Society of London (ZSL) and the considerable number of people who assisted in various ways with sample collection. In particular, we would like to thank John Fataki Bolingo, Bryna Griffin, Terese Hart, Chrysostome Kaghoma, Luaison, John Lukas, Kambale Magloire, Ephrem Mpaka, Stuart Nixon, Linda Penfold, Elise Queslin, Alex Quinn, Rosemary Ruf and Ashley Vosper. Finally, the CRC gratefully acknowledges the structural support of the Flemish Government.

Author Contributions

Conceived and designed the experiments: DWGS MWB. Performed the experiments: DWGS JH. Analyzed the data: DWGS JW JGE MWB. Contributed reagents/materials/analysis tools: DWGS PG PH JS NK. Wrote the paper: DWGS. Commented on manuscript: DWGS JH PG PH JS NK JW JGE MWB.

References

1. Mallon D, Kümpel N, Quinn A, Shurter S, Lukas J, et al. (2013) *Okapia johnstoni*. In: IUCN 2013. IUCN Red List of Threatened Species. Version 2013.2. Available: http://www.iucnredlist.org Accessed 2013 Nov 26.

2. IUCN (2008) IUCN SSC Antelope Specialist Group 2008. *Okapia johnstoni*. In: IUCN 2013. IUCN Red List of Threatened Species. Version 2013.1. Available: http://www.iucnredlist.org Accessed 2013 Oct 2.

3. Lankester ER (1901) On *Okapia johnstoni*. Proceedings of the Zoological Society of London 2: 279–281.

4. Hart JA, Hart TB (1989) Ranging and feeding behaviour of okapi (*Okapia johnstoni*) in the Ituri Forest of Zaire: food limitation in a rain-forest herbivore? Symposium of the Zoological Society of London 61: 31–50.

5. Nixon SC, Lusenge T (2008) Conservation status of okapi in Virunga National Park, Democratic Republic of Congo. The Zoological Society of London, London.

6. Frankham R, Ballou JD, Briscoe DA (2002) Introduction to Conservation Genetics. The Edinburgh Building, Cambridge CB2 2RU, United Kingdom: Cambridge University Press.

7. Harris JM, Solounias N, Geraads D (2010) Giraffoidea. In: Werdelin L, Sanders WJ, editors Cenozoic Mammals of Africa. Berkeley, Los Angeles & London: University of California Press. pp. 797–811.

8. Alpers DL, Van Vuuren BJ, Arctander P, Robinson TJ (2004) Population genetics of the roan antelope (*Hippotragus equinus*) with suggestions for conservation. Molecular ecology 13: 1771–1784.

9. Arctander P, Johansen C, Coutellec-Vreto M (1999) Phylogeography of three closely related African bovids (tribe Alcelaphini). Molecular biology and evolution 16: 1724–1739.

10. Brown DM, Brenneman Ra, Koepfli K-P, Pollinger JP, Milá B, et al. (2007) Extensive population genetic structure in the giraffe. BMC Biology 5: 57–70.

11. Moodley Y, Bruford MW (2007) Molecular biogeography: towards an integrated framework for conserving pan-African biodiversity. PloS one 2: e454.

12. Nersting LG, Arctander P (2001) Phylogeography and conservation of impala and greater kudu. Molecular Ecology 10: 711–719.

13. van Hooft WF, Groen AF, Prins HHT (2002) Phylogeography of the African buffalo based on mitochondrial and Y-chromosomal loci: Pleistocene origin and population expansion of the Cape buffalo subspecies. Molecular Ecology 11: 267–279.

14. Lorenzen ED, Masembe C, Arctander P, Siegismund HR (2010) A long-standing Pleistocene refugium in southern Africa and a mosaic of refugia in East Africa: insights from mtDNA and the common eland antelope. Journal of Biogeography 37: 571–581.

15. Stuart C, Stuart T (1997) Field guide to the larger mammals of Africa. Cornelis Struik House, 80 McKenzie Street, Cape Town 8001: Struik Publishers.

16. Cowling SA, Cox PM, Jones CD, Maslin MA, Peros M, et al. (2008) Simulated glacial and interglacial vegetation across Africa: implications for species phylogenies and trans-African migration of plants and animals. Global Change Biology 14: 827–840.

17. Hamilton AC (1975) A quantitative analysis of altitudinal zonation in Ugandan forests. Plant Ecology 30: 99–106.

18. Livingstone DA (1975) Late Quaternary Climatic Change in Africa. Annual Review of Ecology and Systematics 6: 249–280.

19. Clifford SL, Anthony NM, Bawe-Johnson M, Abernethy Ka, Tutin CEG, et al. (2004) Mitochondrial DNA phylogeography of western lowland gorillas (*Gorilla gorilla gorilla*). Molecular Ecology 13: 1551–1565, 1567.

20. Mboumba JF, Deleporte P, Colyn M, Nicolas V (2011) Phylogeography of *Mus (Nannomys) minutoides* (Rodentia, Muridae) in West Central African savannahs: singular vicariance in neighbouring populations. Journal of Zoological Systematics and Evolutionary Research 49: 77–85.

21. Muwanika VB, Nyakaana S, Siegismund HR, Arctander P (2003) Phylogeography and population structure of the common warthog (*Phacochoerus africanus*) inferred from variation in mitochondrial DNA sequences and microsatellite loci. Heredity 91: 361–372.

22. Eriksson J, Hohmann G, Boesch C, Vigilant L (2004) Rivers influence the population structure of bonobos (*Pan paniscus*). Molecular Ecology 13: 3425–3435.

23. Kawamoto Y, Takemoto H, Higuchi S, Sakamaki T, Hart Ja, et al. (2013) Genetic structure of wild bonobo populations: diversity of mitochondrial DNA and geographical distribution. PloS one 8: e59660.

24. Hey J (2010) The divergence of chimpanzee species and subspecies as revealed in multipopulation isolation-with-migration analyses. Molecular biology and evolution 27: 921–933.

25. Caswell JL, Mallick S, Richter DJ, Neubauer J, Schirmer C, et al. (2008) Analysis of chimpanzee history based on genome sequence alignments. PLoS genetics 4: e1000057.

26. Won Y-J, Hey J (2005) Divergence population genetics of chimpanzees. Molecular biology and evolution 22: 297–307.

27. Kennis J, Nicolas V, Hulselmans J, Katuala PGB, Wendelen W, et al. (2011) The impact of the Congo River and its tributaries on the rodent genus Praomys: speciation origin or range expansion limit? Zoological Journal of the Linnean Society 163: 983–1002.

28. Aitken N, Smith S, Schwarz C, Morin Pa (2004) Single nucleotide polymorphism (SNP) discovery in mammals: a targeted-gene approach. Molecular Ecology 13: 1423–1431.

29. GeneCodes Sequencer. 4.9 ed: Gene Codes Corporation, Ann Arbor, MI USA Available: http://www.genecodes.com. Accessed 2013 Dec 12. pp. Sequence analysis software.

30. Clement M, Posanda D, Crandall KA (2000) TCS: a computer program to estimate gene genealogies. Molecular Ecology 9: 1657–1660.

31. Xia X, Xie Z (2001) DAMBE: Data analysis in molecular biology and evolution. Journal of Heredity 92: 371–373.

32. Librado P, Rozas J (2009) DnaSP v5: A software for comprehensive analysis of DNA polymorphism data. Bioinformatics 25: 1451–1452.

33. Bensasson D, Zhang D-X, Hartl DL, Hewitt GM (2001) Mitochondrial pseudogenes: evolution's misplaced witnesses. Trends in Ecology and Evolution 16: 314–321.

34. Excoffier L, Lischer HEL (2010) Arlequin suite ver 3.5: A new series of programs to perform population genetics analyses under Linux and Windows. Molecular Ecology Resources 10: 564–567.

35. Drummond AJ, Rambaut A (2007) BEAST: Bayesian evolutionary analysis by sampling trees. BMC Evolutionary Biology: 214.

36. Hassanin A, Delsuc F, Ropiquet A, Hammer C, van Vuuren BJ, et al. (2012) Pattern and timing of diversification of Cetartiodactyla (Mammalia, Laurasiatheria), as revealed by a comprehensive analysis of mitochondrial genomes. Comptes Rendus Biologies: 32–50.

37. Posada D (2008) jModelTest: Phylogenetic Model Averaging. Molecular Biology and Evolution: 1253–1256.

38. Guindon S, Gascuel O (2003) A simple, fast, and accurate algorithm to estimate large phylogenies by maximum likelihood. Systems Biology: 696–704.

39. Kass RE, Raftery AE (1995) Bayes Factors. Journal of the American Statistical Association 90: 773–795.

40. Rambaut A, Drummond AJ (2007) Tracer v1.5, Available: http://tree.bio.ed.ac.uk/software/. Accessed 2014 Jun 11.

41. Lopes JS, Balding D, Beaumont M (2009) PopABC: a program to infer historical demographic parameters. Bioinformatics 25: 2747–2749.

42. Stephens M, Smith NJ, Donnelly P (2001) A new statistical method for haplotype reconstruction from population data. American Journal of Human Genetics 68: 978–989.

43. Graur D, Martin W (2004) Reading the entrails of chickens: molecular timescales of evolution and the illusion of precision. Trends in genetics 20: 80–86.

44. Bromham L, Penny D (2003) The modern molecular clock. Nature reviews Genetics 4: 216–224.

45. Ho SYW, Lanfear R, Bromham L, Phillips MJ, Soubrier J, et al. (2011) Time-dependent rates of molecular evolution. Molecular Ecology 20: 3087–3101.

46. Ho SYW, Larson G (2006) Molecular clocks: when times are a-changin'. Trends in genetics 22: 79–83.

47. Johnston AR, Anthony NM (2012) A multi-locus species phylogeny of African forest duikers in the subfamily Cephalophinae: evidence for a recent radiation in the Pleistocene. BMC Evolutionary Biology 12: 120.

48. Dietrich WO (1942) Ältestquartäre Säugetiere aus der Südlichen Serengeti, Deutsch-Ostafrika. Palaeontographica, Abt A 94: 3–22.

49. Van der Made J, Morales J (2012) *Mitilanotherium inexpectatum* (Giraffidae, Mammalia) from Huélago (Lower Pleistocene; Guadix-Baza basin, Granada, Spain) - observations on a peculiar biogegraphic pattern. Estudios Geológicos 67: 613–627.

50. Gentry AW (1990) Ruminant artiodactyls of Pasalar, Turkey. Journal of Human Evolution 19: 529–550.

51. Harris JM (1991) Family Giraffidae. In: Harris JM, editor. Koobi Fora Research Project Volume 3 The fossils ungulates: geology, fossil artiodactyls, and palaeonenvironments. Clarendon Press, Oxford. pp. 93–138.

52. Bonis Ld, Koufos G, Sen S (1997) A giraffid from the Middle Miocene of the island of Chios, Greece. Palaeontology 40: 121–133.

53. Fernandez MH, Vrba ES (2005) A complete estimate of the phylogenetic relationships in Ruminantia: a dated species-level supertree of the extant ruminants. Biological reviews of the Cambridge Philosophical Society 80: 269–302.

54. Maglio VJ (1973) Origin and evolution of the elephantidae. Transactions of the American Philosophical Society 63: 1–149.

55. Rohland N, Pollack JL, Nagel D, Beauval C, Airvaux J, et al. (2005) The population history of extant and extinct hyenas. Molecular biology and evolution 22: 2435–2443.

56. Potts R (2013) Hominin evolution in settings of strong environmental variability. Quaternary Science Reviews 73: 1–13.

57. DeMenocal PB (2004) African climate change and faunal evolution during the Pliocene–Pleistocene. Earth and Planetary Science Letters 220: 3–24.

58. Maslin MA, Christensen B (2007) Tectonics, orbital forcing, global climate change, and human evolution in Africa: introduction to the African paleoclimate special volume. Journal of human evolution 53: 443–464.

59. DeMenocal PB (1995) Plio-Pleistocene African Climate. Science 270: 53–59.

60. Born C, Alvarez N, McKey D, Ossari S, Wickings EJ, et al. (2011) Insights into the biogeographical history of the Lower Guinea Forest Domain: evidence for the role of refugia in the intraspecific differentiation of *Aucoumea klaineana*. Molecular Ecology 20: 131–142.

61. Bowie RCK, Fjeldså J, Hackett SJ, Bates JM, Crowe TM (2006) Coalescent models reveal the relative roles of ancestral polymorphism, vicariance, and dispersal in shaping phylogeographical structure of an African montane forest robin. Molecular phylogenetics and evolution 38: 171–188.

62. Nicolas V, Missoup AD, Denys C, Kerbis Peterhans J, Katuala P, et al. (2011) The roles of rivers and Pleistocene refugia in shaping genetic diversity in *Praomys misonnei* in tropical Africa. Journal of Biogeography 38: 191–207.

63. Plana V (2004) Mechanisms and tempo of evolution in the African Guineo-Congolian rainforest. Philosophical transactions of the Royal Society of London Series B, Biological sciences 359: 1585–1594.

64. Quérouil S, Verheyen E, Dillen M, Colyn M (2003) Patterns of diversification in two African forest shrews: *Sylvisorex johnstoni* and *Sylvisorex ollula* (Soricidae, Insectivora) in relation to paleo-environmental changes. Molecular phylogenetics and evolution 28: 24–37.

65. Colinvaux PA, De Oliveira PE, Bush MB (2000) Amazonian and neotropical plant communities on glacial time-scales: The failure of the aridity and refuge hypotheses. Quaternary Science Reviews 19: 141–169.

66. de Thoisy B, da Silva AG, Ruiz-García M, Ramirez O, et al. (2010) Population history, phylogeography, and conservation genetics of the last Neotropical mega-herbivore, the lowland tapir (*Tapirus terrestris*). BMC Evolutionary Biology 10: 278.

67. Lessa EP, Cook JA, Patton JL (2003) Genetic footprints of demographic expansion in North America, but not Amazonia, during the Late Quaternary. Proceedings of the National Academy of Sciences 100: 10331–10334.

68. Pickles RSA, Groombridge JJ, Zambrana Rojas VD, Van Damme P, Gottelli D, et al. (2011) Evolutionary history and identification of conservation units in the giant otter, *Pteronura brasiliensis*. Molecular phylogenetics and evolution 61: 616–627.

69. Willis KJ, Whittaker RJ (2000) The refugial debate. Science 287: 1406–1407.

70. Fischer A, Prüfer K, Good JM, Halbwax M, Wiebe V, et al. (2011) Bonobos fall within the genomic variation of chimpanzees. PloS one 6: e21605.
71. Anka Z, Séranne M, Lopez M, Scheck-Wenderoth M, Savoye B (2009) The long-term evolution of the Congo deep-sea fan: A basin-wide view of the interaction between a giant submarine fan and a mature passive margin (ZaiAngo project). Tectonophysics 470: 42–56.
72. Braga J (1995) Skeletal variation and measure of divergence among chimpanzees. Comptes Rendus de l'Academie des Sciences 320: 1025–1030.

73. Uchida A (1996) What we don't know about great ape variation. Trends in Ecology and Evolution 16: 163–168.
74. Hart J, Detwiler KM, Gilbert CC, Burrell AS, Fuller JL, et al. (2012) Lesula: A new species of cercopithecus monkey endemic to the Democratic Republic of Congo and implications for conservation of Congo's central basin. PLoS ONE 7: e44271.
75. Moritz C (1994) Applications of mitochondrial DNA analysis in conservation: a critical review. Molecular Ecology 3: 401–411.

2

Phenotypic Variation in Infants, Not Adults, Reflects Genotypic Variation among Chimpanzees and Bonobos

Naoki Morimoto[1]*, Marcia S. Ponce de León[2], Christoph P. E. Zollikofer[2]*

1 Laboratory of Physical Anthropology, Graduate School of Science, Kyoto University, Kyoto, Japan, **2** Anthropological Institute, University of Zurich, Zurich, Switzerland

Abstract

Studies comparing phenotypic variation with neutral genetic variation in modern humans have shown that genetic drift is a main factor of evolutionary diversification among populations. The genetic population history of our closest living relatives, the chimpanzees and bonobos, is now equally well documented, but phenotypic variation among these taxa remains relatively unexplored, and phenotype-genotype correlations are not yet documented. Also, while the adult phenotype is typically used as a reference, it remains to be investigated how phenotype-genotye correlations change during development. Here we address these questions by analyzing phenotypic evolutionary and developmental diversification in the species and subspecies of the genus *Pan*. Our analyses focus on the morphology of the femoral diaphysis, which represents a functionally constrained element of the locomotor system. Results show that during infancy phenotypic distances between taxa are largely congruent with non-coding (neutral) genotypic distances. Later during ontogeny, however, phenotypic distances deviate from genotypic distances, mainly as an effect of heterochronic shifts between taxon-specific developmental programs. Early phenotypic differences between *Pan* taxa are thus likely brought about by genetic drift while late differences reflect taxon-specific adaptations.

Editor: David Caramelli, University of Florence, Italy

Funding: This work was supported by the Swiss National Science Foundation (no. 3100A0-109344/1) and Japan Society for Promotion of Science Research Fellowship for Young Scientists (no. 251133). The funders had no role in study design, data collection and analysis, decision to publish, or preparation of the manuscript.

Competing Interests: The authors have declared that no competing interests exist.

* Email: morimoto@anthro.zool.kyoto-u.ac.jp (NM); zolli@aim.uzh.ch (CPEZ)

Introduction

The ready accessibility of population-wide genotypic and phenotypic data from humans and our closest relatives, the great apes, has spurred a large number of studies investigating the relationship between patterns of genotypic and phenotypic evolution. One central issue is the relative role of neutral versus adaptive evolutionary processes in shaping genotypic and phenotypic variation. A steadily growing number of studies indicates that variation of cranial morphology among modern human populations, and between modern humans and fossil hominins (species related more closely to modern humans than to great apes) largely reflects the effects of genetic drift, while only a small proportion of variation can be attributed to selection [1,2,3,4,5,6,7,8,9,10]. Fossil hominin aDNA now also permits insights into earlier phases of human population and evolutionary history at an unprecedented level of detail [11,12,13,14,15]. These analyses are limited, however, by the "aDNA preservation horizon", which is currently around 50,000 years BP for fossil hominin nDNA, and around 400,000 years BP for mtDNA from temperate zones [16].

One possible solution to investigate genotype-phenotype evolution beyond this horizon is to study living great ape species as a model system. The genus *Pan* represents the best model for this purpose, since it is our closest living relative, its species, subspecies and population structure is now genetically well-documented [17,18,19,20], and population history and genetic diversification

are well understood [18,19,21,22,23]. To date, two *Pan* species, *P. troglodytes* (common chimpanzee) and *P. paniscus* (bonobo) are recognized, and *P. troglodytes* is subdivided into four subspecies (*P. t. troglodytes*, *P. t. schweinfurthii*, *P. t. verus* and *P. t. ellioti*) [19]. Also, these *Pan* taxa have been the subject of detailed anatomical [24,25,26,27,28], morphological [29,30,31,32,33], phylogeographic [17,19,23,34], and behavioral [32,35,36,37,38,39,40] studies.

The extant *Pan* taxa are closely related to each other, which represents several advantages for comparative analyses. First, genotypic differences between taxa are small compared to variation within each taxon, such that the number of genes associated with phenotypic differentiation during (sub-) speciation is expected to be comparatively small [41]. Second, diversity among *Pan troglodytes* taxa represents patterns of incipient speciation, which are not yet blurred by long-term processes of taxon-specific specialization and/or convergence [42,43]. Also, we may note that the estimated time frame of *Pan* speciation [19,23] is comparable to that of our own genus *Homo* (ca. 2 million years).

Despite the increasing knowledge about *Pan* taxa, it still remains to be explored how changes at the level of the genotype are linked to changes at the level of the phenotype during speciation. The first aim of this study is thus to provide new phenotypic data documenting the evolutionary divergence of *Pan* taxa, and to relate this new evidence to the well-established body of genotypic evidence. While evolutionary studies traditionally focus on variation in craniodental features e.g. [44,45], we study here morphological variation of the femoral shaft (= diaphysis). The femur is a functionally highly constrained element of the postcranial skeleton, and can thus be expected to be under strong stabilizing selection.

Most studies exploring genotype-phenotype relationships in great apes and humans have naturally focused on adult morphologies. This is because taxon-specific morphological features are thought to be more clearly expressed in adults than in juveniles. However, there is clear evidence that the phenotypes of early ontogenetic stages, and patterns of developmental change, are highly informative about patterns of evolutionary divergence at the levels of skeletal structure e.g. [46,47,48,49,50,51,52,53], of locomotor behaviors [35,37], and of social interactions [54]. The second aim of this study is thus to expand the scope of genotype-phenotype comparisons by taking into account the perspective of ontogeny. Here we explore how genotype-phenotype relationships change during the development of the femoral diaphysis in the different *Pan* taxa, and relate this information to evolutionary change at the level of the genotype and phenotype. Specifically, we explore when during ontogeny the effects of drift versus selection become evident in taxon-specific phenotypes.

Measuring genotype-phenotype relationships is a complex endeavor, both theoretically and practically, and requires several model assumptions. In the standard model of quantitative population genetics, phenotypic variance V_P is the combination of genetic variance V_G and environmental variance V_E: $V_P = V_E + V_G$. Empirical data and theoretical considerations indicate that, for complex traits, phenotypic variance can be approximated by $V_P = V_E + V_A$, where V_A represents additive genetic variation (the portion of phenotypic variation that can be explained by the cumulative effects of allelic variation) [55]. The question of interest here is how V_P and V_A evolve in segregating populations. In a constant environment (V_E = const.), $V_P = V_A$, such that phenotypic variation reflects additive genotypic variation. Under these basic model assumptions, effects of drift and selection are typically estimated by comparing neutral genotypic distances with non-neutral distances [56,57,58,59,60]. The former distances (F_{ST}: genetic variation within subpopulation relative to total genetic variation [61,62]) are estimated from non-coding genetic markers thought to evolve under no selection such as STRs (short tandem repeats) and non-coding SNPs (single nucleotide polymorphisms) [63]. The latter distances are typically estimated from continuous quantitative genetic traits (Q_{ST}: evaluated in analogy to F_{ST} [64]) assuming additive genetic effects [64]. The question is whether Q_{ST} is equal to, smaller than, or larger than F_{ST}, which indicates neutral evolution, uniform or stabilizing selection, and diversifying selection, respectively [65].

Q_{ST} can be estimated from phenotypic distance P_{ST} [66] using a measure of heritability (h^2, proportion of additive genetic variance to phenotypic variance, V_A / V_P) [66,67,68,69,70]. In wild populations, heritability h^2 is often unknown and needs to be estimated from largely comparable lab studies. Furthermore, h^2 tends to change due to *in-vivo* environmental effects that accumulate during an individual's lifetime, and due to developmental changes in gene activation patterns [71,72,73]. In any case, estimates of h^2 affect the distance measures expressed by Q_{ST}, such that estimating the relative contribution of additive genetic and *in-vivo* environmental effects to P_{ST} remains a challenge [74].

A further challenge of $F_{ST} - Q_{ST}$ comparisons is the practical difficulty in measuring genotypic and phenotypic distances. Genotypic distances have been typically calculated using population-specific allele frequencies [75] (e.g., in Nei's standard distance D_a [76] and Cavalli-Sforza and Edwards chord distance D_{CH}

[77]). One problem is that sample sizes of wild populations are often limited, which makes it difficult to estimate population-specific allele frequencies and within-population variation. Complementary methods have thus been proposed, e.g. Principal Components Analysis (PCA) of genetic data [78,79]. While phenotypic distances have traditionally been evaluated from arrays of linear and angular measurements, geometric morphometrics (GM) offers elegant methods to quantify complex patterns of phenotypic variation [80,81,82]. In GM, biological form is typically measured by the spatial configuration (3D geometry) of anatomical points of reference, so-called landmarks [83,84]. Alternatively, various methods of GM have been developed to quantify the shape of landmark-free biological structures such as outlines [85], endocranial cavities [86] and longbone shafts [46,87]. One key feature of all GM methods is that phenotypic variation can simultaneously be represented in physical (three-dimensional) space by means of graphical interpolation and in multivariate space by means of PCA. PCA thus provides an ideal means to compare multivariate genotypic and phenotypic data independent of underlying population models.

Materials and Methods

Volumetric data of the femora of $N = 146$ *Pan* specimens were acquired with computed tomography (CT) ($N = 50$ *Pan troglodytes troglodytes*, $N = 39$ *P.t. schweinfurthii*, $N = 26$ *P. t. verus*, $N = 31$ *P. paniscus*; see Figs. S1 and S2, Table S1, and Text S1 and S2 for details on sample structure). *P. t. troglodytes* and *P. t. verus* specimens were obtained from the collections of the Anthropological Institute and Museum of the University of Zurich (AIMUZH), *P. t. schweinfurthii* specimens were obtained from the collections of the Royal Africa Museum, Tervuren, Belgium (MRA), and *P. paniscus* specimens were obtained from AIMUZH and MRA (Table S1). Each taxon is represented by four consecutive ontogenetic stages from infancy to adulthood. These were defined according to dental eruption: m2 (second deciduous molar erupted), M1, M2, M3 (first, second, third permanent molars erupted). In *Pan*, m2, M1, M2 and M3 erupt approximately at 0.5–0.83, 3, 7 and 11 years after birth, respectively [88].

Because femoral epiphyses are not yet ossified during the early stages of ontogeny, we focus on diaphyseal morphology. Effects of *in-vivo* bone modification in the femur have been studied in various *Pan* taxa, and it has been shown that ontogenetic changes in femoral morphology reflect an underlying developmental program that is fairly independent of environmental influences [87]. In other words, environmental variance V_E remains approximately constant throughout ontogeny [31,87,89] (see Text S3), which is an important prerequisite to estimate Q_{ST} from P_{ST} [74].

To quantify a specimen's diaphyseal surface morphology the transverse radius of curvature was evaluated for each point of the external (subperiosteal) surface, as specified in ref. [87]. The data of all specimens were then analyzed by means of morphometric mapping (MM) methods [87,90] (Fig. S3 and Text S1). MM is a landmark-free geometric morphometric method that permits dense sampling of data from smooth surfaces. It is thus well suited to quantify even subtle morphological differences in femoral shaft form between different taxa and/or developmental stages [87,91,92,93]. To correct for size differences between specimens, size is normalized by diaphyseal length and the median value of the radius of curvature. Shape variation is then decomposed into statistically independent shape components, which represent multivariate descriptors of the total femoral diaphyseal morphology. Since MM establishes a direct link between femoral geometry and its multivariate representation, patterns of inter- and intra-group variation can be visualized in multivariate shape space ("morphospace"; Fig. 1) as well as in real (physical) space (Fig. 2). To infer the femoral diaphyseal morphology and its developmental pattern in the last common ancestor (LCA) of *Pan* taxa, the phylogenetic tree of *Pan* taxa was projected onto the morphospace using a model of squared-change parsimony under a Brownian motion model [94] for each ontogenetic stage (Fig. S4) using the software package MorphoJ [95]. Also, MM was used to infer the infant and adult femoral diaphyseal morphology of the LCA (Fig. 2).

Mean femoral diaphyseal shape was calculated for each taxon at each ontogenetic stage i, and inter-taxon phenotypic (*i.e.*, morphometric) distance matrices $\mathbf{M}_i$ were calculated for each stage. As a phenotypic distance metric, the Euclidean distance in morphospace was used. Between-taxon quantitative genetic differentiation (Q_{ST}) was also estimated for each ontogenetic stage. To this end, pairwise Q_{ST}s were evaluated from P_{ST}s with the software RMET 5.0 [96,97], using PC scores (PC1–3) and a standard estimation of heritability $h^2 = 0.55$. This procedure resulted in stage-specific distance matrices $\mathbf{Q}_i$.

Genotypic distances between *Pan* taxa (matrices $\mathbf{F}$) were calculated from sequence datasets. The sequence data of 150,000 bp on 15 non-coding autosomal regions in $\mathcal{N} = 74$ *Pan* specimens were obtained from GenBank (accession number: JF725992–727161 [22]). Inter-taxon genotypic distances were evaluated with various methods; Nei's standard distance D_a [76], Cavalli-Sforza and Edwards chord distance D_{CH} [77], and Euclidean distances in Patterson's PC space D_{PPC} [78,79]. Further, F_{ST} and R_{ST} from published sources were also used to construct genotypic distance matrices ([18,19,21,22]; refs. [18] and [19] use the same marker set) (Table S2).

Overall, three kinds of between-taxon distance matrices $\mathbf{F}$ (genotypic), $\mathbf{M}$ (phenotypic) and $\mathbf{Q}$ (quantitative genetic) were evaluated, and these matrices were used for $\mathbf{F}-\mathbf{M}$ and $\mathbf{F}-\mathbf{Q}$ ($F_{ST} - Q_{ST}$ [P_{ST}]) comparisons. The similarity between these distance matrices was evaluated with principal coordinate analysis (PCO), and assessed statistically with the Mantel test and resampling statistics (see Text S1 and Fig. S3 for details on PCO and resampling statistics). In brief, PCO transforms a between-taxon distance matrix into a "taxon constellation" (i.e., locations of taxa relative to each other in multivariate space). To assess the coincidence between genotypic and phenotypic taxon constellations, we used Procrustes analysis. This method superimposes two or more different constellations using a least-squares criterion. The Mantel test was performed using Relethford's MANTEL 3.1 (software programs RMET and MANTEL are available at http://employees.oneonta.edu/relethjh/programs/).

The fact that more than two *Pan* taxa are studied here facilitates rather than complicates F_{ST}–Q_{ST} comparisons. For $K = 2$ groups (populations or taxa), one F_{ST} distance is compared with one Q_{ST} distance. These need to be scaled appropriately with an estimate of h^2 to permit significant implications on neutral versus adaptive evolution, but h^2 is typically unknown. For $K > 2$ groups (this study: $K = 4$), the structures of two $K \times K$ distance matrices (F and Q) are compared, and scaling issues can be addressed with methods of matrix-matrix correlation and multidimensional scaling (MDS) such as the PCO method used here e.g. [2,7,98,99,100]. Assuming that $h^2(i) = $ const. for all groups at a given ontogenetic stage i, MDS will thus scale P_{ST} and Q_{ST} relative to F_{ST} even without explicit estimates of $h^2(i)$ (refs. [10,101]).

These matrix-matrix comparisons permit to assess whether the structure of a phenotypic ($\mathbf{M}$) or quantitative-genetic ($\mathbf{Q}$) distance matrix is similar to, or deviates from, a putatively neutral genotypic distance matrix $\mathbf{F}$. Similarity would imply that $\mathbf{M}$ and $\mathbf{Q}$ are scaled versions of $\mathbf{F}$ (scaling factor h^2). An important assumption is that the genetic markers to estimate F_{ST} follow neutral evolution. This is critical to evaluate the relative role of neutral and adaptive processes from phenotypic data. The genetic markers used here to estimate F_{ST} represent non-coding regions

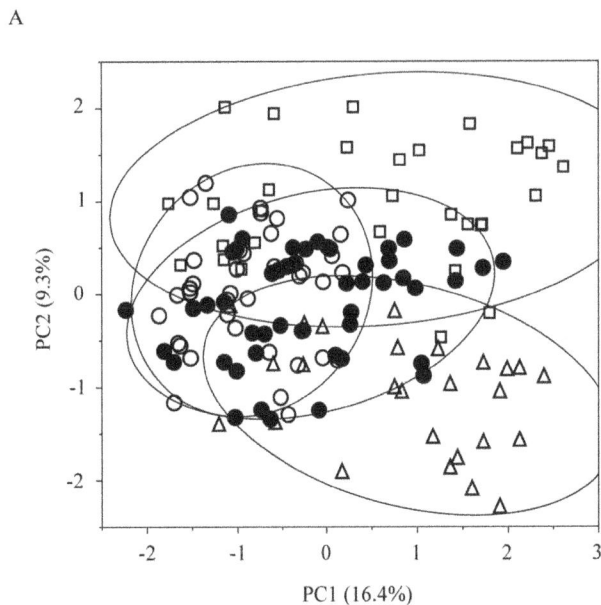

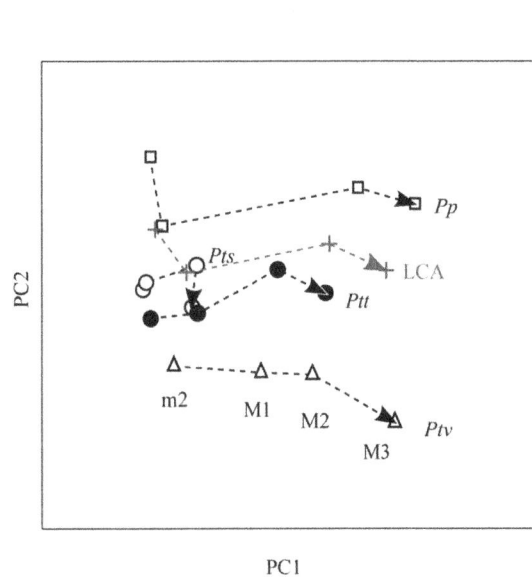

A

B

Figure 1. Femoral diaphyseal shape variation in an ontogenetic sample of *Pan* taxa. A: Variation along the first two principal components of shape, PC1 and PC2 (filled circles: *P.t. troglodytes*, open circles: *P.t. schweinfurthii*, open triangles: *P.t. verus*, open squares: *P. paniscus*). Solid outlines show 95%-density ellipses for each taxon. B: plot of mean shapes at consecutive ontogenetic stages. m2: second deciduous molar erupted; M1/M2/M3: permanent molars 1/2/3 erupted. Gray symbols and dashed line indicate the inferred shape at each ontogenetic stage and ontogenetic trajectory of the last common ancestor.

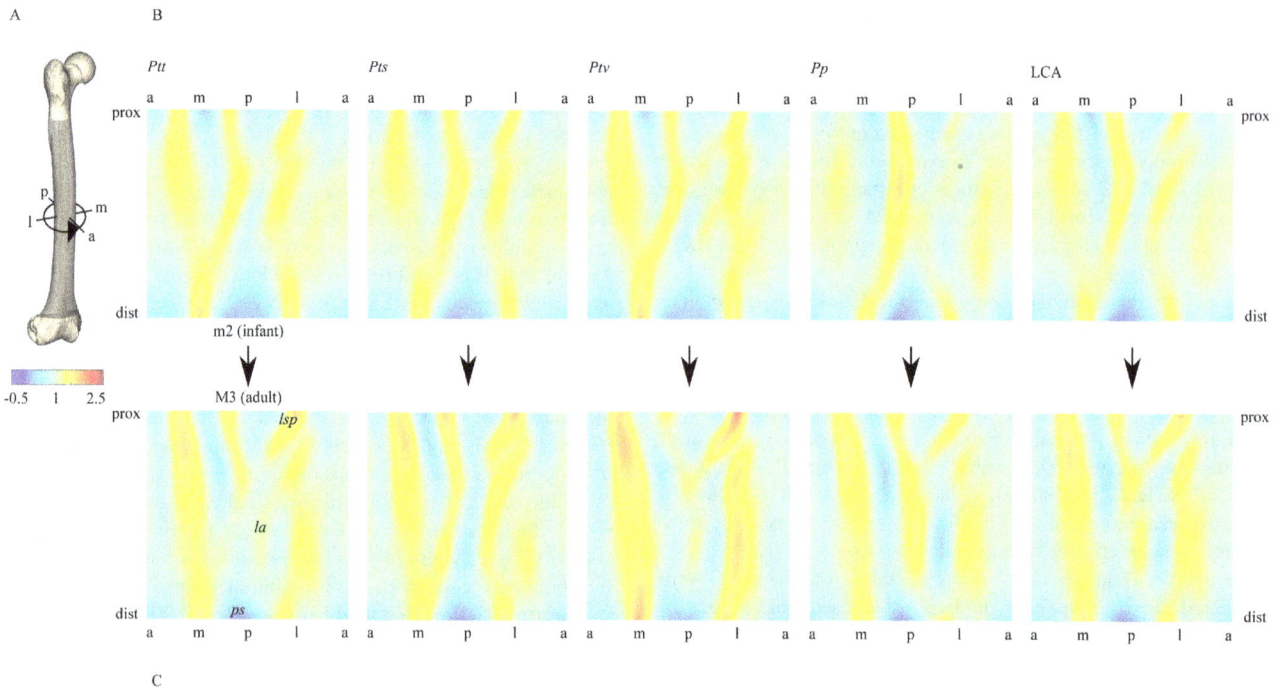

Figure 2. Taxon-specific femoral diaphyseal shapes. A: principle of morphometric map generation (anterior [0°] → medial [90°] → posterior [180°] → lateral [270°] → anterior [360°]). B, C: morphometric maps of taxon-specific morphologies at ontogenetic stages m2 (B, infant) and M3 (C, adult) (false-color images of external surface curvature [relative units]). *la*: linea aspera, *lsp*: lateral spiral pilaster, *ps*: popliteal surface.

[18,19,21,22], so it is reasonable to assume that variation reflects neutral processes.

Results

Fig. 1 shows commonalities and differences in femoral diaphyseal shape and shape variation between *Pan* taxa. The first two principal components represented here (PC1 and PC2) account for 25.7% of the total shape variation in the sample. There is substantial overlap between taxon-specific distributions of *P. t. troglodytes* and *P. t. schweinfurthii*, but almost no overlap between *P. paniscus* and *P. t. verus* (Fig. 1A). At each ontogenetic stage, taxon-specific mean shapes are statistically different from each other (Fig. 1B, Table S3). Furthermore, taxon-specific ontogenetic trajectories (see SI and refs. [102,103]) have statistically similar directions through morphospace (Fig. 1B and Table S4). Trajectories differ from each other, however, in their length (mostly along PC1), and in their location in morphospace (mostly along PC2) (Fig. 1B). Trajectories of *P. t. troglodytes* and *P. t. schweinfurthii* are in close vicinity, but the trajectory of the latter taxon is significantly shorter than that of the former. Compared to these taxa, the trajectory of *P. paniscus* is significantly longer (Fig. 1B, Table S5).

Differences between trajectories are already present at the m2 (infant) stage, indicating that taxon-specific femoral shape is established early during ontogeny. The differences in trajectory length indicate that the shape differences between *Pan* taxa increase toward adulthood. Longer trajectories indicate a larger total amount of femoral shape change during ontogeny, and possibly higher rates of shape change. Fig. 2 visualizes the corresponding real-space patterns of femoral diaphyseal shape change from infant to adult for each taxon. Each stage- and taxon-specific diaphyseal shape is represented here with a morphometric map (MM), which represents surface structures around (x-axis) and

along (y-axis) the femoral diaphysis. MMs visually confirm that taxon-specific femoral shape is present already at the m2 (infant) stage, and that taxon-specific features become more pronounced toward the M3 (adult) stage.

Using methods of squared-change parsimony [94], it is possible to infer the ontogenetic trajectory of the LCA of *Pan* taxa. The LCA trajectory lies between the trajectory of *P. paniscus* and the average trajectory of *P. troglodytes* taxa (Figs. 1, 2, S4). The length of the LCA trajectory is comparable to that of *P. t. troglodytes*, *P. t. verus*, and *P. paniscus*, but is longer than that of *P. t. schweinfurthii*.

All measures of genotypic distances (F_{ST}, D_a, D_{CH}, D_{PPC}) are highly correlated with each other (Table S6; Mantel test). Genotypic distances (F_{ST} and R_{ST}) evaluated from different marker sets [18,19,21,22] (Table S2) are also concordant with each other (Fig. S5), indicating that potential noise due to the small sample sizes of these studies does not greatly affect the results [104]. In all further comparative analyses we use D_{PPC} because evaluation of this distance measure does not presuppose estimation of within-group variance.

To assess the congruence between genotypic and phenotypic distance matrices, we projected the genotypic and phenotypic PCO data into the same multidimensional space and aligned them with Procrustes Analysis. Patterns of phenotypic similarity among *Pan* taxa (P_{ST}) are overall congruent with patterns of genetic similarity (D_{PPC}, F_{ST}) (Figs. 3A, S5, Tables 1, S6, S7). Figs. 3A and S5 show that the match between genotypic and phenotypic data is closest at the m2 (infant) stage (Table 1; $p<0.05$, Mantel test). While taxa advance along their ontogenetic trajectories, patterns of phenotypic variation tend to deviate from the pattern of genetic variation (Fig. 3A, S5). These results are statistically supported by a resampling test (Fig. 3B). **F–M** correlation is highest at the m2 (infant) stage ($R^2 = 0.80$, $p = 0.02$), and is lowest at the M3 (adult) stage ($R^2 = 0.20$, $p = 0.37$). Likewise, the **F–M** correlation between genotypic and phenotypic distances evaluated by a Mantel test is

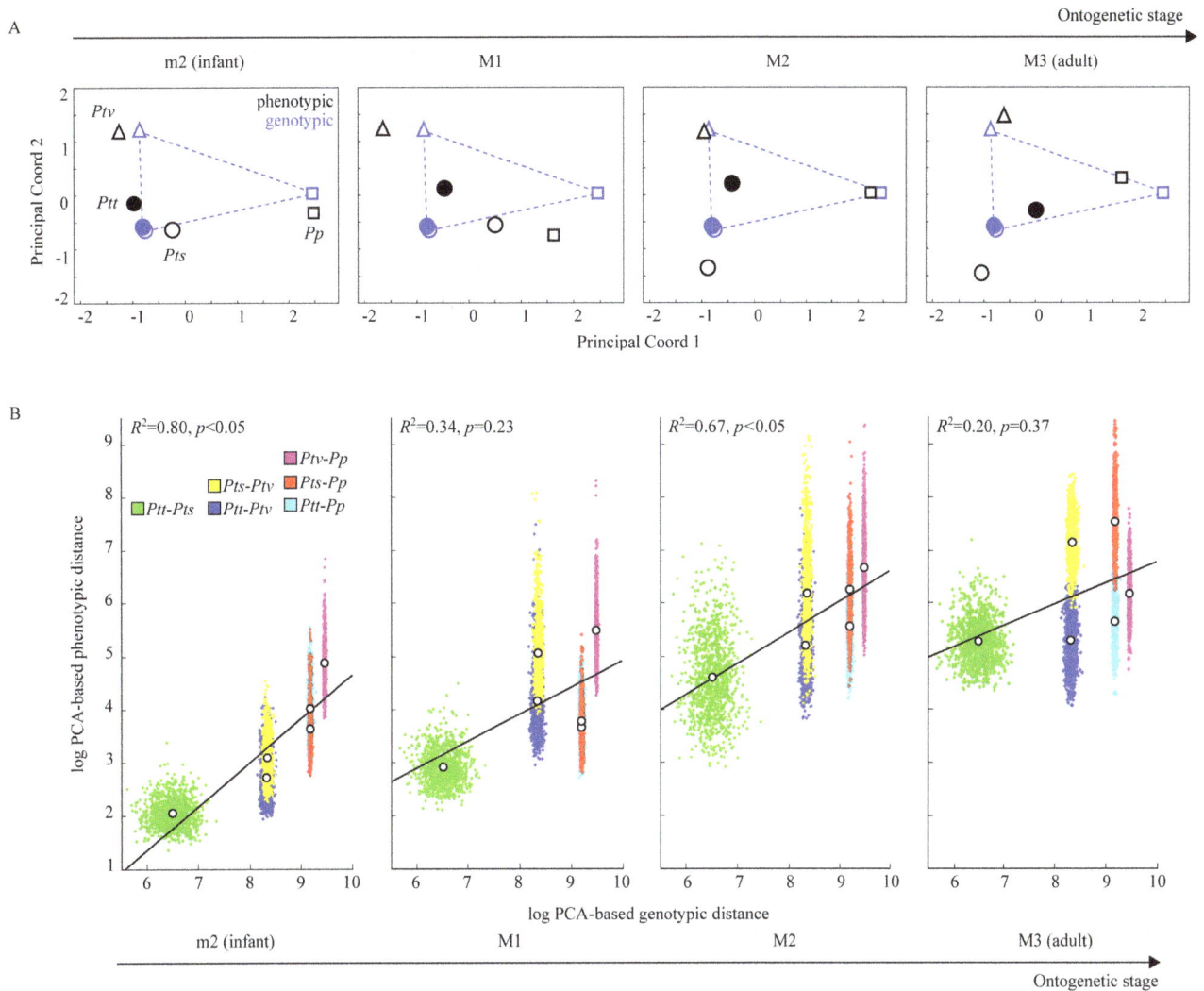

Figure 3. Comparison of genotypic and phenotypic distances between *Pan* taxa. A: Principal Coordinates Analysis (PCO) permits representation of genotypic and phenotypic distance data in the same multivariate space. The four subgraphs show phenotypic data (black dots) for consecutive ontogenetic stages m2, M1, M2, and M3, and genotypic data (same blue dots for all stages). For graphical clarity genotypic data points, which are independent of ontogenetic stage, are connected with dashed lines. Note that during ontogeny the phenotypic distance configuration departs from the neutral genetic distance configuration (see also Fig. S5). B: Correlation between phenotypic and neutral genetic distances between taxa. Each point cloud consists of 1000 randomly sampled phenotypic and genotypic distances between individuals belonging to different *Pan* taxa (resampling procedures are explained in Text S1). Correlation of phenotypic and neutral genetic distances is highest at the m2 (infant) stage and declines towards adulthood (M3). Genetic and phenotypic distances are normalized by their respective median values. Note overall increase of phenotypic distance between taxa toward adulthood.

highest at the m2 stage (Tables 1 and S7). **F–M** correlation is also significant at the M2 stage, but to a lesser extent than at the m2 stage. The decline in **F–M** correlation from infancy to adulthood thus follows a non-monotonous pattern.

The results of **F–Q** comparisons (i.e., standard $F_{ST}–Q_{ST}$ tests) are similar to the results obtained with PCA/PCO analyses (Table 1). The correlation between F_{ST} and Q_{ST} [P_{ST}] is highest at the m2 (infant) stage ($R^2 = 0.72$, $p<0.01$), and lowest at the M3 (adult) stage ($R^2 = 0.10$, $p = 0.35$). The finding that correlation between genotypic and phenotypic markers decreases during ontogeny is thus independent of the method of genotypic and phenotypic distance measurement.

Discussion

Investigating the evolutionary divergence between populations and/or closely related taxa at the level of genes and phenes, and inferring underlying processes of selection and drift, has become an important research topic in primatology and anthropology [1,2,3,4,5,6,7,8,9,10]. Progress in this field is fostered by the availability of ever-increasing volumes of genomic and phenomic data, and sophisticated analytical tools to compare patterns of genotypic and phenotypic variation. While DNA sequence data provide *static* structural information about the genome, data at any level above the DNA (from the transcriptome to morphology) provide *dynamic* structural information about the phenotype, which changes during ontogeny. Interestingly, the effect of ontogenetic time on correlations between genotypic and phenotypic variation is still relatively unexplored. For example, ontogenetic time does

Table 1. Correlation between genotypic and genotypic distance matrices.

		m2 (infant)	M1	M2	M3 (adult)
genotypic distance[1]–phenotypic distance[2] (Mantel[3])	R^2	**0.84**	0.15	**0.64**	0.18
	p	**<0.01**	0.2609	**<0.01**	0.3478
genotypic distance–phenotypic distance (resampling[4])	R^2	**0.80**	0.34	**0.67**	0.20
	p	**0.015**	0.23	**0.045**	0.37
F_{ST}–Q_{ST} test[5] (Mantel)	R^2	**0.72**	0.40	**0.67**	0.10
	p	**<0.01**	0.087	**<0.01**	0. 3478

[1]Euclidean distance in Patterson's PC space.
[2]Euclidean distance in morphospace (shape PCs).
[3]correlation (R^2) and significance levels (p) evaluated with Mantel test (1000 permutations).
[4]evaluated with resampling statistics (see methods; Fig. S3C).
[5]estimate of heritability h^2: 0.55.

not appear as an explicit variable in the standard equations relating V_P to V_A, nor is it typically considered explicitly in F_{ST}–Q_{ST} comparisons.

To fill this gap, we studied femoral diaphyseal shape change in the genus *Pan* and compared patterns of phenotypic divergence (both during development and evolution) with patterns of genotypic divergence. The results presented here yield several new insights into evolutionary and developmental links between genotypic and phenotypic diversification in *Pan*. Before any general inferences can be drawn, it should be reminded, however, that the genotypic and phenotypic data sets studied here represent subsets of the total genotypic/phenotypic evidence that is potentially available for such studies.

The close correspondence between genotypic and phenotypic distances at the earliest ontogenetic stage analyzed here (the m2 stage) gives rise to two alternative hypotheses; H0: if the molecular markers of refs. [18,19,21,22] track neutral evolution then the observed pattern of phenotypic evolution is "neutral-like" within the constraints imposed by stabilizing selection (often described as "wandering around an adaptive optimum" [105,106,107]); H1: if the pattern of phenotypic distances between taxa is the result of selection and adaptation, then the molecular markers are non-neutral and carry an adaptive signal. Given the good evidence for neutrality in the molecular markers [108] used here, hypothesis H1 is less likely. Also, the congruence of the genotypic distance patterns evaluated from different marker types (Fig. S5) suggests that H1 is less likely, since one would expect that selection acts differently on different marker types. Our data thus support hypothesis H0, which implies that morphological variation of the femoral diaphysis in infant *Pan* reflects neutral evolutionary diversification between taxa rather than taxon-specific adaptation.

While phenotypic distances between *Pan* taxa at the m2 stage are in good concordance with genotypic distances ($R^2 = 0.8$; Fig. 3B), correlations are lower at later ontogenetic stages, and reach a value of $R^2 = 0.2$ at adulthood (Figs. 3, S5; Table 1). As already reported in earlier studies [74,109,110], correlations between molecular and phenotypic markers are typically low, and this has been interpreted in two ways: (1) that (non-coding) molecular marker variation does not adequately represent the quantitative genetic variation of coding genes that becomes manifest in the phenotype, and (2) that environmental variation has a significant influence on V_P, and hence on Q_{ST}.

The ontogenetic data presented in this study provide an empirical basis to test these hypotheses. The high correlation ($R^2 = 0.80$) between inter-taxon molecular and phenotypic varia-tion at the m2 stage (Fig. 3B) indicates that, during early ontogeny, molecular marker variation indeed represents quantitative genetic variation. Departure from genotypic-phenotypic correspondence during later ontogenetic stages might indicate *in-vivo* modification of the femoral shaft morphology, indicating an increasing contribution of V_E to V_P over ontogenetic time. Given the evidence from earlier studies investigating *in-vivo* effects on femoral shaft morphology [31,87,89,111], however, this interpretation is unlikely, and V_E remains fairly constant from infancy to adulthood [87]. Another possible explanation is size allometry, implying that the observed pattern of phenotypic divergence reflects differences in adult body mass among *Pan* taxa. Since direct data on body mass are available for only few specimens in this study, we use the taxon-specific body masses reported in the literature [112] to test this hypothesis. Taxon-specific means of PC scores at adulthood are not correlated with adult body masses of *Pan* taxa (Fig. S6, Table S8). It is thus unlikely that the observed pattern of divergence is due to allometry.

After excluding major environmental and allometric effects, it appears most likely that phenotypic divergence is caused by genetically determined taxon-specific developmental programs. This implies that the genetic variance V_G changes during ontogenetic time t: $V_P(t) = V_E + V_G(t)$. In the present case, it is not known whether $V_G(t)$ can be approximated by additive genetic variance $V_A(t)$ alone, or whether non-additive effects have to be taken into account. Several alternative hypotheses must thus be considered to explain the observed pattern of phenotypic divergence. Under the additive genetic variance model [$V_P(t) = V_E + V_A(t)$], our hypothesis is that the genes mediating early ontogeny (up to the m2 stage) evolved by neutral processes ($Q_{ST} \sim F_{ST}$), whereas the genes mediating late ontogeny (from m2 to adulthood) evolved under selection ($Q_{ST} > F_{ST}$), probably as an adaptation to taxon-specific locomotor regimes. An alternative hypothesis is that non-additive effects V_N are a function of developmental time: $V_P(t) = V_E + V_A(t) + V_N(t)$. With the currently available empirical evidence, we cannot decide between these hypotheses. In any case, the molecular markers used here to estimate V_G are unlikely to represent variation in the actual coding genes that cause V_P to increase over ontogenetic time [110].

In spite of these uncertainties, our data permit inferences on the developmental mechanisms that cause taxon-specific differences in femoral diaphyseal shape, and to speculate on their genetic basis. As shown in Fig. 1B, taxon-specific ontogenetic trajectories set out at similar locations along PC1, but differ in their length. This pattern indicates differences in taxon-specific *rates* of development from the m2 stage onward, resulting in significant differences between adult morphologies. Evolutionary divergence via differential developmental rates is well-known as heterochrony. It thus appears that heterochronic shifts played a major role in the development of the adult femoral morphologies of *Pan* taxa. Such shifts might be effected by changes in a small number of developmental genes [113,114], which are difficult to trace with standard molecular markers, but might be further investigated with whole-genome comparisons [23].

It has been shown that a marked paedomorphic pattern is expressed in the skull relative to the postcranial skeleton in bonobos (*P. paniscus*) compared to common chimpanzees (*P. troglodytes*) [33,115,116]. The present study shows that the femur also exhibits heterochronic variation among *Pan* taxa. It is interesting to note that the femoral diaphysis of bonobos exhibits peramorphic development compared to common chimpanzees. This mosaic structure of evolutionary developmental modification is in concordance with the observation made earlier that *P. paniscus* is not just a paedomorphic chimpanzee [116,117]. It remains to be elucidated whether cranial and postcranial ontogenies are governed by the same set of "heterochrony genes", which have different local effects, or whether different sets of heterochrony genes are expressed locally [113,118].

Currently, we can only speculate about the adaptive significance of taxon-specific heterochronic modifications of femoral development, since more comparative field data are necessary to specify the diversity of locomotor behaviors and their ontogeny in all *Pan* taxa. The inferred femoral diaphyseal morphology and developmental trajectory of the *Pan* LCA indicates that the peramorphic pattern as in *P. paniscus*, *P. t. troglodytes* and *P. t. verus* represents the primitive state whereas the paedomorphic (rate hypomorphic) pattern as in *P. t. schweinfurthii* represents the derived state. The inferred femoral diaphyseal morphology of the LCA at the adult stage is relatively close to the morphology of adult *P. paniscus* and *P. t. troglodytes*. The locomotor repertoire of the LCA might thus have been close to that of adult *P. paniscus* and *P. t. troglodytes*.

The data presented here provide empirical insights into the role of neutral and adaptive evolutionary mechanisms at the level of genes and phenes. In the system studied here, it appears that – among the closely related *Pan* taxa – early developmental genes evolve mostly neutrally and produce neutral taxon-specific phenotypes, while selection acts on late developmental genes (most likely on those involved in the regulation of developmental rates) and produces adaptive phenotypes.

Evidence for this pattern of evolution has also been found in the hominin clade. For example, the pattern of genotypic and phenotypic divergence between *Homo sapiens* and *H. neanderthalensis* is concordant with a model of neutral evolution by mutation and drift [6,8]. Also, parallel ontogenetic trajectories and heterochronic divergence during late ontogeny are reported for *Homo sapiens* and *H. neanderthalensis* [51]. Likewise, it appears that genetic and phenotypic divergence in early *Homo* and between modern human populations is governed to a large extent by neutral processes [1,3,5,10,119,120]. Our data indicate that this pattern of evolution might be more general than currently thought and characteristic not only for *Homo* but also for the taxa descending from the last common ancestor of humans and chimpanzees. It remains to be tested whether the observed patterns of developmental diversification in *Pan* also characterize the developmental diversification in other great ape taxa.

As a general outcome of this study, we may state that the phenotype of early developmental stages conveys a better neutral phylogenetic signal than the adult phenotype. This finding is in contrast with the traditional notion that the fully-developed adult phenotype is most significant for taxonomy and phyletic inference. The close match between patterns of neutral molecular and phenotypic variation during early ontogeny, however, indicates that immature individuals are of special relevance to infer phylogenetic relationships, although taxon-specific features are less expressed in early stages of ontogeny (Fig. 2B) compared to late stages (Fig. 2C). Femoral diaphyseal morphology of hominoids provides a good example. While adult-based studies often show similarities of femoral diaphyseal morphology among great apes to the exclusion of humans e.g. [121,122,123], at an early developmental stage humans and chimpanzees are grouped together to the exclusion of gorillas [46]. Furthermore, our data may explain why previous meta-analyses showed a generally low correlation of F_{ST} and Q_{ST} in adult phenotypes [74,110,124]. Generalizing our findings to hominoid (and hominin) evolution, the comparison of immature and adult phenotypes will permit a better discrimination between phyletic and adaptive signals in the phenotype.

Supporting Information

Figure S1 Geographical distribution and taxonomy of *Pan* (modified from ref. [22]).

Figure S2 Sample structure by taxon and age class. A, distribution of femoral diaphyseal length (measured as the linear distance between proximal and distal epiphyseal lines). B: distribution of femoral diaphyseal cross-sectional area (measured as the median of cross-sectional areas between proximal and distal epiphyses). Filled circles: *P.t. troglodytes*, open circles: *P.t. schweinfurthii*, open triangles: *P.t. verus*, open squares: *P. paniscus*. Age classes: m2: second deciduous molar erupted; M1/M2/M3: permanent molars 1/2/3 erupted. Each symbol represents a specimen; black lines/whiskers indicate mean and range; red boxes and whiskers indicate first/third quartiles and median.

Figure S3 Principle of morphometric mapping. A, 3D representation of the right femur. B, principle of cylindrical projection (anterior [0°] → medial [90°] → posterior [180°] → lateral [270°] → anterior [0°]).

Figure S4 Phylogenetic tree in morphospace. The phylogenetic tree (blue lines; diamonds indicate the inferred state of last common ancestor at each ontogenetic stage) of the genus *Pan* is projected onto the shape space using a model of squared-change parsimony. A: m2 (infant), B: M1, C: M2, D: M3 (adult) stage. Gray symbols and line indicate the inferred ontogenetic trajectory of the last common ancestor.

Figure S5 Phenetic and genetic similarity between *Pan* taxa. Principal Coordinates Analysis (PCO) of phenetic and genetic distance data. Phenetic data (black) are given for consecutive ontogenetic stages (connected with dashed lines). Genetic data (color) are from ref. [18] (blue), ref. [19] (green), ref. [21] (red), and ref. [22] (magenta). Note that during ontogeny the phenetic distance configuration departs from the genetic distance configuration.

Figure S6 Correlation of taxon-specific means of adult body weight and PC scores. Taxon-specific means of adult body weight was calculated as a mean of male and female body weight taken from the literature [112].

Table S1 Specimen list. The following specimens are used in this study. AIMUZH: Anthropological Institute and Museum of University of Zurich. MRA: Royal Africa Museum, Tervuren, Belgium.

Table S2 Genetic distances between _Pan_ taxa (F_{ST} and R_{ST}).

Table S3 Phenetic distances between taxon-specific mean shapes.

Table S4 Divergence of ontogenetic vector.

Table S5 F-test on taxon-specific variance along PC1.

Table S6 Correlation of genetic and phenetic distances.

Table S7 Correlation between phenetic and genetic distance matrices.

Table S8 Correlation between PC scores and taxon-specific adult body masses.

Text S1 Materials and methods.

Text S2 Habitats of _Pan_ taxa.

Text S3 _In-vivo_ bone modification in the femur of _Pan_ taxa.

Acknowledgments

We thank P. Jans, E Gillissen and W Coudyzer for help with sample preparation and CT scanning. The comments of H. Bagheri, M. Kobayashi, and T. Marques-Bonet are greatly acknowledged. We are also grateful to the anonymous reviewers for their valuable comments and suggestions.

Author Contributions

Conceived and designed the experiments: NM MSPDL CPEZ. Performed the experiments: NM. Analyzed the data: NM. Contributed reagents/materials/analysis tools: NM MSPDL CPEZ. Contributed to the writing of the manuscript: NM MSPDL CPEZ.

References

1. Ackermann RR, Cheverud JM (2004) Detecting genetic drift versus selection in human evolution. Proc Natl Acad Sci U S A 101: 17946–17951.
2. Roseman CC (2004) Detecting interregionally diversifying natural selection on modern human cranial form by using matched molecular and morphometric data. Proceedings of the National Academy of Sciences of the United States of America 101: 12824–12829.
3. Roseman CC, Weaver TD (2004) Multivariate apportionment of global human craniometric diversity. American Journal of Physical Anthropology 125: 257–263.
4. Harvati K, Weaver TD (2006) Human cranial anatomy and the differential preservation of population history and climate signatures. Anat Rec A Discov Mol Cell Evol Biol 288: 1225–1233.
5. Roseman CC, Weaver TD (2007) Molecules versus morphology? Not for the human cranium. Bioessays 29: 1185–1188.
6. Weaver TD, Roseman CC, Stringer CB (2007) Were neandertal and modern human cranial differences produced by natural selection or genetic drift? Journal of Human Evolution 53: 135–145.
7. Smith HF, Terhune CE, Lockwood CA (2007) Genetic, geographic, and environmental correlates of human temporal bone variation. American Journal of Physical Anthropology 134: 312–322.
8. Weaver TD, Roseman CC, Stringer CB (2008) Close correspondence between quantitative- and molecular-genetic divergence times for Neandertals and modern humans. Proc Natl Acad Sci U S A 105: 4645–4649.
9. von Cramon-Taubadel N, Weaver TD (2009) Insights from a quantitative genetic approach to human morphological evolution. Evolutionary Anthropology 18: 237–240.
10. Betti L, Balloux F, Hanihara T, Manica A (2010) The relative role of drift and selection in shaping the human skull. American Journal of Physical Anthropology 141: 76–82.
11. Reich D, Green RE, Kircher M, Krause J, Patterson N, et al. (2010) Genetic history of an archaic hominin group from Denisova Cave in Siberia. Nature 468: 1053–1060.
12. Krause J, Fu Q, Good JM, Viola B, Shunkov MV, et al. (2010) The complete mitochondrial DNA genome of an unknown hominin from southern Siberia. Nature 464: 894–897.
13. Green RE, Krause J, Briggs AW, Maricic T, Stenzel U, et al. (2010) A draft sequence of the Neandertal genome. Science 328: 710–722.
14. Hawks J (2013) Significance of Neandertal and Denisovan genomes in human evolution. Annual Review of Anthropology 42: 433–449.
15. Sankararaman S, Patterson N, Li H, Pääbo S, Reich D (2012) The date of interbreeding between Neandertals and modern humans. PLoS Genetics 8: e1002947.
16. Meyer M, Fu Q, Aximu-Petri A, Glocke I, Nickel B, et al. (2014) A mitochondrial genome sequence of a hominin from Sima de los Huesos. Nature 505: 403–406.
17. Gonder MK, Disotell TR, Oates JF (2006) New genetic evidence on the evolution of chimpanzee populations and implications for taxonomy. International Journal of Primatology 27: 1103–1127.
18. Becquet C, Patterson N, Stone AC, Przeworski M, Reich D (2007) Genetic structure of chimpanzee populations. PLoS Genetics 3: e66.
19. Gonder MK, Locatelli S, Ghobrial L, Mitchell MW, Kujawski JT, et al. (2011) Evidence from Cameroon reveals differences in the genetic structure and histories of chimpanzee populations. Proceedings of the National Academy of Sciences of the United States of America 108: 4766–4771.
20. Auton A, Fledel-Alon A, Pfeifer S, Venn O, Ségurel L, et al. (2012) A fine-scale chimpanzee genetic map from population sequencing. Science 336: 193–198.
21. Fischer A, Pollack J, Thalmann O, Nickel B, Pääbo S (2006) Demographic history and genetic differentiation in apes. Current Biology 16: 1133–1138.
22. Fischer A, Prufer K, Good JM, Halbwax M, Wiebe V, et al. (2011) Bonobos fall within the genomic variation of chimpanzees. PLoS ONE 6: e21605.
23. Prado-Martinez J, Sudmant PH, Kidd JM, Li H, Kelley JL, et al. (2013) Great ape genetic diversity and population history. Nature 499: 471–475.
24. Champneys F (1871) On the muscles and nerves of a chimpanzee (_Troglodytes niger_) and a _Cynocepalus anubis_. J Anat Lond 6: 176–211.
25. Crass E (1952) Musculature of the hip and thigh of the chimpanzee: a comparison to man and other primates. PhD thesis, Univ Wisconsin.
26. Sigmon BA (1974) A functional analysis of pongid hip and thigh musculature. Journal of Human Evolution 3: 161–185.
27. Stern JT (1972) Anatomical and functional specializations of human gluteus maximus. American Journal of Physical Anthropology 36: 315–338.
28. Morimoto N, Zollikofer CPE, Ponce de León MS (2011) Femoral morphology and femoropelvic musculoskeletal anatomy of humans and great apes: a comparative virtopsy study. Anatomical Record 294: 1433–1445.
29. Bourne GH, editor (1969) The Chimpanzee, Vol. 1: Anatomy, Behavior, and Diseases of Chimpanzees. Basel/New York: Karger.
30. Bourne GH, editor (1971) The Chimpanzee, Vol. 4: Behavior, Growth, and Pathology of Chimpanzees. Basel/New York: Karger.
31. Carlson KJ, Doran-Sheehy DM, Hunt KD, Nishida T, Yamanaka A, et al. (2006) Locomotor behavior and long bone morphology in individual free-ranging chimpanzees. Journal of Human Evolution 50: 394–404.
32. Doran DM (1993) Comparative locomotor behavior of chimpanzees and bonobos - the influence of morphology on locomotion. American Journal of Physical Anthropology 91: 83–98.
33. Lieberman DE, Carlo J, Ponce de León MS, Zollikofer CP (2007) A geometric morphometric analysis of heterochrony in the cranium of chimpanzees and bonobos. J Hum Evol 52: 647–662.
34. Morin PA, Moore JJ, Chakraborty R, Jin L, Goodall J, et al. (1994) Kin selection, social-structure, gene flow, and the evolution of chimpanzees. Science 265: 1193–1201.
35. Doran DM (1992) The ontogeny of chimpanzee and pygmy chimpanzee locomotor behavior: a case-study of paedomorphism and its behavioral-correlates. Journal of Human Evolution 23: 139–157.
36. Doran DM (1996) Comparative positional behabior of the African apes. In: McGrew MC, Marchant LF, Nishida T, editors. Great Ape Societies. Cambridge: Cambridge University Press.

37. Doran DM (1997) Ontogeny of locomotion in mountain gorillas and chimpanzees. J Hum Evol 32: 323–344.

38. Doran DM, Jungers WL, Sugiyama Y, Fleagle J, Heesy C (2002) Multivariate and phylogenetic approaches to understanding chimpanzee and bonobo behavioral diversity. In: Boesch C, Hohmann G, Marchant LF, editors. Behavioural diversity in Chimpanzees and Bonobos. Cambridge: Cambridge University Press. 14–34.

39. Goodall J (1986) The Chimpanzees of Gombe. Cambridge: Harvard University Press.

40. McGrew MC, Marchant LF, Nishida T, editors (1996) Great Ape Societies. Cambridge: Cambridge University Press.

41. Nei M (2007) The new mutation theory of phenotypic evolution. Proc Natl Acad Sci U S A 104: 12235–12242.

42. West-Eberhard MJ (2005) Developmental plasticity and the origin of species differences. Proc Natl Acad Sci U S A 102 Suppl 1: 6543–6549.

43. Shaw KL, Mullen SP (2011) Genes versus phenotypes in the study of speciation. Genetica 139: 649–661.

44. Collard M, Wood B (2000) How reliable are human phylogenetic hypotheses? Proceedings of the National Academy of Sciences 97: 5003–5006.

45. Strait DS, Grine FE (2004) Inferring hominoid and early hominid phylogeny using craniodental characters: the role of fossil taxa. Journal of Human Evolution 47: 399–452.

46. Morimoto N, Zollikofer CPE, Ponce de León MS (2012) Shared human-chimpanzee pattern of perinatal femoral shaft morphology and its implications for the evolution of hominin locomotor adaptations. PLoS ONE 7: e41980.

47. Geiger M, Forasiepi AM, Koyabu D, Sanchez-Villagra MR (2013) Heterochrony and post-natal growth in mammals - an examination of growth plates in limbs. Journal of Evolutionary Biology 20: 12279.

48. Koyabu D, Endo H, Mitgutsch C, Suwa G, Catania KC, et al. (2011) Heterochrony and developmental modularity of cranial osteogenesis in lipotyphlan mammals. Evodevo 2: 21.

49. Wilson LAB, Sánchez-Villagra MR (2011) Evolution and phylogenetic signal of growth trajectories: the case of chelid turtles. Journal of Experimental Zoology Part B: Molecular and Developmental Evolution 316B: 50–60.

50. Sánchez M (2012) Embryos in Deep Time: The Rock Record of Biological Development: University of California Press. 265 p.

51. Ponce de León MS, Zollikofer CPE (2001) Neanderthal cranial ontogeny and its implications for late hominid diversity. Nature 412: 534–538.

52. Ackermann RR (2005) Ontogenetic integration of the hominoid face. Journal of Human Evolution 48: 175–197.

53. Gunz P, Neubauer S, Golovanova L, Doronichev V, Maureille B, et al. (2012) A uniquely modern human pattern of endocranial development. Insights from a new cranial reconstruction of the Neandertal newborn from Mezmaiskaya. Journal of Human Evolution 62: 300–313.

54. Palagi E, Cordoni G (2012) The right time to happen: play developmental divergence in the two *Pan* species. PLoS ONE 7: e52767.

55. Hill WG, Goddard ME, Visscher PM (2008) Data and theory point to mainly additive genetic variance for complex traits. PLoS Genetics 4: e1000008.

56. Morgan TJ, Evans MA, Garland T Jr, Swallow JG, Carter PA (2005) Molecular and quantitative genetic divergence among populations of house mice with known evolutionary histories. Heredity 94: 518–525.

57. Whitlock MC (2008) Evolutionary inference from Q_{ST}. Molecular Ecology 17: 1885–1896.

58. Smith HF (2009) Which cranial regions reflect molecular distances reliably in humans? Evidence from three-dimensional morphology. American Journal of Human Biology 21: 36–47.

59. Brommer JE (2011) Whither P_{ST}? The approximation of Q_{ST} by P_{ST} in evolutionary and conservation biology. Journal of Evolutionary Biology 24: 1160–1168.

60. Edelaar PIM, Burraco P, Gomez-Mestre I (2011) Comparisons between Q_{ST} and F_{ST}—how wrong have we been? Molecular Ecology 20: 4830–4839.

61. Wright S (1969) Evolution and the Genetics of Populations, Vol. II. The Theory of Gene Frequencies. Chicago: University of Chicago Press.

62. Wright S (1978) Evolution and the Genetics of Populations, Vol. IV. Variability Within and Among Natural Populations. Chicago: University of Chicago Press.

63. Holsinger KE, Weir BS (2009) Genetics in geographically structured populations: defining, estimating and interpreting FST. Nature Reviews Genetics 10: 639–650.

64. Spitze K (1993) Population-structure in *Daphnia obtusa*: quantitative genetic and allozymic variation. Genetics 135: 367–374.

65. Leinonen T, McCairns RJS, O'Hara RB, Merila J (2013) Q_{ST}–F_{ST} comparisons: evolutionary and ecological insights from genomic heterogeneity. Nature Reviews Genetics 14: 179–190.

66. Leinonen T, Cano JM, MÄKinen H, MerilÄ J (2006) Contrasting patterns of body shape and neutral genetic divergence in marine and lake populations of threespine sticklebacks. Journal of Evolutionary Biology 19: 1803–1812.

67. Merilä J, Björklund M, Baker AJ (1997) Historical demography and present day population structure of the greenfinch, carduelis chloris-an analysis of mtDNA control-region sequences. Evolution 51: 946–956.

68. Storz JF (2002) Contrasting patterns of divergence in quantitative traits and neutral DNA markers: analysis of clinal variation. Molecular Ecology 11: 2537–2551.

69. Saint-Laurent R, Legault M, Bernatchez L (2003) Divergent selection maintains adaptive differentiation despite high gene flow between sympatric rainbow smelt ecotypes (*Osmerus mordax* Mitchill). Molecular Ecology 12: 315–330.

70. Slate J (2013) From beavis to beak color: a simulation study to examine how much qtl mapping can reveal about the genetic architecture of quantitative traits. Evolution 67: 1251–1262.

71. Atchley WR (1984) Ontogeny, timing of development, and genetic variance-covariance structure. American Naturalist 123: 519–540.

72. Charmantier A, Perrins C, McCleery RH, Sheldon BC (2006) Age-dependent genetic variance in a life-history trait in the mute swan. Proceedings of the Royal Society B-Biological Sciences 273: 225–232.

73. Lesser KJ, Paiusi IC, Leips J (2006) Naturally occurring genetic variation in the age-specific immune response of *Drosophila melanogaster*. Aging Cell 5: 293–295.

74. Pujol B, Wilson AJ, Ross RIC, Pannell JR (2008) Are Q_{ST}–F_{ST} comparisons for natural populations meaningful? Molecular Ecology 17: 4782–4785.

75. Kalinowski ST (2002) Evolutionary and statistical properties of three genetic distances. Molecular Ecology 11: 1263–1273.

76. Nei M, Tajima F, Tateno Y (1983) Accuracy of estimated phylogenetic trees from molecular data. II. Gene frequency data. Journal of Molecular Evolution 19: 153–170.

77. Cavalli-Sforza LL, Edwards AW (1967) Phylogenetic analysis. Models and estimation procedures. Am J Hum Genet 19: 233–257.

78. Patterson N, Price AL, Reich D (2006) Population structure and eigenanalysis. PLoS Genetics 2: e190.

79. Price AL, Patterson NJ, Plenge RM, Weinblatt ME, Shadick NA, et al. (2006) Principal components analysis corrects for stratification in genome-wide association studies. Nat Genet 38: 904–909.

80. Slice DE (2007) Geometric morphometrics. Annual Review of Anthropology 36: 261–281.

81. Mitteroecker P, Gunz P (2009) Advances in geometric morphometrics. Evolutionary Biology 36: 235–247.

82. Zollikofer CPE, Ponce de León MS (2005) Virtual Reconstruction. A Primer in Computer-Assisted Paleontology and Biomechanics. Hoboken: NJ: John Wiley & Sons.

83. Bookstein F (1991) Morphometric Tools for Landmark Data: Geometry and Biology. Cambridge: Camnridge University Press.

84. Gunz P, Mitteroecker P, Bookstein FL (2005) Semilandmarks in Three Dimensions. In: Slice DE, editor. Developments in Primatology: Progress and Prospects. New York: Springer.

85. Kuhl F, Giardina C (1982) Elliptic Fourier features of a closed contour. Computer graphics and image processing 18: 236–258.

86. Specht M, Lebrun R, Zollikofer CPE (2007) Visualizing shape transformation between chimpanzee and human braincases. Visual Computer 23: 743–751.

87. Morimoto N, Zollikofer CPE, Ponce de León MS (2011) Exploring femoral diaphyseal shape variation in wild and captive chimpanzees by means of morphometric mapping: a test of Wolff's Law. Anatomical Record 294: 589–609.

88. Bolter DR, Zihlman AL (2011) Brief communication: dental development timing in captive *Pan paniscus* with comparisons to *Pan troglodytes*. American Journal of Physical Anthropology 145: 647–652.

89. Carlson K, Sumner D, Morbeck M, Nishida T, Yamanaka A, et al. (2008) Role of nonbehavioral factors in adjusting long bone siaphyseal atructure in free-ranging *Pan troglodytes*. International Journal of Primatology 29: 1401–1420.

90. Zollikofer CPE, Ponce de León MS (2001) Computer-assisted morphometry of hominoid fossils: the role of morphometric maps. In: De Bonis L, Koufos G, Andrews P, editors. Phylogeny of the Neogene Hominoid Primates of Eurasia. Cambridge: Cambridge University Press. 50–59.

91. Bondioli L, Bayle P, Dean C, Mazurier A, Puymerail L, et al. (2010) Technical note: Morphometric maps of long bone shafts and dental roots for imaging topographic thickness variation. American Journal of Physical Anthropology 142: 328–334.

92. Puymerail L (2013) The functionally-related signatures characterizing the endostructural organisation of the femoral shaft in modern humans and chimpanzee. Comptes Rendus Palevol.

93. Puymerail L, Ruff CB, Bondioli L, Widianto H, Trinkaus E, et al. (2012) Structural analysis of the Kresna 11 *Homo erectus* femoral shaft (Sangiran, Java). Journal of Human Evolution 63: 741–749.

94. Maddison WP (1991) Squared-change parsimony reconstructions of ancestral states for continuous-valued characters on a phylogenetic tree. Systematic Biology 40: 304–314.

95. Klingenberg CP (2011) MorphoJ: an integrated software package for geometric morphometrics. Molecular Ecology Resources 11: 353–357.

96. Relethford JH, Blangero J (1990) Detection of differential gene flow from patterns of quantitative variation. Human Biology 62: 5–25.

97. Relethford JH, Crawford MH, Blangero J (1997) Genetic drift and gene flow in post-famine Ireland. Human Biology 69: 443–465.

98. Sæther SA, Fiske P, Kålås JA, Kuresoo A, Luigujõe L, et al. (2007) Inferring local adaptation from Q_{ST}–F_{ST} comparisons: neutral genetic and quantitative trait variation in European populations of great snipe. Journal of Evolutionary Biology 20: 1563–1576.

99. Chapuis E, Martin G, Goudet J (2008) Effects of selection and drift on G matrix evolution in a heterogeneous environment: a multivariate Q_{st}-F_{st} test with the freshwater snail *Galba truncatula*. Genetics 180: 2151–2161.

100. Martin G, Chapuis E, Goudet J (2008) Multivariate Q_{st}–F_{st} comparisons: a neutrality test for the evolution of the G matrix in structured populations. Genetics 180: 2135–2149.

101. Relethford JH (2004) Boas and beyond: Migration and craniometric variation. American Journal of Human Biology 16: 379–386.

102. Zollikofer CPE, Ponce de León MS (2006) Neanderthals and modern humans - chimps and bonobos: similarities and differences in development and evolution. In: Harvati K, Harrison T, editors. Neanderthals Revisited: New Approaches and Perspectives. New York: Springer. 71–88.

103. Penin X, Berge C, Baylac M (2002) Ontogenetic study of the skull in modern humans and the common chimpanzees: neotenic hypothesis reconsidered with a tridimensional Procrustes analysis. Am J Phys Anthropol 118: 50–62.

104. Willing E-M, Dreyer C, van Oosterhout C (2012) Estimates of genetic differentiation measured by Fst do not necessarily require large sample sizes when using many SNP markers. PLoS ONE 7: e42649.

105. Hunt G (2007) The relative importance of directional change, random walks, and stasis in the evolution of fossil lineages. Proceedings of the National Academy of Sciences of the United States of America 104: 18404–18408.

106. Haller BC, Hendry AP (2014) Solving the paradox of stasis: squashed stabilizing selection and the limits of detection. Evolution 68: 483–500.

107. Wagner GP (1996) Apparent stabilizing selection and the maintenance of neutral genetic variation. Genetics 143: 617–619.

108. Kirk H, Freeland JR (2011) Applications and implications of neutral versus non-neutral markers in molecular ecology. Int J Mol Sci 12: 3966–3988.

109. McKay JK, Latta RG (2002) Adaptive population divergence: markers, QTL and traits. Trends in Ecology & Evolution 17: 285–291.

110. Reed DH, Frankham R (2001) How closely correlated are molecular and quantitative measures of genetic variation? A meta-analysis. Evolution 55: 1095–1103.

111. Carlson KJ, Lublinsky S, Judex S (2008) Do different locomotor modes during growth modulate trabecular architecture in the murine hind limb? Integrative and Comparative Biology 48: 385–393.

112. Smith RJ, Jungers WL (1997) Body mass in comparative primatology. Journal of Human Evolution 32: 523–559.

113. Somel M, Franz H, Yan Z, Lorenc A, Guo S, et al. (2009) Transcriptional neoteny in the human brain. Proceedings of the National Academy of Sciences of the United States of America 106: 5743–5748.

114. Somel M, Liu X, Tang L, Yan Z, Hu H, et al. (2011) MicroRNA-driven developmental remodeling in the brain distinguishes humans from other primates. PLoS Biology 9: e1001214.

115. Mitteroecker P, Gunz P, Bookstein FL (2005) Heterochrony and geometric morphometrics: a comparison of cranial growth in *Pan paniscus* versus *Pan troglodytes*. Evolution & Development 7: 244–258.

116. Shea BT (1983) Pedomorphosis and neoteny in the pygmy chimpanzee. Science 222: 521–522.

117. Shea BT (1983) Allometry and heterochrony in the African apes. American Journal of Physical Anthropology 62: 275–289.

118. Khaitovich P, Enard W, Lachmann M, Pääbo S (2006) Evolution of primate gene expression. Nature Reviews Genetics 7: 693–702.

119. Lynch M (1989) Phylogenetic hypotheses under the assumption of neutral quantitative-genetic variation. Evolution 43: 1–17.

120. Roseman CC (2004) Detecting interregionally diversifying natural selection on modern human cranial form by using matched molecular and morphometric data. Proc Natl Acad Sci U S A 101: 12824–12829.

121. Lovejoy CO, Meindl RS, Ohman JC, Heiple KG, White TD (2002) The Maka femur and its bearing on the antiquity of human walking: Applying contemporary concepts of morphogenesis to the human fossil record. American Journal of Physical Anthropology 119: 97–133.

122. Ruff CB (2002) Long bone articular and diaphyseal structure in old world monkeys and apes. I: Locomotor effects. American Journal of Physical Anthropology 119: 305–342.

123. Carlson KJ (2005) Investigating the form-function interface in African apes: Relationships between principal moments of area and positional behaviors in femoral and humeral diaphyses. Am J Phys Anthropol 127: 312–334.

124. Leinonen T, O'Hara RB, Cano JM, Merilä J (2008) Comparative studies of quantitative trait and neutral marker divergence: a meta-analysis. Journal of Evolutionary Biology 21: 1–17.

3

Genetic Signatures for Enhanced Olfaction in the African Mole-Rats

Sofia Stathopoulos[1]*, Jacqueline M. Bishop[2], Colleen O'Ryan[1]

1 Department of Molecular and Cell Biology, University of Cape Town, Cape Town, Western Cape, South Africa, 2 Department of Biological Sciences, University of Cape Town, Cape Town, Western Cape, South Africa

Abstract

The Olfactory Receptor (OR) superfamily, the largest in the vertebrate genome, is responsible for vertebrate olfaction and is traditionally subdivided into 17 OR families. Recent studies characterising whole-OR subgenomes revealed a 'birth and death' model of evolution for a range of species, however little is known about fine-scale evolutionary dynamics within single-OR families. This study reports the first assessment of fine-scale OR evolution and variation in African mole-rats (Bathyergidae), a family of subterranean rodents endemic to sub-Saharan Africa. Because of the selective pressures of life underground, enhanced olfaction is proposed to be fundamental to the evolutionary success of the Bathyergidae, resulting in a highly diversified OR gene-repertoire. Using a PCR-sequencing approach, we analysed variation in the OR7 family across 14 extant bathyergid species, which revealed enhanced levels of functional polymorphisms concentrated across the receptors' ligand-binding region. We propose that mole-rats are able to recognise a broad range of odorants and that this diversity is reflected throughout their OR7 gene repertoire. Using both classic tests and tree-based methods to test for signals of selection, we investigate evolutionary forces across the mole-rat OR7 gene tree. Four well-supported clades emerged in the OR phylogeny, with varying signals of selection; from neutrality to positive and purifying selection. Bathyergid life-history traits and environmental niche-specialisation are explored as possible drivers of adaptive OR evolution, emerging as non-exclusive contributors to the positive selection observed at OR7 genes. Our results reveal unexpected complexity of evolutionary mechanisms acting within a single OR family, providing insightful perspectives into OR evolutionary dynamics.

Editor: Hiroaki Matsunami, Duke University, United States of America

Funding: This study was supported by the University of Cape Town Research Committee and the South African National Research Foundation. The funders had no role in study design, data collection and analysis, decision to publish, or preparation of the manuscript.

Competing Interests: The authors have declared that no competing interests exist.

* E-mail: sofiastathopoulos@hotmail.com

Introduction

Highly developed olfaction and odour discrimination underpin a number of fitness-related behaviours in mammals, from foraging and predator avoidance, to individual recognition, mate choice and maternal care [1–3]. In vertebrates, odour molecules are detected by seven trans-membrane G-protein-coupled receptors (7-TM GPCRs) encoded by the olfactory receptor (OR) gene family - the largest in the vertebrate genome [4]. From available genome data, it is clear that the extent of the vertebrate OR repertoire varies considerably, ranging from ~100 genes in fish [5], to 400–1000 ORs in tetrapods (from 388 functional ORs in humans to 1259 functional genes in rats; [5–7]), where the expansion of OR gene repertoires is thought to reflect the shift from aquatic to terrestrial environments in the Middle Devonian, some 395 MYA [8]. As with the evolution of most multi-gene families, dynamic and rapid evolution via the birth-and-death model has been proposed for the OR gene family. Here, new OR genes arise through duplication and then either diversify in function in response to selection, lose function via pseudogenization, or are lost from the genome [9–11]. Thus, the extent of any OR repertoire (i.e. number of genes and the diversity among these genes) depends on diverse evolutionary forces, as well as the extent of duplication and inactivation events that characterise the evolution of a species' genome [11].

Vertebrate ORs are predominantly expressed in the sensory neurons of the main olfactory epithelium (MOE) [12]; further evidence also supports their expression in the rodent vomeronasal organ (VNO) and septal organ of Masera [13].

Genetic variation within OR genes is concentrated in the ligand-binding pockets of the receptors, spanning trans-membrane domains 2–7 (TM 2–7) [12,14,15]. High levels of polymorphism in this region are associated with the recognition of a wide range of chemicals, including both odorants and semiochemicals [16–18]. While the overall structure of ORs is maintained by strong purifying selection, a signal of positive selection in the ligand-binding region is reported in a diverse range of species, from fish to rodents [19–21]. This is consistent with the evolutionary pressure to generate and maintain adaptive binding properties at ORs, for the recognition of ecologically important odorants across species and habitats [22].

Olfactory acuity in vertebrates is commonly measured using the number of 'functional' OR genes in a species genome, together with the ratio of functional OR genes: pseudogenes [18,23,24]. Functional OR gene number is thought to be proportional to the range of scents that can be detected and discriminated between [17,18]. On the other hand, the ratio of OR genes:pseudogenes depends on the evolutionary forces that have shaped the OR repertoire of a species. Accordingly, these two measures vary across species, as a result of both lineage age and the selective environment in which they have evolved. For example, a number of extant rodent species, known to rely on highly developed olfaction for fitness-related tasks, have a large proportion of functional OR genes in their repertoires. In contrast, in species where olfaction has regressed, there is a higher fraction of OR pseudogenes. For example, in primates the evolution of full trichromatic vision is proposed to have influenced loss of OR diversity [25] (but see [26]).

Increasing evidence supports a role for ecological niche adaptations in the evolution of the vertebrate OR repertoire. A recent comparative survey of mammalian OR subfamily diversity, proposed a significant role for ecological niches in the evolution of OR functional diversity [22]. Similarly, the loss of OR functionality in cetaceans appears directly related to the evolution of an aquatic lifestyle [27]. Noteworthy, is the higher proportion of functional ORs reported in baleen whales (Mysticeti), which have a complex olfactory bulb, in comparison to toothed whales (Odontoceti), implying greater olfactory ability in mysticetes. This increased olfactory sensitivity is hypothesised to enable mysticetes to orientate more successfully toward aggregations of their dominant food source, krill [28]. Likewise, elapid snakes, viviparous species that have recently adapted to a marine lifestyle (Subfamily Hydrophiinae, ~8 MYA; [29]), have also experienced extensive OR pseudogenisation in comparison to both oviparous aquatic snake species, which still require land-based nests for their eggs, and fully terrestrial species [30]. In birds, a larger OR repertoire is found in a number of nocturnal species, that are known to rely on olfactory cues, as compared to their closest diurnal relatives [31]. Thus, the physical environment clearly influences functional diversification and size of this multi-gene family [27,30,32].

Here we explore OR diversity and evolution within a single OR family, namely OR7, in the African mole-rats. These burrowing rodents of the family Bathyergidae are endemic to sub-Saharan Africa, and most notable for their broad range of social strategies [33]. Whilst they do disperse above-ground, mole-rats essentially live permanently underground and have evolved an array of morphological, physiological and behavioural adaptations [33,34]. All species are poorly equipped for utilisation of the visual field [35] and exhibit little neuro-anatomical or molecular evidence of adaptation for low-light vision [36–38]. Whilst light/dark discrimination has been reported, the bathyergid central visual system is significantly reduced [37,39] and, in the absence of visual cues, all species exhibit enhanced olfactory sensitivity [33,36]. Olfactory cues direct mole-rats digging towards food resources, thus minimising the energy investment necessary for successful foraging [40,41]. For example, naked mole-rats, *Heterocephalus glaber*, recruit colony members to food sources by laying down odour trails [42], and similarly use olfactory cues during colony interactions [43–46]. Furthermore, complex scent marking rituals are used in common nesting and latrine areas within the extensive burrow systems of all the social mole-rats [33,47]. This chemo-communication in naked mole-rats is perhaps surprising, given that they lack a functional vomeronasal organ (VNO) [48]. Thus, pheromonal communication in naked mole-rats may be mediated by the MOE in a similar manner to that hypothesised for humans [49]. Other examples of chemo-communication in bathyergids are reported in species of the social genus *Cryptomys*, where individuals are able to discriminate between kinspecific and heterospecific odours using a proposed "self-referent matching" mechanism [50–

52]; this information is used to both reinforce individual and group recognition rituals and to limit incestuous mating [53].

Given the socio-ecological significance of odour discrimination in the Bathyergidae, we examined OR7 diversity across all genera of extant mole-rats and present the first assessment of OR gene diversity and evolution in a subterranean mammal. Useing PCR and sequencing methods, we characterise representative OR7 diversity across 14 bathyergid species and classify bathyergid OR genes, based on phylogenetic relationships together with a range of published OR subgenomes. We hypothesize that well-developed olfaction in Bathyergidae is the result of an expansion within the OR multi-gene family, resulting in increased divergence among OR7 genes. We also test whether patterns of OR7 variation in the amino acids involved in ligand-binding, are consistent with a scenario of adaptive functional variability across the Bathyergidae. In this context, we use phylogenetic-based methods to test whether adaptive evolution has operated differentially across bathyergid OR7 clades. Finally, we investigate a role for sociality and environmental niche specialisation in determining OR7 gene diversity in mole-rats and interpret our results within the framework of Nei's 'birth-and-death' model of evolution for multi-gene families [9].

Results

Olfactory Receptor Diversity in African Mole-rats

The Bathy-OR1/Bathy-OR2 primer pair were designed in this study and yielded unambiguous amplification of OR7 loci in all 14 African mole-rat species This produced a final alignment of 178 unique OR7 sequences (GenBank accession numbers KF453235–KF453412), and a BLAST search confirmed the sequence identity as OR7 genes for all sequences in the dataset. A 'conserved domains' search revealed the presence of typical GPCRs features in all sequences [54], whilst known OR motifs were confirmed by eye from the amino acid alignment [23,55].

Consistent with published studies, mole-rat OR sequences were considered to be pseudogenes if they had mutations that disrupted the 7TM receptor structure; these mutations included stop codons and frameshift mutations [8,31,56]. Using these criteria, 97 of the 178 bathyergid OR sequences were classified as pseudogenes. However, this may be a potential underestimation of the number of pseudogenes because of additional mutations outside the amplified region (TM 2–7), or mutations in promoter regions that were not amplied [25,57].

After allelic variants were merged, 119 unique OR7 genes were identified from the original pool of 178 OR7 gene candidates, including 51 putatively functional ORs and 68 OR pseudogenes. Interestingly, alleles of the same OR7 gene (as well as identical alleles) were identified across a number of mole-rat species and tentatively supports the idea that OR7 diversification may have preceded speciation in Bathyergidae.

The distribution of amino acid diversity across Bathyergidae OR7 genes was assessed based on Katada et al.'s molecular model of the mouse mOR-EG receptor [15]. The topological distribution of conserved and variable sites in mole-rat receptors is analogous to that of mOR-EG [15], with 73% of highly conserved residues shared, and 88% of variable residues occupying the same locations (Figure 1). High levels of both nucleotide and amino acid sequence polymorphism were detected in mole-rat OR7 sequences, and variability is concentrated in the region between TM3 and TM6, which corresponds to the predicted core of the ligand-binding pocket of ORs (Figure 1) [14,15]. Interestingly, 19 of the 26 amino acid residues predicted to be involved in ligand-binding [14,15], are variable in bathyergids. If residues in TM domains 2 and 7 are

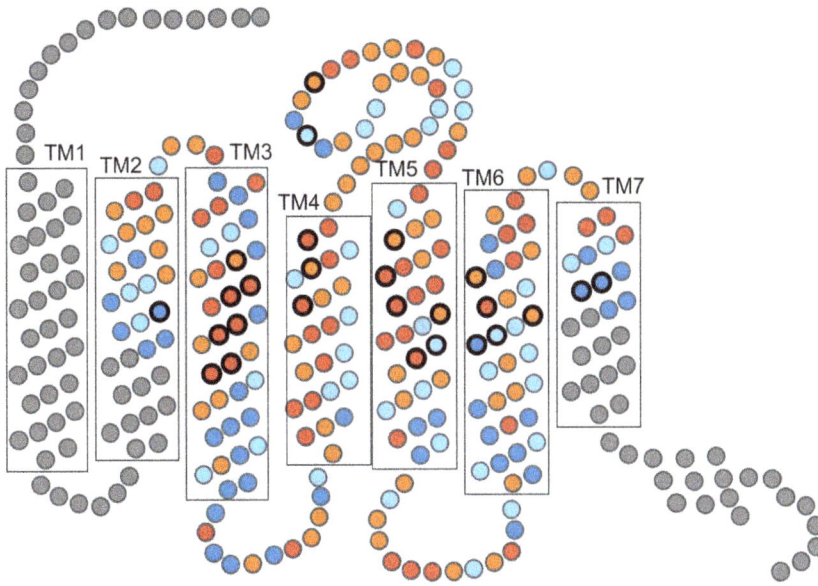

Figure 1. Functional variability across mole-rat OR7 receptors (redrawn from [15]**).** Functional variation is colour coded based on the number of different amino-acids presents at each position: red – highly variable (≥5); orange – variable (3–4); light blue – conserved (2); dark blue – highly conserved (1). Amino-acid positions involved odorant-binding are circled in black [14,15]; these are predominantly variable in our dataset, as expected. Abbreviations stand for the following: TM trans-membrane domain, EC extra-cellular and IC intracellular domain.

excluded, 83% of the alleged odorant-binding sites in mole-rat OR7 genes are polymorphic, consistent with a role in odorant recognition [14].

In keeping with other published studies, recombination is not a significant mechanism for the generation of sequence variability across mole-rat OR7 loci [10]. Tests for linkage disequilibrium did not indicate significant pairwise associations between polymorphic sites ($ZZ = 0.006$). This is consistent with the widely accepted idea that variability across OR genes is predominantly the result of gene duplication events and nucleotide substitution driven by positive selection, rather than recombination [10]. This result means that recombinant PCR artefacts are unlikely to have obscured the signal in our dataset [58].

Phylogenetic Relationships among Mole-rat OR7 Sequences

Phylogenetic reconstruction revealed four well-supported clades of closely-related OR7 genes (bootstrap support ≥97%). The four clades were named clades A–D, and an isolated gene (BJ4_A12), that is a sister lineage to clades A and B, was also observed. Identical phylogenetic topology was recovered when only a single representative sequence for each putative OR7 gene was analysed. The numbers and ratios of functional OR7 genes and pseudogenes across clades A–D are reported in Table S1.

The four clades do not cluster in a species-specific way. Instead, sequences in each clade were found to share functional motifs across the ligand-binding sites. Of the 23 amino-acid positions involved in odorant-binding across TM3-6 [14,15], only three were found to be conserved across all clades, whilst the remaining 20 sites displayed clade-specific motifs. This is consistent with OR7 genes in each clade having different binding properties. Furthermore, there is a striking prevalence of hydrophobic amino-acids (92% of all amino-acids involved in odorant-binding) across the putative ligand-binding domain of OR7 genes in all clades. This result supports Katada et al.'s hypothesis [15], that the interaction

between ORs and odorant ligands occur primarily via hydrophobic and van der Waals interactions [59].

Classification and Evolution of Mole-rat OR7 Genes

Using genetic similarity criteria, mammalian OR genes are subdivided into Class I and Class II genes [60,61], and these classes are further partitioned into 17 families. There are four Class I families, families 51, 52, 55 and 56, and 13 Class II families, families 1 to 13 [61,62]. Although the differential functions of these families and the range of odorants they can recognise is poorly understood [5], it has been mooted that each family might detect a particular class of odorant molecules [63].

In order to identify the OR genes amplified in our study, we inferred phylogenetic relationships between mole-rat OR genes and representative OR sequences from the entire OR repertoires of 18 different mammalian species [22]. The resulting phylogeny reveals strong support for Bathyergidae OR genes clustering together with Family 7 OR (OR7) genes from a number of mammalian species (Figure 2).

Family 7 OR genes represent a polyphyletic family of Class II OR genes in mammals, and are classified as part of the larger grouping of families 1/3/7 [22]. However, OR genes from families 1 and 3 appear to group independently from family 7, in strongly supported clades in our tree (Figure 2), and are more distantly related to the mole-rat OR genes characterized in this study.

The evolution of OR7 Bathyergidae genes was inferred by phylogenetic analyses of all the available mammalian OR7 sequences from Hayden et al.'s dataset [22]. Again, African mole-rat OR7 genes clustered into four strongly supported clades, which correspond to clades A–D in the Bathyergidae phylogenetic tree (with the exception of two genes; Figure S2). Interestingly, clades A, B and D appear to be Bathyergidae-specific clades, whilst clade C included OR7 genes from other mammalian species. Other family-specific clades are highlighted in the tree by a colour-coded classification of mammalian OR7 genes (Figure S2).

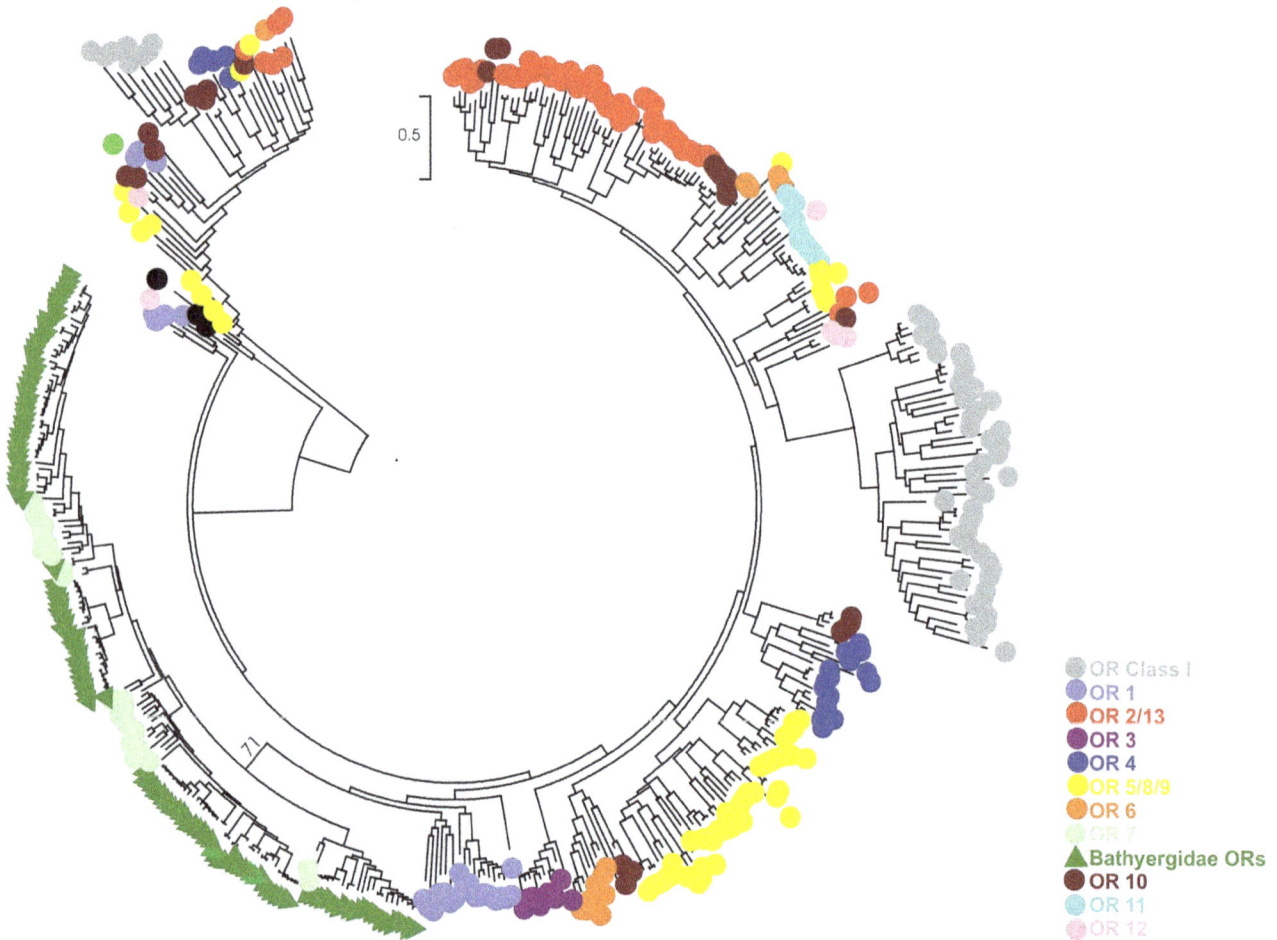

Figure 2. Mammalian OR family structure. Maximum likelihood tree obtained with Tamura-Nei substitution model (1000 bootstrap) using representative sequences of all OR Families from the available Mammalian database [22], together with the Bathyergidae OR genes characterised in this study. OR families are colour-coded as reported on the right. All Bathyergidae ORs appear to cluster together with mammalian Family 7 OR genes (indicated in green, together with the bootstrap support value for that branch).

Signatures of Selection in the African Mole-rat OR7 Tree

Tests for differential positive selection across bathyergid OR7 clades revealed a number of evolutionary patterns. Likelihood ratio tests (LRT) for ongoing positive selection were performed on the functional OR7 genes from clades A, C and D, while clade B was excluded from this analysis due to insufficient sample size. The LRT results reveal significant positive selection for clade A only, although numerous codons were found to be evolving under positive selection in both clades A and C (Figure 3.II).

Notably, no amino-acid sites in clade D were characterised by dN/dS>1; instead, all sites within this clade were characterised by dN/dS ratios <1, which is consistent with purifying selective forces acting along the OR genes of this clade. A codon-based Z-test was then used and a strong signal of purifying selection was confirmed (p<0.0001).

To identify episodic events of adaptive evolution across specific Bathyergidae OR7 lineages, we used a branch-site test of positive selection across all branches of the African mole-rat OR gene tree. Six branches in the tree support a signal of positive selection in the corresponding lineages (p<0.05). However, when Q-values are taken into account, positive selection can only be inferred unequivocally for two branches (75 and 34, Q-value <0.001; Figure 3.I). The next two branches (# 27 and 63 Figure 3.I) are

only mildly significant (Q-value = 0.13), whereas from the fifth branch the Q value jumps to 0.54. Results from a Bayes Empirical Bayes (BEB) [64,65] analysis to identify which amino-acid sites are evolving under adaptive evolution, revealed that the number and location of positively selected sites vary among these lineages (reported in Figure 3.II and Table S2).

Divergent, lineage-specific evolutionary forces in mole-rat OR7 genes are revealed when considering signals of both current and episodic selection and the proportion of functional genes in of the four clades. Firstly, within clade A, significant adaptive selection was detected both at ancestral branches (branch #34, Figure 3.I) and at the tips of the tree. Clade A is also characterised by a relatively higher OR7 genes: pseudogenes ratio, when compared to other clades. Along branch #34, six of the eight amino-acid sites that were identified as evolving under positive selection, based on the BEB analysis, lie in TM3-6 region (Figure 3.II). This is suggestive of selection acting predominantly on the odorant-binding region in this gene lineage, presumably to generate novel binding properties. The second branch that carries a mild signal of positive selection in this clade (# 27 Figure 3.I), leads to a subset of *H. glaber* OR7 pseudogenes that have only one positively selected codon within TM2 (Table S2). In addition, only pseudogenes are present in the sub-clade derived from branch #27, further

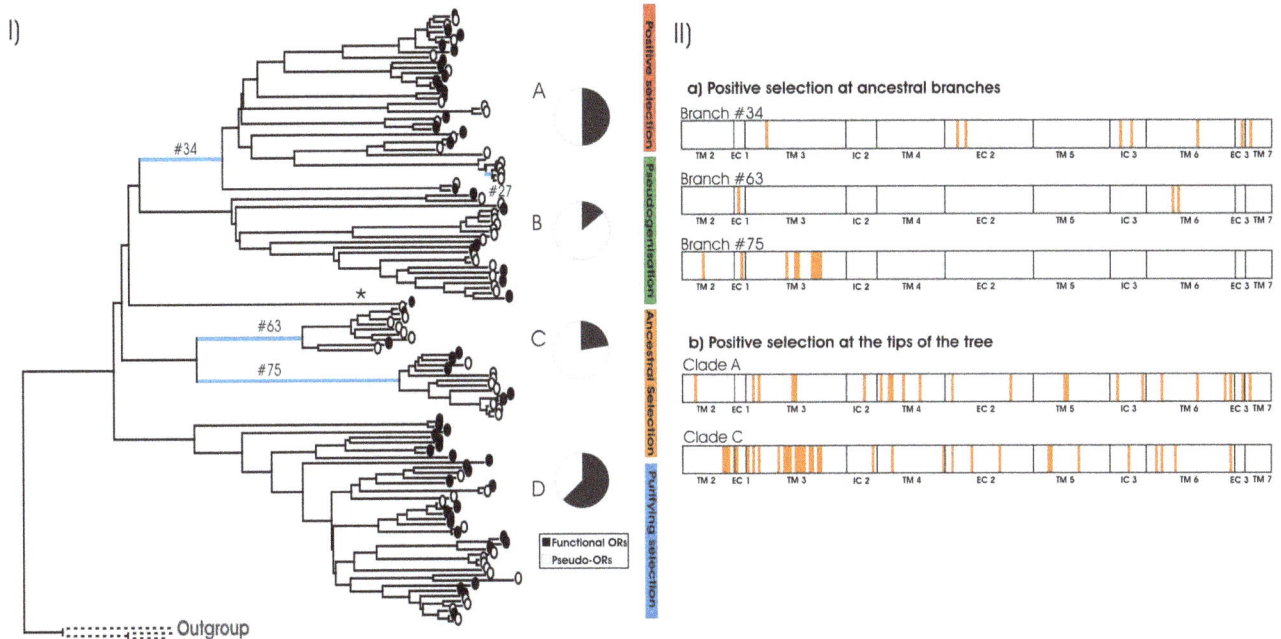

Figure 3. I Divergent evolutionary forces on the Bathyergidae OR7 gene tree. Simplified schematic view of the maximum likelihood tree (GTR, 1000 bootstrap) constructed using a single representative sequence for each putative Bathyergid OR gene; three rhodopsin-like GPCRs were used to root the tree (accession numbers NP_001287.2, NP_005292.2, NP_037014.2). Black filled circles at branch tips represent the putatively functional OR genes; empty circles represent OR pseudogenes. Pie charts represent the proportions of functional OR genes (black) and OR pseudogenes (white) in clades A–D; only one isolated gene falls out of these clades and is indicated with an asterisk. Positively selected lineages, according to branch-sites analysis, are coloured in blue; # branch numbers correspond to those assigned by CodeML [107,108]. A summary of the selective forces acting on Bathyergidae OR7 gene family, based on ancestral and ongoing selection, as well as on the ratios and numbers of functional OR genes in each clade (Table S1), is represented in vertical colour bars. Figure 3.II Positively selected residues in Bathyergidae OR7 lineages. a) Results from the Bayes Empirical Bayes analysis reveal a prevalence of positively selected sites across the odorant-binding region of ORs (TM3-6); branch numbers match labelled branches in Figure 3.I. Amino-acid positions and location domains were assigned based on Katada et al.'s molecular model [15]. b) An analysis of ongoing selection on functional nucleotide alignments from clades A and C identifies a number of amino-acid positions characterised by dN/dS>1 (indicated in orange), whilst no such sites are found across clade D (clade B was excluded from the analysis for insufficient sample size).

strengthening the idea that the mild signal of positive selection detected by the branch-site test may be a consequence of pseudogenisation, rather than adaptive evolution. Secondly, Clade B supports a small number of putatively functional OR7 genes, together with the highest proportion of pseudogenes in the dataset (86%, Figure 3.II). This is possibly indicative of the fact that ORs within this clade may be secondary for bathyergid olfaction and thus more susceptible to mutation, resulting in pseudogenes. Accordingly, the branch-site test of positive selection failed to detect any episodic events of positive selection across this clade. Thirdly, Clade C is characterised by a strong signal of positive selection on one ancestral branch (#63) and mild positive selection on the other ancestral branch (#75), with selection concentrated on the ligand-binding region of the genes (Figure 3 and Table S2). Although several codons at the tips of the tree are under positive selection, the LRT did not identify an unambiguous signal of selection. This, together with a relatively lower proportion of functional genes in this clade (23% functional OR7 genes in clade C *versus* 50% in clade A), is consistent with an ancestral pulse of adaptive evolution on clade C OR7 genes, perhaps indicating a phase when new OR functionalities were acquired within this gene lineage. Finally, strong purifying selection has maintained an unaltered pool of OR7 genes, over a long period of evolutionary time, in Clade D. Since the divergence of the major mole-rat genera *Bathyergus, Georychus, Cryptomys* and *Fukomys* ~ 15–17MYA [66] (Figure S1) and throughout the phylogeny, no periods of

adaptive evolution are detected in this clade. Interestingly, the highest proportion of functional OR7 genes (62.5%), as well as the greatest number of putatively functional genes, are found in clade D (Figure 3.II). These results are consistent with a scenario where odorant chemicals, that carry fundamental information for Bathyergidae fitness, are recognised by clade D ORs and are therefore actively maintained unchanged over time.

The Roles of Sociality and Environment in Shaping OR7 Evolution

We tested whether episodic positive selection has acted differentially on OR7 genes across specific bathyergid lineages. The bathyergid OR7 gene phylogeny was partitioned between solitary and social species and explored with a branch-site test of positive selection (following Ramm et al. [67]). No significant correlation was found between social phenotypes and positive selection (LnL difference = 0, p = 1).

The role of the environment in shaping OR7 diversity was also explored, by comparing OR genes of families 1/3/7 in mole-rats with a suite of mammalian species occupying the full spectrum of ecological habitats. Following Hayden et al. [22], the different proportions of OR 1/3/7 pseudogenes were calculated for each ecological habitat or 'ecogroup', and we introduced the mole-rat 'Subterranean' group to the analysis. Proportions of OR pseudogenes within ecogroups are reported in Figure 4.

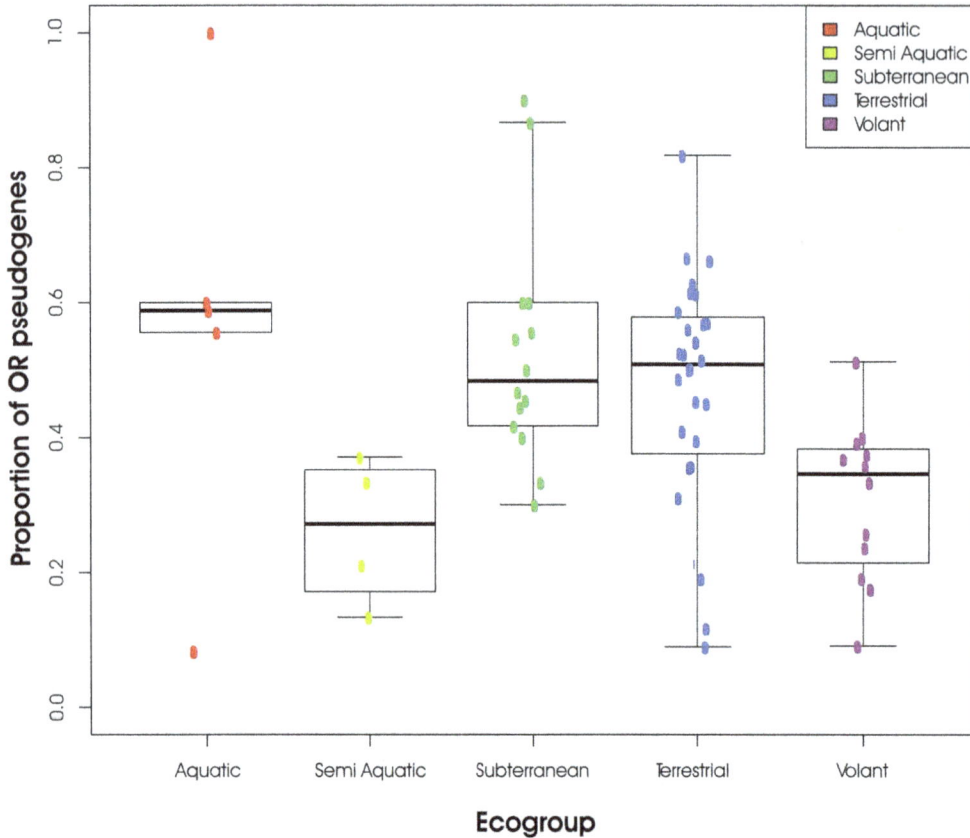

Figure 4. Proportions of OR 1/3/7 pseudogenes across Ecogroups. The mean percentage of pseudogenes and standard error are indicated for each Ecogroup.

A Wilcoxon rank-sum test was used to identify whether ecogroups differed significantly with respect to their proportions of (non)functional OR1/3/7. Significant differences were found between Subterranean and Semi-aquatic (p = 0.036), Subterranean and Volant (p = 0.011), and Terrestrial and Volant ecogroups (p = 0.014). A nearly significant value differentiates the Terrestrial from the Semi-Aquatic ecogroup (p = 0.06, Table 1). Thus, whilst the subterranean environment undoubtedly contributes to the evolution of the observed differences in OR 1/3/7 ratios across the broad range of mammals analysed, our data set is not significantly differentiated from the Terrestrial and Aquatic ecotypes.

Discussion

This study reports the first assessment of OR gene diversity in the African mole-rats and represents the first study of OR gene evolution in a subterranean mammal. Phylogenetic inference of a range of mammalian OR subgenomes identified the majority of sequences we recovered as belonging to the OR7 subfamily [22,61]. We report evidence for a large number of functional polymorphisms that translate into diverse binding properties, as well as the presence of OR polymorphisms conserved across mole-rat species, indicating an ancient origin for some aspects of bathyergid OR7 diversification [68]. Our analysis of signatures of selection on mole-rat OR7 loci revealed evidence for clade-specific evolution of olfactory receptor genes. Our results are discussed in the context of the possible evolutionary drivers of OR7 diversification, and provide insight into the complex evolutionary

Table 1. Pairwise comparisons between Ecogroups using Wilcoxon rank sum test (p-value adjustment method: BH).

	Aquatic	Semi-Aquatic	Subterranean	Terrestrial
Semi-Aquatic	0.317	–	–	–
Subterranean	0.505	**0.036***	–	–
Terrestrial	0.483	0.06	0.81	–
Volant	0.127	0.518	**0.011***	**0.014***

*Ecogroups that differ significantly (p<0.05).

history of a gene family that may be linked to individual fitness in this unusual mammalian lineage.

Functional Variation across Bathyergid OR7 Genes

Four strongly supported OR7 lineages (clades A–D) were consistently recovered from all the OR phylogenies inferred in this study. Functional OR7 genes exhibited clade-specific motifs across the amino-acid sites involved in odorant-binding [14,15], and we propose that ORs in each clade may have different binding properties. Although polymorphisms characterise these sites across clades, their chemical properties are similar, with a remarkable prevalence of hydrophobic residues across the putative ligand-binding OR domain. This finding is consistent with Katada et al.'s hypothesis [15], that the binding of odorant molecules in the odorant-binding pocket of ORs is mediated by hydrophobic interactions and van der Waals forces [59]. The OR binding-pocket spans TM3-6 and constitutes a binding environment that is broad i.e. able to recognise a range of odorants, but also selective for the shape, size and length of odorant ligands [15]. With the exception of a few known odorant-OR dyads, the functional characterisation of ORs and their respective ligands remains a major challenge [69]. Nevertheless, in humans single nucleotide polymorphisms in specific OR genes have been found to determine whether a specific odorant is detected or not [70]. Therefore, functional variability of the magnitude observed in our data set is consistent with a scenario where diverse binding properties have been selected for at OR7 loci.

Given the direct association between functional OR diversity and olfactory ability [17,18], together with the role of olfaction in the socio-ecological success of bathyergids, we predicted that positive Darwinian selection has played a fundamental role in the evolution of variability at OR7 loci in the African mole-rats. Whilst we did find evidence for positive selection within our dataset, we also, somewhat unexpectedly, found strong evidence for divergent patterns of selection across the OR7 phylogenetic tree.

A Role for Positive Selection in the Evolution of Bathyergid OR7 Genes

Positive selection is proposed to maintain functional variability at vertebrate OR loci, particularly in the ligand-binding region of ORs, while the overall receptor structure typical of GPCRs is thought to evolve under purifying selection [19,20,71]. Signatures of positive selection in clade A are similar to those reported for other vertebrate species [19–21], with selection acting predominantly on the ligand-binding domain of mole-rat OR7 genes. Similarly, two ancestral branches in clade C carry a signal of adaptive evolution across the receptors' ligand-binding region and are likely the result of a historic pulse in positive selection on these loci. Based on such signals of positive selection, we suggest that functional variation at mole-rat OR7 loci was generated in response to selective pressures for enhanced sensitivity to the range of odorants recognised by mole-rats, and/or to optimise the recognition of crucial odorants. From this perspective, the detection of such odorant molecules may be directly related to fitness in mole-rats. Consistent with this scenario, adaptive evolution is likely an indicator of intra-specific competition for olfactorily-mediated resources [20]. Emes et al. [20] present the hypothesis that OR gene duplication and sequence diversification, driven by positive selection, are the result of intense competition between individuals, e.g. for food or predator avoidance. Unfortunately, there is limited information on specific ORs and their odorant ligands and it is therefore difficult to establish an explicit link between fitness and OR diversity at specific loci [63].

A theoretical association between OR variation and fitness is nonetheless indisputable, since ORs need to recognise odorants from an ever-changing environment, in a way that is perhaps comparable to the co-evolution of MHC receptors and the pathogen environment [71].

In this study we tested the possible roles of the subterranean environment and of the different levels of sociality in selecting for enhanced functional OR7 variation in mole-rats. The contribution of sociality (or solitariness), in shaping OR7 diversity, was explored using a tree-based method (following Ramm et al. [67]), but the analysis failed to indicate a significant correlation between the social phenotype of species and positive selection. While odour detection is of primary importance for mole-rats, the comparable degrees of selection detected in both social and solitary species may be the result of selection for functional diversification very early in the evolution of the bathyergid lineage. We integrated our dataset into a broad analysis of orthologous mammalian OR genes, to explore how environmental niche-specialisation may have influenced OR7 diversification in mole-rats. Proportions of (non)functional ORs across OR1/3/7 gene families, reveal that the Subterranean ecogroup differs significantly from the Volant and Semi-aquatic groups, but is not significantly different from the Terrestrial and Aquatic groups (Figure 4, Table 1). The lack of a significant difference between Terrestrial and Subterranean ecogroups may be biased by the heterogeneous taxonomic coverage in the two datasets analysed, together with the different ages of the taxa being compared. Species coverage in the Terrestrial ecogroup spans four superorders of mammals, with 28 species from more than 20 different families, and extremely variable lineage ages (e.g. Muridae 31 MY, Canidae 12 MY) [72,73]. In contrast, only a single, relatively ancient mammalian family represents the Subterranean group (Bathyergidae 49 MY) [74,75]. Ideally, a more balanced species coverage across ecogroups, considering only those taxa with similar ages, could be used to test more accurately for the role of environment. Because continuous 'birth and death' evolution theoretically leads to an increase of OR pseudogenes, which are essentially neutral [76,77], older species may have accumulated a greater proportion of pseudogenes simply as a function of time. Even though OR pseudogenes will eventually become unidentifiable due to accumulated mutations, some ORs classified as 'non-functional' may still play a regulatory role in gene expression. Zhang et al. [78] report that 67% of human pseudogenes are in fact transcribed and this may explain the persistence of OR 'pseudogenes' in the genome over long periods of time. A further caveat, given our methodology, is that our data set is unlikely to be fully representative of the true pattern in the Bathyergidae, and analysis of the recently published naked mole-rat genome (http://naked-mole-rat.org) will provide valuable insight into this question in future studies. Nevertheless, a role for sociality and the environment in shaping and/or maintaining OR variation in mole-rats cannot be excluded. Undoubtedly, olfactory requirements will differ between solitary and social bathyergid species because of the fundamental differences in lifestyles. For example, social species require a mechanism to optimise kin recognition and use this behaviour to avoid incestuous matings and maintain colony cohesion [44,53,79]. The observed tendency of the subterranean environment to influence OR7 diversity, is only partly consistent with Hayden et al.'s conclusions [22] that natural selection, via niche-specific adaptation, shapes OR subgenomes. Nonetheless, it is reasonable to propose that the olfactory requirements of species that inhabit such diverse ecogroups are different and may be reflected in other OR gene families. The necessity to detect either airborne or water-soluble odorants is the most logical reason why the OR repertoires of terrestrial and aquatic species differ [5,17,80]. The subterranean environment, on the other hand, presents unique challenges. These include the absence of visual cues and limited auditory cues, requiring fossorial species to compensate with enhanced olfaction and hence a diversification in OR genes.

Purifying Selection and Ancient OR7 Variation

Based on the occurrence of allelic variants of the same OR7 genes, as well as identical OR7 sequences across mole-rat species, we suggest that a proportion of the variability observed in the Bathyergidae might be of ancient origin. This idea is also supported by the clustering of sequences into distinct OR7 lineages, rather than in a species-specific manner (Figure S1). It is worth noting that the majority of functional "ancient" bathyergid OR7 alleles, i.e. conserved alleles from a single OR gene that are present across numerous mole-rat species, occur within clade D (70%) only. This clade was identified as evolving under strong purifying selection and supports the highest number and proportion of functional OR7 genes. Conserved OR7 alleles across this clade may represent allelic variants that maintain precise binding properties among bathyergid species, enabling them to detect primary olfactants e.g. plant exudates released from roots of edible plants [40,41].

The occurrence of ancient OR loci has also been reported in a comparative study of whole Mouse and Rat OR subgenomes, where the presence of conserved OR loci across species is proposed to be the result of 'slow OR evolution' [81]. The authors propose a scenario where the genes have evolved under neutrality, such that loci shared between the two species are a consequence of relatively recent divergence; *Mus* and *Rattus* are estimated to have diverged ~23 MYA [72]. We argue that purifying selection in the African mole-rats has ensured the persistence of conserved OR loci across all the major generic divergence events in the family i.e. *Bathyergus*, *Georychus*, *Cryptomys* and *Fukomys* onwards (~17–15 MYA) [66]. Under a scenario of neutral evolution, alleles are predicted to become species-specific only when species have diverged for more than 4Ne generations (where Ne is the effective population size; [82]). Information on bathyergid effective population sizes is not available, but even calculations based on educated estimates suggest highly unrealistic population sizes would be required to account for the retention of conserved OR7 loci due to 'slow OR evolution'. Conserved OR7 alleles across bathyergid species are therefore most likely the result of strong purifying selection for alleles that confer significant fitness benefits. Indeed, one could envisage that at least some of the conserved OR loci identified by Zhang et al. [81], are a result of selective pressures acting to maintain specific capacities of odour recognition among sympatric Muridae, rather than a by-product of neutral evolution, especially given the divergence time between the two species [72].

A New Method for Characterising OR Subfamilies

Based on our analyses, the subpool of Bathyergidae OR genes isolated and characterised in this study belongs to a single family of ORs, namely OR7. In order to perform fine-scale classification of the complex vertebrate OR gene superfamily, a number of recent studies subdivide OR families into 'subfamilies' based on 'pattern' i.e. setting sequence similarity cut-offs of generally 60% [23,61]. On inspection of average pairwise distances, based on the number of nucleotide differences between functional OR7 genes from clades A–D, we observe that between 62–68% of the sequence similarity occurs across clades. Therefore, if we were to classify Bathyergidae OR7 genes into subfamilies according to 'pattern',

and using the cut-off limit of 60% [61], the observed clade structure would not reflect subfamily structure. This is because all OR7 genes would fall into a single subfamily. Nevertheless, the results presented here are consistent with the clustering of ORs into clades that have evolved under unrelated selective forces, potentially reflecting their underlying biological significance. Despite the high percentage of between-clade sequence similarity, there appears to be strong functional association between genes belonging to each clade. Thus, it is tempting to speculate that from a functional viewpoint each clade may represent a distinct OR7 subfamily. The above discussion on classification of OR genes into families based on sequence similarity cut-offs, raises the debate of the appropriateness of this practice that has been common in many large-scale OR studies. From our results it is clear that analysis of the evolutionary mechanisms that shape OR genetic diversity across clades can be used as an additional, novel and potentially more accurate method in classifying OR genes, informed by 'process' rather than 'pattern' alone.

Using a recent dated phylogeny, based on 66 genes and over 2000 mammalian species, the OR7 gene family is thought to have diversified after the Placental-Marsupial split ~147 MYA [18,83]. High levels of gene duplication in humans have resulted in OR7 being the largest family of the OR subgenome, occurring as OR7-specific clusters scattered across a number of genomic locations [8]. Although the function of OR7 remains poorly understood, some OR7 genes have played a significant role in recent mammalian evolution, e.g. OR7D4 in humans binds the steroid compounds androstenone and androstadienone [84]. Interestingly, these two compounds were classified as human 'pheromone candidates' after they were found to influence both brain function and, more recently, endocrine balance in humans [85–87]. These studies suggest that in those species where the VNO is considered to be a 'nonchemosensory vestige', like *Homo sapiens* or indeed the naked mole-rat [49], pheromonal communication may still occur, possibly mediated by ORs in the MOE and possibly including loci in the OR7 subfamily.

Conclusion

This study represents the first assessment of OR7 diversity for a family of subterranean mammals. In exploring the mechanisms shaping the evolution of the African mole-rat olfactory repertoire, we reveal that olfaction in mole-rats has been subject to a spectrum of evolutionary forces. Positive selection emerges as the foremost evolutionary process shaping functional OR7 variability in the family; nonetheless, neither the divergent social strategies of mole-rats nor the specialised subterranean environment emerge as clear drivers of this process. In addition to classic features of 'birth and death' evolution [10], an important role for purifying selection also emerges in the evolution of OR7 genes in mole-rats. The 'clade structure' observed in the Bathyergidae OR7 gene tree is consistent with a 'subfamily structure' based on OR7 functional properties, and likely reflects the broad range of odorant ligands that mole-rat OR7 genes can recognise. These findings challenge the commonly accepted theory that closely related ORs necessarily share functional properties [56], and reveal the intricate mechanisms of OR evolution at a 'microscopic' single-OR family scale, thus offering a valuable perspective on the breadth and complexity of OR evolution at the subgenome level.

Materials and Methods

Ethics Statement

Mole-rat tissue samples were collected as part of a previous study carried out by Deuve et al. [88–90] with full ethics approval by the University of Stellenbosch, Ethics Clearance Certificate # 2006B01006.

Olfactory Receptor Gene Isolation and Identification

Genomic DNA was extracted from fresh muscle tissue using a standard phenol-chloroform protocol [91]. Species sampled include representative taxa from all currently recognised genera in the Bathyergidae: *Bathyergus janetta* (BJ), *Bathyergus suillus* (BS), *Cryptomys hottentotus hottentotus* (CHH), *Cryptomys hottentotus natalensis* (CHN), *Cryptomys hottentotus pretoriae* (CHP), *Fukomys mechowi* (CM), *Fukomys amatus* (CA), *Fukomys anselli* (CAN), *Fukomys bocagei* (CB), *Fukomys damarensis* (CDM), *Fukomys darlingi* (CD), *Georychus capensis* (GC), *Heliophobius argentocinereus* (HA), *Heterocephalus glaber* (HG).

Vertebrate olfactory receptors display a conserved overall structure typical of GPCRs, with variability concentrated across the ligand binding pockets, spanning transmembrane (TM) domains 3–6 [12,15]. To provide a reference sequence for the development of bathyergid-specific PCR primers, we used the degenerate PCR primers A4/B6 described in Buck and Axel [4] to amplify TM 2–7 from a single *C. damarensis* individual; conditions followed those reported in Buck and Axel [4]. PCR products were gel purified using the Wizard SV Gel and PCR Clean-up System (Promega) and cloned into *E. coli* DH5α CaCl$_2$-competent cells using pGEM-T-Easy Vector System (Promega). Insert-containing clones were sequenced using a BigDye Terminator v.3.1 Cycle Sequencing Kit (Applied Biosystems). Post-sequencing purification was performed using Centrisep Columns (Princeton) and DNA sequences were analysed on an ABI 3130 Genetic Analyser using 3130 Genetic Analyser Data Collection software v.5.2.

Bathyergid-specific OR primers were designed to TM domains 2–7 (approx. size 645 bp): Bathy-OR1 5′- GCG GAC ATC YGT TTC AC - 3′; Bathy-OR2 5′- GTG ACC ACA GTG TAC ATC –3′. The Bathy-OR1/Bathy-OR2 primer pair successfully amplified unambiguous PCR products in all 14 mole-rat species, using the following conditions: 95°C for 1 min, 54°C for 3 min, 72°C for 3 min (35 cycles). Each 40 ul reaction contained between 50–100 ng genomic DNA, 2 pmol/ul of each Bathy-OR1 and Bathy-OR2 primers, 1.5 mM MgCl$_2$, 0.2 mM dNTPs, 1.25 U Super-Therm *Taq* DNA polymerase and 1x corresponding *Taq* reaction buffer. PCR products were gel-purified, cloned and sequenced as described previously. Between 10 and 50 clones were sequenced in both direction using using both the forward and reverse-primers for 1 to 3 individuals for each species; this produced 402 OR sequences. Forward and reverse sequences were aligned and checked for ambiguities by eye in Bioedit v7.0.8.0 [92], resulting in 201 putative OR sequences. The sequences were then aligned using Clustal W v2.0 [93,94] and translated in Bioedit v7.0.8.0 [92]; identical sequences were identified by pairwise comparisons in MEGA v5 [95] and the final data set comprised 178 unique OR sequences.

A BLAST search was performed against the nucleotide collection data, available on NCBI (www.ncbi.nlm.nih.gov), to assign identity to both nucleotide and amino acid OR sequences.

OR Sequence Identification

The role of recombination in generating sequence variation in any dataset, either *in vivo* or *in vitro*, was evaluated by calculating the level of linkage disequilibrium between polymorphic sites as a function of their physical distance, using Rozas et al's [96] ZZ value in DnaSP4.5 [97]. This test reduces the possibility of *in vitro* recombination generating false OR variability, since recombination is established as only a minor source of OR variation *in vivo* [10].

Following Steiger et al. [31], OR sequences were classified as pseudogenes if they contained stop codons or frame-shift mutations that disrupted the overall receptor structure. Sequences that translated into putatively functional OR genes, but that differed in length, were considered to be functional only if they maintained the known features of ORs (e.g. the MAYDRFVAIC and KAFSTCASH motifs in TM domains 3 and 6, respectively), and if the variability mapped to the ligand-binding pockets of ORs [14,15].

In order to identify allelic variants of OR genes, pairwise comparisons were performed across all unique sequences in our dataset using MEGA v5 [95]. Allelic pairs based on the pairwise comparison matrix generated in MEGA were then identified using *alleles.R* (R. Gaujoux, unpublished) developed in R (R Development Core Team, 2008, http://www.R-project.org). The criteria for allele identification described by Kishida [18] were applied to the bathyergid dataset using the following cut-off limits: within a species, sequences that shared 99% sequence similarity were considered to be alleles of the same gene; across species, the cut-offs were 98% within the same genus and 96% across genera. Single base-pair differences as well as two base-pair differences were assumed to represent identical sequences due to PCR or sequencing errors. Similarly, when two allelic variants shared more than 99% sequence similarity across species (i.e. between 3–5 base pair differences), they were considered to represent identical alleles. When more than two putative alleles of the same OR gene were found in an individual, given the defined cut-offs, two copies of that particular gene were assumed to be present. Similarly, when two presumed alleles were of different functional status i.e. one putatively functional and one pseudogene, they were considered to belong to two different OR genes, the result of a duplication event followed by pseudogenisation. Whenever the percentage sequence similarity led to ambiguous results e.g. when transitivity was not applicable (A = B, B = C but A≠C, with ' = ' meaning 'alleles' based on sequence similarity), phylogenetic relationships (described below) were used to allocate alleles to different OR genes. Once identified, alleles of the same OR gene were collapsed down to a single consensus sequence for each putative gene, and used in subsequent analyses.

Phylogenetic Analyses

Evolutionary relationships among Bathyergidae OR7 genes were explored using maximum likelihood (ML) [98], based on the general time-reversible model (GTR) [99] as determined jModeltest [100], and constructed in MEGA v5 [95]. Tree topology was inferred using all unique mole-rat OR sequences identified, with three non-OR GPCR genes used as an outgroup; robustness of the tree topology was tested using 1000 bootstrap replicates [101]. The resulting tree was used in combination with the pairwise comparison matrix to determine allelic relationships amongst sequences. If sequence similarity led to uncertain allelic allocation, alleles were considered to be sister taxa in the phylogenetic tree. A further ML tree was then constructed (GTR, 1000 bootstrap) using only a single representative sequence for each putative OR gene.

In the most comprehensive survey of mammalian ORs to date, Hayden et al. [22] used a combination of sequence similarity and phylogenetic criteria for OR gene classification. Their dataset analysed the entire OR subgenomes of 50 mammalian species, consisting of ~50,000 OR sequences. One or two representative sequences per species, for each of the 17 OR families from Hayden et al. [22], were aligned together with the bathyergid OR dataset, using the online Clustal W alignment tool from the European Bioinformatics Institute (available at www.ebi.ac.uk). Aligned sequences were then imported into Bioedit v7.0.8.0 [92] and corrections to the alignment were made by eye. An ML tree was then constructed using the Tamura-Nei substitution model [102] in MEGA v5 [95] with 1000 bootstrap replications. Phylogenetic analysis based on all nucleotide sites included 312 representative sequences from all OR gene families across 18 different mammalian species, as well as the 119 Bathyergidae ORs. All Bathyergidae sequences clustered together with known Family 7 ORs [22]. Family information was included in Bathyergidae sequence nomenclature (GenBank accession numbers KF453235–KF453412).

All the available OR sequences belonging to Family 7 from Hayden et al's dataset [22], representing the entire Family 7 OR subgenome of 18 different mammalian species, were then aligned with the 119 mole-rat sequences using the online Clustal W tool. Aligned sequences were corrected by eye in Bioedit v7.0.8.0 [92], and a final ML tree (Tamura-Nei) was constructed in MEGA v5 and tested using 1000 bootstraps replicates. Positions containing alignment gaps were eliminated from the pairwise sequence comparisons (pairwise deletion option), resulting in 805 nucleotide positions in the final dataset.

Signatures of Selection on Bathyergid OR7 Genes

Because olfactory receptors display a highly conserved overall structure, with variability limited to a set of amino acid residues involved in the binding of odorant molecules, an average of substitution rates across the entire OR gene provides neither an accurate nor informative test of positive selection [12,15]. Therefore, tests for positive selection were applied to different codon sites in the dataset using the SELECTON server (available at http://selecton.tau.ac.il/) [103,104]. Estimates of the ratio of non-synonymous (dN) to synonymous (dS) substitutions were obtained for each codon, and significance assessed via a LRT. The LRT compared two nested models for each codon: a null model (M8a), which assumes no selection, and an alternative model (M8) which allows positive selection to occur. Three sets of bathyergid OR7 genes, corresponding to clades A, C and D, identified in the phylogenetic analysis, were analysed separately using codon-based multiple sequence alignment (MSA); pseudogenes were not included in the analysis. Clade B displayed only three putatively functional OR genes, and was therefore excluded from this analysis due to insufficient sample size. Across clade D, several codons with dN/dS ratios <1 were identified, while none appeared to have dN/dS >1. A codon-based-Z test, to test for purifying selection (overall average), was performed on clade D's functional ORs using the Nei-Gojobori method [105] implemented in MEGA5 [95], with 1000 bootstrap replicates to determine significance of purifying selection across this clade.

To investigate whether positive selection may have acted along specific bathyergid OR7 lineages, a branch-site test (test 2 in [106]) was carried out in PAML [107,108] using CodeML and based on the ML tree of African mole-rat OR7 genes. CodeML was used to estimate the dN/dS ratio (ω) on codon (nucleotide) alignments across the topology of the trees. Two nested models, null and alternative, were computed and compared using a LRT. In the null model, codons along all branches are either under purifying selection (ω <1) or under neutral evolution (ω = 1), and the foreground branch may have different proportions of sites under neutral selection than the background branches (i.e. relaxed purifying selection). In the alternative model, some sites on the foreground branch may be under positive selection (ω >1). Following Yang [109], stop codons and alignment gaps were excluded from the alignment used to construct a ML tree (as previously described), and the resulting tree maintained the same

tree topology as the original bathyergid OR tree. In the branch-site test of positive selection, each branch of the OR gene tree was labelled in turn as foreground; a LRT was performed on all pairs of nested models and compared to a χ^2 distribution to determine significance. Furthermore, the Q-value, a measure of the false discovery rate (FDR) due to multiple testing, was calculated for each branch using the 'Q-value' software available at http://genomics.princeton.edu [110–112]. When the LRT remained significant after the correction for multiple testing (i.e. both p- and Q-values<0.05), the posterior probability of sites being under positive selection (dN/dS >1) was calculated using the BEB method [64,65] implemented in CodeML.

Testing the Role of Sociality in OR7 Variation

To test the role of sociality in shaping OR7 variation, we partitioned the OR7 gene tree between Solitary and Social bathyergid species (following Ramm et al. [67]), and performed a branch-site test of positive selection as previously described. The tree was partitioned by labelling the terminal branches of the phylogeny according to the social status of the corresponding species. The analysed data comprised all 119 unique bathyergid OR genes identified in this study, including both functional OR7 genes and pseudogenes. Stop codons from OR pseudogenes and alignment gaps were excluded from the OR7 alignment. OR7 genes from the genera *Bathyergus*, *Georychus* and *Heliophobius* were labelled as Solitary, while those belonging to *Cryptomys*, *Fukomys* and *Heterocephalus* were labelled Social [33].

We did not hypothesise *a priori* which social system (i.e. solitary or social) would be subject to positive selection, and therefore conducted two branch-site analyses. In the first analysis we tested whether social lineages carried a signal of increased selection when compared to the solitary ones, labelling all the terminal branches of the OR7 gene tree that belonged to social bathyergid as 'foreground'. In the second analysis, the test was performed with the 'solitary leaves' of the tree labelled as foreground. With these branches defined as foreground, a LRT was performed on all pairs of nested models (null and alternative) and compared to a χ^2 distribution to determine significance. A Q-value was then calculated for each branch using the 'Q-value' software as before. When the LRT was significant with a FDR below 5%, the posterior probability of sites being under positive selection (dN/dS >1) was then calculated using the BEB method in CodeML.

The Subterranean Ecogroup as a Driver of Bathyergid OR7 Diversification

The role of the subterranean environment as a driver of OR evolution across the Bathyergidae, was explored by comparing the ratios of functional OR genes:pseudogenes across ecotypes (following [22]). Hayden et al.'s dataset [22], comprising ratios of OR functional genes:pseudogenes from whole-OR-subgenome data for 50 mammalian species, covered a range of environmental niches, namely: Terrestrial, Aquatic, Semi-aquatic and Volant (i.e. bats). Hayden et al. [22] performed a Bayesian phylogenetic analysis to classify OR genes into gene families, and the 17 'traditional' OR families were recovered [61]; the following families were found to group together; OR 2/13; OR 1/3/7; OR 5/8/9. Data for this part of the study is available at http://genome.cshlp.org/content/20/1/1/suppl/DC1.

The authors used a principal component analysis, based on the different proportions of functional ORs and pseudogenes across gene families, to then compare data from the different ecogroups, and in so doing identify the OR families that explain most of the variation between these groups.

Following Hayden et al. [22], OR7 belongs to the broader mammalian OR grouping of OR 1/3/7. Here, we used the numbers of functional OR 1/3/7 genes and pseudogenes, as well as their relative proportions as reported in [22], together with data from the 14 Bathyergidae species analysed in this study, to test our hypothesis that the subterranean niche has contributed to the diversification of functional variability in the mole-rat OR genome. In this context, species used in the analysis were classified according to five ecogroups: Aquatic, Semi-aquatic, *Subterranean*, Terrestrial and Volant.

OR1/3/7 ratios were plotted across all species within each ecogroup, and the mean percentage of pseudogenes and associated standard error were calculated in R (R Development Core Team, 2008, http://www.R-project.org). To test for pairwise differences in the distributions of pseudogene proportions between each ecogroup, we applied a non-parametric Wilcoxon-test [113], as well as the Benjamini & Hochberg 'BH' correction for multiple testing [114].

Supporting Information

Figure S1   Bathyergid OR7 gene tree. Maximum-likelihood tree (GTR, 1000 bootstrap) constructed using all 178 unique Bathyergid OR sequences; three rhodopsin-like GPCRs are used as outgroups (accession numbers NP_001287.2, NP_005292.2, NP_037014.2). The four main OR clades are indicated (A–D); only one isolated gene (BJ4_A12) falls out of these clades and is labelled with an asterisk. Abbreviations correspond to gene names in Genbank accession numbers KF453235–KF453412 and contain species information as follows: *Bathyergus janetta* (BJ), *Bathyergus suillus* (BS), *Cryptomys hottentotus hottentotus* (CHH), *Cryptomys hottentotus natalensis* (CHN), *Cryptomys hottentotus pretoriae* (CHP), *Fukomys mechowi* (CM), *Fukomys amatus* (CA), *Fukomys anselli* (CAN), *Fukomys bocagei* (CB), *Fukomys damarensis* (CDM), *Fukomys darlingi* (CD), *Georychus capensis* (GC), *Heliophobius argentocinereus* (HA), *Heterocephalus glaber* (HG).

Figure S2   Mammalian OR7 gene tree. Maximum likelihood tree (Tamura-Nei, 1000 boostrap) constructed with all the available mammalian OR7 genes [22]. Each circle dot corresponds to an OR7 gene belonging to family 7; ORs from different taxonomic families are colour-coded as indicated on the figure. Rhodopsin-like non-OR GPCRs are used as an outgroup (accession numbers NP_001287.2, NP_005292.2, NP_037014.2). Bathyergidae ORs from clades A–D are indicated in green; bootstrap values are reported for the main bathyergid clades.

Table S1   Numbers of functional ORs and pseudogenes in clades A–D.

Table S2   Positively selected residues in Bathyergidae OR7 lineages.

Acknowledgments

We thank Nigel Bennett and Terry Robinson for providing mole-rat tissue samples for this study. Renaud Gaujoux provided extensive advice and invaluable help with the data analysis.

Author Contributions

Conceived and designed the experiments: SS JMB COR. Performed the experiments: SS. Analyzed the data: SS. Contributed reagents/materials/analysis tools: SS COR. Wrote the paper: SS JMB COR.

References

1. Firestein S (2001) How the olfactory system makes sense of scents. Nature 413: 211–218.
2. Brennan PA, Kendrick KM (2006) Mammalian social odours: attraction and individual recognition. Philos Trans R Soc B 361: 2061–2078.
3. Isogai Y, Si S, Pont-Lezica L, Tan T, Kapoor V, et al. (2011) Molecular organisation of vomeronasal chemoreception. Nature 478: 241–245.
4. Buck L, Axel R (1991) A novel multigene family may encode odorant receptors: a molecular basis for odor recognition. Cell 65: 175–187.
5. Nei M, Niimura Y, Nozawa M (2008) The evolution of animal chemosensory receptor gene repertoires: roles of chance and necessity. Nature Rev Genet 9: 951–963.
6. Niimura Y, Nei M (2005b) Evolutionary dynamics of olfactory receptor genes in fishes and tetrapods. Proc Natl Acad Sci U S A 102: 6039–6044.
7. Niimura Y (2009) On the origin and evolution of vertebrate olfactory receptor genes: comparative genome analysis among 23 chordate species. Genome Biol Evol 1: 34–44.
8. Glusman G, Yanai I, Rubin I, Lancet D (2001) The complete human olfactory subgenome. Genome Res 11: 685–702.
9. Nei M, Gu X, Sitnikova T (1997) Evolution by the birth-and-death process in multigene families of the vertebrate immune system. Proc Natl Acad Sci U S A 94: 7799–7806.
10. Nei M, Rooney AP (2005) Concerted and birth-and-death evolution of multigene families. Annu Rev Genet 39: 121–152.
11. Niimura Y, Nei M (2007) Extensive gains and losses of olfactory receptor genes in mammalian evolution. PLoS ONE 8: e708.
12. Gaillard I, Rouquier S, Pin JP, Mollard P, Richerd S, et al. (2002) A single olfactory receptor specifically binds a set of odorant molecules. Eur J Neurosci 15: 409–418.
13. Keller A, Vosshall LB (2008) Better smelling through genetics: mammalian odor perception. Curr Opin Neurobiol 18: 364–369.
14. Man O, Gilad Y, Lancet D (2004) Prediction of the odorant binding site of olfactory receptor proteins by human–mouse comparisons. Protein Sci 13: 240–254.
15. Katada S, Hirokawa T, Oka Y, Suwa M, Touhara K (2005) Structural basis for a broad but selective ligand spectrum of a mouse olfactory receptor: mapping the odorant-binding site. J Neurosci 25: 1806–1815.
16. Dulac C, Wagner S (2006) Genetic analysis of brain circuits underlying pheromone signaling. Annu Rev Genet 40: 449–467.
17. Niimura Y, Nei M (2006) Evolutionary dynamics of olfactory and other chemosensory receptor genes in vertebrates. J Hum Genet 51: 505–517.
18. Kishida T (2008) Pattern of the divergence of olfactory receptor genes during tetrapod evolution. PLoS ONE 3: e2385.
19. Alioto TS, Ngai J (2005) The odorant receptor repertoire of teleost fish. BMC Genomics 6: 173.
20. Emes RD, Beatson SA, Ponting CP, Goodstadt L (2004) Evolution and comparative genomics of odorant- and pheromone-associated genes in rodents. Genome Res 14: 591–602.
21. Kondo R, Kaneko S, Sun H, Sakaizumi M, Chigusa SI (2002) Diversification of olfactory receptor genes in the Japanese medaka fish, Oryzias latipes. Gene 282: 113–120.
22. Hayden S, Bekaert M, Crider TA, Marini S, Murphy WJ, et al. (2010) Ecological adaptation determines functional mammalian olfactory subgenomes. Genome Res 20: 1–9.
23. Godfrey PA, Malnic B, Buck LB (2004) The mouse olfactory receptor gene family. Proc Natl Acad Sci U S A 101: 2156–2161.
24. Ache BW, Young JM (2005) Olfaction: diverse Species, conserved principles. Neuron 48: 417–430.
25. Gilad Y, Wiebe V, Przeworski M, Lancet D, Pääbo S (2004) Loss of olfactory receptor genes coincides with the acquisition of full trichromatic vision in primates. PLoS Biol 2: e5.
26. Matsui A, Go Y, Niimura Y (2010) Degeneration of Olfactory Receptor gene repertoires in primates: no direct link to full trichromatic vision. Mol Biol Evol 27: 1192–1200.
27. Kishida T, Kubota S, Shirayama Y, Fukami H (2007) The olfactory receptor gene repertoires in secondary-adapted marine vertebrates: evidence for reduction of the functional proportions in cetaceans. Biology Lett 3: 428–430.
28. Thewissen JGM, George J, Rosa C, Kishida T (2011) Olfaction and brain size in the bowhead whale. Mar Mammal Sci 27: 282–294.
29. Sanders KL, Lee MSY, Leys R, Foster R, Keogh JS (2008) Molecular phylogeny and divergence dates for Austral-asian elapids and sea snakes (hydrophiinae): evidence from seven genes for rapid evolutionary radiations. J Evol Biol 21: 682–695.
30. Kishida T, Hikida T (2010) Degeneration patterns of the olfactory receptor genes in sea snakes. J Evol Biol 23: 302–310.

31. Steiger SS, Fidler AE, Kempenaers B (2009) Evidence for increased olfactory receptor gene repertoire size in two nocturnal bird species with well-developed olfactory ability. BMC Evol Biol 9: 117.

32. Niimura Y, Nei M (2005a) Comparative evolutionary analysis of olfactory receptor gene clusters between humans and mice. Gene 346: 13–21.

33. Bennett NC, Faulkes CG (2000) African mole-rats ecology and eusociality. Cambridge: Cambridge University Press. 273 p.

34. Lacey EA, Patton JL, Cameron GN (2000) Life Underground: The Biology of Subterranean Rodents. Chicago and London: University of Chicago Press. 457 p.

35. Kott O, Šumbera R, Němec P (2010). Light perception in two strictly subterranean rodents: life in the dark or blue? PLoS ONE 5: e11810.

36. Eloff G (1958) The structural and functional degeneration of the eye of South African rodent moles Cryptomys bigalkei and Bathyergus maritimus. S Afr J Sci 54: 293–302.

37. Nemec P, Cveková P, Benada O, Wielkopolska E, Olkowicz S, et al. (2008) The visual system in subterranean African mole-rats (Rodentia, Bathyergidae): retina, subcortical visual nuclei and primary visual cortex. Brain Res Bull 75: 356–364.

38. Zhao HB, Ru BH, Teeling EC, Faulkes CG, Zhang SY, et al. (2009) Rhodopsin molecular evolution in mammals inhabiting low light environments. PLoS ONE 4: e8326.

39. Crish SD, Rice FL, Park TJ, Comer CM (2003) Somatosensory organization and behavior in naked mole-rats I: vibrissa-like body hairs comprise a sensory array that mediates orientation to tactile stimuli. Brain Behav Evolut 62: 141–151.

40. Heth G, Todrank J, Begall S, Koch R, Zilbiger Y, et al. (2002a) Odours underground: subterranean rodents may not forage "blindly". Behav Ecol Sociobiol 52: 53–58.

41. Lange S, Neumann B, Hagemeyer P, Burda H (2005). Kairomone-guided food location in subterranean Zambian mole-rats (Cryptomys spp., Bathyergidae). Folia Zool 54: 263–268.

42. Judd TM, Sherman PW (1996) Naked mole-rats recruit colony mates to food sources. Anim Behav 52: 957–969.

43. Faulkes CG (1990) Social suppression of reproduction in the naked mole-rat Heterocephalus glaber. PhD thesis (unpublished), University of London, United Kingdom.

44. O'Riain MJ, Jarvis JUM (1997) Colony member recognition and xenophobia in the naked mole rat. Anim Behav 53: 487–498.

45. Reeve HK, Sherman PW (1991) Intracolonial aggression and nepotism by the breeding female naked mole-rat. In: Sherman PW, Jarvis JUM, Alexander RD, editors. The Biology of the Naked Mole-Rat. Princeton: Princeton University Press. 337–357.

46. Jarvis JUM (1991) Reproduction of naked mole-rats. In: Sherman PW, Jarvis JUM, Alexander RD, editors. The Biology of the Naked Mole-Rat. Princeton: Princeton University Press. 384–425.

47. Jarvis JUM, Sherman PW (2002) Heterocephalus glaber. Mammal Sp 706: 1–9.

48. Smith TD, Bhatnagar KP, Dennis JC, Morrison EE, Park TJ (2007) Growth deficient vomeronasal organs in the naked mole-rat (Heterocephalus glaber). Brain Res 1132: 78–83.

49. Bhatnagar KP, Smith TD (2010) The human vomeronasal organ. Part VI: A nonchemosensory vestige in the context of major variations of the mammalian vomeronasal organ. Curr Neurobiol 1: 1–9.

50. Holmes WG, Sherman PW (1983) Kin recognition in animals. Am Sci 71: 46–55.

51. Heth G, Todrank J, Burda H (2002) Individual odor similarities within colonies and across species of Cryptomys mole rats. J Mammal 83: 569–575.

52. Heth G, Todrank J, Begall S, Wegner RE, Burda H (2004) Genetic relatedness discrimination in eusocial Cryptomys anselli mole-rats, Bathyergidae, Rodentia. Folia Zool 53: 269–278.

53. Burda H (1995) Individual recognition and incest avoidance in eusocial common mole-rats rather than reproductive suppression by parents. Experientia 51: 411 413.

54. Terakita A (2005) The Opsins. Genome Biol 6: 213.

55. Zhang X, Firestein S (2002) The olfactory receptor gene superfamily of the mouse. Nature Neurosci 5: 124–133.

56. Malnic B, Godfrey PA, Buck L (2004) The Human Olfactory receptor gene family. Proc Natl Acad Sci U S A 101: 2584–2589.

57. Rouquier S, Blancher A, Giorgi D (2000) The olfactory receptor gene repertoire in primates and mouse: Evidence for reduction of the functional fraction in primates. Proc Natl Acad Sci U S A 97: 2870–2874.

58. Meyerhans A, Vartanian J-P, Wain-Hobson S (1990) DNA recombination during PCR. Nucleic Acids Res 18: 1687–1691.

59. Parsegian VA (2006) van der Waals Forces: a Handbook for Biologists, Chemists, Engineers, and Physicists. Cambridge: Cambridge University Press.

60. Freitag J, Krieger J, Strotmann J, Breer H (1995) Two classes of olfactory receptors in Xenopus laevis. Neuron 15: 1383–1392.

61. Glusman G, Bahar A, Sharon D, Pilpel Y, White J, et al. (2000) The olfactory receptor gene superfamily: Data mining, classification, and nomenclature. Mamm Genome 11: 1016–1023.

62. Warren WC, Hillier LW, Marshall Graves JA, Birney E, Ponting CP, et al. (2008) Genome analysis of the platypus reveals unique signatures of evolution. Nature 453: 175–183.

63. Zarzo M (2007) The sense of smell: Molecular basis of odorant recognition. Biol Rev Camb Philos 82: 455–479.

64. Nielsen R, Yang Z (1998) Likelihood models for detecting positively selected amino acid sites and applications to the HIV-1 envelope gene. Genetics 148: 929–936.

65. Yang Z, Wong WSW, Nielsen R (2005) Bayes empirical Bayes inference of amino acid sites under positive selection. Mol Biol Evol 22: 1107–1118.

66. Ingram CM, Burda H, Honeycutt RL (2004) Molecular phylogenetics and taxonomy of the African mole-rats, genus Cryptomys and the new genus Coetomys Gray, 1864. Mol Phylogenet Evol 31: 997–1014.

67. Ramm SA, Oliver PL, Ponting CP, Stockley P, Emes RD (2008) Sexual selection and the adaptive evolution of mammalian ejaculate proteins. Mol Biol Evol 25: 207–219.

68. Klein J, Sato A, Nagl S, O'hUigín C (1998) Molecular Trans-Species Polymorphism. Annu Rev Ecol Syst 29: 1–21.

69. Saito H, Chi Q, Zhuang H, Matsunami H, Mainland JD (2009) Odor coding by a mammalian receptor repertoire. Science Sign 2: ra9.

70. Menashe I, Abaffy T, Hasin Y, Goshen S, Yahalom V, et al. (2007). Genetic elucidation of human hyperosmia to isovaleric acid. PLoS Biol 5: e284.

71. Kambere MB, Lane RP (2007) Co-regulation of a large and rapidly evolving repertoire of odorant receptor genes. BMC Neurosci 8: S2. doi: 10.1186/1471-2202-8-S3-S2.

72. Adkins RM, Gelke EL, Rowe D, Honeycutt RL (2001) Molecular phylogeny and divergence time estimates for major rodent groups: evidence from multiple genes. Mol Biol Evol 18: 777–791.

73. Bardeleben C, Moore RL, Wayne RK (2005) A molecular phylogeny of the Canidae based on six nuclear loci. Mol Phylogenet Evol 37: 815–831.

74. Nedbal MA, Allard MW, Honeycutt RL (1994) Molecular systematics of hystricognath rodents: evidence from the mitochondrial 12S rRNA gene. Mol Phylogenet Evol 3: 206–220.

75. Blanga-Kanfi S, Miranda H, Penn O, Pupko T, DeBry RW, et al. (2009) Rodent phylogeny revised: analysis of six nuclear genes from all major rodent clades. BMC Evol Biol 9: 71.

76. Li WH, Gojobori T, Nei M (1981) Pseudogenes as a paradigm of neutral evolution. Nature 292: 237–239.

77. Gilad Y, Bustamante CD, Lancet D, Paabo S (2003) Natural selection on the olfactory receptor gene family in humans and chimpanzees. Am J Hum Genet 73: 489–501.

78. Zhang X, De la Cruz O, Pinto JM, Nicolae D, Firestein S, et al. (2007) Characterizing the expression of the human olfactory receptor gene family using a novel DNA microarray. Genome Biol 8: R86.

79. Spinks AC, O'Riain MJ, Polakow DA (1998) Intercolonial encounters and xenophobia in the common mole-rat, Cryptomys hottentotus hottentotus (Bathyergidae): the effects of aridity, sex and reproductive status. Behav Ecol 69: 224–234.

80. Freitag J, Ludwig G, Andreini I, Rossler P, Breer H (1998) Olfactory receptors in aquatic and terrestrial vertebrates. J Comp Physiol 183: 635–650.

81. Zhang X, Zhang X, Firestein S (2007) Comparative genomics of odorant and pheromone receptor genes in rodents. Genomics 89: 441–450.

82. Kimura M, Ohta T (1969) The average number of generations until fixation of a mutant gene in a finite population. Genetics 61: 763–771.

83. Bininda-Emonds ORP, Cardillo M, Jones KE, MacPhee RDE, Beck RMD, et al. (2007) The delayed rise of present-day mammals. Nature 446: 507–512.

84. Keller A, Zhuang H, Chi Q, Vosshall LB, Matsunami H (2007) Genetic variation in a human odorant receptor alters odour perception. Nature 449: 468–472.

85. Jacob S, Kinnunen LH, Metz J, Cooper M, McClintock MK (2001) Sustained human chemosignal unconsciously alters brain function. Neuroreport 12: 2391–2394.

86. Jacob S, Hayreh DJ, McClintock MK (2001) Context-dependent effects of steroid chemosignals on human physiology and mood. Physiol Behav 74: 15–27.

87. Wyart C, Webster WW, Chen JH, Wilson SR, McClary A, et al. (2007) Smelling a single component of male sweat alters levels of cortisol in women. J Neurosci 27: 1261–1265.

88. Deuve JL, Bennett NC, O'Brien PCM, Ferguson-Smith M, Faulkes CG, et al. (2006) Complex evolution of X and Y autosomal translocations in the Giant mole-rat, Cryptomys mechowi (Bathyergidae). Chromosome Res 14: 681–69.

89. Deuve JL, Bennett NC, Britton-Davidian J, Robinson TJ (2008) Chromosomal Phylogeny and evolution of the African mole-rats (Bathyergidae). Chromosome Res 16: 57–74.

90. Deuve JL, Bennett NC, Ruiz-Herrera A, Waters PD, Britton-Davidian J, et al. (2008) Dissection of a Y-autosome translocation in an African mole-rat, Cryptomys hottentotus (Rodentia: Bathyergidae) and implications for the evolution of a meiotic sex chromosome chain. Chromosoma 117: 211–217.

91. Sambrook J, Frisch E, Maniatis T (1989) Molecular Cloning: A Laboratory Manual. Cold Spring Harbor : Cold Spring Harbor Laboratory Press.

92. Hall TA (1999) BioEdit: a user-friendly biological sequence alignment editor and analysis program for Windows 95/98/NT. Nucl Acid S 41: 95–98.

93. Thompson JD, Higgins DG, Gibson TJ (1994) CLUSTAL W: improving the sensitivity of progressive multiple sequence alignment through sequence weighting, position-specific gap penalties and weight matrix choice. Nucl Acid S 22: 4673–4680.

94. Larkin MA, Blackshields G, Brown NP, Chenna R, McGettigan PA, et al. (2007) Clustal W and Clustal X version 2.0. Bioinformatics Appl Note 23: 2947–2948.

95. Tamura K, Peterson D, Peterson N, Stecher G, Nei M, et al. (2011) MEGA5: molecular evolutionary genetics analysis using maximum likelihood, evolutionary distance, and maximum parsimony methods. Mol Biol Evol 28: 2731–2739.

96. Rozas J, Gullaud M, Blandin G, Aguadé M (2001) DNA variation at the rp49 gene region of Drosophila simulans: evolutionary inferences from an unusual haplotype structure. Genetics 158: 1147–1155.

97. Librado P, Rozas J (2009) DnaSP v5: A software for comprehensive analysis of DNA polymorphism data. Bioinformatics 25: 1451–1452.

98. Felsenstein J (1981) Evolutionary trees from DNA sequences: A maximum likelihood approach. J Mol Evol 17: 368–376.

99. Tavaré S (1986) Some probabilistic and statistical problems in the analysis of DNA sequences. Lectures on Mathematics in the Life Sciences (American Mathematical Society) 17: 57–86.

100. Posada D (2008) jModelTest: Phylogenetic model averaging. Mol Biol Evol 25: 1253–1256.

101. Felsenstein J (1985) Confidence limits on phylogenies: An approach using the bootstrap. Evolution 39: 783–791.

102. Tamura K, Nei M (1993) Estimation of the number of nucleotide substitutions in the control region of mitochondrial DNA in humans and chimpanzees. Mol Biol Evol 10: 512–526.

103. Doron-Faigenboim A, Stern A, Mayrose I, Bacharach E, Pupko T (2005) Selecton: a server for detecting evolutionary forces at a single amino-acid site. Bioinformatics 21: 2101–2103.

104. Stern A, Doron-Faigenboim A, Erez E, Martz E, Bacharach E, et al. (2007) Selecton 2007: advanced models for detecting positive and purifying selection using a Bayesian inference approach. Nucleic Acids Res 35: 506–511.

105. Nei M, Gojobori T (1986) Simple methods for estimating the numbers of synonymous and nonsynonymous nucleotide substitutions. Mol Biol Evol 3: 418–426.

106. Zhang J, Nielsen R, Yang Z (2005) Evaluation of an improved branch-site likelihood method for detecting positive selection at the molecular level. Mol Biol Evol 22: 2472–2479.

107. Yang Z (1997) PAML: a program package for phylogenetic analysis by maximum likelihood. Comput Appl Biosci 13: 555–556.

108. Yang Z (2007) PAML 4: a program package for phylogenetic analysis by maximum likelihood. Mol Biol Evol 24: 1586–1591.

109. Yang Z (2009) PAML manual. Version 4.3.

110. Storey JD (2002) A direct approach to false discovery rates. J Roy Stat Soc B 64: 479–498.

111. Storey JD (2003) The positive false discovery rate: A Bayesian interpretation and the q-value. Ann Stat 31: 2013–2035.

112. Storey JD, Taylor JE, Siegmund D (2004) Strong control, conservative point estimation, and simultaneous conservative consistency of false discovery rates: A unified approach. J Roy Stat Soc B 66: 187–205.

113. Wilcoxon F (1945) Individual comparisons by ranking methods. Biometrics Bull 1: 80–83.

114. Benjamini Y, Hochberg Y (1995) Controlling the false discovery rate: a practical and powerful approach to multiple testing. J Roy Stat Soc B 57: 289–300.

4

Divergence and Selectivity of Expression-Coupled Histone Modifications in Budding Yeasts

Yaron Mosesson, Yoav Voichek, Naama Barkai*

Department of Molecular Genetics, Weizmann Institute of Science, Rehovot, Israel

Abstract

Various histone modifications are widely associated with gene expression, but their functional selectivity at individual genes remains to be characterized. Here, we identify widespread differences between genome-wide patterns of two prominent marks, H3K9ac and H3K4me3, in budding yeasts. As well as characteristic gene profiles, relative modification levels vary significantly amongst genes, irrespective of expression. Interestingly, we show that these differences couple to contrasting features: higher methylation to essential, periodically expressed, 'DPN' (Depleted Proximal Nucleosome) genes, and higher acetylation to non-essential, responsive, 'OPN' (Occupied Proximal Nucleosome) genes. Thus, H3K4me3 may generally associate with expression stability, and H3K9ac, with variability. To evaluate this notion, we examine their association with expression divergence between the closely related species, *S. cerevisiae* and *S. paradoxus*. Although individually well conserved at orthologous genes, changes between modifications are mostly uncorrelated, indicating largely non-overlapping regulatory mechanisms. Notably, we find that inter-species differences in methylation, but not acetylation, are well correlated with expression changes, thereby proposing H3K4me3 as a candidate regulator of expression divergence. Taken together, our results suggest distinct evolutionary roles for expression-linked modifications, wherein H3K4me3 may contribute to stabilize average expression, whilst H3K9ac associates with more indirect aspects such as responsiveness.

Editor: Robert Feil, CNRS, France

Funding: This work was supported by the ERC, the Israel Science Foundation, and the Helen and Martin Kimmel award for Innovative Investigations. The funders had no role in study design, data collection and analysis, decision to publish, or preparation of the manuscript.

Competing Interests: The authors have declared that no competing interests exist.

* Email: naama.barkai@weizmann.ac.il

Introduction

A dynamic portfolio of gene transcripts shapes cellular phenotype, but much about the regulation of transcription remains to be understood. Clearly, multiple, convergent regulatory mechanisms, converge at individual genes, including recruitment of selective transcription factors (TFs), organization of nucleosomes, and post-translational modification of key components such as the RNA polymerase II complex (Pol II) and histone proteins. These mechanisms collectively govern different phases of the transcriptional cycle, such as initiation and elongation. Therefore, by assimilation of regulatory inputs, cells may construct different kinetic strategies of transcription, suited to the requirements of a particular gene [1].

Histone modifications, including acetylation, methylation, and ubiquitylation, have emerged as pivotal features of transcriptional regulation. Various forms, targeting specific histone residues, have been extensively associated with active gene expression in genome-wide studies throughout eukaryotes [2–6]. Acetylation levels are governed by the net activity of histone acetyl transferases (HATs) and histone deacetylases (HDACs). Prominent transcription-linked HATs, including Gcn5 and Esa1, selectively target lysine clusters mainly within the amino-terminal regions of histone 3 (H3) and histone 4 (H4). Gcn5, in turn, forms part of the SAGA complex, a multi-modular assembly extensively coupled to the transcription machinery [7]. Accordingly, SAGA is recruited to active promoters, at levels that reflect gene activity [8], and to coding regions, in patterns that mirror Pol II occupancy [9,10]. Closely

spaced acetyl marks are thought to promote the loosening or eviction of nucleosomes, both via collective neutralization of histone/DNA interactions [11], and by direct engagement of nucleosome remodeling complexes [12]. This facilitates binding of initiation factors at the promoter, and presumably, paves a path for Pol II progress across coding regions. Hence, histone acetylation is often synonymous with active transcription, and may influence the kinetics of both initiation and elongation. HDACs, in contrast, generally associate with repression. As with HATs, several HDACs, such as the prototypical Rpd3 complex, are enlisted at different regions along the gene body [13,14], through various mechanisms, such as via direct engagement by various histone methyl marks, and recruitment by global repressor proteins (reviewed in [15]). Hence, by countering nucleosome pliability, deacetylation serves to prevent or retard transcriptional activity.

Histone methylation encompasses a variety of forms (mono-, di-, tri-, at several acceptor residues), and associates with both activated and repressed gene expression. Tri-methylation of H3K4 (H4K4me3), generated by Set1 methyltransferase, is most commonly correlated with active expression (reviewed in [16]). Set1, a component of the COMPASS complex, couples to several core regulators of transcription. In particular, COMPASS binding at active genes, and subsequent methylation, depends on mono-ubiquitylation of histone H2B (H2Bub) [17]. Indeed, both modifications are dependent on the Pol II-binding PAF complex, a global regulator that serves a platform for various histone-modifying activities during Pol II progression [18]. These associations implicate H2Bub and H3K4me 3 (as well as other methyl marks) in the control of elongation. As well as a methylation-deacetylation cross talk, H3K4me3 marks may also promote acetylation via methyl-binding domains within components of HAT complexes, including SAGA [19]. SAGA also incorporates a de-ubiquitylation module targeting H2Bub [20]. On aggregate, therefore, histone modifications are richly inter-linked both positively and negatively, in line with a model wherein orchestrated turnover of histone marks facilitates the disassembly and reassembly of nucleosomes during Pol II passage.

Interactions between histone-modifying enzymes raise an important issue; namely, to what degree are their corresponding marks connected in the context of expression? Are these generically utilized within a transcriptional cycle, or are their distinct chemical attributes exploited in order to forge customized kinetic programs? An extensive regulatory repertoire associated with each modification suggests that, in evolutionary terms, a degree of functional independence will have been encoded. Notably, by deletion of individual chromatin regulators in budding yeast, only limited changes in steady-state expression were observed, although functional inter-relationships between regulators could be reconstructed based on their gene targets [21]. This indicates the likely importance inter-complex cooperativity, but also attests to gene selectivity of regulatory complexes.

To investigate the selective nature of histone marks, we focus on the relationship between H3K9ac (generally representative of Gcn5-dependent acetylation of several other lysines within H3) and H3K4me3 in budding yeasts. Two complementary approaches are taken: firstly, we compare genome-wide profiles of both modifications in *S. cerevisiae*, and analyze their concordance at individual genes. We ask: can we distinguish different modification patterns with respect to nucleosome and gene preferences? Can we infer functional attributes from differences in gene targets? Second, we explore divergence of H3K9ac and H3K4me3 between the closely related species, *S. cerevisiae* and *S. paradoxus*. This approach offers a valuable means to evaluate the evolutionary association between sets of parameters, and therefore the degree to which these share a common genetic basis, and in turn, mechanisms of regulation. Hence, it allows us to interrogate the level of inter-dependence between H3K9ac and H3K4me3. Likewise, we ask how divergence of each modification correlates with expression divergence, and therefore whether or not either may serve as an evolutionary candidate to modulate expression changes between related species. Previous association studies in different yeast strains suggest that a large fraction of expression differences may be linked to polymorphisms affecting chromatin regulators [22,23]. On the other hand, features such as nucleosome occupancy and TF binding sites may only weakly explain expression divergence between species [24,25]. Hence, we evaluate the evolutionary utility of these histone marks in this context.

Our analyses reveal widespread differences between acetylation and methylation. Interestingly, each mark is found enriched at contrasting types of genes: higher H3K9ac at genes often variably transcribed (non-essential, responsive and OPN genes), and residual H3K4me3 at those more rigidly expressed (essential, periodic and DPN genes). Comparisons between species suggest largely independent regulation of each modification, and moreover, reveal significantly better agreement of methylation and expression divergence. Taken together, our results indicate that selective utilization of these chromatin features in different transcriptional programs may be highly prevalent, and further,

propose that acetylation engenders variable transcription while H3K4me3 contributes to a more stable program.

Materials and Methods

Yeast strains and antibodies

S. cerevisiae (BY4743) and a homozygous diploid *S. paradoxus* (generated from the CBS432 strain by transient HO inactivation) were used. ChIP antibodies to acetyl K9 of histone H3 (anti-H3K9ac), tri-methyl K4 of histone H3 (anti-H3K4me3), and an unmodified histone H3 were from Abcam (ab4441, ab8580, and ab1791, respectively).

Chromatin immunoprecipitation and mRNA extraction

ChIP was performed as previously described (Weinberger et al. 2012). Briefly, cells growing at mid-log phase (OD~0.7) were fixed, lysed and sonicated to generate DNA fragments with average size of 150-300 bp. Antibodies were added for an overnight incubation and precipitated using Protein A agarose beads. Following extensive washes, antibody-bound DNA was eluted in SDS and cross-links were reversed by incubation at 65°C. Proteins and RNA were then degraded, and the DNA obtained using the Qiagen PCR purification kit. To verify specific enrichment of DNA, multiplex PCR was performed at several loci selected according to previously published data (Liu et al. 2005). mRNA was generated in parallel from aliquots taken prior to cell fixation. Equal amounts of polyA-containing RNA from *S. cerevisiae* and *S. paradoxus* were taken before reverse transcription and library preparation for high throughput sequencing.

ChIP-seq and RNA-seq analysis

High-throughput sequencing was performed using the Illumina GAIIx system. To minimize experimental artifacts between samples, ChIP samples using the same antibody in the two yeast species were combined and processed together. Likewise, RNA samples were also combined. Raw data (40 base reads) were aligned to reference genomes of both *S. cerevisiae* (from Saccharomyces Genome Database) and *S. paradoxus* (obtained from the SGRP at the Sanger Institute), using the Illumina pipeline (Casava) together with Bowtie software [26]. Given significant sequence divergence between the species (~10% non-identical), almost all reads (>95%) could be unequivocally matched to either one or the other reference. To more accurately score each base, we estimated the characteristic length of the DNA fragments for each sequenced sample. This was done by generating separate profiles for the forward and reverse strands (scoring the first sequenced base), shifting the profile of one strand a base at a time, and calculating the overall correlation of the shifted profile with that of the other strand. The offset giving maximal correlation was considered as the average fragment length (typically, c. 125 bp, with rank correlation between strands of ~0.70). To obtain a profile per base pair, the forward and reverse strand profiles were shifted towards each other (by half the optimal offset) and then summed. Because the individual samples did not yield the same total number of reads, all profiles were normalized to that with the lowest total (~6.5 million reads).

For individual genes, we took the modification profile around the transcription start site (TSS) including the promoter region (−500 bp to +900 bp relative to the TSS), and that around the transcription termination site (TTS; −400 bp to +200 bp relative to the TTS). Binned profiles per gene around these regions were then obtained by taking the log$_2$-transformed mean signal over 20 bp intervals. In order to derive modification levels at individual nucleosomes in *S. cerevisiae*, we utilized consensus coordinates in assembled by Jiang et al. from several high-resolution maps of nucleosome occupancy (Jiang and Pugh 2009). Here, nucleosomes per gene were also numbered according to their position relative to the largely consensual TSS nucleosome: -2 and -1 for promoter-residing nucleosomes, $+1$ for the TSS nucleosome, and $+2$, $+3$, etc., for successive downstream nucleosomes (Jiang and Pugh 2009). Mean modification levels encompassing individual nucleosomes were taken and $\log_2$ transformed. For nucleosome occupancy profiles around the TSS and TTS, we employed previously generated MNase-seq data for *S. cerevisiae* and *S. paradoxus* from our laboratory [24]. Sequenced reads were binned as above. For RNA-seq experiments, reads were mapped to the reference genomes and the number of reads residing between the gene TSS and TTS summed. Gene expression levels were determined by normalizing for gene length and for the total number of reads mapped, and after taking an average of two biological repeats.

Because the *S. paradoxus* reference genome was incompletely annotated, we localized the orthologous genes by aligning the *S. cerevisiae* gene sequences with the appropriate chromosome of *S. paradoxus*. Most of genes (c. 90%) were aligned, and the orthologous TSS and TTS coordinates identified. Thereafter, binned modification profiles around the TSS and TTS were obtained as above. To facilitate interspecies comparisons, we also extracted average modification levels at three regions along the gene: the promoter (denoted 'prom': -320 to -160 relative to the TSS), the regions with highest abundance for each modification (denoted 'peak': 0 to $+140$ for H3K9ac, $+100$ to $+580$ for H3K4me3) and the end of genes (denoted 'end': -260 to $+60$ relative to the TTS). Furthermore, for interspecies comparisons at these regions, genes lacking data for either one of the modifications, mostly corresponding to short genes, were excluded from the analysis (e.g. leaving c. 4900 genes for analysis of 'peak' H3K9ac and H3K4me3).

Normalization of H3K9ac and H3K4me3 ChIP-seq data for Histone H3

In order to normalize H3K9ac and H3K4me3 levels in *S. cerevisiae* for nucleosome occupancy, we performed ChIP-seq using an anti-H3 antibody. The same experimental protocol described was employed, including similar growth conditions, sonication procedure, and Illumina pipeline. We obtained a lower number of mapped sequencing reads ($\sim$2.2 million) due to lower antibody efficacy. H3 levels were smoothed using a moving window (of 140 bases), and profiles per gene then generated as above, taking the mean signal over 20 bp intervals. For normalization, we subtracted H3 from modification levels ($\log_2$-transformed), either at each binned position along a gene, or after taking mean modification and H3 levels, either across the respective 'peak' regions or across the proximal open reading frame (-60 to $+580$ relative to the TSS).

Gene classifications and statistical analyses

Classifications of DPN and OPN genes were taken from a previous study [27]. Likewise, annotation of TATA and TATA-less genes was from previous work [28]. For periodic genes, we utilized a data set generated in Pramila et al., comprising the probabilities for genes to be periodically transcribed, which were calculated from measurements of transcript levels in synchronized cells at high temporal resolution [29]. We took the top 800 ranked genes for our analysis. A measure of expression responsiveness was calculated from a compendium of >1500 expression data sets for all genes, under a variety of conditions, including environmental stresses, mutations and developmental transitions [30]. We took

the standard deviation of expression per gene as a measure of responsiveness: the top and bottom ranked genes were assigned as responsive and non-responsive, respectively (1000 genes each). For classification according to gene ontology, we used the GO slim annotations, comprising 83 groupings (www.geneontology.org). Genes were also grouped according their tendency to be co-expressed with other genes; here, we employed previous data from our lab, comprising 26 pre-defined, overlapping transcriptional modules, which were constructed thorough analysis of numerous expression data sets [30]. Enrichment or depletion was calculated via a hypergeometric test in Matlab, or as a percentage of the abundance expected for the null hypothesis.

Results

Distinct genomic patterns of H3K9ac and H3K4me3 in *Saccharomyces cerevisiae*

Despite common ties to gene transcription, the extent to which H3K9ac and H3K4me3 patterns coincide within the genome remains unclear. To address this, we generated high-resolution maps in *S. cerevisiae*: cells growing in rich conditions (YPD) were subjected to chromatin immunoprecipitation (ChIP) with specific antibodies and subsequent high-throughput sequencing using the Illumina platform (see Materials and Methods for details). Sequenced reads were mapped, and to facilitate comparison between modifications, normalized for total read counts and $\log_2$ transformed. For a general perspective, we assessed their relative abundance at gene promoters and at open reading frames (ORFs; taking the mean signal across these regions). As shown, acetylation is clearly more enriched at gene promoters compared to methylation (Figure S1A in File SI), suggesting their differential function within this region. Both modifications are similarly prevalent within the ORF. We then plotted average modification profiles after aligning genes by their transcription start or end sites (TSS and TTS, respectively). As depicted, H3K9ac peaks at the TSS and sharply declines at downstream nucleosomes, while H3K4me3 exhibits a broader plateau encompassing several nucleosomes within the proximal ORF, which decreases thereafter (Figure 1A). These distinct profiles in *S. cerevisiae* are in accord with recent ChIP-seq studies [14,31,32]. Employing consensus genome-wide nucleosome positions compiled from several high-resolution maps [33], we inquired on modification patterns at individual nucleosomes. Here, nucleosomes are numbered according to their location along a gene: '-2nuc' and '-1nuc' encompass promoter-residing nucleosomes, '+1nuc' generally lies at the TSS, and nucleosomes further downstream are denoted '+2nuc', '+3nuc', etc. As shown, distributions for successive nucleosomes confirm highest abundance of H3K9ac marks at +1nuc, and that a significant fraction of genes also show high acetylation at -1nuc, while H3K4me3 is most prevalent at +2nuc and +3nuc (Figure S1B in File SI).

Given these differences, we assessed the influence of nucleosome occupancy in shaping the distinct profiles. To address this, average acetylation and methylation at individual nucleosomes were plotted after sorting by increasing occupancies. Interestingly, both modifications at -1nuc tended to mirror occupancy, while no association could be observed at nucleosomes that harbor highest H3K9ac and H3K4me3 levels (Figure S1C in File SI); accordingly, a positive correlation between modification and occupancy levels within the promoter ($r = 0.2$ to 0.4) declines to zero within the transcribed region (Figure 1B). This indicates that profiles within the gene body are actively determined, either by depositing/removing enzymes or as a consequence of nucleosome turnover,

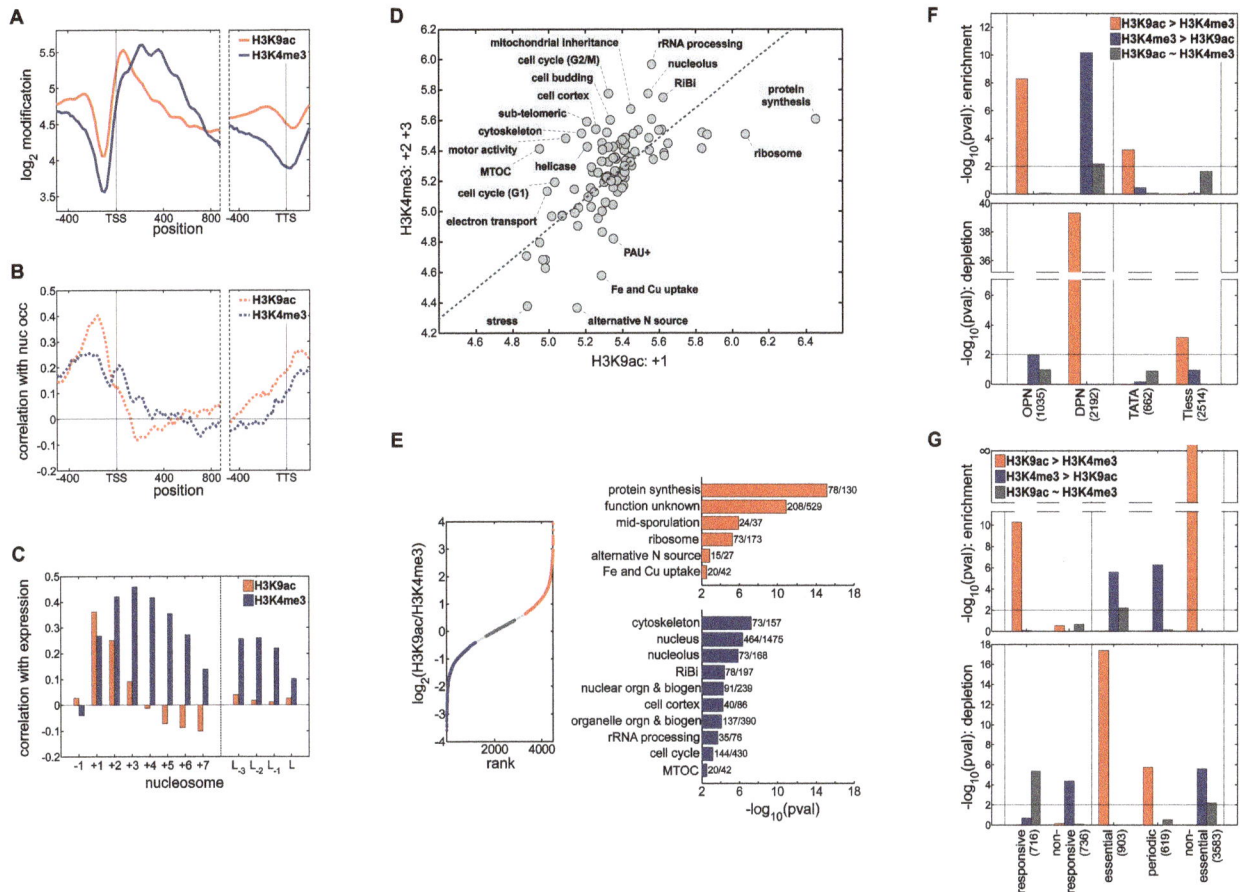

Figure 1. Biased distributions of expression-associated histone modifications in *S. cerevisiae*. (A) Average H3K9ac (*red*) and H3K4me3 (*blue*) profiles around the transcription start and termination sites of genes (*TSS* and *TTS*, respectively). **(B)** Traces depicting the correlation of H3K9ac and H3K4me3 profiles with nucleosome occupancy, around the TSS and TTS of genes. **(C)** Bar graphs showing the correlation between mRNA expression and modifications at individual nucleosomes across genes, as indicated. **(D)** Levels of H3K9ac (at +1nuc) and H3K4me3 (mean of +2nuc and +3nuc levels) were obtained for all genes. Genes were then grouped by gene ontology ('GO slim' categories, www.geneontology.org) or by pre-defined transcriptional modules [30], and the mean H3K9ac and H3K4me3 per category plotted against each other. Gene groups exhibiting higher average levels of one modification over the other are indicated. **(E)** Modification levels at specific nucleosomes for all genes were taken as in (D), and the ratio between H3K9ac and H3K4me3 calculated (log₂(H3K9ac/H3K4me3)). Genes were ranked according to this ratio, and three sectors (1200 genes each) were considered: H3K4me3 > H3K9ac (*blue*), H3K4me3 ≈ H3K9ac (*grey*), and H3K4me3 < H3K9ac (*red*). Enrichment of various categories of genes (as in (D)) within each of these sectors was then assessed using a hypergeometric test. Significantly enriched categories (−1*log₁₀(pval) > 2) are depicted in the bar graphs. **(F)** Genes were classified according to their promoter nucleosome architecture (occupied proximal nucleosome, '*OPN*'; depleted proximal nucleosome, '*DPN*'), or according to whether or not they incorporate a TATA-box within the promoter (TATA-containing, '*TATA*'; or TATA-deficient, '*Tless*'). Thereafter, enrichment (*top panel*) or depletion (*bottom panel*) of these classes amongst the sectors defined in terms of the genic H3K9ac/H3K4me3 ratio (as in (E)) was calculated using a hypergeometric test (*top panel*). Calculated p values are shown as -1*log10(pval). The number of genes in each subgroup is indicated. **(G)** As in (F) but assessing the enrichment (*top panel*) and depletion (*bottom panel*) of genes classified according to several features of expression: '*responsive*' and '*non-responsive*' genes, defined by their expression variance across a large compendium of conditions [30]; '*periodic*' genes, which show cyclical expression between successive cell cycles (800 genes; as defined in [29]); '*essential*' and '*non-essential*' genes (defined according to the viability in rich media of their respective deletion mutants).

whereas regulation of promoter modification levels may be less dynamic.

In order to directly account for the contribution of nucleosome occupancy, we performed ChIP-seq using an antibody to Histone H3, employing the same conditions as above. We then normalized for occupancy by subtracting log-transformed H3 from modification levels for each position (see Materials and Methods). Examples of H3K9ac and H3K4me3 profiles at individual genes before and after H3 normalization are depicted (Figure S2A in File SI). To gauge the general effect of occupancy, we plotted the average H3-normalized modification levels along a gene. As shown, the characteristic modification profiles are similar after normalizing for H3 levels: H3K9ac clearly peaks at the TSS

nucleosome, is also prevalent at the promoter, and sharply diminishes at internal ORF nucleosomes; H3K4me3 remains maximal at +2nuc and +3nuc (Figure S2B in File SI). Hence the distinct nucleosome preferences for these modifications are underscored after taking nucleosome occupancy into consideration.

Within this context, the positional association of modifications with gene expression was assessed. We generated RNA-seq data in cells growing under similar conditions, and expression levels were assigned by averaging mapped read counts across a gene. As expected, after classifying genes by expression level, average H3K9ac and H3K4me3 profiles unravel accordingly; that is, genic peak heights directly reflect the median expression of respective groups (Figure S1D in File SI). Examined per nucleosome, it is clear that association with expression varies along a gene: maximal correlation with H3K9ac occurs at +1nuc (r≈0.35), and with H3K4me3, at +2nuc to +4nuc (r≈0.45; Figure 1C), in accord with previous results [4,14,31]. Respectively, these comprise the nucleosomes with highest signals, which are also independent of occupancy. While promoter modifications showed no correlation with expression, we found a weak negative association with acetylation at distal coding-region nucleosomes, perhaps due to active de-acetylation. To corroborate these results, we also tested expression correlation after taking nucleosome occupancy into account. As above, average H3-normalized H3K9ac and H3K4me3 profiles for genes grouped by expression were well separated according to the group's mean expression level (Figure S2C in File SI, upper and middle panels). Further, the different positional correlations across a gene for H3K9ac and H3K4me3 were also maintained (Figure S2C in File SI, lower panel).

In summary, although mutually coupled to expression, H3K9ac and H3K4me3 concentrate to different nucleosomes along a gene. Conceivably, this may demarcate distinct functional domains during transcription: enrichment of acetylation at the TSS may drive initiation [34], while broader downstream methylation could serve a role in the fidelity of incipient transcription [35]. We note that these steady state profiles are the net outcome of extensive crosstalk between multiple histone modifying enzymes, hence, dynamic modification at other nucleosomes may be highly significant. Still, the question arises as to whether or not H3K9ac and H3K4me3 are similarly invoked at a particular gene during transcription.

Disparity between H3K9ac and H3K4me3 levels prevails at specific classes of genes

Mutual links to gene expression ensure a minimal coordination between acetylation and methylation, but to what degree does this occur? According to one possibility, if both marks contribute equivalently within a transcriptional cycle, then one might expect synchrony at all genes. Alternatively, these marks may have been deployed unevenly, for instance, in order to exploit different activities that each presents. To address this, intragenic modification levels were compared. Here, we examined only those nucleosomes that best correlate with expression, as we considered that incorporating other regions would likely detract from focusing on expression-linked phenomena. Hence, taking H3K9ac at +1nuc and average H3K4me3 across +2nuc and +3nuc, we calculate a Pearson correlation of ~0.3 (Figure S1E in File SI, left panel), which indicates that for given expression, relative modification levels may fluctuate significantly. Furthermore, after normalizing for expression, the correlation between modifications that can be attributed to expression is weaker (Figure S1E in File SI, right panel; r≈0.3-0.15). Hence, in the context of expression, the perceived lack of coordination may either reflect redundant functions of these modifications, or else be concealing deliberate asymmetry of acetylation-vs-methylation at a particular gene.

To examine this issue, we surveyed the gene pool by drawing on pre-assembled categories: 83 groups based on gene ontologies (GO slim database; www.geneontology.org), and an additional 26 groups comprising transcriptional modules re-constructed from multiple expression data sets [30]. Average acetylation and methylation per group were then compared. As shown, most display similar relative levels, indeed reflecting a much higher correlation coefficient compared to that for individual genes (~0.65; Figure 1D). Predictably, gene sets that are highly/poorly expressed in rich media contain high/low amounts of both marks on average (e.g. 'Golgi apparatus' or 'cytoplasmic vesicle' vs

'peroxisome' or 'meiosis', respectively). We estimated the fraction of individual genes showing concordant modification levels, by taking the $\log_2$(H3K9ac/H3K4me3) ratio and defining a nominal cutoff for modification differences of ±0.5 from the median (~1.4-fold; data not shown). By these criteria, c. 45% of genes exhibit similar modification levels, corresponding to the indicated correlation coefficient of ~0.3 (Figure S1E in File SI). For a minority of groups, however, disparity between H3K9ac and H3K4me3 persists across the constituent genes. For instance, on average, 'protein synthesis' genes appear hyper-acetylated, while 'rRNA processing' genes are, on average, methylation-rich (Figure 1D), suggesting that these biases may result from shared regulatory features at such genes.

To adopt a more systematic approach, we calculated the ratio of modifications ($\log_2$ (H3K9ac/H3K4me3)) at individual genes, and then ranked them accordingly. For this analysis, we considered genes above a minimal gene length (700 bases; ~4500 genes); because methylation, on average, concentrates at several nucleosomes within the proximal ORF, we reasoned that this constraint allows a more suitable comparison between modifications per gene. Three classes (1200 genes each) were then taken - the highest (H3K9ac > H3K4me3) and lowest (H3K4me3 > H3K9ac) ranked, and those intermediately ranked (H3K9ac ≈ H3K4me3) - and subsequently analyzed for enrichment of our pre-defined categories. Indeed, genes engaged in protein synthesis, in particular, were highly enriched within the first class (hypergeometric test, $p≈10^{-15}$), while higher methylation was significant for cytoskeletal ($p≈10^{-7}$) and nucleolar genes ($p≈10^{-6}$), as well as genes involved in biogenesis/assembly of ribosomes ('RiBi'), organelles and the nucleus ($p≈10^{-4}$; Figure 1E). Examples of gene contingents with tendencies towards with a higher acetylation or methylation are shown (Figure S1F in File SI). No categories were found to preferentially contain concordant acetylation and methylation levels.

A possible caveat of the analysis relates to the comparison of modifications at different nucleosomes along a gene (+1nuc versus +2/3nucs). As described, these regions were taken because we reasoned these to be the most informative in the context of active expression, based on their different positional correlations with expression (Figure 1C). However, variations in nucleosome occupancies at these positions may directly impact on relative H3K9ac and H3K4me3 amounts at individual genes, and so affect our analysis. Hence, to address this issue, we employed H3-normalized modification levels at these positions. Here, we took the mean $\log_2$-transformed H3 and modification signal across the respective regions (−60 to +140 for H3K9ac, +100 to +580 for H3K4me3, relative to the TSS; Figure S3A in File SI), and subtracted H3 from modification values. Genes were then ranked, and enrichment of gene groups tested assessed as before. As shown, similar gene ontologies recurred in this more stringent analysis, albeit manifesting with generally lower p-values (Figure S3B in File SI); for example, higher H3K9ac remained significant for protein synthesis and ribosome genes ($p≈10^{-10}$ and 10^{-5}, respectively), and higher H3K4me3 for nucleolar and cytoskeletal genes ($p≈10^{-6}$ and 10^{-5}, respectively). Another possible issue is that prevalence of modifications across individual nucleosomes may vary significantly from gene to gene, notwithstanding the characteristic average profiles. Accordingly, we also considered a less stringent definition, wherein we took a broader region encompassing the proximal ORF (−60 to +580, essentially nucleosomes +1 to +4) for both H3K9ac and H3K4me3 (Figure S4A in File SI). For this analysis, we tested H3-normalized, as well as the original modification data. Indeed, both definitions yielded similar results, and strongly supported previous observations;

recurrent gene ontologies were enriched for differential modification, and further, additional significant groups were apparent amongst the higher H3K4me3 category (e.g. 'vesicle-mediated transport', 'helicase activity'; Figure S4B in File SI, and data not shown). Overall, by accounting for nucleosome occupancy, and by taking a broader definition of H3K9ac and H3K4me3 levels, modification biases amongst particular gene groupings are reaffirmed.

These observations suggest unique functional properties of each modification. Therefore, we asked whether or not their disparity associates with structural and other attributes of genes. Patterns of promoter nucleosome occupancy have been strongly implicated in differential regulation of transcription. Genes with nucleosomes positioned over their promoter region ('occupied proximal nucleosome'; OPN), often occluding TF binding sites, tend to exhibit 'plasticity' of expression between cellular states. This includes non-uniform expression within an isogenic population, expression responsiveness to environmental changes, and divergent expression between related species [27,36]. Presumably, the opportunity for such variability stems from the need to mobilize a multifaceted regulatory program, which includes de-occlusion of TF binding sites, in order for transcription to ensue productively. At the other end are genes characterized by nucleosome-free promoter regions, flanked by well-positioned nucleosomes ('depleted proximal nucleosome'; DPN). This design likely facilitates assembly of the general transcription machinery, and may reflect a more homogeneous, less sensitive regulatory regime. Likewise, promoters that incorporate a TATA-box are often found in variable genes, while TATA-less genes are usually less variable [28]. Interestingly, we found that these contrasting features revealed clear preferences: OPN genes were significantly enriched in genes with higher H3K9ac (hypergeometric test, $p\approx10^{-8}$; fold enrichment over the expected abundance (fe) $\approx5\%$), and DPN genes were more prevalent in the class defined by higher H3K4me3 ($p\approx10^{-10}$; fe$\approx15\%$) and markedly under-represented in the higher H3K9ac class ($p\approx10^{-38}$; fe$\approx-25\%$; Figure 1F, and data not shown). Despite considerable overlap with DPN genes, TATA-less genes were not enriched in any class, while TATA-containing genes showed similar, but less significant, tendencies as OPN genes. Notably, these observations are well corroborated by further analyses employing H3-normalized modification levels. As shown, the corresponding enrichment/depletion of OPN and DPN genes is maintained, while TATA status appears not to be linked (Figure S3C in File SI, left panels). Moreover, when taking average H3K9ac and H3K4me3 levels over the proximal ORF, as before, the modification preferences at OPN and DPN genes are generally strengthened (e.g. fe$\approx45\%$ and fe$\approx25\%$, for higher H3K9ac and higher H3K4me3, respectively; Figure S4C in File SI, left panels). Taken together, these results suggest that nucleosome architecture may play a role in selective engagement of histone marks.

Other aspects of gene expression were also analyzed within this context, namely, responsiveness/plasticity and periodicity. 'Responsiveness' describes the potential expression range available to particular gene, and therefore, an indication of its sensitivity to regulatory inputs. To quantify responsiveness, we employed a compendium of transcription profiles at all genes in a variety of conditions (>1500 conditions, including environmental stresses, mutations and developmental transitions [30], and calculated for each gene the average magnitude of expression modulation. Periodicity, on the other hand, specifies the regularity with which genes are expressed between successive cell cycles; presumably, certain classes of genes, such as those with a 'housekeeping' role, encode an ability to be consistently and robustly transcribed.

Highly periodic/cyclical genes were taken from previous studies employing finely resolved temporal expression profiles in synchronized cells [29]. In a similar vein, essential and non-essential genes were also analyzed. Clear differences also emerged on the basis of these classifications: higher acetylation relative to methylation was highly favored for responsive ($p\approx10^{-10}$; fe$\approx35\%$) and for non-essential genes ($p > 10^{-100}$; fe$\approx10\%$; Figure 1G, and data not shown). Essential and periodically expressed genes, on the other hand, frequently displayed higher methylation ($p\approx10^{-5}$ and 10^{-6}; fe$\approx25\%$ and 30%, respectively), and furthermore, clearly disfavored higher acetylation (respectively, $p\approx10^{-17}$ and 10^{-6}; fe$\approx -40\%$ and 25%; Figure 1G, and data not shown). As previously, we re-appraised these findings using the alternative measures for modification levels described (H3-normalized, and mean over the proximal ORF). Here, preferences for differential acetylation-to-methylation clearly recapitulated for essential and non-essential genes, but responsiveness was only significant in the broader analysis (Figure S3C and Figure S4C in File SI, right panels).

To test the relative influence these parameters, we asked how their combination with nucleosome architecture affects the association with modification disparity. As shown, OPN/DPN status was generally a major determinant, as the respective biases were mostly retained amongst various sub-groups (Figure S5A in File SI, and data not shown). Periodicity and essentiality, however, associated with higher methylation, appeared to over-ride the preference of OPN genes for higher acetylation, which suggests that other regulatory features in addition to nucleosome architecture may guide selective use of modifications. Responsiveness *per se*, although initially associated with higher H3K9ac, did not affect the H3K4me3 bias at DPN or essential genes (Figure S5A in File SI, and data not shown), perhaps indicating the importance of H3K4me3-linked mechanisms at such genes. In further examples, we re-examined candidate GO categories (ribosome, RiBi, cytoskeletal and nuclear genes, and genes with unknown function). As shown, despite a tendency to fall within particular modification classes, further sub-division on the basis of OPN/DPN status augmented or diminished their enrichment accordingly, again underscoring the influence of nucleosome structure (Figure S5B in File SI).

Taken together, these results raise an interesting hypothesis; namely, that expression-linked histone modifications are not equivalently or redundantly used, but rather may be mobilized in a selective manner to create suitable programs of transcription. Higher H3K4me3, recapitulated at essential, periodic and DPN genes, may encompass a 'stable' strategy, wherein H3K4me3 guides reliable gene transcription to pre-programmed levels, without necessarily disclosing the temporal itinerary. Higher H3K9ac marks at intrinsically variable OPN genes may embody a more dynamic strategy; conceivably, residual steady-state TSS acetylation reflects rapid initiation of transcription, serving to meet changeable expression targets on demand (see Discussion).

Divergence of H3K9ac and H3K4me3 patterns in related yeasts

Our observations suggest that each modification associates with a distinct feature of transcriptional regulation. In order to test our hypothesis, we adopted an evolutionary perspective, and analyzed divergence patterns amongst related yeast species. This offers two advantages: first, by evaluating the evolutionary coordination between acetyl and methyl marks, we may question whether or not their respective machineries are governed by common genetic ground, and therefore, the likelihood or not that they are functionally autonomous. Second, a similar analysis with expression divergence discloses the extent to which genetic regulation of transcription encompasses that of each histone mark. A strong evolutionary relationship would indicate a direct functional association (either causal or consequential) with average expression levels, while lack of correlation would point to an indirect role, and/or links to processes other than gene expression. Hence, if we propose that H3K4me3 is involved in the stabilizing expression between successive cell generations, then one might expect good agreement with expression divergence between evolved states. Conversely, with a putative role in the kinetics of transcription rather than yield, weaker evolutionary coordination of H3K9ac changes might be anticipated.

Hence, we compared *S. cerevisiae* and *S. paradoxus*. Having diverged over 5my from a common ancestor, and exhibiting 80–90% sequence identity, these species are sufficiently close that the repertoire and chromosomal order of their genes is well preserved (shared synteny), yet sufficiently distant for widespread divergence of gene expression and its regulation [37,38]. Genome-wide acetylation and methylation maps, as well as mRNA levels per gene, were generated for *S. paradoxus* as previously, and juxtaposed to those obtained for *S. cerevisiae*. Our analysis included c. 6000 pairs (90%) of genes, and sequencing reads were normalized for total read counts and log_2 transformed. First, orthologous genes were compared by matching modification levels at corresponding loci (mean signal per 20 bp window); as shown, both H3K9ac and H3K4me3 were well correlated ($r \approx 0.6$ and $r \approx 0.55$, respectively; Figure 2A), indicating significant local agreement between species. Accordingly, average profiles for all genes were highly similar (Figure 2B, upper panels), supporting the notion that the characteristic genic distributions of H3K9ac and H3K4me3 are a recurrent design feature. To determine the degree of concordance across a gene, paired levels at each position were assessed for all genes. This revealed that the regions most enriched for acetylation (essentially +1nuc) and methylation (essentially +2nuc and +3nuc) were very well correlated between species ($r \approx 0.6$ and $r \approx 0.7$, respectively; Figure 2B, lower panel). Further, we detected remarkable agreement at the promoter particularly between H3K9ac levels ($r \approx 0.7$ and $r \approx 0.5$, respectively; Figure 2B, lower panel). However, given highly similar nucleosome occupancies between these species [24], significant correlation of promoter modifications may be an indirect consequence.

In the absence of well-defined consensus nucleosome positions for *S. paradoxus*, we considered three regions: the promoter ('prom': −320 to −160 relative to the TSS), the regions with highest abundance for each modification ('peak': 0 to +140 for H3K9ac, +100 to +580 for H3K4me3) and the end of genes ('end': −260 to +60 relative to the TTS). The extent of divergence at each region was then assessed; as shown, 'peak' acetylation and methylation (in particular) were best conserved, with only c.10% of genes displaying more than two-fold difference between the species despite high absolute levels within these regions (Figure 2C). H3K9ac differences at the end of genes were also restricted, likely due to lower signals in general, while promoter methylation was the most variable (almost 40% of genes with > 2-fold difference; Figure 2C). Hence, evolutionary conservation across a gene largely mirrors abundance and correlation with expression, reaffirming the functional significance of the respective patterns.

Disparity of H3K9ac and H3K4me3 at orthologous genes is often maintained, yet global patterns of inter-species divergence between modifications appear largely independent

Given considerable interspecies agreement between individual modifications, we asked whether this level of conservation

encompasses the intragenic disparity in *S. cerevisiae* (Figure 1). To address this, acetylation-to-methylation ratios were compared between species; indeed, they were well correlated ($r \approx 0.5$; Figure 2D), indicating that a significant proportion of genes retain disparity of modifications in evolution. We examined some of the strongest examples noted in *S. cerevisiae*; as shown, the prominent acetylation of protein synthesis genes and methylation of cytoskeletal genes are maintained, and a similar dichotomy survives for 'ribosome' vs 'RiBi' genes (Figure 2E). Overall, the marked evolutionary correlation for individual modifications is largely upheld in the context of their ratios, indicating the likely utility of asymmetric modifications. Predictably, similarly utilized orthologous genes are governed by similar regulatory constraints.

Apart from this basic association, however, the direct question of whether acetyl and methyl marks diverge together remains open. To approach this, we calculated a regression curve from the interspecies scatter plot for each modification, and extracted those genes farthest from the curve in either direction (Cer>Par and Par>Cer, c. 800 genes each). Coincidence of acetylation and methylation differences was then tested. As shown, overlap of genes showing both higher H3K9ac and H3K4me3 in the same species was only marginally significant (hypergeometric test, $p \approx 0.03$), and clearly, the overwhelming majority of genes (c. 85%) did not concur (Figure 2F). That is, divergence of one modification gave no indication for divergence of the other. We extended this analysis to all genes and plotted H3K9ac changes against H3K4me3 changes; as expected, there was only a nominal correlation between changes (Figure 2G; $r \approx 0.1$). Significant changes were then defined using a threshold of 1.3-fold (0.4 in log_2 scale), encompassing c. 50% genes. Using this threshold, we noted that H3K9ac or H3K4me3 co-varied at only 17% of genes, while 22% showed no change in both modifications. Genes with differences only in acetylation were more prevalent than those only varying in methylation (27% and 21% of genes, respectively), and indeed, a significant fraction showed opposite changes (Figure 2G). Hence, transitions of intragenic H3K9ac and H3K4me3 levels between evolved states appear largely uncoordinated, with almost 50% of genes exhibiting variation in only one of the modifications. In this context, the retention of modification disparity at a fraction of genes may correspond to some level of mutual constraint, wherein acetylation levels vary within boundaries set by methylation levels, and *vice versa*. This may arise, for instance, by selective engagement of cross-talk mechanisms. However, the overwhelming lack of correlation indicates that acetyl and methyl levels are governed by largely independent sets of genetic determinants, which in turn allow considerable functional autonomy.

Changes in H3K4me3 coincide with expression divergence more often than do H3K9ac changes

In light of their dissimilar evolutionary profiles, the question as to which histone mark diverges more closely with expression is raised; here, genes with significant changes between *S. cerevisiae* and *S. paradoxus* in both mRNA levels and modifications were considered (at least 1.3-fold). As shown, expression differences correlated significantly better with changes in 'peak' H3K4me3, as compared to 'peak' H3K9ac ($r \approx 0.4$ and $r \approx <0.2$, respectively; Figure 3A). Examining this further, all genes were classified according to whether they vary in expression, modifications (average signal for the 'peak' regions) or both. The proportion of genes that co-varied with expression was similar for both modifications (c.20%), but H3K9ac, relative to H3K4me3, showed a higher fraction of changes not coordinated with

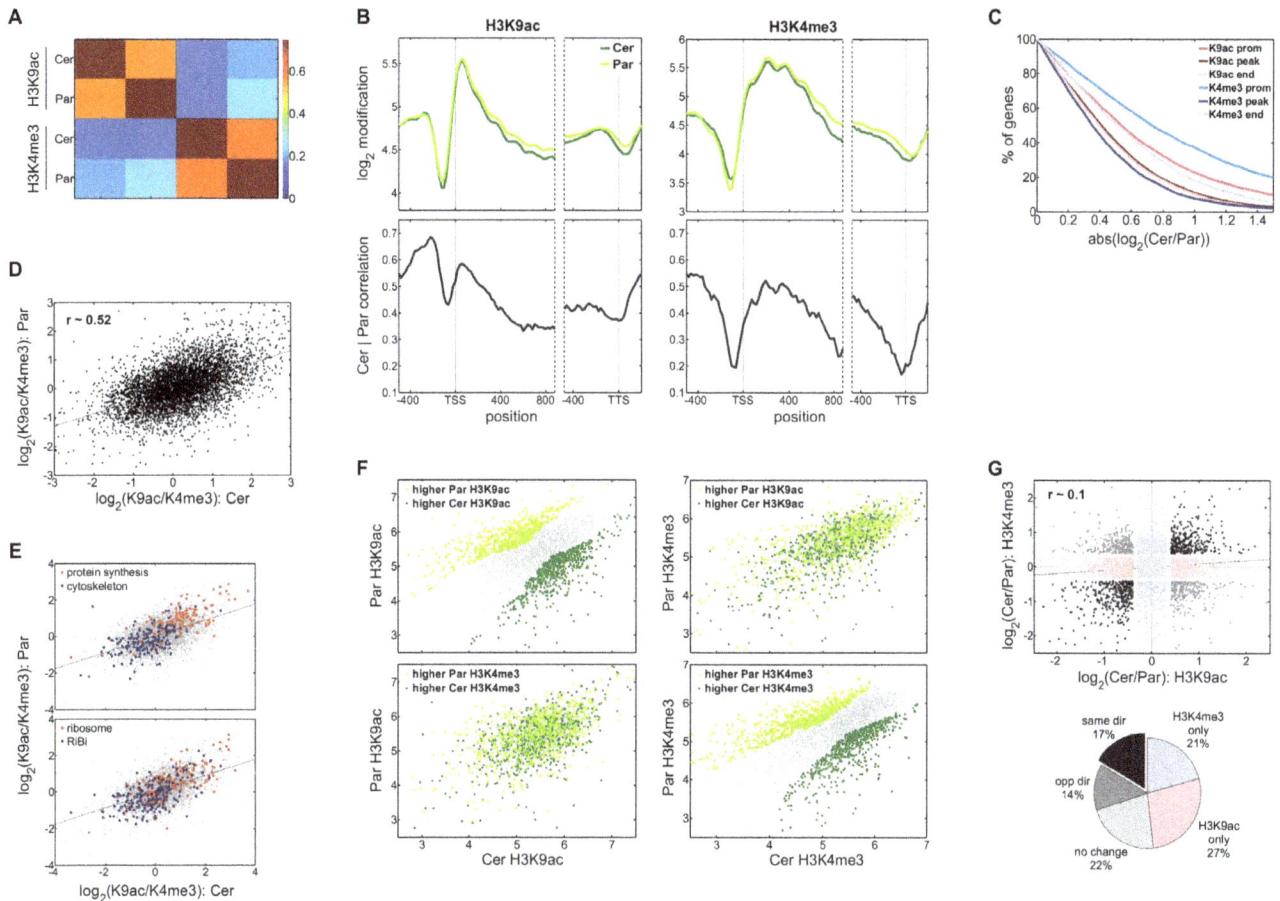

Figure 2. Divergence of H3K9ac or H3K4me3 patterns between closely related yeasts. (A) Overall correlation of H3K9ac and H3K4me3 patterns between *S. cerevisiae* and *S. paradoxus*. Heatmap of Pearson correlations calculated after juxtaposing modification profiles (mean signal over 20bp intervals) at c.6000 orthologous genes (including the promoter and coding regions). (B) Average H3K9ac (*upper left panel*) and H3K4me3 (*upper right panel*) profiles around the TSS and TTS, for *S. cerevisiae* (*dark green*) and *S. paradoxus* (*light green*). Interspecies correlations for H3K9ac and H3K4me3 at the given positions along a gene are shown in the lower panels. (C) Interspecies differences (log$_2$(Cer/Par)) in modification levels for different regions along a gene (*upper panel*). Mean levels across three regions were taken: the promoter ('*prom*': −320 to −160 relative to the TSS), the respective loci with highest prevalence on average ('*peak*': 0 to +140 for H3K9ac, +100 to +580 for H3K4me3) and around the TTS ('*end*': −260 to + 60 relative to the TTS), as indicated. The percentage of genes with absolute differences at these regions exceeding an increasing threshold (log$_2$ scale) is depicted. (D) Comparison of intragenic differences in modification levels (log$_2$(H3K9ac/H3K4me3)) between *S. cerevisiae* and *S. paradoxus* at orthologous genes. The interspecies correlation, and a linear fit of the data are shown. (E) Relative modification levels at orthologous genes for selected ontological groups. Shown are interspecies comparisons for 'protein synthesis' and 'cytoskeleton' genes (*upper panel*), and 'ribosome' and 'RiBi' genes (*lower panel*). (F) Scatter plots comparing 'peak' H3K9ac and 'peak' H3K4me3 between species. For each modification, the most divergent genes were extracted by applying the Lowess method to the interspecies plot, and selecting those genes farthest from the regression curve in either direction (c. 800 genes each; *upper left* and *lower right* panels). Genes with higher H3K9ac in *S. cerevisiae* (*dark green*) or *S. paradoxus* (*light green)* were overlaid onto the interspecies H3K4me3 plot (*upper right*), or vice versa (*lower left*). (G) Plot of interspecies differences in H3K9ac against H3K4me3. Genes showing consistent changes for both modifications (*black*), opposite changes (*grey*), no changes in either modification (*light grey*), changes in only in H3K9ac (*pink*), and changes only in H3K4me3 (*light blue*) are marked. Absolute differences above 0.4 (log$_2$ scale) were considered as significant. The Pearson correlation and a linear fit for the data are shown. Relative proportions of each group are indicated in the adjacent pie chart.

expression (modification changes only or opposite changes: 38% vs 30%; Figure 3B and 3C).

On further analysis, the better predictive value of 'peak' methylation over 'peak' acetylation is clarified. First, correlation with methylation differences consistently rises with an increasing threshold for absolute changes (e.g. r≈0.6 for genes with greater than two-fold difference), but association with acetylation plateaus at significantly lower correlations (maximally, r≈0.3; Figure 3D, top panel). Accordingly, for genes with significant expression differences (at least 1.3-fold), the percentage of genes with consistent H3K4me3 changes (that is, in the same direction) outweighs those for H3K9ac at all taken thresholds (e.g. 75% compared to <60% with consistent changes, respectively, for a

two-fold threshold; Figure 3D, middle panel). Further, we selected the most and least divergent genes with respect to expression (1000 genes each), and compared their presence for a range of modification differences. Clearly, the fraction of divergent genes far outweighed that of non-divergent genes for methylation, but not for acetylation, across the scale of difference thresholds tested (Figure 3D, bottom panel); for instance, a two-fold threshold for H3K4me3 differences incorporated a 4.5-fold enrichment in the fraction of divergent over non-divergent genes, while H3K9ac coupled to only a 1.5-fold enrichment. Hence, between closely related yeasts, methylation downstream of the TSS often varies together with mRNA levels, but association with TSS acetylation changes is significantly weaker.

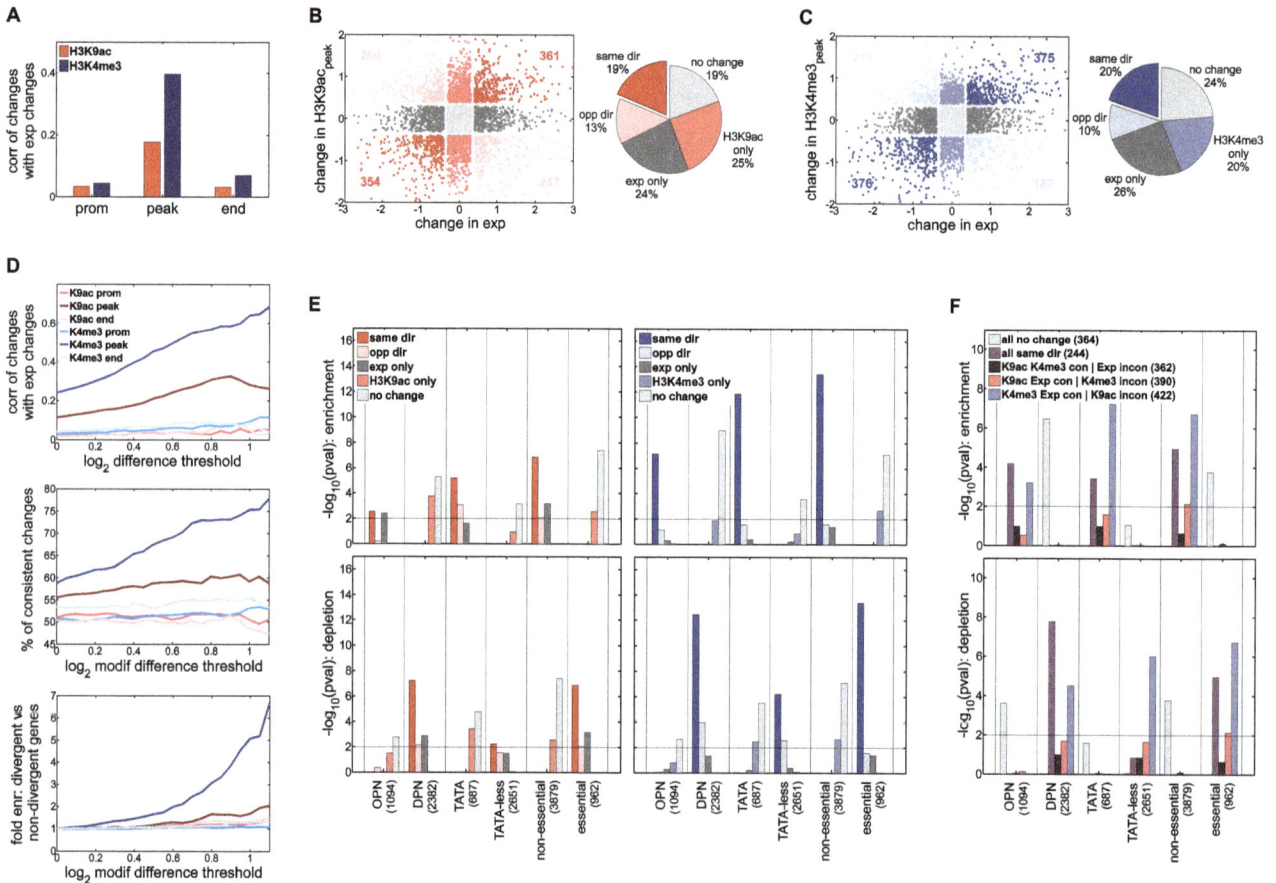

Figure 3. Interspecies differences in H3K4me3, compared to H3K9ac, better predict expression divergence. (A) Correlation between interspecies expression differences and either H3K9ac (*red*) or H3K4me3 (*blue*) changes, at different regions along a gene, as indicated. Correlations were calculated for significant changes in expression and modifications (absolute differences greater than 0.4 ($\log_2$ scale)). **(B)** Scatter plot of interspecies differences in mRNA levels against changes in 'peak' H3K9ac. Genes showing consistent changes between *S. cerevisiae* and *S. paradoxus* (*dark red*), opposite changes (*pink*), no changes in either parameter (*light grey*), changes in only in H3K9ac (*light red*), and changes only in expression (*dark grey*) are marked. The numbers of genes showing consistent or opposite changes are noted, and the relative proportions of each group are indicated in the adjacent pie chart. **(C)** As in (B), but for H3K4me3 changes. Genes showing consistent changes (*dark blue*), opposite changes (*light blue*), no changes in either parameter (*light grey*), changes in only in H3K9ac (*blue*), and changes only in expression (*dark grey*) are marked. **(D)** *Upper panel*, Interspecies correlation between expression differences and H3K9ac (*red*) or H3K4me3 (*blue*) differences at different regions along a gene, as indicated, calculated at increasing absolute thresholds (for both expression and modification changes). *Middle panel*, Graph depicting the proportion of genes for which interspecies changes in expression and modifications (at different regions) are consistent (in the same direction). Percentages were calculated for an increasing threshold of absolute modification differences, and considering significant expression differences (> 0.4). *Lower panel*, Genes with the largest and smallest differences in expression *S. cerevisiae* and *S. paradoxus* ('divergent' and 'non-divergent'; 1000 genes each) were taken. Thereafter, for an increasing threshold of absolute differences in modifications at different regions, the fold enrichment of divergent over non-divergent genes was calculated at each threshold. **(E)** Enrichment (*upper panels*) or depletion (*lower panels*) of sets of genes defined according to patterns of H3K9ac/mRNA changes (*left*) or H3K4me3/mRNA changes (*right*) amongst the indicated architectural or phenomenological gene classes; *p* values were calculated using a hypergeometric test. **(F)** Genes were classified based on interspecies variation in all three parameters (H3K9ac, H3K4me3, expression): those showing no change in either (*light grey*), consistent changes amongst all three (*purple*), consistent changes between H3K9ac and H3K4me3 but not expression (*dark grey*), consistent changes between H3K9ac and expression but not H3K4me3 (*light red*), and those showing consistent changes between H3K4me3 and expression but not H3K9ac (*blue*). Gene sets were subsequently analyzed as in (E).

Particular gene attributes favor coordination of H3K4me3 and expression divergence

Interspecies variation in methylation, in particular, may disclose the degree of expression divergence, but this relationship is only upheld for a fraction of the gene pool (c.40%). Hence, we asked whether certain gene attributes might be favored amongst these genes, and tested association with nucleosome organization (OPN vs DPN) and promoter sequence (TATA-containing vs TATA-less). Amongst genes for which mRNA and H3K4me3 changes were consistent (in the same direction between species), both TATA-containing and OPN genes were highly enriched

($p \approx 10^{-12}$ and $p \approx 10^{-7}$, respectively), while DPN and TATA-less genes were significantly depleted ($p \approx 10^{-12}$ and $p \approx 10^{-6}$, respectively; Figure 3E). As there is considerable overlap between OPN and TATA-containing genes, we looked at intersecting gene groups to determine the dominant feature; as depicted, evolutionary coordination of expression with methylation is favored when both features are present (Figure S6 in File SI). This is to be expected given that OPN and TATA-containing genes are often divergently expressed in evolution. However, it is notable that these genes were not significant when expression changes were not accompanied by consistent H3K4me3 changes (Figure 3E, and Figure S6 in File SI). Furthermore, H3K9ac marks did not recapitulate these biases; for instance, TATA-containing genes showed both consistent and opposite changes with regards to expression divergence ($p \approx 10^{-5}$ and $p \approx 10^{-3}$; respectively), and OPN genes were equally present whether or not H3K9ac levels co-varied with expression ($p \approx 10^{-2}$; Figure 3E).

Essential and non-essential genes were similarly analyzed, and based on their evolutionary proximity, we assumed equivalent gene cohorts in *S. cerevisiae* and *S. paradoxus*. Similar to OPN and TATA-containing genes, non-essential genes were highly prevalent when methylation and mRNA levels varied together ($p \approx 10^{-13}$), but not amongst genes with uncoordinated differences. For H3K9ac, non-essential genes also distributed amongst those with inconsistent changes. Differences between the modifications are even more pronounced when assessing H3K9ac, H3K4me3 and mRNA changes concurrently. For instance, TATA-containing genes appear enriched amongst those that exhibit consistent methylation and expression changes between species, regardless of whether or not acetylation changes are in agreement, but not vice versa (Figure 3F). In sum, variation in methylation, but not acetylation, appears to be well coordinated with expression divergence at TATA-containing, OPN and non-essential genes. At such genes, evolution of expression might have entailed modulation of H3K4me3 levels.

For a functional perspective, genes grouped by gene ontology or 'transcriptional modules' (as described earlier) were also examined. This revealed that gene sets with coordinated changes are largely common to both acetylation and methylation. For instance, higher modification and expression in *S. cerevisiae* was significant for amino acid biosynthesis genes; for *S. paradoxus*, higher modification/expression levels were prevalent at mitochondrial genes, in line with the differential respiratory strategies of the species under rich conditions (Figure S7 in File SI). In contrast, several sets were inconsistent with respect to acetylation, but not methylation. For instance, genes engaged in protein synthesis tended to be differentially expressed without variation in H3K9ac levels; ribosomal RNA-associated genes, which were invariantly expressed, were also inclined towards invariant H3K4me3, but not invariant H3K9ac. Other gene sets (e.g. bud-neck, nuclear, nucleolar and cell cycle genes) also tended towards variable acetylation with no change in expression (Figure S7 in File SI). Hence, examination of functionally coordinated groups also highlights a propensity for incoherent acetylation and expression differences, whereas methylation tends to concur for a greater number. Overall, these results raise the notion that, at divergently expressed orthologous genes, H3K4me3 may have been enlisted as a means to stabilize interspecies differences in mRNA levels. Poorer evolutionary coordination with H3K9ac, on the other hand, indicates indirect involvement, conceivably, in the temporal regulation of transcription.

Discussion

Histone acetylation and methylation are central players in the transcriptional process, yet as distinct chemical moieties, each is likely to embody specialized functions. In this study, we compared the distributions of H3K9ac and H3K4me3 marks at high resolution in budding yeasts, in order to discover distinguishing features in the context of gene expression. Differences were manifested at several levels. First, each mark exhibited characteristic gene patterns: acetylation concentrated at TSS nucleosomes, and methylation, at proximal ORF nucleosomes. Second, we found that these marks were asymmetrically distributed amongst the gene pool: H3K9ac appeared more prominently at non-essential, responsive genes and an OPN architecture, while higher

H3K4me3 was enriched at essential, periodically expressed and DPN genes (Figure 4). Third, both acetylation and methylation patterns were well conserved between *S. cerevisiae* and *S. paradoxus*, but their divergence at orthologous genes appeared largely uncoordinated with each other, indicating largely independent evolutionary trajectories of their respective machineries. Finally, expression divergence, particularly at OPN and TATA-containing genes, was well explained by changes in H3K4me3, but not H3K9ac, suggesting methylation as a plausible evolutionary candidate to tune transcriptional output.

In our initial survey of gene ontology/transcriptional modules, it was interesting to note that higher methylation, relative to acetylation, appeared to manifest preferentially at gene cohorts with 'housekeeping' or structural roles, such as the biogenesis and/or maintenance of organelles and the cytoskeleton (Figure 1D and 1E, and Figure S1F in File SI). A similar preference was also found when classifying genes in terms of essentiality and periodicity between cell cycles (Figure 1G). Therefore, we considered that higher aggregate H3K4me3 levels might relate to the need to ensure that such genes are to be infallibly and consistently expressed between successive cell generations. Indeed, a growing body of work from across the eukaryotic divide associates the H3K4me3 mark with stable transcription. Using live imaging to follow transcription events at a single housekeeping gene in individual *Dictyostelium* cells, Muramoto et al. showed that frequency of transcriptional pulses tended to be inherited between successive cell generations, and that H3K4me3 marks were important in maintaining the memory of active states [39]. In a similar vein, diverse studies attribute to H3K4me3 a role in preserving and/or reflecting the characteristic expression profiles of differentiated cells. These include key involvement in persistent expression of somatic genes in nuclear transplantation experiments [40], and maintaining homeostasis in adult cardiomyocytes [41]. Interestingly, a recent report, charting the proximity of various histone modifications with single nucleotide polymorphisms (SNPs) linked to complex human traits, found H3K4me3 to be the most informative for identifying the participant cell types [42], again, coupling this mark to the phenotype of specialized cells.

In line with conferring memory of expression, the dynamics of histone methylation are believed to be markedly slower than those of acetylation [43]; turnover rates for acetylation are typically in the order of minutes [44]. H3K4me3 marks, on the other hand, are likely more dependent on DNA replication for their erasure, and may linger for hours beyond the residence time of an actively transcribing Pol II [45–47]. Further, persistent methylation has been found to obstruct reactivation of recently transcribed genes [48], thereby curbing transcriptional output. Indeed, increasing evidence attributes to H3K4me3 repressive roles in gene expression, through several mechanisms. For instance, elevated gene expression upon deletion of the methylase, Set1, has been recently linked to de-repression of attendant antisense transcription [49–51]. Further, methylation marks, including H3K4me2/3 and H3K36me2/3, are thought to decrease accessibility of the local chromatin structure, often via direct association with various HDACs. Chromatin compaction may serve various ends, including reducing the processivity of an elongating Pol II, protecting against cryptic initiation of transcription, and preventing basal gene activation [15]. Plausibly, such buffering mechanisms converge on a common evolutionary objective – to impart fidelity, both in terms of Pol II accuracy and in delimiting mRNA output boundaries.

If H3K4me3 marks contribute to constrained but reliable gene transcription, then the synchrony with expression changes that we observed between related yeasts (Figure 3) may be readily

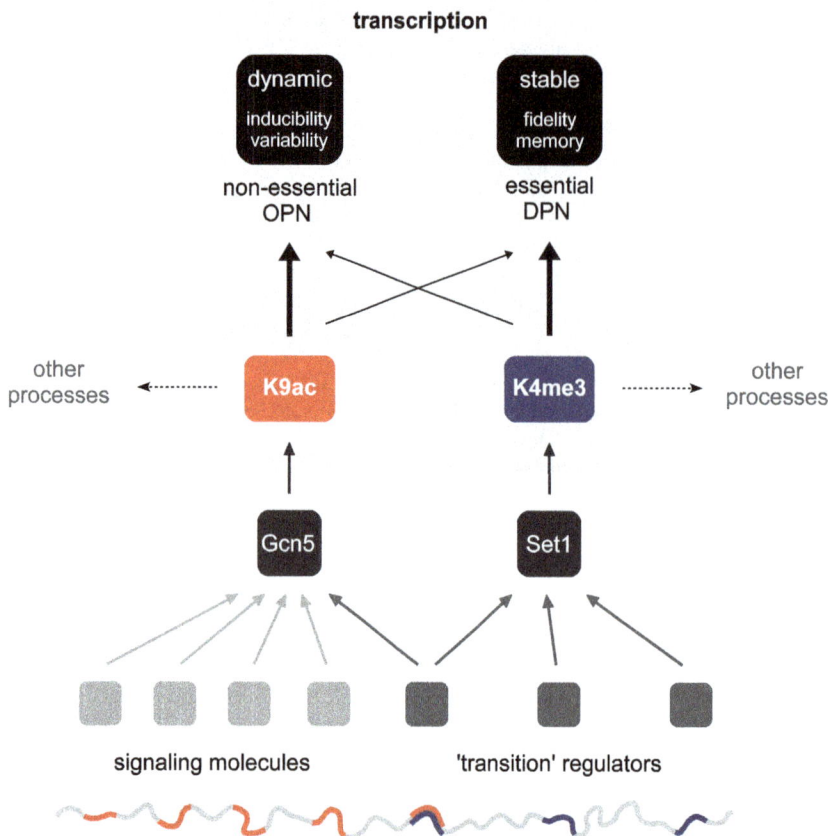

Figure 4. Selective utilization and evolutionary divergence of transcription-linked histone modifications. Scheme depicting the differential distribution of H3K9ac and H3K4me3 marks amongst the gene pool in budding yeast. Higher acetylation manifests at non-essential OPN genes, while methylation appears enriched at essential DPN genes. We propose that these patterns reflect disparate blueprints of transcriptional regulation. That is, selective engagement of acetylation at the TSS may enable a dynamic program, varying both temporally and quantitatively, through control of transcription induction. H3K4me3 may engender higher fidelity and consistency of transcription, and so contribute to a more stable program. This division of labor is corroborated in comparisons between related yeast species. In particular, H3K9ac and H3K4me3 marks appear to vary independently of each other, yet interspecies changes in steady-state expression are well explained by differences in methylation, rather than acetylation. Hence, evolutionary regulation of these marks likely involves genetic changes at largely non-overlapping sets of *trans* factors. Conceivably, mutations at a variety of signaling molecules that react to environmental perturbations could impinge on Gcn5 acetylase activity, whereas modulation of Set1 may involve changes at a more select set of regulators engaged in large-scale phenotypic transitions (such as cell division or differentiation).

explained. Conceivably, evolution has enlisted H3K4me3 as a means to stabilize gene output at levels suitable to the adapted species: that is, robust interspecies transitions in the expression of orthologous genes may have been partly achieved by modulating histone methylation. It is interesting to note that other features of the transcriptional process have not been found to concur evolutionarily. For instance, sequence divergence at TF binding sites explains only a small fraction of interspecies expression differences [25]. Likewise, although nucleosome positioning at gene promoters often correlates with expression changes between different environments and cell types [5,52,53], this relationship is not upheld in evolutionary timescales [24]. Namely, the genetic basis of expression divergence does not appear to encompass regulation of nucleosome positioning, but may rather involve mutations that impinge on the H3K4me3 marks, amongst other features to be identified. Hence, this class of chromatin features may serve as a rheostat for appropriate gene yield, whilst overriding discrepancies of other features such as nucleosome positioning and TF binding sites.

Acetylation presents a different picture in our analyses. Unlike H3K4me3, H3K9ac marks concentrated at TSS nucleosomes (Figure 1A), which co-localize with the transcription initiation machinery. Moreover, H3K9ac was generally prevalent at non-essential genes, as well as at responsive genes intrinsically capable of variable expression in different environments (Figure 1G). Protein synthesis and ribosomal protein (RP) genes, which are highly expressed in response to nutrient signals in exponentially growing cells, were also targets for hyper-acetylation (Figure 1E, and Figure S1F in File SI). Further clues into the utility of excess acetylation emerge from considering OPN/DPN status. In general, this classification of genes has proved insightful in gene regulation: OPN genes tend to incorporate greater regulatory complexity, including gene-tailored TFs, and presumably, promoter nucleosome-remodeling activities, which, in turn, enables inducible, conditional gene expression [27,36]. The enrichment of H3K9ac over H3K4me3 at OPN-structured genes, and its relative dearth at DPN genes (Figure 1F), may therefore be a consequence of selective recruitment of acetylation enzymes. Indeed, although likely integral to general transcription, the HAT-bearing SAGA complex is often specifically engaged by multiple transcription factors in response to inductive signaling [54]. Further, several components of SAGA may substitute for components of the general transcription machinery (TFIID), particularly during transcription of stress-induced genes [55]. Hence, where H3K4me3 marks are perhaps more generically employed and less amenable to change, acetylation machineries appear highly sensitive to signaling, linking H3K9ac (and other histone acetyl marks) with timely induction of transcription [34].

Acetylation dynamics that are far above the rate of histone turnover, by definition, implies the continuous engagement of both HAT and HDAC activities. Interestingly, this reversibility is thought to be important to leverage responsiveness. For instance, deletion of a component of the RPD3L deacetylase complex, which is recruited to gene promoters, impaired both induction and repression of ESR genes in response to stress [56]. Early studies showed that the coding region deacetylase complex, SET3, was required for efficient activation of the GAL gene cluster [57]. Hence, dynamic acetylation at activated genes may serve to reset chromatin template between successive rounds of transcription, thereby creating conditions for efficient re-initiation.

In this light, the poorer association we found between divergence of H3K9ac marks and expression in related yeasts (Figure 3) may be rationalized. If acetylation impinges primarily on the efficiency/kinetics of gene induction, then it is reasonable to expect that its links to steady-state expression dissipate during evolution. For example, several groups of genes presented differences in H3K9ac without varying in expression (Figure S7 in File SI); such genes may be encountering different efficiencies of transcription, according to given conditions in the given species, but the cumulative outcome on expression levels may ultimately be unchanged. Accordingly, when inspecting simultaneous modifications, the presence of H3K9ac marks did not improve the predictive power of H3K4me3 changes in expression divergence (Figure 3F). Genes pertaining to oxidative phosphorylation were a clear exception in this regard (Figure S7 in File SI), but given that the preferred respiratory programs of S. cerevisiae and S. paradoxus (anaerobic vs aerobic respiration, respectively) are most likely hard-wired through signaling, interspecies differences in the efficiency of gene induction, in addition to output, might be expected. Our hypothesis concurs well with a study examining natural variation in H3K14 acetylation between different S. cerevisiae strains; inter-strain differences were also not indicative of expression divergence, but H3K14ac changes were enriched at responsive genes [58].

It is interesting to speculate on the evolutionary aspects of our findings. At orthologous genes, H3K9ac changes were only nominally correlated with H3K4me3 changes ($r \approx 0.1$; Figure 2F and 2G). This implies that interspecies genetic differences that impinge on each mark are largely non-overlapping, and have probably evolved independently. On the other hand, higher coordination of H3K4me3 and steady-state mRNA changes, especially at TATA-containing and OPN genes (Figure 3A and 3E), indicates that the methylation and transcription machineries are to some extent affected by a common pool of mutations. What then is the nature of such mutations? Given general pervasiveness of chromatin features, and their specific utility as modulators of gene expression, it seems unlikely that their genetic regulation involves many individual changes in the proximity of target genes (cis effects). Rather, mutations that affect the activity/expression of regulatory factors (trans effects) are more plausible evolutionarily. Previous work employing an interspecific hybrid strain to distinguish cis/trans contributions to expression divergence strongly corroborates this assumption: although cis effects were generally

more prevalent, trans regulation predominated at genes most affected by deletion of chromatin regulators [38]. Likewise, expression evolution of intrinsically divergent OPN and TATA-containing genes was also better explained by trans rather than cis effects [23,38]. Obvious candidates for evolutionary selection are core components of the H3K4 methylation complex (COMPASS). However, other factors that exert leverage over this machinery are also likely to have been enlisted. The recurrent association of H3K4me3 marks with steady-state phenotype, together with their relative stability, perhaps predicts that potential evolutionary targets will have been limited to a select type; namely, those capable of driving large-scale cellular transitions, such as division or differentiation (Figure 4).

What might be the genetic basis of H3K9ac marks? As with other chromatin regulators, trans effects that originate at the enzymatic machinery are likely. However, compared to methylation, a variety of ancillary factors might be expected. Rapid dynamics, together with a posited role in transcription induction, predicts that acetyl marks are sensitive to multiple signaling pathways, including those that mediate prompt adaptation to environmental perturbations. Accordingly, a large collection of mutations at upstream signaling molecules may have contributed to H3K9ac divergence (Figure 4), which is in line with a weaker association with expression divergence (Figure 3). Furthermore, H3 acetylation and H3K4 methylation are engaged in other DNA-centered processes, including replication and recombination [59,60], indicating additional complexity into the genetic regulation of these marks in the context of transcription.

Conclusions

Overall, our results show that distinct histone modifications appear to be deployed selectively across the gene pool, in a manner that associates with expression variability on the one hand (higher H3K9ac at the TSS) and stability on the other (higher H3K4me3 at proximal ORF nucleosomes). Our comparative inter-species analysis corroborates their independent regulation, and proposes H3K4me3, rather than acetylation, as a possible evolutionary means to control expression divergence. In future work, it will be interesting to test cis and trans contributions to expression regulation by profiling an interspecific hybrid. Moreover, current high-throughput ChIP-seq technologies make it feasible to extend this comparative approach to multiple chromatin features in parallel, including other histone marks, nucleosome remodelers and histone chaperones. This will help to unravel the interfaces that link mutations, target molecules, functional pleitropy, and cellular phenotype.

Supporting Information

File S1 Figure S1, Genome-wide features of H3K9ac and H3K4me3 marks in S. cerevisiae. **Figure S2**, H3-normalization of H3K9ac and H3K4me3 levels in S. cerevisiae. **Figure S3**, Disparity of H3-normalized H3K9ac or H3K4me3 levels per gene: analysis of expression-correlated regions. **Figure S4**, Disparity of H3-normalized H3K9ac or H3K4me3 levels per gene: analysis across the proximal ORF. **Figure S5**, Analysis of H3K9ac/H3K4me3 disparity amongst various gene subsets. **Figure S6**, Association of gene architecture with evolutionary coordination between histone modifications and expression. **Figure S7**, Co-divergence of histone modifications with expression in the context of gene ontology.

Acknowledgments

We thank Itay Tirosh for insightful comments and members of our group for helpful discussions. This work was supported by the ERC, the Israel Science Foundation, and the Helen and Martin Kimmel award for Innovative Investigations.

Author Contributions

Conceived and designed the experiments: YM NB. Performed the experiments: YM. Analyzed the data: YM YV. Wrote the paper: YM NB.

References

1. Larson DR (2011) What do expression dynamics tell us about the mechanism of transcription? Curr Opin Genet Dev 21: 591–599. doi:10.1016/j.gde.2011.07.010.

2. Santos-Rosa H, Schneider R, Bannister AJ, Sherriff J, Bernstein BE, et al. (2002) Active genes are tri-methylated at K4 of histone H3. Nature 419: 407–411. doi:10.1038/nature01080.

3. Kurdistani SK, Tavazoie S, Grunstein M (2004) Mapping global histone acetylation patterns to gene expression. Cell 117: 721–733. doi:10.1016/j.cell.2004.05.023.

4. Liu CL, Kaplan T, Kim M, Buratowski S, Schreiber SL, et al. (2005) Single-nucleosome mapping of histone modifications in S. cerevisiae. PLoS Biol 3: e328. doi:10.1371/journal.pbio.0030328.

5. Pokholok DK, Harbison CT, Levine S, Cole M, Hannett NM, et al. (2005) Genome-wide map of nucleosome acetylation and methylation in yeast. Cell 122: 517–527. doi:10.1016/j.cell.2005.06.026.

6. Barski A, Cuddapah S, Cui K, Roh T-Y, Schones DE, et al. (2007) High-resolution profiling of histone methylations in the human genome. Cell 129: 823–837. doi:10.1016/j.cell.2007.05.009.

7. Koutelou E, Hirsch CL, Dent SY (2010) Multiple faces of the SAGA complex. Curr Opin Cell Biol 22: 374–382. doi:10.1016/j.ceb.2010.03.005.

8. Robert F, Pokholok DK, Hannett NM, Rinaldi NJ, Chandy M, et al. (2004) Global position and recruitment of HATs and HDACs in the yeast genome. Mol Cell 16: 199–209. doi:10.1016/j.molcel.2004.09.021.

9. Govind CK, Zhang F, Qiu H, Hofmeyer K, Hinnebusch AG (2007) Gcn5 promotes acetylation, eviction, and methylation of nucleosomes in transcribed coding regions. Mol Cell 25: 31–42. doi:10.1016/j.molcel.2006.11.020.

10. Weake VM, Dyer JO, Seidel C, Box A, Swanson SK, et al. (2011) Post-transcription initiation function of the ubiquitous SAGA complex in tissue-specific gene activation. Genes Dev 25: 1499–1509. doi:10.1101/gad.2046211.

11. Protacio RU, Li G, Lowary PT, Widom J (2000) Effects of histone tail domains on the rate of transcriptional elongation through a nucleosome. Mol Cell Biol 20: 8866–8878.

12. Carey M, Li B, Workman JL (2006) RSC exploits histone acetylation to abrogate the nucleosomal block to RNA polymerase II elongation. Mol Cell 24: 481–487. doi:10.1016/j.molcel.2006.09.012.

13. Robyr D, Suka Y, Xenarios I, Kurdistani SK, Wang A, et al. (2002) Microarray deacetylation maps determine genome-wide functions for yeast histone deacetylases. Cell 109: 437–446.

14. Weinberger L, Voichek Y, Tirosh I, Hornung G, Amit I, et al. (2012) Expression Noise and Acetylation Profiles Distinguish HDAC Functions. Mol Cell 47: 193–202. doi:10.1016/j.molcel.2012.05.008.

15. Smolle M, Workman JL (2013) Transcription-associated histone modifications and cryptic transcription. Biochim Biophys Acta 1829: 84–97. doi:10.1016/j.bbagrm.2012.08.008.

16. Shilatifard A (2012) The COMPASS family of histone H3K4 methylases: mechanisms of regulation in development and disease pathogenesis. Annu Rev Biochem 81: 65–95. doi:10.1146/annurev-biochem-051710-134100.

17. Lee J-S, Shukla A, Schneider J, Swanson SK, Washburn MP, et al. (2007) Histone crosstalk between H2B monoubiquitination and H3 methylation mediated by COMPASS. Cell 131: 1084–1096. doi:10.1016/j.cell.2007.09.046.

18. Krogan NJ, Dover J, Wood A, Schneider J, Heidt J, et al. (2003) The Paf1 complex is required for histone H3 methylation by COMPASS and Dot1p: linking transcriptional elongation to histone methylation. Mol Cell 11: 721–729.

19. Bian C, Xu C, Ruan J, Lee KK, Burke TL, et al. (2011) Sgf29 binds histone H3K4me2/3 and is required for SAGA complex recruitment and histone H3 acetylation. EMBO J 30: 2829–2842. doi:10.1038/emboj.2011.193.

20. Henry KW (2003) Transcriptional activation via sequential histone H2B ubiquitylation and deubiquitylation, mediated by SAGA-associated Ubp8. Genes Dev 17: 2648–2663. doi:10.1101/gad.1144003.

21. Lenstra TL, Benschop JJ, Kim T, Schulze JM, Brabers NACH, et al. (2011) The Specificity and Topology of Chromatin Interaction Pathways in Yeast. Mol Cell 42: 536–549. doi:10.1016/j.molcel.2011.03.026.

22. Lee S-I, Pe'er D, Dudley AM, Church GM, Koller D (2006) Identifying regulatory mechanisms using individual variation reveals key role for chromatin modification. Proceedings of the National Academy of Sciences 103: 14062–14067. doi:10.1073/pnas.0601852103.

23. Choi JK, Kim Y-J (2008) Epigenetic regulation and the variability of gene expression. Nat Genet 40: 141–147. doi:10.1038/ng.2007.58.

24. Tirosh I, Sigal N, Barkai N (2010) Divergence of nucleosome positioning between two closely related yeast species: genetic basis and functional consequences. Mol Syst Biol 6: 365. doi:10.1038/msb.2010.20.

25. Zhang Z, Gu J, Gu X (2004) How much expression divergence after yeast gene duplication could be explained by regulatory motif evolution? Trends in Genetics 20: 403–407. doi:10.1016/j.tig.2004.07.006.

26. Langmead B, Trapnell C, Pop M, Salzberg SL (2009) Ultrafast and memory-efficient alignment of short DNA sequences to the human genome. Genome Biol 10: R25. doi:10.1186/gb-2009-10-3-r25.

27. Tirosh I, Barkai N (2008) Two strategies for gene regulation by promoter nucleosomes. Genome Res 18: 1084–1091. doi:10.1101/gr.076059.108.

28. Tirosh I, Weinberger A, Carmi M, Barkai N (2006) A genetic signature of interspecies variations in gene expression. Nat Genet 38: 830–834. doi:10.1038/ng1819.

29. Pramila T, Wu W, Miles S, Noble WS, Breeden LL (2006) The Forkhead transcription factor Hcm1 regulates chromosome segregation genes and fills the S-phase gap in the transcriptional circuitry of the cell cycle. Genes Dev 20: 2266–2278. doi:10.1101/gad.1450606.

30. Ihmels J, Friedlander G, Bergmann S, Sarig O, Ziv Y, et al. (2002) Revealing modular organization in the yeast transcriptional network. Nat Genet 31: 370–377. doi:10.1038/ng941.

31. Zhang L, Ma H, Pugh BF (2011) Stable and dynamic nucleosome states during a meiotic developmental process. Genome Res 21: 875–884. doi:10.1101/gr.117465.110.

32. Maltby VE, Martin BJE, Brind'Amour J, Chruscicki AT, McBurney KL, et al. (2012) Histone H3K4 demethylation is negatively regulated by histone H3 acetylation in Saccharomyces cerevisiae. Proceedings of the National Academy of Sciences 109: 18505–18510. doi:10.1073/pnas.1202070109.

33. Jiang C, Pugh BF (2009) A compiled and systematic reference map of nucleosome positions across the Saccharomyces cerevisiae genome. Genome Biol 10: R109. doi:10.1186/gb-2009-10-10-r109.

34. Clayton AL, Hazzalin CA, Mahadevan LC (2006) Enhanced histone acetylation and transcription: a dynamic perspective. Mol Cell 23: 289–296. doi:10.1016/j.molcel.2006.06.017.

35. Morillon A, Karabetsou N, Nair A, Mellor J (2005) Dynamic lysine methylation on histone H3 defines the regulatory phase of gene transcription. Mol Cell 18: 723–734. doi:10.1016/j.molcel.2005.05.009.

36. Choi JK, Kim Y-J (2009) Intrinsic variability of gene expression encoded in nucleosome positioning sequences. Nat Genet 41: 498–503. doi:10.1038/ng.319.

37. Thompson DA, Regev A (2009) Fungal regulatory evolution: cis and trans in the balance. FEBS Lett 583: 7–7. doi:10.1016/j.febslet.2009.11.032.

38. Tirosh I, Reikhav S, Levy AA, Barkai N (2009) A Yeast Hybrid Provides Insight into the Evolution of Gene Expression Regulation. Science 324: 659–662. doi:10.1126/science.1169766.

39. Muramoto T, Müller I, Thomas G, Melvin A, Chubb JR (2010) Methylation of H3K4 Is required for inheritance of active transcriptional states. Curr Biol 20: 397–406. doi:10.1016/j.cub.2010.01.017.

40. Ng RK, Gurdon JB (2008) Epigenetic memory of an active gene state depends on histone H3.3 incorporation into chromatin in the absence of transcription. Nat Cell Biol 10: 102–109. doi:10.1038/ncb1674.

41. Stein AB, Jones TA, Herron TJ, Patel SR, Day SM, et al. (2011) Loss of H3K9 methylation destabilizes gene expression patterns and physiological functions in adult murine cardiomyocytes. J Clin Invest 121: 2641–2650. doi:10.1172/JCI44641DS1.

42. Trynka G, Sandor C, Han B, Xu H, Stranger BE, et al. (2013) Chromatin marks identify critical cell types for fine mapping complex trait variants. Nat Genet 45: 124–130. doi:10.1038/ng.2504.

43. Barth TK, Imhof A (2010) Fast signals and slow marks: the dynamics of histone modifications. Trends Biochem Sci: 1–9. doi:10.1016/j.tibs.2010.05.006.

44. Katan-Khaykovich Y, Struhl K (2002) Dynamics of global histone acetylation and deacetylation in vivo: rapid restoration of normal histone acetylation status upon removal of activators and repressors. Genes Dev 16: 743–752. doi:10.1101/gad.967302.

45. Ng HH, Robert F, Young RA, Struhl K (2003) Targeted recruitment of Set1 histone methylase by elongating Pol II provides a localized mark and memory of recent transcriptional activity. Mol Cell 11: 709–719.

46. Armstrong JA, Radman-Livaja M, Quan TK, Valenzuela L, Van Welsem T, et al. (2012) A key role for chd1 in histone h3 dynamics at the 3' ends of long genes in yeast. PLoS Genet 8: e1002811. doi:10.1371/journal.pgen.1002811.

47. Le Martelot G, Canella D, Symul L, Migliavacca E, Gilardi F, et al. (2012) Genome-wide RNA polymerase II profiles and RNA accumulation reveal kinetics of transcription and associated epigenetic changes during diurnal cycles. PLoS Biol 10: e1001442. doi:10.1371/journal.pbio.1001442.

48. Zhou BO, Zhou J-Q (2011) Recent transcription-induced histone H3 lysine 4 (H3K4) methylation inhibits gene reactivation. J Biol Chem 286: 34770–34776. doi:10.1074/jbc.M111.273128.

49. Seila AC, Calabrese JM, Levine SS, Yeo GW, Rahl PB, et al. (2008) Divergent transcription from active promoters. Science 322: 1849–1851. doi:10.1126/science.1162253.

50. Margaritis T, Oreal V, Brabers N, Maestroni L, Vitaliano-Prunier A, et al. (2012) Two distinct repressive mechanisms for histone 3 lysine 4 methylation through promoting 3'-end antisense transcription. PLoS Genet 8: e1002952. doi:10.1371/journal.pgen.1002952.

51. Weiner A, Chen HV, Liu CL, Rahat A, Klien A, et al. (2012) Systematic Dissection of Roles for Chromatin Regulators in a Yeast Stress Response. PLoS Biol 10: e1001369. doi:10.1371/journal.pbio.1001369.

52. Lee C-K, Shibata Y, Rao B, Strahl BD, Lieb JD (2004) Evidence for nucleosome depletion at active regulatory regions genome-wide. Nat Genet 36: 900–905. doi:10.1038/ng1400.

53. Shivaswamy S, Bhinge A, Zhao Y, Jones S, Hirst M, et al. (2008) Dynamic remodeling of individual nucleosomes across a eukaryotic genome in response to transcriptional perturbation. PLoS Biol 6: e65. doi:10.1371/journal.pbio.0060065.

54. Weake VM, Workman JL (2012) SAGA function in tissue-specific gene expression. Trends Cell Biol 22: 177–184. doi:10.1016/j.tcb.2011.11.005.

55. Pugh BF, Huisinga KL (2004) A genome-wide housekeeping role for TFIID and a highly regulated stress-related role for SAGA in Saccharomyces cerevisiae. Mol Cell 13: 573–585.

56. Alejandro-Osorio AL, Huebert DJ, Porcaro DT, Sonntag ME, Nillasithanukroh S, et al. (2009) The histone deacetylase Rpd3p is required for transient changes in genomic expression in response to stress. Genome Biol 10: R57. doi:10.1186/gb-2009-10-5-r57.

57. Wang A, Kurdistani SK, Grunstein M (2002) Requirement of Hos2 histone deacetylase for gene activity in yeast. Science 298: 1412–1414. doi:10.1126/science.1077790.

58. Nagarajan M, Veyrieras J-B, De Dieuleveult M, Bottin H, Fehrmann S, et al. (2010) Natural single-nucleosome epi-polymorphisms in yeast. PLoS Genet 6: e1000913. doi:10.1371/journal.pgen.1000913.

59. Lucas I, Grunstein M, Rubbi L, Brewer BJ, Vogelauer M (2002) Histone acetylation regulates the time of replication origin firing. Mol Cell 10: 1223–1233.

60. Sommermeyer V, Béneut C, Chaplais E, Serrentino ME, Borde V (2013) Spp1, a member of the Set1 Complex, promotes meiotic DSB formation in promoters by tethering histone H3K4 methylation sites to chromosome axes. Mol Cell 49: 43–54. doi:10.1016/j.molcel.2012.11.008.

5

Evolution of Proliferation and the Angiogenic Switch in Tumors with High Clonal Diversity

Scott T. Bickel[1], Joseph D. Juliano[2], John D. Nagy[1,3]*

1 Department of Life Sciences, Scottsdale Community College, Scottsdale, Arizona, United States of America, **2** Department of Chemistry and Biochemistry, Arizona State University, Tempe, Arizona, United States of America, **3** School of Mathematical and Statistical Sciences, Arizona State University, Tempe, Arizona, United States of America

Abstract

Natural selection among tumor cell clones is thought to produce hallmark properties of malignancy. Efforts to understand evolution of one such hallmark—the angiogenic switch—has suggested that selection for angiogenesis can "run away" and generate a hypertumor, a form of evolutionary suicide by extreme vascular hypo- or hyperplasia. This phenomenon is predicted by models of tumor angiogenesis studied with the techniques of adaptive dynamics. These techniques also predict that selection drives tumor proliferative potential towards an evolutionarily stable strategy (ESS) that is also convergence-stable. However, adaptive dynamics are predicated on two key assumptions: (i) no more than two distinct clones or evolutionary strategies can exist in the tumor at any given time; and (ii) mutations cause small phenotypic changes. Here we show, using a stochastic simulation, that relaxation of these assumptions has no effect on the predictions of adaptive dynamics in this case. In particular, selection drives proliferative potential towards, and angiogenic potential away from, their respective ESSs. However, these simulations also show that tumor behavior is highly contingent on mutational history, particularly for angiogenesis. Individual tumors frequently grow to lethal size before the evolutionary endpoint is approached. In fact, most tumor dynamics are predicted to be in the evolutionarily transient regime throughout their natural history, so that clinically, the ESS is often largely irrelevant. In addition, we show that clonal diversity as measured by the Shannon Information Index correlates with the speed of approach to the evolutionary endpoint. This observation dovetails with results showing that clonal diversity in Barrett's esophagus predicts progression to malignancy.

Editor: Juan F. Poyatos, Spanish National Research Council (CSIC), Spain

Funding: This research is supported in part by Scottsdale Community College and Arizona State University's Barrett Honors College. The funders had no role in study design, data collection and analysis, decision to publish, or preparation of the manuscript.

Competing Interests: The authors have declared that no competing interests exist.

* E-mail: jdnagy@asu.edu

Introduction

Natural selection has long been recognized as the ultimate driver of cancer progression and pathogenesis (see [1] for a recent review; see also [2]). In early stages of tumor progression, heterogeneous populations of malignant and healthy cells compete for available resources. Tumor cell clones that have acquired, via mutation and epigenetic effects, malignant "hallmark" phenotypes [3,4] gain proliferative and (or) survival advantages relative to other lineages in their tumor microenvironment. Eventually the hallmark-carrying mutant clones come to dominate the tumor and destroy tissue homeostasis. If this interpretation is correct, then the mechanism causing malignancy—heritable variation conferring advantages to particular clonal lineages—is precisely evolution by natural selection.

Explaining the adaptive significance of most cancer hallmarks is straightforward. However, angiogenesis—the ability of tumors to generate their own vascular infrastructure—presents a difficult case. Angiogenesis is coordinated directly and indirectly by cancer cells using a variety of signaling molecules, including vascular endothelial growth factor (VEGF), angiopoietins, fibroblastic growth factors (FGFs), platelet-derived growth factor (PDGF), epidermal growth factor (EGF), transforming growth factors (TGFα and -β), and thrombospondin-1 (TSP-1), among others. These factors act in a variety of ways on vascular endothelial cells

and (or) their precursors. Target cell responses include proliferation, chemotaxis and differentiation into functional microvessel endothelial cells [3–6]. The balance between pro- and anti-angiogenic molecules in the local milieu define the "angiogenic signal" [7]. In hypoxic tissues, this balance tips in favor of angiogenesis [5,6]. Cancer is often characterized by derangement of this signaling system, generating among certain tumor clones more-or-less constitutive production of pro-angiogenic signals and receptors, a condition referred to as the angiogenic switch [3,4,7–12]. The intensity of this switch varies among tumors even of the same histological type and tissue of origin [4,13].

Angiogenesis clearly benefits tumors. In addition to nutrient delivery and waste removal, tumor microvessels provide routes for metastasis. However, all tumor cells receive the benefits of angiogenesis whether or not they participate in producing the signal. Therefore, the signal is a public good. As is well known from decades of research into the "free-rider" problem in economics and evolutionary biology, public goods are susceptible to exploitation by free-riders. In this context, free-riders would be clones that, by mutation or epigenetic alteration, decrease or stop their own production of proangiogenic signals. Since metabolic energy is required to produce the angiogenic signal, free-riders eliminate one drain on internal energy reserves with no immediate detriment. However, they gain an immediate advantage—saved energy reserves can be committed to proliferation and maintenance metabolism. Free-rider clones would therefore be expected to expand more rapidly than angiogenic clones due to their inherited advantage. The obvious fact that the tumor, and the free-riders themselves, would suffer hypoxia once free-riding becomes dominant is irrelevant. Natural selection does not act to benefit the tumor. Selection simply favors clones with the highest growth and survival potential once the chains of kin selection and other evolutionary forces compelling cooperation have been broken. In any environment, even severely hypoxic ones, free-rider clones will always have an advantage over angiogenic clones, all else being equal, because they have less demand for energy to produce a public good. Any angiogenic clone will certainly benefit from being angiogenic. But the free-rider benefits equally. The fact that cancer cells tend to disperse from unfavorable environments does not eliminate the problem. It simply spreads it. If the hallmarks of cancer are consequences of evolution, it is not immediately clear why the angiogenic switch persists in malignant tumors.

Indeed, modeling studies initiated by one of us (JDN) predict that such nonangiogenic free-riders can damage or perhaps destroy all or part of a growing tumor [14–16]. This predicted "tumor-on-a-tumor" phenomenon has a conceptual sister, *viz.* hyperparisitism—one parasite exploiting another. Therefore, the parallel term "hypertumor" was suggested to describe it [14], and it has since been recognized as a form of evolutionary suicide [17,18].

The early hypertumor models were limited by the fact that costs associated with hallmark phenotypes could not be investigated because the models lacked a proper description of energetic trade-offs. That limitation was addressed in a recent study by Nagy and Armbruster [17] in which the original models were extended to include an energetic "opportunity cost." This extension required the addition of a submodel describing intracellular adenylate dynamics to the existing tissue-level model of angiogenesis. The result was a multiscale system with three distinct spatial and temporal levels: intracellular energy metabolism on scales of µm and seconds to minutes; tissue-level on scales of mm and hours to days; and evolutionary, with scales of cm and months to years (Fig. 1). In this formulation, ATP, the primary energy currency in cells, is partitioned among three major energetically demanding programs: proliferation, cell maintenance and secretion.

This model confirms that evolution of hallmarks acting on proliferation rate differs markedly from evolution of the angiogenic switch. In particular, selection drives energy allocation for proliferation to an intermediate state that balances evolutionary benefits of reproduction with opportunity costs of shunting reducing power away from cell maintenance. This attracting state is an evolutionarily stable strategy (ESS) *sensu* Maynard Smith and Price [19,20]; that is, it is a strategy that, when used by almost all residents of a population, cannot be invaded by any possible mutant. This ESS is also convergence stable (essentially an evolutionary attractor; see [21] for a review).

In contrast, angiogenic potential does not evolve to an intermediate state in this model. As predicted by the free-rider argument, direct selection on angiogenesis signaling is powerless to produce the angiogenic switch via the benefits of eventual increased perfusion. However, this model predicts an indirect evolutionary pathway to the angiogenic switch caused by an interesting property of the adenylate homeostasis mechanism. Both modeling and *in vivo* studies suggest that intracellular equilibrium ATP concentration is a unimodal function of overall cell metabolic rate [11,23]. To the left of the mode (relatively low metabolic rates), cells respond to slight increases in ATP consumption by excessively ramping up *de novo* adenylate synthesis,

resulting in a paradoxical increase in equilibrium ATP concentration. Therefore, under conditions favoring such overcompensation, mutants that increase production of the angiogenic signal, which requires ATP for protein synthesis and secretion, can gain more ATP for proliferation. This pleiotropic effect on both proliferation and angiogenic potential confers to the mutant a selective advantage. In the model, the evolutionary picture is complicated by an interaction between this phenomenon and the neovascularization in response to the angiogenic signal. This interaction generates an ESS that is always evolutionarily repelling (details in [17]). That is, if clones differ only in ATP allocation to angiogenesis, clones slightly more (or less) committed to angiogenesis than the ESS are vulnerable to invasion by a mutant clone with higher (respectively, lower) energy allocation to angiogenesis. Therefore, this model predicts that, given enough time, selection will run away either to vascular hyper- or hypoplasia, eventually reaching a tumor inviability region. The latter possibility is the original hypertumor prediction, while the former represents a novel form of evolutionary suicide [14,16,17].

The open questions we address here are the following: (i) what are the likely trajectories tumors traverse through their evolutionary "strategy spaces" as angiogenic and proliferative potentials evolve? (ii) what variation in these evolutionary trajectories can be expected? and (iii) how rapidly will the traverse occur? Answers to these questions are required before practically testable predictions from the model can be distilled, but they could not be addressed in the previous modeling attempts. In these studies, the evolutionary analysis relied on the techniques of adaptive dynamics [21,24,25], which require the assumption that mutation dynamics are much slower than ecological dynamics. At most only two competing clones can exist in a tumor. One arises as a rare mutant within a tumor populated almost exclusively by a resident clone. Competitive exclusion is the rule in these pairwise bouts; either the mutant invades the tumor and eliminates the resident clone, or the resident eliminates the mutant. Either way, "ecological" dynamics of competition are assumed to reach their endpoint before a new mutant arises, so ecological and evolutionary timescales decouple. Also, all mutations are assumed to have a small effect; therefore, the difference between phenotypes of resident and mutant clones is always small [25]. Given the genomic instability characteristic of many malignant tumors [4], these assumptions are likely to be violated in real tumors. So, here we repeat the evolutionary analysis with the adaptive dynamics assumptions relaxed. To achieve this, we first define a stochastic simulation analogue of the multiscale evolutionary model in [17]. The equations governing intracellular adenylate, tumor growth and angiogenesis are unchanged. The only alteration we introduce is at the evolutionary scale. In particular, we allow an indefinite number of clones to compete at the same time, and mutant clones arise at random times independent of the current state of the system. Here we show that relaxation of the adaptive dynamics assumptions has no effect on predicted evolutionary endpoints from the original adaptive dynamics analysis. However, the simulations predict that evolutionary dynamics of both angiogenesis and proliferative capacity is dominated by mutational history. Practically, this prediction suggests that the disease is on an evolutionary transient throughout its clinical course—that is, an attracting ESS is rarely if ever approached—and the tumor's evolutionary tempo and trajectory are largely determined by phenotypes of early mutants, which in practice are likely to resist prediction. We refer to this prediction as the historical contingency effect, following [26].

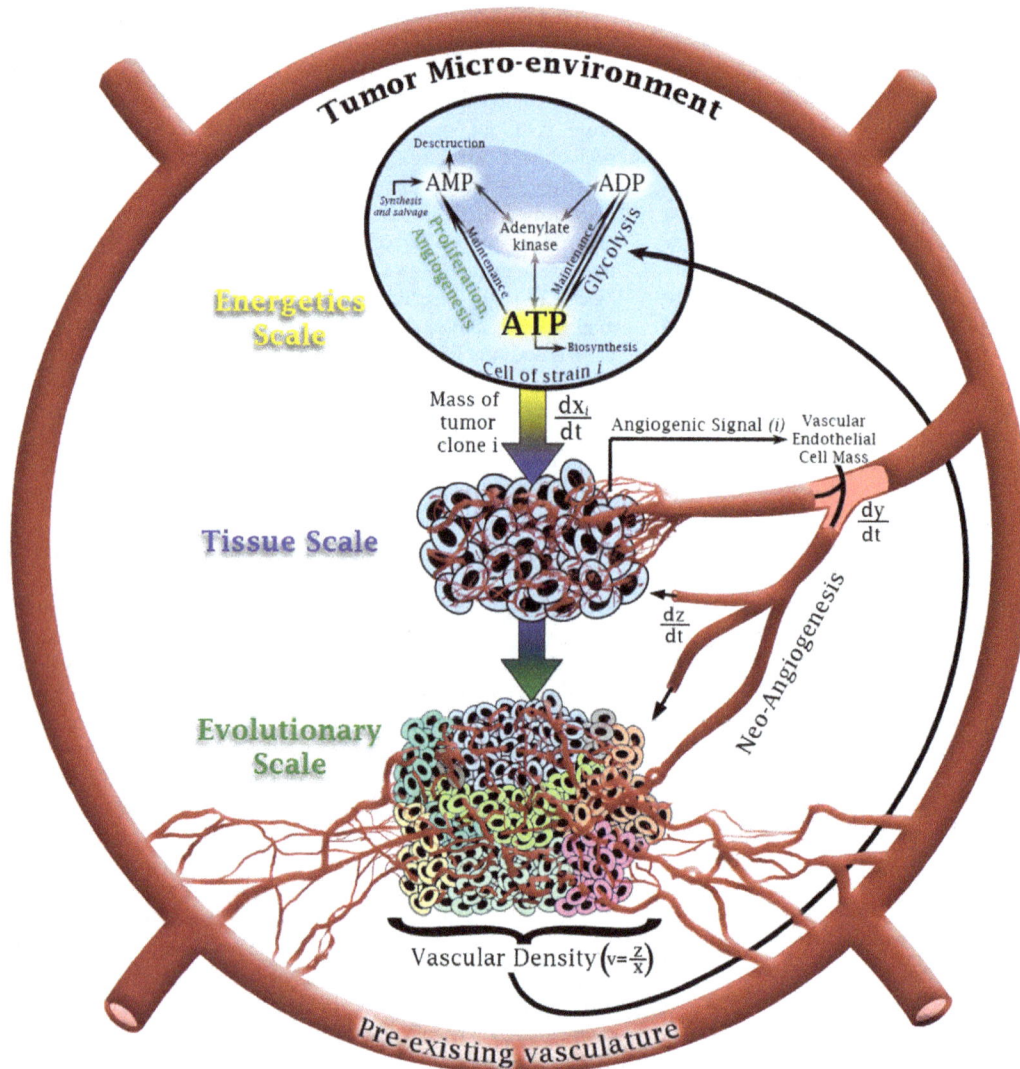

Figure 1. Schematic representation of the multiscale model. The "energetic scale" submodel [equations (1) through (10)] governs dynamics of adenylate (AMP, ADP and ATP). Interconversions among the three species occur via maintenance metabolism (e.g., biosynthesis, volume control), chemical energy to support proliferation and angiogenesis signal production, glycolysis and, most importantly, the adenylate kinase reaction, among others. Sources for adenylate include *de novo* synthesis of AMP and salvage from nucleic acid catabolism. Adenylate sinks include AMP destruction by deaminases and nucleotidases and ATP loss to nucleic acid synthesis. Clonal expansion or regression at the tissue scale [model (11)] depends primarily on mean tumor microvessel density, which is controlled in part by angiogenic factor secretion by existing clones. Blood vessels grow from existing vasculature via chemotaxis and maturation of vascular endothelial cell precursors in and near the tumor. At the evolutionary scale, angiogenic and proliferative potential varies among clones (different colored cell subpopulations) as they compete for resources delivered by microvessels. Evolutionary scale dynamics are handled in the simulation (see "Simulation Methods" above).

Methods

The deterministic model underlying our simulations [17] comprise two distinct systems of ordinary differential equations (ODEs) governing dynamics at three time and spatial scales (Fig. 1). The first system, which extends the pioneering work of Martinov, Ataullakhanov, Vitvitsky and colleagues [22,23], tracks intracellular adenylate dynamics, with adenylate concentrations scaled in fmol and time scaled in mins ("Energetics Scale" in Fig. 1). The second system, operating at tissue and evolutionary scales, describes growth dynamics of the tumor, dynamics of its vascular infrastructure and competition among clones within the tumor. The tumor is assumed to comprise some number of genetically and phenotypically distinct clones, a collection of immature

vascular endothelial cell precursors (VECPs) and patent, functional microvessels. In the original formulation, the number of competing clones was limited to 2, but here we allow the tumor to house an arbitrary number, S, of distinct clones. Dynamics of a single clone ("Tissue Scale" in Fig. 1) is governed by two ODEs—one for clonal mass (in g) and one for mean tumor microvessel density. Time is scaled in days. Overall tumor dynamics ("Evolutionary Scale" in Fig. 1) is therefore determined by $S+2$ ODEs, one for each clone, plus equations for VECPs and microvessels, with dependencies on the energetic states of cells determined by equations at the energetics scale. This tissue-level model is derived directly from previous work of Nagy and colleagues [14,16].

Interactions among all three scales revolve around tumor perfusion, measured as microvessel length density. In the model, hypoxic tumor cells secrete a chemical signal composed of a variety of tumor angiogenesis factors (TAF) to which VECPs respond by proliferating, maturing and integrating themselves into functional microvessels. The interaction between intracellular and tissue levels arises as cells partition their available chemical potential energy, primarily in the form of ATP, among three energy-dependent activities: maintenance metabolism (cell volume regulation, maintenance protein production and other life-support physiology), proliferation and, potentially, secretion of TAF. In turn, these three activities feed into tissue-level phenomena of blood vessel growth and clone-specific expansion. These growth phenomena then feed back to the intracellular scale because relative growth rates of vessels and tumors determine perfusion and therefore nutrient delivery, which in turn sets the cellular energy charge and ATP regeneration rate, as detailed below. Tumor vascular dynamics depend on both the strength of the angiogenic signal and rate of tumor growth. Tumor vascular density determines rate of ATP synthesis. Since malignant tumors are often characterized by dampened oxidative metabolism (Warburg effect [27,28]), vascular feedback on ATP synthesis primarily occurs via delivery of glucose for glycolysis (Fig. 1). Although an exhaustive derivation of the model can be found in [17] (see also [14]), we provide a detailed outline below for completeness. All parameters in the model have been estimated carefully from data when possible or from biological first principles and model behavior when not. Details of the parameterizations are outside the scope of this paper but are available in references [14,16,17].

Cell Energetics Scale Model

Let $A_{1i}(t)$, $A_{2i}(t)$ and $A_{3i}(t)$ represent mean intracellular concentrations of adenylate 5′ mono-, di- and triphosphate, respectively, in clonal lineage $i \in \{1,2,\ldots,S\}$ at time $t \in [0,\infty)$. (Variables and parameters in this model are summarized in Tables 1 and 2, respectively.) We assume that each clone acts independently of all others, and all cells in a given lineage are identical. Adenylate concentrations are to be understood as mean-field or ensemble averages within clone i.

Intracellular adenylate dynamics are governed by the following system of ODEs:

$$\begin{cases} \dfrac{dA_{1i}}{dt} = \alpha_a + k(A_{2i}^2 - A_{1i}A_{3i}) + a_i A_{3i} - f(A_{1i},A_{3i}), \\[2mm] \dfrac{dA_{2i}}{dt} = 2k(A_{1i}A_{3i} - A_{2i}^2) + bA_{3i} - G(\phi_i,v)A_{2i}, \\[2mm] \dfrac{dA_{3i}}{dt} = G(\phi_i,v)A_{2i} - c_i A_{3i} + k(A_{2i}^2 - A_{1i}A_{3i}), \\[2mm] \qquad\qquad i \qquad\qquad\qquad\qquad\qquad \in \{1,2,\ldots,S\}. \end{cases} \quad (1)$$

Table 1. Dependent variables studied in this model.

Variable	Meaning
t	Time
$A_j(t)$	Concentration of adenylate 5′ j-phosphate
$x_i(t)$	Mass of the ith clone's parenchymal cells
$\mathbf{x}(t)$	Total parenchymal mass
$y(t)$	Total mass of precursor VECs
$z(t)$	Total length of mature microvessels
$v(t)$	Tumor vascularization ($= z/\mathbf{x}$)

Total adenylate in cells is controlled primarily via synthesis and destruction of AMP. In the model, AMP appears *de novo* at constant rate $\alpha_a > 0$, representing mainly synthesis from inosine monophosphate and salvage from nucleic acid recycling. The function $f(A_{1i},A_{3i})$ represents irreversible AMP recycling by enzymes primarily in the 5′ nucleotidase and AMP deaminase families. Specifically,

$$f(A_{1i},A_{3i}) := M_1\left(\frac{A_{1i}}{k_1 + A_{1i}}\right)^4 + \frac{M_2 A_{1i}(k_2 + A_{3i})}{k_3 + k_4 A_{1i} + A_{1i}^2}, \quad (2)$$

which is an empirical model of AMP destruction suggested by Martinov et al. [23], who also provided empirical estimates of (positive) parameters M_1, M_2, k_1, k_2 and k_3. The first and second terms in equation (2) represent the actions of AMP deaminases and 5′ nucleotideases, respectively.

In resting mammalian cells, adenylate dynamics are dominated by the adenylate kinase reaction,

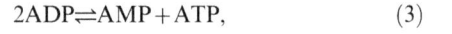

$$2ADP \rightleftharpoons AMP + ATP, \quad (3)$$

which is represented in the model by the second, first and third terms in each equation of model (1), respectively. In most cells this reaction has approximately equal forward and backward rates [29], which we denote here as the positive constant k.

Besides the adenylate kinase reaction (3), interconversion among adenylate species revolves around either ATP hydrolysis or synthesis. The former is primarily governed by positive parameters a_i, b and c_i, which themselves are sums of constants representing metabolism supporting proliferation, TAF secretion and maintenance. In particular,

$$a_i = \beta_1 + \gamma_a + \eta_i + \zeta_i(v), \quad (4)$$

$$b = \beta_2 + \gamma_a, \quad (5)$$

$$c_i = a_i + b + \mu - \gamma_a, \quad (6)$$

for $i \in \{1,2,\ldots,S\}$. Parameters $\beta_1, \beta_2 > 0$ are basic maintenance metabolism rates. The former is the ATP→AMP conversion rate, e.g., for biosynthesis (amino acid adenylation) and to power the phosphoribosyl pyrophosphate synthetase reaction, among others [17]. The latter (β_2) represents the ATP→ADP conversion rate, primarily, but not exclusively, for cell volume control. A second pathway from ATP to both ADP and AMP exists via the adenosine kinase reaction,

$$ATP + Adenosine \rightarrow ADP + AMP,$$

which we assume occurs at base rate $\gamma_a > 0$. Parameter $\mu > 0$ is the per-ATP rate of nucleic acid synthesis (assumed to be an irreversible sink for ATP). All these parameters are assumed constant across clones.

The key evolutionary parameters are η_i and $\hat{\zeta}_i$, both positive constants representing mean per-molecule rates at which clone i cells allocate ATP to proliferation and angiogenesis secretion, respectively. The dependence of ζ_i on v arises because clones vary the intensity of their angiogenic signal as a function of vascular density. Specifically, we assume that

Table 2. Parameters and default values representing a resting cell (from [17]).

Parameter	Meaning	Default	Units
Evolutionary			
λ	Mutation rate parameter	0.1	hr^{-1}
η_i	Proliferation secretion effort of clone i	1, max = 12	min^{-1}
$\hat{\zeta}_i$	Basic TAF secretion effort of clone i	0.11	g/U/min
Energetic			
α_a	de novo AMP synthesis rate	5.725×10^{-5}	fmol/min
k	Adenylate kinase rate parameter	10^6	1/fmol/min
β_1	Maintenance ATP to AMP rate	4	min^{-1}
β_2	Maintenance ATP to ADP rate	4	min^{-1}
γ_a	Adenosine kinase rate	0.01	min^{-1}
μ	ATP destruction rate	0.01	min^{-1}
ξ	Nutrient sensitivity of TAF secretion	10/3	g/U
s_{max}	Physiological max ATP regeneration rate	390	min^{-1}
M_1	AMP deaminase parameter	0.4	fmol/min
M_2	Nucleotidase parameter	9.167×10^{-7}	fmol/min
k_1	AMP deaminase parameter	0.5	fmol
k_2	Nucleotidase parameter	5×10^{-3}	fmol
k_3	Nucleotidase parameter	2.5×10^{-10}	$fmol^2$
k_4	Nucleotidase parameter	5×10^{-5}	fmol
Tissue			
p	Basic parenchyma proliferation rate	0.072	hr^{-1}
k_s	Proliferation sensitivity parameter	2	fmol/min
m	Parenchyma mortality parameter	0.0698	fmol/hr
α	Max VEC response to TAF	0.1	hr^{-1}
k_v	Sensitivity of VECs to TAF	0.0115	fmol/min
β	VEC death/maturation rate	0.04	hr^{-1}
γ	VEC maturation rate	3	U/g/hr
δ	Microvessel remolding rate	4×10^{-3}	g/U/hr

$$\zeta_i(v) = \hat{\zeta}_i v e^{-\xi v}, \quad i \in \{1,2,\ldots,S\}, \qquad (7)$$

where v is vascular density (defined in the Tissue Scale section below). This functional form, adapted from [14], assumes that TAF signaling rate is a unimodal function of vascular density, v. It qualitatively mirrors observed increases in secretion of angiogenesis-promoting growth factors as cells become hypoxic [5] with the added assumption that cells suffering extreme hypoxia lose the ability to produce the signal. Parameter $\xi > 0$ is constant across clones and determines the vascular density at which angiogenesis secretion peaks. Parameter $\hat{\zeta}_i \geq 0$ varies among clones, is constant within a clone, and measures the general intensity of the angiogenic signal.

Finally, ATP is "regenerated" from ADP primarily via glycolysis in cancer cells. This conversion is governed in the model by the function $G(\phi, v)$. Here, ϕ denotes the "energy charge" of a cell, which is a sort of weighted average of "high energy" phosphoryl groups in adenylate. Specifically,

$$\phi_i := \frac{A_{3i} + (1/2)A_{2i}}{A_{1i} + A_{2i} + A_{3i}} \qquad (8)$$

is the mean energy charge of cells in the ith clone. Mean glycolysis rate is assumed to vary among clones only as they vary in mean energy charge; in particular,

$$G(\phi_i, v) := 4s(v)\phi_i(1 - \phi_i), \qquad (9)$$

where

$$s(v) := \frac{s_{max} v}{0.1 + v}, \qquad (10)$$

and $s_{max} > 0$ is constant. The particular forms and parameterizations for $G(\phi, v)$ [equation (9)] and $s(v)$ [equation (10)] were chosen to fit data from [22] (see [17] for details).

Tumor Tissue Scale Model

Let $x_i(t)$, $i \in \{1,2,\ldots,S\}$, be the mass (in g) of clone i at time $t \in [0,\infty)$. Also let $y(t)$ and $z(t)$ be VECP mass (in g) and total length of microvessels, respectively, within the tumor at time t. Microvessel length is scaled such that $z = 1$ when total length of microvessels in the tumor equals that of 1 g of healthy tissue in the tumor's site of origin [14,17]. We assume that mean proliferation and angiogenesis signal production for cells of clone i depend on the clone's mean intracellular ATP concentration. However, mean ATP concentration depends on vascular density, v, the clone's energy commitment to proliferation, η_i and its commitment to angiogenesis, $\zeta_i(t)$ (see previous section). Therefore, there is a continuous feedback between energetic and tissue scale dynamics. Since adenylate dynamics of model (1) equilibrate very rapidly on the time scale of the tissue model (11) (ref. [17]), we further assume that mean ATP concentration in all clones is locked in quasi-equilibrium for the adenylate model. We denote the ATP quasi-equilibrium for the ith clone as $\bar{A}_{3i}$.

These assumptions lead to our tissue-level model of tumor growth, adapted from [14,17]:

$$
\begin{cases}
\dfrac{dx_i}{dt} = \left(\dfrac{p\eta_i \bar{A}_{3i}}{k_s + \eta_i \bar{A}_{3i}} - \dfrac{m}{\bar{A}_{3i}} \right) x_i, & i \in \{1,2,\ldots,S\}, \\[2ex]
\dfrac{dy}{dt} = \left(\dfrac{\sum\limits_{i=1}^{S} \psi_i(v,\bar{A}_{3i}) x_i}{x} - \beta, \right) y, \\[2ex]
\dfrac{dz}{dt} = \gamma y - \delta v z, \\[2ex]
\psi_i(v,\bar{A}_{3i}) = \dfrac{\alpha \zeta_i(v)\bar{A}_{3i}}{k_v + \zeta_i(v)\bar{A}_{3i}}, & i \in \{1,2,\ldots,S\}, \\[2ex]
v = \dfrac{z}{x}, \quad x = \sum\limits_{i=1}^{S} x_i.
\end{cases}
\tag{11}
$$

(Note that we have suppressed the time arguments of dependent variables for clarity.) Here, $v(t)$ denotes tumor microvessel length density (in microvessel units/g), and z is scaled such that $v=1$ when tumor vascular density equals that of surrounding healthy tissue. The mean ATP hydrolysis rate in support of proliferation in clone i is $\eta_i \bar{A}_{3i}$. Mean per-cell proliferation rate is a monotonically increasing, saturating function of $\eta_i \bar{A}_{3i}$, which we represent with a Michaelis-Menten form in which $p > 0$ and $k_s > 0$ are maximum proliferation rates and half-saturation constants, respectively. We also assume that a clone's mean per capita mortality rate is inversely proportional to $\bar{A}_{3i}$, with constant $m > 0$.

Mean energetic commitment to angiogenic signal production in clone i takes a similar form to that for proliferation, viz. $\zeta_i(v)\bar{A}_{3i}$. As with proliferation, ATP invested in angiogenesis gives diminishing returns, as represented by another Michaelis-Menten function, with maximum angiogenic signal production α and half-saturation constant k_v [see equation for $\psi_i(v,\bar{A}_{3i})$ in system (11)]. Overall angiogenic signal is the average signal strength of all clones weighted by clone density, and we assume that per capita proliferation of VECPs is proportional to the strength of this signal [first term, second equation in system (11)]. Mean per capita VECP mortality and maturation rates, combined into parameter $\beta > 0$, are assumed to be fixed. As VECP cells mature, they integrate themselves into functional tumor microvessels at rate γ, composed of both the rate constant and a unit conversion factor. There is evidence that tumors actively maintain their vascular infrastructure even after its initial construction [5,6,12,30]. Viewing this maintenance as the tumor provisioning microvessels with a resource, which may simply be space, we assume that this resource is proportional to tumor mass, say $R=c_1 x$, where $c_1 > 0$ is constant. Resource availability per microvessel unit is therefore the ratio $R/z = c_1 x / z = c_1 / v$. We assume that per capita microvessel remodeling rate is inversely proportional to this ratio; that is, it is proportional to v with proportionality constant δ

(which includes c_1; second term, z equation of system (11); see also ref. [14]).

This modeling approach implicitly assumes that the average conditions in the tumor are predictive of tumor dynamics. In particular, vascularization, $v(t)$, is interpreted as mean vessel density throughout the tumor at any given time (or, alternatively, the ensemble average of many similar tumors). However, since mutant clones are initially localized and vary in angiogenic efficacy, one should question the assumption that all clones are equally vascularized on average. On the other hand, cancer cells are characterized by their ability to infiltrate surrounding tissues, and many if not most exhibit positive chemotaxis up nutrient gradients and therefore tend to move towards areas of locally high vascularization [31]. So how violated the averaging assumption is, and the consequences of that level of violation, remain open questions to be addressed in subsequent approximations.

Evolutionary Scale Model

The tumor's vascular support, measured by v, reacts to changes in the clonal composition of angiogenic phenotypes, their prevalence and overall abundance. In turn, clonal composition is determined by selection pressures generated by a particular vascular environment. This interaction dictates dynamics at the evolutionary scale. Here we follow the prevalence of each phenotype within the tumor, where the phenotype of clone i is defined as $\{\eta_i, \hat{\zeta}_i\}$; that is, the phenotype is the clone's energetic commitment to angiogenesis and proliferation. We make no explicit hypotheses about how these phenotypes relate to genotypes except for the general assumptions that these traits have high heritability, that they are polygenic, and that new phenotypes may arise by mutation. However, we allow the possibility that multiple genotypes can generate the same phenotype. We also leave open the possibility that phenotype could result from a persistent epigenetic change. For simplicity, however, we refer to new phenotypes as "mutants." In any case, phenotype is fixed for all cells in the clone, although angiogenesis signaling and proliferation rates in a single clone are not fixed since these depend on vascularization, which is dynamic.

Evolutionary dynamics of clonal phenotypes in this model were initially analyzed using adaptive dynamics [17]. The technique is founded on the question, can a rare mutant strategy invade an otherwise monomorphic population using a different strategy (the "resident" strategy) in the resident's equilibrium environment [21,24,25]? The set of all possible strategies is referred to as the strategy space. One analyzes the ability of any possible mutant phenotype to invade any resident strategy in an environment set by the resident. Such an analysis produces a good, if not complete, picture of the evolutionary dynamics, including the existence and location in strategy space of evolutionarily important points or sets. In particular, adaptive dynamics identifies evolutionarily stable strategies (ESS), can be used to determine whether these points are evolutionarily attracting (continuously stable strategies [32]) or repelling and assess potential for evolutionary suicide [18] and evolutionary branching [21], among other things.

However, adaptive dynamics is limited by two fundamental assumptions. First, interarrival times between mutations must be long compared to population dynamics so that fate of the mutant is determined before the next mutant arises. Although this assumption improves analytical tractability, it removes mutational dynamics from the evolutionary picture, rendering transient evolutionary dynamics invisible. We can only see the potential evolutionary endpoints. Also, this assumption is almost certainly violated in most cancers, which are well-known to be genomically and genetically heterogeneous [1,33]. Second, most adaptive dynamics analyses rely on the assumption that mutations have small phenotypic effects. Although this assumption is not strictly required, relaxing it typically compounds analytical complexity. But again, in cancer this assumption has dubious validity because even minor mutations in both coding and control regions of genes can have massive effects on cell phenotype. A relevant example here would include a mutation in the control region of *HIF1A*, the gene for the α subunit of hypoxia-inducible factor 1, which could generate an enormous alteration of a clone's angiogenic potential [34]. The main goal of this paper is to assess the effects these assumptions have on the predictions of the coupled models (1) and (11). Therefore, we relax these assumptions, at the cost of sacrificing analytical tractability, which leads to the simulations described below.

Simulation Methods

In concept, our simulations operate as follows. Initial tumors are assumed to be small (10 mg) and monomorphic with vascular density equal to that of surrounding healthy tissue ($z(0)=1$). Therefore, simulation initial conditions were always the following: $x_1(0)=0.01$ g, $x_i(0)=0$ for all $i\in\{2,3,\ldots,S\}$, $y(0)=0.001$ g, $z(0)=0.01$, $A_{j1}(0)=\bar{A}_{j1}, j\in\{1,2,3\}$, where $\bar{A}_{j1}$ are the equilibrium concentrations in a tissue with $v=1$, and all A_{ji} for $j\in\{1,2,3\}$ and $i\in\{2,3,\ldots,S\}$ are left undefined until they arise via mutation. Mutations occur as discrete events, with one new mutant clone introduced at each event. Biologically, all mutation events except the first are assumed to be independent of time, the composition of the tumor and the number of previous mutation events. Mathematically we therefore assume that, if $\{T_n\}_{n=0}^{S-1}$ is the set of arrival times for the S mutations defining new clones (assuming that $T_0=0$), then $T_1, T_2-T_1, \ldots, T_{S-1}-T_{S-2}$ are independent, identically distributed exponential random variables with parameter λ (mean λ^{-1}; biologically, the mean interarrival times of mutations).

On the time intervals $[T_{n-1},T_n), n\in\{1,2,\ldots,S-1\}$, tumors grow according to models (1) and (11) with $x_i=0$ and A_{ji} ignored for all $i\in\{n+1,\ldots,S\}$ and $j\in\{1,2,3\}$. Mutant clones enter the model when they have grown to a size large enough to be buffered from stochastic extinction; therefore, a mutation event represents the arrival of an already sizable mutant clone, which is assumed to have initial mass of 0.1 mg and initial adenylate concentrations equal to the quasi-equilibrium for that clone at the current vascular density. So, at arrival times T_n, $n\in\{0,1,\ldots,S-1\}$, the mass of the new mutant clone, $x_{i+1}(T_n)$, is set instantly to 0.1 mg, and the three adenylate species for the new clone instantly take on their quasi-equilibrium values for that clone given $v(T_n)$; that is, $A_{j(n+1)}(T_n)=\bar{A}_{j(n+1)}(v(T_n)), j\in\{1,2,3\}$. In the terminal time period, $[T_{S-1},\infty)$, the tumor grows according to models (1) and (11), with all variables strictly positive. In practice, simulations ended at $t=4$ "years." This horizon ensures that tumors, if viable, will grow beyond the model's design, given that the model is meant to represent tumors still growing in their exponential phase. In addition, it allows vascular density to equilibrate so that the evolutionarily dominant clone can more easily be defined (see below).

In any given simulation, only one parameter, either η or $\hat{\zeta}$, was allowed to evolve—all others were fixed at their default values. Values of η for each clone were drawn from the interval $[1,12]$ and for $\hat{\zeta}$ from the interval $[0,4]$, which includes essentially all biologically feasible values for these parameters [17]. The probability distributions for the draws were uniform over those intervals because, as mentioned earlier, single mutations may have large phenotypic effects, and a comprehensive mapping between

pathological genetic and epigenetic alterations and phenotypic effects is unavailable. Given this uncertainty, a uniform probability distribution is the proper prior assumption. Further uncertainty arises in the mutation rate parameter, λ. We explore the consequences of this uncertainty by fixing λ in each simulation, but varying it between 10 and 50 hours among simulations. In each case, however, the number of (eventually) competing clones, S, was fixed at 50. Each simulation scenario was then repeated 1000 times with the same initial conditions.

We evaluated the evolutionary outcome of the simulation using two measures. The first represents what would appear be the evolutionary "winner" in a histopathology study. In this case, we defined the "evolutionarily dominant," or just dominant, clone to be the clone with the largest mass at the end of the simulation (4 years). The second measure of evolutionary success conforms more closely to an evolutionary biologist's notion of evolutionary advantage. In this case, we define the dominant clone as that which has the highest per capita growth rate at simulation's end. These two methods frequently identified different evolutionarily favored clones. For instance, many simulated tumors had small, but very rapidly growing, clones within them at the end of the simulation. These clones were clearly outcompeting all others, but because they arose late the the natural history of the tumor, they had not had time to impact the tumor's histology. This phenomenon lead us to fix $S=50$ mutants—mutants that arise beyond the 50th almost always remain pathologically irrelevant.

A measure of clonal diversity within a tumor provides a concise description of a tumor's evolutionary state with, potentially, clinically relevant predictive power [1,35,36]. We therefore measure phenotypic diversity with the Shannon Diversity Index, or Shannon Information measure [37], which quantifies the mean relative abundance of tumor clones. Maley et al. [35] used it to measure clonal diversity in Barrett's esophagus, and we adopt it here for the same reason they did—it does not overemphasize the most common clone. Given that selection acts on clones of all sizes, and that rate of evolution depends on availability of adaptive phenotypes [38], Shannon's diversity index is a better measure of evolutionary potential than are many other commonly used indexes, like Simpson's. In the model, the practical difficulties of estimation from field data [39] are eliminated. The index, H, is defined as

$$H := -\sum_{i=1}^{S} q_i \ln q_i, \quad \sum_{i=1}^{S} q_i = 1, \qquad (12)$$

where q_i is the proportion of the tumor mass contributed by the ith clone. The Shannon index thus varies between 0 (a monomorphic tumor) and $\ln S$ (all S clones contribute equally to tumor mass). If the evolutionary endpoint is a monomorphic tumor, the Shannon index may be used as a rough measure of how close the tumor is to its climax histology (in the sense of the ecologist's "climax ecosystem").

Results and Discussion

Evolution of Proliferation

Our simulations show that relaxation of restrictive adaptive dynamics assumptions do not alter the predictions of the adaptive dynamics analysis. No matter which definition one uses for "evolutionary dominance," either the histopathologist's (most mass) or the evolutionary biologist's (largest per-capita growth rate), the mean dominant η from our simulations agrees very well with the CSS (evolutionary attracting strategy) predicted by adaptive dynamics; for example, with default parameter values, the predicted ESS for proliferation effort was $\eta \approx 3.67 \text{ min}^{-1}$ (approximately 5.5 fmol or 7×10^9 ATP molecules per minute in a resting cell, about half the rate measured in mouse LS cell culture [40]) which agrees very well with our simulations (Fig. 2). Nevertheless, the evolutionarily dominant clone at the close of any given simulation varies, sometimes significantly, from the theoretical ESS from adaptive dynamics. These deviations appear to be caused by mutational contingency—although the ESS is a deterministic evolutionary endpoint, by chance, the mutational history [26] fails to direct the system towards the optimal η before the tumor grows out of bounds. This picture predicts that tumors of the same type, even in the same genetic background, will vary in evolutionary strategy in a symmetric distribution around the ESS (Fig. 2). However, at the whole-tumor level, the weighted mean proliferation effort, defined as

$$\bar{\eta} := \sum_{i=1}^{S} \frac{x_i \eta_i}{x}, \qquad (13)$$

also tends towards the predicted ESS, as does the ensemble average of many simulations (Fig. 3).

Historical contingency has a greater impact on the histopathological picture of tumor natural history than it does on the evolutionary view. If mutant clones nearest the ESS arise late, they will tend to contribute little to the tumor mass as it approaches lethality; the clone favored by selection has no time to become histologically significant. However, these favored clones' per capita expansion rates are large. Therefore, the variance in "dominant phenotype" will tend to be greater in the histopathological rather than the evolutionary view (Fig. 2). This effect is magnified as mutation rate decreases—longer mutation interarrival times gives early-arising, suboptimal clones more time to gain bulk before

more well-adapted mutants crop up. As a result, if the number of mutations is fixed, tumors with lower mutation rates have greater variance in "dominant phenotype" at the histological level (Fig. 2, blue plots) and take longer to approach the ESS at the whole tumor level (Fig. 3). In contrast, since time of arrival has little effect on per capita growth rate, mutation rate has little to no effect on the "dominant" clone as defined by per capita growth rate (Fig. 2, black plots).

These results suggest that, in clinical applications, simple measures of clonal diversity that fail to take clonal abundance into account, like mean phenotype or the ecologist's "species diversity" measure [39], will be inferior to metrics like the Shannon diversity index, which magnifies the relative contribution of rare clones to overall tumor diversity. We explore the consequences of this suggestion by evaluating the dynamics of diversity in our simulations, taking proliferation commitment, η, as the phenotype (Fig. 4). In most simulations, transient dynamics in H are longer than are transients in mean η (compare Figs. 3 and 4). Interestingly, the Shannon index tends to increase with increasing mutation rate even though the total number of mutations remains the same (Fig. 4). Historical contingency is the culprit here, too. When interarrival times are relatively long, new mutant clones have a diminishing impact on this diversity measure since previously successful clones have more time to gain mass before they are challenged by new mutants. In addition, H exhibits a large variance both among tumors in the ensemble and within a given tumor over time well after mutations have stopped (Fig. 4, purple curves).

How this variation relates to clinical prognosis is an interesting open question. Maley et al. [35] addressed a similar issue in Barrett's esophagus. Their study suggested that higher Shannon diversity index in cell ploidy predicts progression from the premalignant state to adenocarcinoma. These authors suggest that this correlation is causative since higher genetic diversity

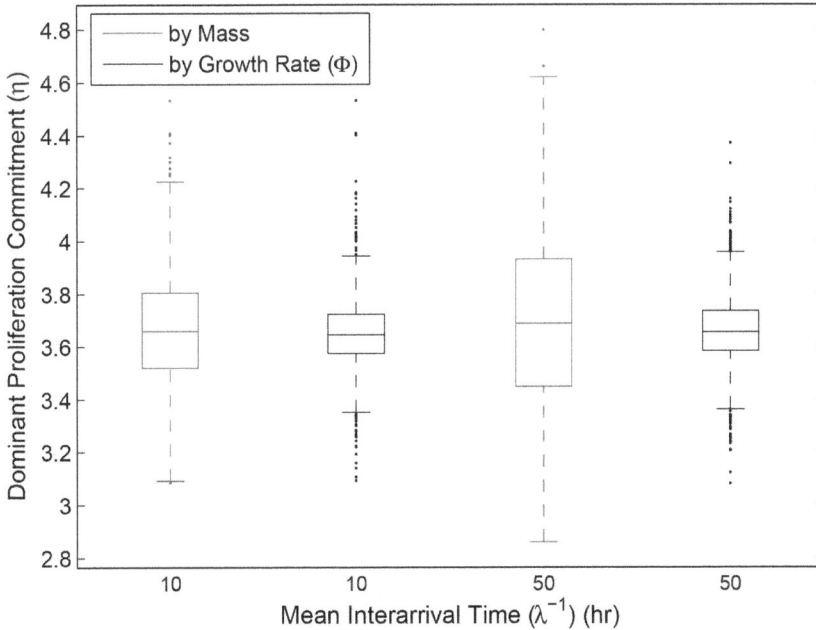

Figure 2. Evolutionarily "dominant" proliferation commitment (η) in 1000 simulations for each of two mutation rates: $\lambda = 0.1$ and **0.02 (plotted as mean interarrival time between mutations 10 and 50 hours, respectively).** All parameters except η were set to the defaults in [17]. Shown are distributions of the histopathologist's "dominant" clone (clone with the most mass; blue) and the evolutionary biologist's dominant clone (clone with the largest per capita growth rate; black).

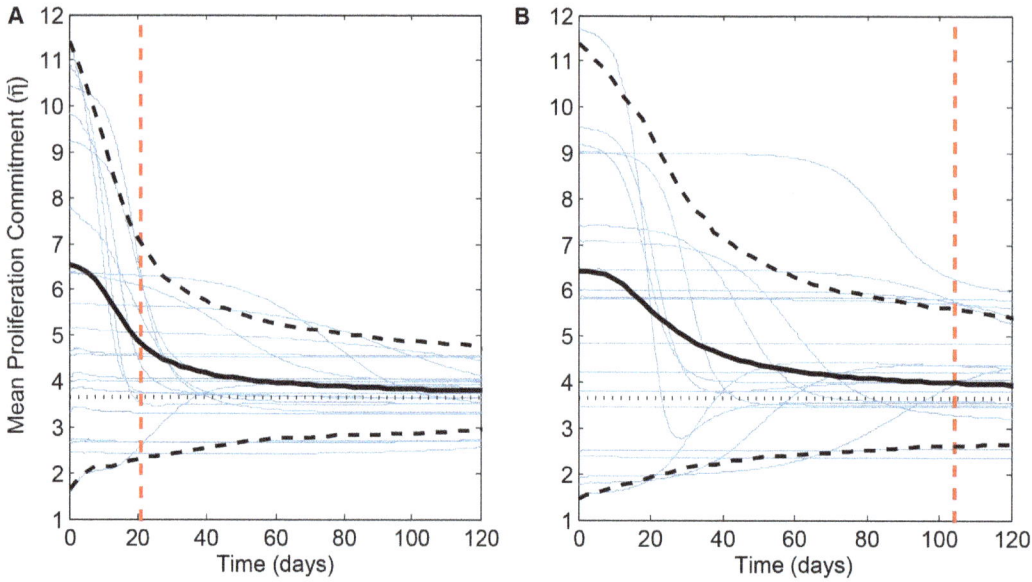

Figure 3. Change over time in weighted mean proliferation effort, $\bar{\eta}$ (see equation (13)). Blue curves in both panels represent the first 20 of the 1000 simulations plotted in Fig. 2. Solid black curves are the ensemble averages of all 1000 runs, dashed black curves mark the inner 95th percentile range for all runs, and dotted black lines represent the ESS predicted by adaptive dynamics theory ($\eta = 3.67$ min^{-1}). Dashed red lines represent mean time of the final mutation (S/λ). (A) $\lambda = 0.1$; (B) $\lambda = 0.02$.

generally leads to more rapid evolution. In the context of the current study, the evolutionary endpoint is the ESS. Therefore, the same reasoning suggests that higher diversity should lead to more rapid convergence to that endpoint. Since diversity increases with mutation rate even for a fixed number of mutations, we therefore expect more rapidly mutating tumors to more rapidly approach the ESS. Indeed, the simulations predict precisely this pattern (Fig. 3). The significance of this is two-fold. Clonal diversity not only helps predict probability of cancer progression, it can also be used to assess the evolutionary potential of tumors that are already malignant. Therefore, we suggest that phenotypic diversity can be a clinically relevant measure that can complement genomic and genetic profiles which are at times so complicated by underlying genetic instability that they can obscure our understanding of tumor drivers.

Evolution of the Angiogenic switch

As described in the introduction, the Nagy-Armbruster model [17] predicts that any ESS for angiogenesis effort ($\hat{\zeta}$) is an

evolutionary repeller—rare mutant strategies further from the ESS are favored. Selection therefore pushes mean phenotype away from the ESS, resulting in runaway selection for either vascular hyper- or hypoplasia. This phenomenon is caused by a complex interaction between adenylate metabolism, energy charge homeostasis and vascular response to angiogenesis signaling. The angiogenesis ESS is also highly sensitive to proliferative effort. For example, if $\eta = 1$, representing proliferative effort in a healthy, homeostatic tissue, the ESS $\hat{\zeta} \approx 0.0524$ min^{-1} (about 0.08 fmol or 10^8 ATP molecules per min per cell). However, as η changes towards its ESS of 3.67 (assuming it does so independently of selection on angiogenesis potential), the ESS for $\hat{\zeta}$ jumps past 90 (more than 135 fmol or 1.8×10^{11} ATP molecules per minute per cell), beyond what is physiologically reasonable.

Here, as in the previous section, we relax the adaptive dynamics assumptions of one mutant challenger at a time and small mutational effect, although we retain the assumption that selection acts only on angiogenesis effort. Here again, simulations agree

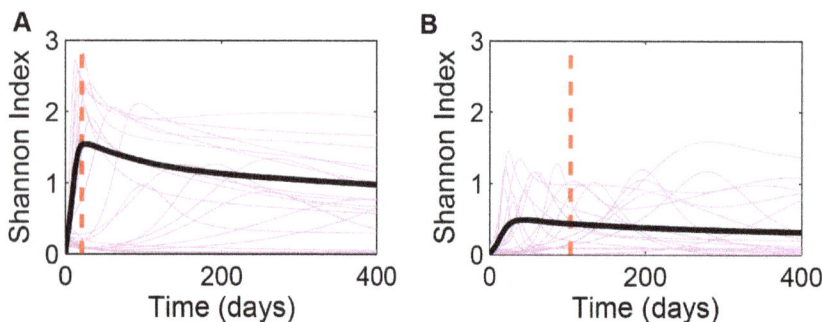

Figure 4. Dynamics of the Shannon diversity index, H, of individual tumors evolving in proliferation commitment, η. Purple curves in both panels represent the first 20 of the 1000 simulations plotted in Fig. 2. Solid black curves are the ensemble averages of all 1000 runs. Dashed red lines represent mean time of the final mutation (S/λ). (A) $\lambda = 0.1$; (B) $\lambda = 0.02$.

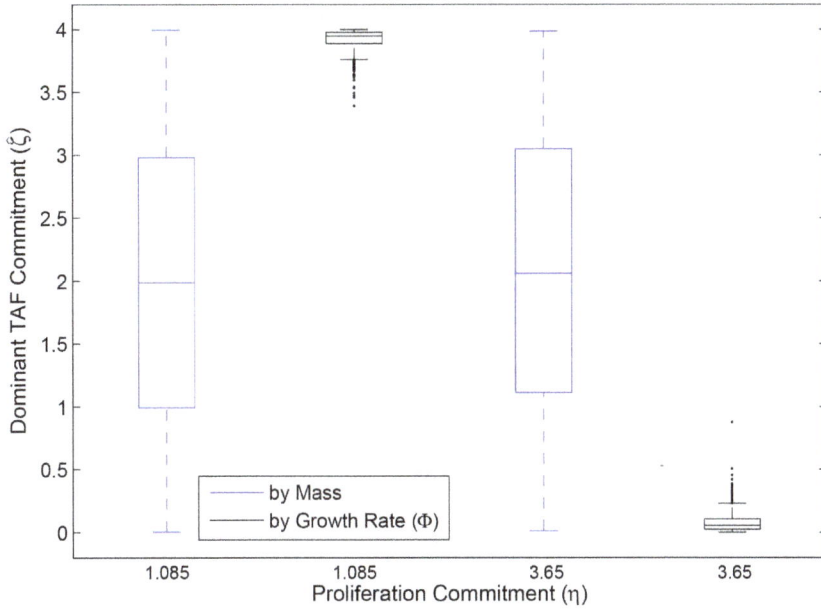

Figure 5. Evolutionarily dominant tumor angiogenesis factor commitment ($\hat{\zeta}$) in 1000 simulations for two values of proliferative effort ($\eta = 1.085$ min^{-1} and 3.65 min^{-1}). In both cases, $\lambda = 0.1$. All other parameters except $\hat{\zeta}$ were set to defaults from [17]. (Compare Fig. 2.)

with the adaptive dynamics analysis. There exists an evolutionarily repelling ESS, so angiogenic commitment evolves to one extreme or the other, as can be seen in Fig. 5. If proliferation commitment is low, the angiogenesis ESS is also relatively low. In consequence, many mutants have an angiogenic commitment above the ESS. Because the ESS is repelling, those clones with the highest angiogenic commitment have the greatest advantage, and so $\hat{\zeta}$ evolves to its highest possible value (Fig. 5, left-hand black box). However, if proliferation commitment is high, then the angiogenesis ESS is beyond what is physiologically possible but is still

repelling. Therefore, mutants with the lowest angiogenic commitment are most favored, and $\hat{\zeta}$ evolves to extremely low values (Fig. 5, right-hand black box).

Nevertheless, these evolutionary forces remain clinically insignificant because historical contingency completely dominates the dynamics. By the end of the simulations, when tumors are well beyond lethal size, strategies that dominate by mass vary greatly in angiogenic phenotype, $\hat{\zeta}$ (Fig. 5, blue boxes). Therefore, even detailed histopathology studies would reveal no evolutionary pattern unless specific markers for angiogenesis were correlated

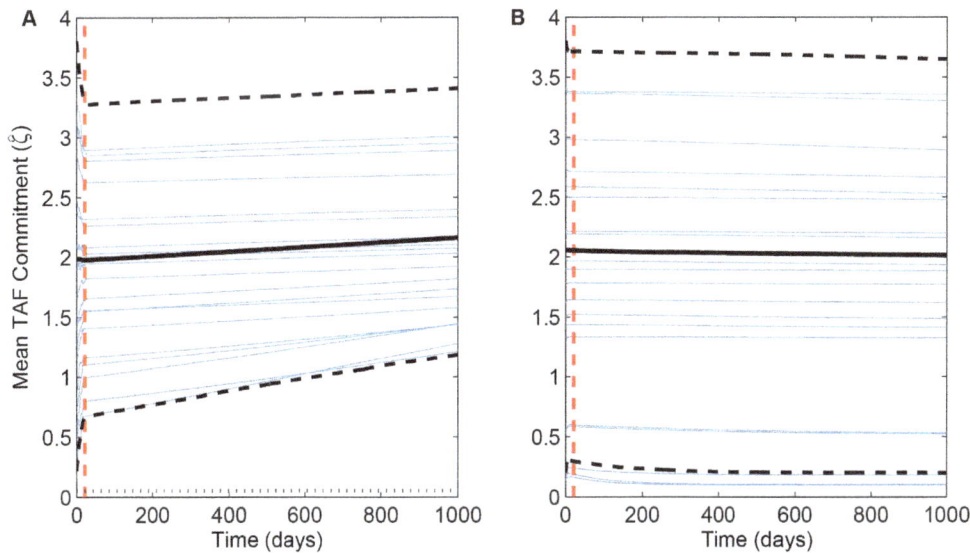

Figure 6. Change over time in weighted mean angiogenesis effort, $\hat{\zeta}$. Blue curves in both panels represent the first 20 of 1000 simulations plotted in Fig. 5. Solid black curves are ensemble averages of all 1000 runs, dashed black curves are the inner 95th percentile of all runs, and dashed red lines are mean time of the last mutation. (Compare Fig. 3.) (A) $\eta = 1.085$ min^{-1}; dotted horizontal line is the ESS from adaptive dynamics (≈ 0.0524 min^{-1}). (B) $\eta = 3.65$ min^{-1}; the ESS value of $\hat{\zeta} > 90$ min^{-1} is not shown.

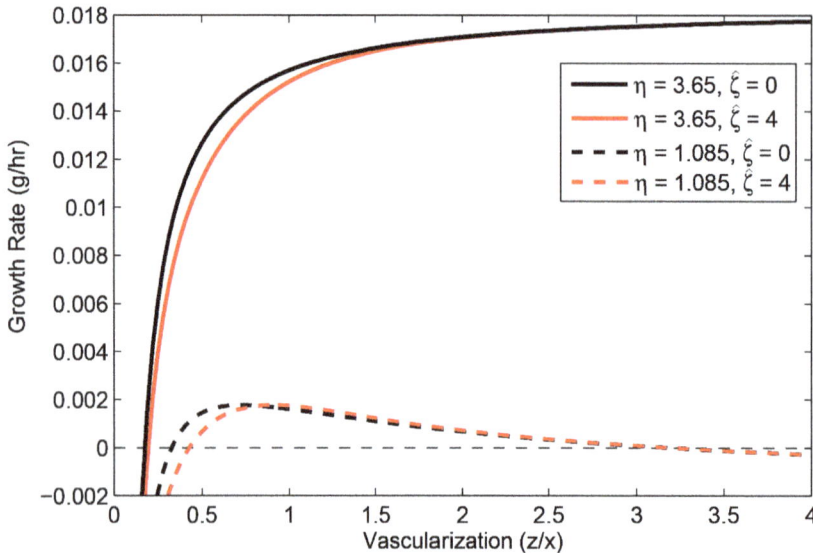

Figure 7. Per capita growth rate of various strategies against tumor vascularization. Gray, dashed horizontal line represents zero growth rate. Solid lines represent clones with high proliferation commitment ($\eta = 3.65$ min^{-1}), while dashed lines represent clones with low proliferation commitment ($\eta = 1.085$ min^{-1}). Black lines represent the lowest angiogenic clones ($\hat{\zeta} = 0$ min^{-1}), red lines; the highest angiogenic clones ($\hat{\zeta} = 4$ min^{-1}), which encompasses the possible curves of all intermediate angiogenic clones.

with proliferation rate. This situation arises because the selection gradient for angiogenesis effort is extremely shallow [17]. Although extreme values of $\hat{\zeta}$ are favored, selective benefits of these extremes are tiny compared to disfavored strategies. Therefore, tumors tend to be dominated histologically by the initial clones that arise because the time required by any selectively favorable clone to overcome these early clones extends well beyond the time required for the tumor to grow to lethal size. Nevertheless, the general evolutionary trend towards "hypertumors" predicted by adaptive dynamics is still evident. All simulations show a clear trajectory towards either vascular hyper- or hypoplasia. In particular, tumors with low proliferation commitment ($\eta = 1.085$ min^{-1}) always evolve towards higher angiogenic potential (Fig. 6A), while tumors with high proliferation commitment ($\eta = 3.65$ min^{-1}) always creep towards failing angiogenesis (Fig. 6B).

In the former case, tumor necrosis occurs once tumor vascular density exceeds 3.2 times normal tissue vascularization (Fig. 7, dashed curves). The necrosis arises as selection continues to favor the most angiogenic clones in the (hypervascular) tumor; the energy wasted by massive angiogenic factor secretion in an environment that promises no more proliferative advantages to cells from more microvessel density ends up causing the tumor's downfall. In contrast, the latter case represents a "classical" hypertumor [14] as selection favors the least angiogenic clones, eventually forcing the vascular density below that required to sustain the cells. Specifically, tumors with proliferation commitment anywhere between 0 and 4 min^{-1} become necrotic once tumor vascularization drops below 0.2 (20% of normal vascular density; Fig. 7, solid curves).

As tumors progress towards their evolutionary endpoint—the ESS in the case of proliferative potential and extreme hypo- or hyperplasia for angiogenic ability—clonal diversity will tend to decrease. Therefore, the diversity index provides at least a rough measure of how close the tumor system is to its evolutionary endpoint. However, the picture is muddied because selection pressures on proliferation and on angiogenic capacities interact. Weak selective pressures on proliferation allow for increasing

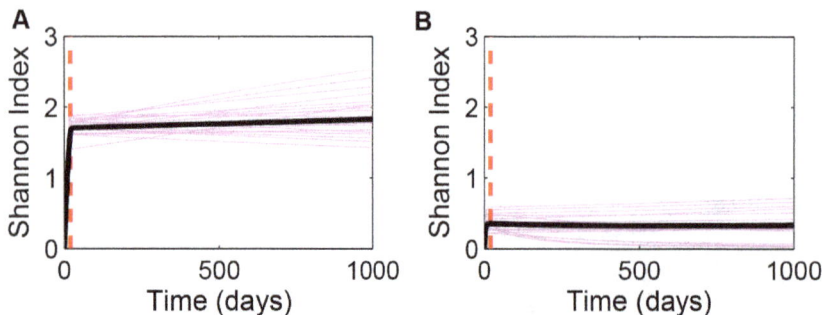

Figure 8. Dynamics of the Shannon diversity index, H, of individual tumors evolving in angiogenic commitment ($\hat{\zeta}$) for two different constant proliferation commitments: (A) $\eta = 1.085$ min^{-1}**; (B)** $\eta = 3.65$ **min^{-1}.** Purple curves in both panels represent the first 20 of the 1000 simulations plotted in Fig. 5. Solid black curves are the ensemble averages of all 1000 runs. Dashed red lines represent mean time of the final mutation (S/λ), with $\lambda = 0.01$.

diversity in angiogenic capacity, whereas tumors evolving rapidly to a high proliferation commitment tend to be significantly less diverse (Fig. 8). Indeed, a small subset of highly proliferative tumors exhibit rapid declines in tumor diversity (Fig. 8B); no similar behavior was observed in tumors with low proliferation commitment (Fig. 8A).

Given the dominant role historical contingency plays in evolution of angiogenesis capacity in these simulations, the clinical significance of hypertumors remains an open question. Evolutionary suicide may be the ultimate endpoint of angiogenic tumors, but they may also tend to kill the host before that endpoint is approached in many, perhaps most, cases. However, the simulations as formulated here cannot be used to assess this suggestion since the parameterization is focused on early tumor growing in their exponential phase.

Conclusions

Tumors exist not as homogeneous entities with universal properties or traits, but rather as diverse collections of heterogeneous cell lineages competing for resources with one another and surrounding healthy cells within the tumor stroma. Thus, viewing tumor progression as an evolutionary process is vital to understanding and eventually treating tumors so that resistance does not evolve. Previous adaptive dynamics modeling has shown that selection acts on cells' commitments to proliferation and TAF secretion potential based on their costs and benefits, defined primarily by their effects on metabolism and per capita growth rate [17]. However, the analytical techniques used in that study,

based on adaptive dynamics, assumes certain biological constraints not commonly observed in cancer growing *in vivo*, including low phenotypic diversity and small mutational effects. Using numerical simulations, we relax these assumptions and show that diverse, complex tumors still adhere to the evolutionary pathways predicted by adaptive dynamics. In particular, (i) the ultimate evolutionary attractor for proliferative commitment is a finite ESS that is also convergence-stable, (ii) selection on angiogenic potential generates an ESS that becomes an evolutionary repeller, and therefore (iii) selection on angiogenesis potential produces vascular instability ultimately leading to evolutionary suicide (hypertumor) by inducing either vascular hypo- or hyperplasia. However, evolutionary trajectories in these simulations are so highly influenced by the tumor's specific mutational history that the predicted evolutionary endpoints may be largely irrelevant to tumor natural history *in vivo*.

Acknowledgments

The authors thank two anonymous reviewers for insightful comments that helped us sharpen the argument and generally improve the manuscript.- This research is supported in part by Scottsdale Community College and Arizona State University's Barrett Honors College.

Author Contributions

Conceived and designed the experiments: JDN STB JDJ. Performed the experiments: STB JDJ. Analyzed the data: JDN STB JDJ. Contributed reagents/materials/analysis tools: JDN. Wrote the paper: STB JDN JDJ.

References

1. Greaves M, Maley CC (2012) Clonal evolution in cancer. Nature 481: 306–313.
2. Weinberg RA (2007) The Biology of Cancer. New York: Garland.
3. Hanahan D, Weinberg RA (2000) The hallmarks of cancer. Cell 100: 57–70.
4. Hanahan D, Weinberg RA (2011) Hallmarks of cancer: The next generation. Cell 144: 646–674.
5. Holash J, Maisonpierre PC, Compton D, Boland P, Alexander CR, et al. (1998) Vessel cooperation, regression and growth in tumors mediated by angiopoietins and VEGF. Science 221: 1994–1998.
6. Neufeld G, Cohen T, Gengrinovitch S, Poltorak Z (1999) Vascular endothelial growth factor (VEGF) and its receptors. FASEB Journal 13: 9–22.
7. Baeriswyl V, Christofori G (2009) The angiogenic switch in carcinogenesis. Seminars in Cancer Biology 19: 329–337.
8. Benjamin LE, Hemo I, Keshet E (1998) A plasticity window for blood vessel remodelling is defined by pericyte coverage of the preformed endothelial network and is regulated by PDGF-B and VEGF. Development 125: 1591–1598.
9. Dunn IF, Heese O, Black PM (2000) Growth factors in glioma angiogenesis: FGFs, PDGF, EGF, and TGFs. Journal of Neuro-Oncology 50: 121–137.
10. Hanahan D, Folkman J (1996) Patterns and emerging mechanisms of the angiogenic switch during tumorigenesis. Cell 86: 353–364.
11. Plate KH, Breier G, Weich HA, Risau W (1992) Vascular endothelial growth factor is a potent tumour angiogenesis factor in human gliomas in vivo. Nature 359: 845–848.
12. Vajkoczy P, Farhadi M, Gaumann A, Heidenreich R, Erber R, et al. (2002) Microtumor growth initi-ates angiogenic sprouting with simultaneous expression of VEGF, VEGF receptor-2 and angopietin-2. Journal of Clinical Investigation 109: 777–785.
13. Bergers G, Benjaimin LE (2003) Tumorigenesis and the angiogenic switch. Nature Reviews Cancer 3: 401–410.
14. Nagy JD (2004) Competition and natural selection in a mathematical model of cancer. Bulletin of Mathematical Biololgy 66: 663–687.
15. Nagy JD (2005) The ecology and evolutionary biology of cancer: A review of mathematical models of necrosis and tumor cell diversity. Mathematical Biosciences and Engineering 2: 381–418.
16. Nagy JD, Victor EM, Cropper JH (2007) Why don't all whales have cancer? A novel hypothesis resolving peto's paradox. Integrative and Comparative Biology 47: 317–328.
17. Nagy JD, Armbruster D (2012) Evolution of uncontrolled proliferation and the angiogenic switch in cancer. Mathematical Biosciences and Engineering 9: 843–876.
18. Parvinen K (2005) Evolutionary suicide. Acta Biotheoretica 53: 241–264.
19. Smith JM (1982) Evolution and the Theory of Games. Cambridge: Cambridge University Press.

20. Smith JM, Price GR (1973) The logic of animal conict. Nature 246: 15–18.
21. Geritz SAH, Kisdi E, Meszéna G, Metz JAJ (1998) Evolutionarily singular stategies and the adap-tive growth and branching of the evolutionary tree. Evolutionary Ecology 12: 35–57.
22. Ataullakhanov FI, Vitvitsky VM (2002) What determines the intracellular ATP concentration. Bioscience Reports 22: 501–511.
23. Martinov MV, Plotnikov AG, Vitvitsky VM, Ataullakhanov FI (2000) Deficiencies of glycolytic enzymes as a possible cause of hemolytic anemia. Biochimica et Biophysica Acta 1474: 75–87.
24. Dieckmann U (1997) Can adaptive dynamics invade? Trends in Ecology and Evolution 12: 128–131.
25. Metz JAJ, Nesbit R, Geritz SAH (1992) How should we define 'fitness' for general ecological scenarios? Trends in Ecology and Evolution 7: 198–202.
26. Vermeij GJ (2006) Historical contingency and the purported uniqueness of evolutionary innovations. Proceedings of the National Academy of Sciences, USA 103: 1804–1809.
27. Cairns RA, Harris IS, Mak TW (2011) Regulation of cancer cell metabolism. Nature Reviews Cancer 11: 85–95.
28. Upadhyay M, Samal J, Kandpal M, Singh OV, Vivekanandan P (2013) The Warburg effect: Insights from the past decade. Pharmacology and Therapeutics 137: 318–330.
29. Hardie DG, Carling D, Carlson M (1998) The AMP-activated/SNF1 protein kinase subfamily: Metabolic sensors of the eukaryotic cell? Annual Review of Biochemistry 67: 821–855.
30. Colombo M, Lombardi L, Melani C, Parenza M, Baroni C, et al. (1996) Hypoxic tumor cell death and modulation of endothelial adhesion molecules in the regression of granulocyte colony stimulating factor-transduced tumors. American Journal of Pathology 148: 473–483.
31. Rong Y, Durden DL, Van Meir EG, Brat DJ (2006) 'Pseudopalisading' necrosis in glioblastoma: A familiar morphologic feature that links vascular pathology, hypoxia and angiogenesis. Journal of Neuropathology and Experimental Neurology 65: 529–539.
32. Eshel I, Motro U (1981) Kin selection and strong evolutionary stability of mutual help. Theoretical Population Biology 19: 420–433.
33. Kops GJPL, Weaver BAA, Cleveland DW (2005) On the road to cancer: Aneuploidy and the mitotic checkpoint. Nature Reviews Cancer 5: 773–785.
34. Kaelin WG, Ratcliffe PJ (2008) Oxygen sensiiing by metazoans: The central role of the HIF hydroxylase pathway. Molecular Cell 30: 393–402.
35. Maley CC, Galipeau PC, Finley JC, Wongsurawat VJ, Li X, et al. (2006) Genetic clonal diversity predicts progression to esophageal adenocarcinoma. Nature Genetics 38: 468–473.
36. Merlo LM, Pepper JW, Reid BJ, Maley CC (2006) Cancer as an evolutionary and ecological process. Nature Reviews Cancer 6: 924–935.

37. Shannon CE, Weaver W (1962) The Mathematical Theory of Communication. Urbana, IL: U. Illinois Press.
38. Hartl DL, Clark AG (2006) Principles of Population Genetics. Sunderland, MA: Sinauer Associates, fourth edition.
39. Lande R (1996) Statistics and partitioning of species diversity, and similarity among multiple communities. Oikos 76: 5–13.
40. Kilburn DG, Lilly MD, Webb FC (1969) The energetics of mammalian cell growth. Journal of Cell Science 4: 645–654.

6

Geographic Variation of Melanisation Patterns in a Hornet Species: Genetic Differences, Climatic Pressures or Aposematic Constraints?

Adrien Perrard[1,2]*, Mariangela Arca[3,4], Quentin Rome[1], Franck Muller[1], Jiangli Tan[5], Sanjaya Bista[6], Hari Nugroho[7], Raymond Baudoin[8], Michel Baylac[1], Jean-François Silvain[3,4], James M. Carpenter[2], Claire Villemant[1]

1 UMR 7205 ISYEB, Muséum National d'Histoire Naturelle, Paris, France, 2 Division of Invertebrate Zoology, American Museum of Natural History, New York, New York, United States of America, 3 Unité de Recherche IRD 072-UPR9034 CNRS, Laboratoire Evolution Génomes et Spéciation, Gif-sur-Yvette, France, 4 Université Paris-Sud 11, Orsay, France, 5 School of Life Sciences, Northwest University, Xi'an, Shaanxi, China, 6 Entomology Division, Nepal Agricultural Research Council (NARC), Lalitpur, Nepal, 7 Museum Zoologicum Bogoriense, Indonesian Institute of Sciences, Cibinong, Bogor, Indonesia, 8 CBNBP, Muséum National d'Histoire Naturelle, Paris, France

Abstract

Coloration of stinging insects is often based on contrasted patterns of light and black pigmentations as a warning signal to predators. However, in many social wasp species, geographic variation drastically modifies this signal through melanic polymorphism potentially driven by different selective pressures. To date, surprisingly little is known about the geographic variation of coloration of social wasps in relation to aposematism and melanism and to genetic and developmental constraints. The main objectives of this study are to improve the description of the colour variation within a social wasp species and to determine which factors are driving this variation. Therefore, we explored the evolutionary history of a polymorphic hornet, *Vespa velutina* Lepeletier, 1836, using mitochondrial and microsatellite markers, and we analysed its melanic variation using a colour space based on a description of body parts coloration. We found two main lineages within the species and confirmed the previous synonymy of *V. auraria* Smith, 1852, under *V. velutina*, differing only by the coloration. We also found that the melanic variation of most body parts was positively correlated, with some segments forming potential colour modules. Finally, we showed that the variation of coloration between populations was not related to their molecular, geographic or climatic differences. Our observations suggest that the coloration patterns of hornets and their geographic variations are determined by genes with an influence of developmental constraints. Our results also highlight that *Vespa velutina* populations have experienced several convergent evolutions of the coloration, more likely influenced by constraints on aposematism and Müllerian mimicry than by abiotic pressures on melanism.

Editor: Alexandre Roulin, University of Lausanne, Switzerland

Funding: This work was funded by two ATM ("Formes possibles, formes réalisées" and "Biodiversité actuelle et fossile") from the Museum National d'Histoire Naturelle, the Bourse Eole of the Franco-Netherland alliance and the Bourse Germaine Cousin of the Société Entomologique de France. The funders had no role in study design, data collection and analysis, decision to publish, or preparation of the manuscript.

Competing Interests: The authors have declared that no competing interests exist.

* E-mail: perrard@mnhn.fr

Introduction

Geographic variation of coloration is one of the most striking aspects of diversity within many species [1], [2]. Numerous factors could influence this colour variation, including evolutionary drift [3], but also various selection forces triggered by environmental differences between populations [1], [4]. One of the factors of selection on coloration is aposematism, the use of a highly visible signal to warn predators of unpalatability [4–6], such as the black and light stripes of wasps [7–9]. Efficiency of aposematic signals relies on the frequency of this signal in the habitats and its probability to be experienced by predators [4], [10], [11]. Therefore, it may seem counter-intuitive to find colour diversity under aposematic constraints. However, colour variation occurs in many aposematic species [11–13]. In these species, selection on aposematic signals can also interact with other selective pressures for driving the colour variation in the species. For example, the

main colour variation within wasp species is a change in the degree of dark pigmentation, or melanisation, inducing light- or dark-coloured morphs and confusing the taxonomy of the group [14], [15], [16]. This melanism is known to be of adaptive importance in insects: it has been related to crypsis, to thermoregulation and to the resistance to pathogens [17–20]. Selection on melanism can thus interact with selection on aposematic patterns of coloration [21]. Surprisingly, the potential relations of intra-specific colour variation with aposematism and melanism have been rarely studied in social wasps (but see [16]). In this study, our aim is to determine which factors influenced the geographic variation of coloration within a species of wasp.

In many organisms, melanism is expressed by a roughly homogeneous increase of the melanin pigmentation over the body [19], [20]. On the contrary, in aposematic insects such as wasps and ladybirds, contrasted colours are a major component of the warning signal. In these organisms, melanism occurs through an increase of the area of delimited melanic patterns on the body [22], [23], [19]. We thus described the variation in coloration of the species by the variation of these melanic patterns.

Warning signal in wasps includes the colour patterns of the different body segments. In order to maintain a recognizable warning signal among the different colour variants, the variation of body parts may follow an integrated process over the entire body, *i.e.* the different segments would vary in a coordinated way, thus maintaining the unity of the general pattern [24], [25]. Furthermore, the different body parts are not likely to play an equivalent role in the warning signal. The coloration of some body parts may be either affected by stronger selective pressures or be under the influence of the same developmental pathways. These phenomena would result in modularity [26], *i.e.* some body parts co-varying more between them than with the other parts, forming colour "modules". Integration and modularity may enhance the quality of the warning signal of the different colour variants by conserving a coherent pattern between the different body parts involved in this signal.

The evolutionary history of the species is also a major component in understanding the colour variation. In case of highly convergent variation in distantly related populations, the melanism variation could be induced by selective pressures caused either by abiotic factors related to climate, or biotic factors related to aposematism. If melanism convergence was caused by abiotic factors, melanisation should present geographical or altitudinal clines related to climatic proximity [27]. On the other hand, the geographical variation of coloration driven by aposematic pressures would more likely reflect a mosaic of locally selected phenotypes than a cline [28].

We studied the colour variation of the yellow-legged hornet *Vespa velutina* Lepeletier, 1836 which presents a dozen distinct colour morphs across its distribution in South-East Asia [15], [22], [29] (Fig. 1). Its coloration varies among populations from almost entirely yellow or orange to extensively black [15], [30], and can be labile within populations. This phenomenon underlay previous distinction of a second species, *Vespa auraria* Smith, 1852, because of the apparent sympatry in different localities of this colour form with another form called *nigrithorax* [14], [31].

In order to assess the evolutionary history of the species across its distribution and to disentangle potential taxonomic differences that could bias the analysis, we first explored the genetic relatedness of *Vespa velutina* populations using mitochondrial DNA sequences and microsatellite markers. Then, to quantify the colour variation of the individuals, we built a "colour space" from the gradual colour variation between populations based on the measure of melanic patterns of the different body parts. We tested the integration of the melanic variation and its modular nature through the correlation of the body parts vectors in the colour space. Finally, we tested whether the colour variation across the species was congruent with the genetic variation, geographic clines or a spatial mosaic by comparing geographic, genetic and phenotypic distances. These analyses allowed us to answer the following questions:

(1) Is *auraria* a lineage genetically divergent from the other colour morphs of *V. velutina*?

(2) Is the colour variation of hornet well depicted by a "colour space" based on melanic patterns?

(3) Was the coloration of an aposematic species influenced by developmental processes such as integration and modularity?

(4) Was the colour variation driven by genetic variation or differences in climatic niches?

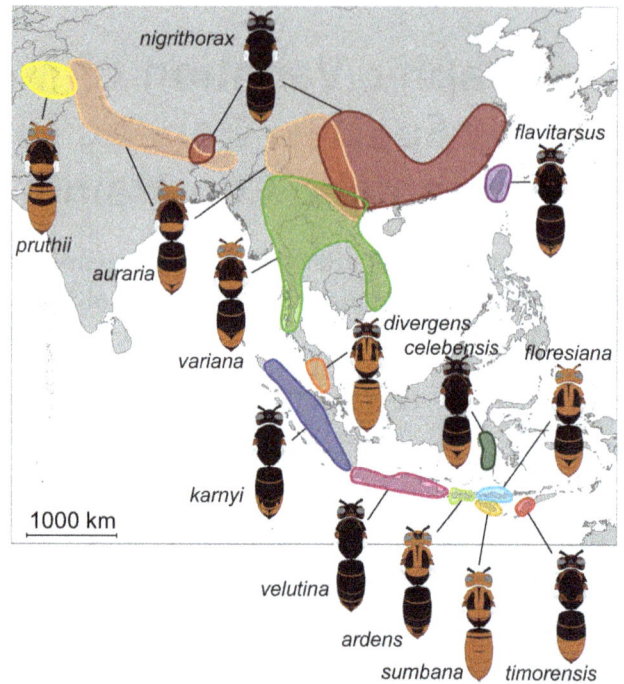

Figure 1. Known distribution of the different colour morphs of *Vespa velutina* across south-east Asia.

We then discussed the implications of our results for the understanding of the development and variation of colour patterns in social wasps.

Material and Methods

A total of 448 specimens of *Vespa velutina* from 216 localities were studied, including 125 recently collected specimens from 69 different localities from Nepal, China, Laos, Thailand, Vietnam and six Indonesian islands. Other specimens used in this study were assembled from public and private collections. No endangered or protected species were sampled for this study. Specimens sampled in Indonesian and Vietnam were collected partially in national parks with the corresponding authorizations: the research permits were obtained from the head office of the forest protection and nature conservation in Indonesia and from the Tropical Institute for Biology in Vietnam; the permits to collect in National Parks (NP) were delivered by the local authorities of Gunung Rinjani NP in Lombok (Indonesia), of Laiwangi Wanggameti NP in Sumba (Indonesia) and of the Bidoup-Nui Ba NP in Vietnam. The localities sampled in other countries were not protected in any way. No specific permits were required for these collections made in collaboration with local researchers and authorizations of the land owners. Indonesian vouchers were stored at the Museum Zoologicum Bogoriense. Other vouchers were housed in the Museum National d'Histoire Naturelle in Paris. The study of public collection specimens was allowed by curators Y. Gérard and A. Drumont (Institut Royal des Sciences Naturelles, Brussels), J. van Achterberg (Nationaal Natuurhistorisch Museum Naturalis, Leiden), G. Broad (Natural History Museum, London) and F. Gusenleitner (Oberösterreichischen Landesmuseen, Linz). The study of private collection specimens was granted by their owners J. Gusenleitner, J. Haxaire, J.-L. Renesson and P. Tripotin.

Over the 216 studied localities, populations were delineated based on colour morphs, geographic distances and ecological environments. We defined ten continental populations (Kashmir, Nepal, Yunnan, Zhejiang, Guangdong, north of Vietnam, south of Vietnam, Thailand, Kra and Malaysia), and eight insular populations from Taiwan and nine Indonesian islands (Sumatra, Java with Bali, Lombok with Sumbawa, Flores, Sumba, Sulawesi and Timor) (Fig. 2, Table 1). Four populations (Kashmir, Nepal, Thailand and Yunnan) each encompass two to three colour morphs for which colour distributions are known to overlap in these areas (Fig. 1). The delineation of these populations was based on the geographic distances of collected individuals and altitude. For example Yunnan specimens were defined as specimens from the mountains while Thailand specimens were defined as located at lower altitude in the North Indo-Burman valley. Three closely-related species were used as outgroup: *V. bicolor* Fabricius, 1787, *V. simillima* Smith, 1868, and *V. vivax* Smith, 1870 [32].

DNA extraction, Polymerase Chain Reaction, Sequencing and Genotyping

Recently collected specimens were preserved in 95% ethanol. Genomic DNA was extracted from legs using QIAGEN 'DNeasy tissue Kit'.

A 658 bp sequence of the mitochondrial gene cytochrome C oxidase subunit I (CO1) was amplified for 119 specimens of *V. velutina*, one of *V. bicolor*, one of *V. vivax* and one of *V. simillima* using universal primer sequences HCO and LCO [33]. DNA amplification followed the standard Polymerase Chain Reaction (PCR) protocol of the Canadian Centre for DNA Barcoding [34]. PCR products were checked on a 2% agarose gel. The purified PCR products were sequenced in both directions. BIOEDIT 7.0.5.3 and CodonCode Aligner V.3.5 were used to align both strands of DNA [35]. Sequences were truncated to the same length (658 bp) to avoid missing data. No insertions, deletions or stop codons were found in the alignment.

Genotypes of *V. velutina* populations were assessed using 11 of the 15 microsatellite loci previously developed for the analysis of the origin of the invasion of *V. velutina* in France: D2-185, R4-100,

R4-114, D3-15, R1-36, R1-75, R1-77, R4-33, R1-137, R1-169 and D2-142 [36], as well as two supplementary loci (List 2015 and List 2020B) [37]. PCR protocols and genotype scoring are detailed by Arca [38]. Because of high frequencies of missing values in the sample caused by low-quality template DNA, the other loci used in Arca *et al.* [36] were not analyzed.

Genetic analyses

Haplotype network and diversity among CO1 sequences were calculated using NETWORK 4.6.1.0 software [39]. An evolutionary tree based on CO1 sequences was computed using the Maximum Likelihood criterion (ML) under a GTR model with PhyML 2.4.4 [40]. Bootstrap supports were calculated from 1000 resamplings.

We built a Neighbor Joining (NJ) tree [41] of individuals using (microsatellite) shared allele distances (DAS; [42]) with the R software and "ape" library [43], and performed a Principal Coordinate Analysis (PCoA) on these distances. Population structure was explored using Bayesian clustering through STRUCTURE 2.0 software [44]. We used admixture model and correlated frequencies. Length of the burnin and the MCMC were 10.000 and 100.000 respectively. Simulations were iterated 10 times for each number of clusters from K = 2 to K = 15. Specimen assignment to a cluster was defined by the probability of the specimen to belong to this cluster with a threshold of 0.5.

Descriptive statistics of populations based on microsatellite data as the genetic diversity estimate θ were obtained using Arlequin 3.01 [45].

Colour pattern analyses

The curved nature of most body segments of wasps makes a standardized quantification of colour over the insect body difficult, due to reflectance and luminosity problems caused by the 3D structure. Furthermore, colour of specimens in natural history collections can be altered by the different collecting methods and conservation conditions. Direct colour quantification using common photograph and pixel-based methods are hardly suitable to such analyses. On the other hand, a semi-quantitative characterisation of melanin patterns can be applied.

We identified the melanic patterns, or modalities, of the different body parts using the original descriptions of *V. velutina* colour morphs and the observation of 448 collection specimens. Body parts with constant coloration (*e.g.* clypeus, propodeum), rare colour variation present in less than 5% of the specimens studied (*e.g.* middle- and hind-femora) or variation without clear patterns (*e.g.* coxae) were not included in the analysis.

We found 23 recurrent colour variations over the body parts, four concerning the head, five the mesosoma, three the legs, seven the metasomal dorsum and four the metasomal sterna (Table 2, Fig. 3). These characters were coded in two to five modalities, giving a total of 73 patterns depicted in Figure 4 (cf. Appendix S1). For each analysed body part, extreme melanic patterns were coded as two binary characters of the presence/absence of the darkest and the lightest modalities. Intermediate modalities were coded following a fuzzy coding between these two extreme characters [46]. In practice, it is equivalent to coding the modalities following their ranking in an ordination of the modalities from one extreme to the other as:

$$\frac{r}{(n-1)}$$

where *r* is the rank of the coded modality, starting from 0 for the extreme modality not described by the character, for example the

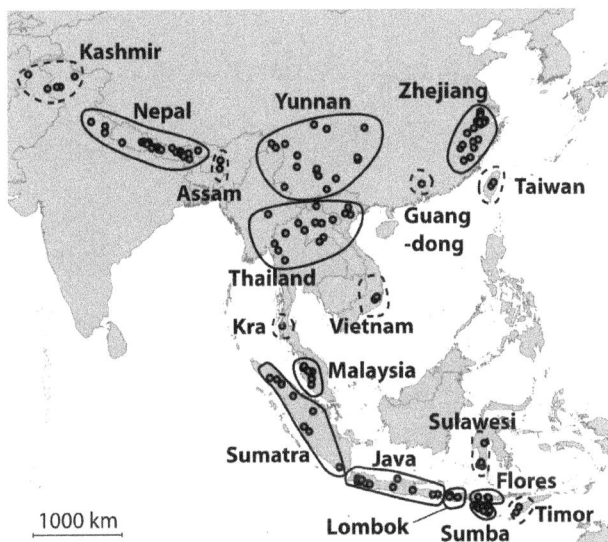

Figure 2. Sampling of *Vespa velutina* across the distribution of the species. Dotted populations are represented by less than 10 specimens.

Table 1. Population sampling.

Name	Distribution	Colour form	Sampling	Molecular
Kashmir	Kashmir, India, Afghanistan	*auraria, pruthii*	9	0/0
Nepal	Nepal, India, Bhutan	*auraria, nigrithorax*	68	29/31
Assam	India	*auraria*	3	0/0
Yunnan	Bhutan, China (Yunnan), Myanmar	*auraria, nigrithorax*	33	13/15
Guangdong	China (Guangdong)	*nigrithorax*	2	0/0
Zhejiang	China (Zhejiang, Shanghai, Jiangsu, Jiangxi, Fujian)	*nigrithorax*	63	29/23
Taiwan	Taiwan	*flavitarsus*	2	0/0
Thailand	Myanmar, Thailand, Laos, north of Vietnam	*auraria, variana*	34	2/2
Vietnam	South of Vietnam	*variana*	10	7/8
Kra	Thailand (Kra Isthmus)	*variana*	1	0/0
Malaysia	Malaysia	*divergens*	34	0/0
Sumatra	Indonesia (Sumatra)	*karnyi*	25	0/0
Java	Indonesia (Java, Bali)	*velutina*	54	4/4
Lombok	Indonesia (Lombok, Sumbawa)	*ardens*	40	5/4
Flores	Indonesia (Flores)	*floresiana*	25	4/5
Sumba	Indonesia (Sumba)	*sumbana*	27	0/1
Sulawesi	Indonesia (Sulawesi)	*celebensis*	10	2/3
Timor	Indonesia (Timor)	*timorensis*	8	0/0

Sampling column refers to number of specimens studied for the coloration and molecular column refers to specimens that provided CO1 / microsatellite data.

Table 2. List and descriptions of colour characters.

N°	Body part	Character	Number of modalities	Illustration (Figure 4)
01	Head	Upper gena	3	A
02	Head	Vertex	3	A
03	Head	Ocellar area	3	A
04	Head	Dorsal margin of scape	3	A
05	Mesosoma	Prothorax	4	B
06	Mesosoma	Mesoscutum	5	B
07	Mesosoma	Scutellum	3	B
08	Mesosoma	Metanotum	4	B
09	Mesosoma	Mesepisternum	2	C
10	Legs	Profemora	3	C
11	Legs	Protibia	3	C
12	Legs	Metatibia apex	2	C
13	Metasoma	1st metasomal tergum	4	D
14	Metasoma	Basal area of the 2nd metasomal tergum	2	D
15	Metasoma	Apical margin of the 2nd metasomal tergum	3	D
16	Metasoma	3rd metasomal tergum	4	D
17	Metasoma	4th metasomal tergum	4	D
18	Metasoma	5th metasomal tergum	3	D
19	Metasoma	6th metasomal tergum	2	D
20	Metasoma	2nd metasomal sternum	4	E
21	Metasoma	3rd metasomal sternum	3	E
22	Metasoma	4th metasomal sternum	3	E
23	Metasoma	5th & 6th metasomal sterna	2	E

Characters' location on the organism is illustrated in Figure 3. Modalities of each colour character are depicted in Figure 4.

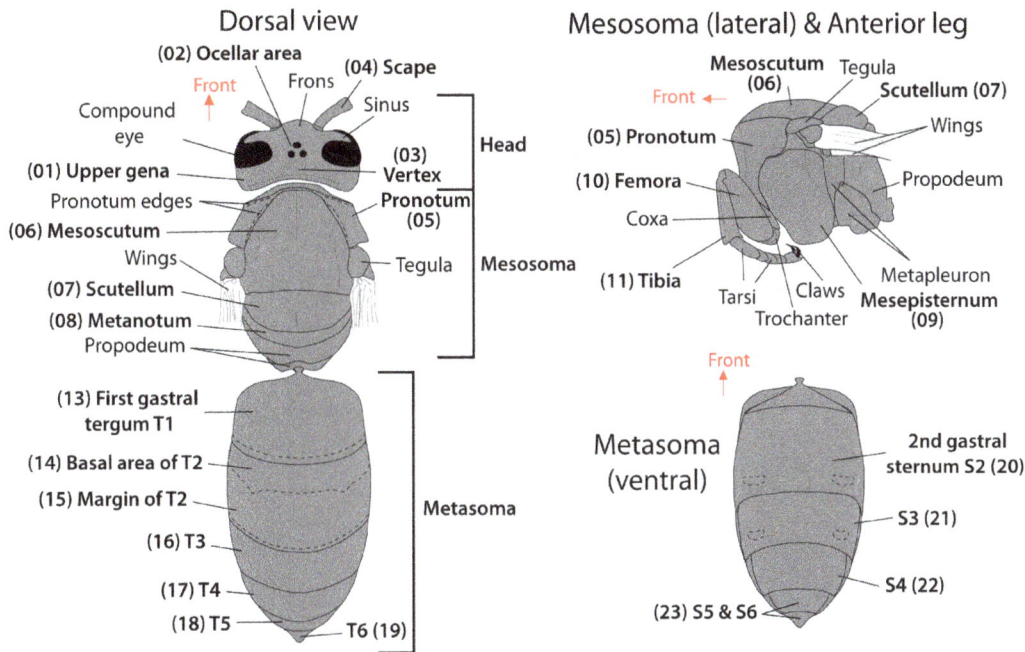

Figure 3. Terminology of the studied body parts of *Vespa velutina.* Characters coding variation of melanisation are in bold and numbered from one to 23. The twelfth character (spot at the apex of the hind-tibia) is not depicted.

darkest modality for a character of presence/absence of the lightest modality, and n the total number of modalities.

Only workers were coded in order to avoid bias caused by social colour dimorphism within the species [15].

We analysed the resulting matrix of 46 variables ranging from 0 to 1 using a correspondence analysis (CA). The coding in two opposed variables for each character induced a marginal value identical for every individual, independently from their melanisation level, and so the same weight in the CA. In such analysis, colour variation is estimated by semi-quantitative variables describing the level of melanisation of each body part. Because the modalities of a given body part could also be considered as independent discrete states between populations, we also computed a multiple correspondence analysis (MCA) on a matrix considering each modality as a variable. The second analysis being based on 73 variables, its description of colour variation was seemingly more precise than the one using the fuzzy coding. However, the CA with fuzzy coding has the advantage of describing the variation of melanic patterns of each body part by a linear direction in the resulting multivariate "colour space". In the MCA, the variation of a single body part is described by a succession of vectors, due to the different modalities.

In order to quantify the accuracy of the colour spaces from the two analyses, we compared the distribution of individuals in these spaces to the classes of colour morphs distinguished by systematists. We tested for the separation of individuals from these different morphs in the colour spaces using cross-validation from a canonical variate analyses.

Both light and dark characters of a body part being vectors with opposite directions from the origin in the CA, we estimated the directions of variation of melanic patterns in the colour space by focusing only on variables describing the lightest modalities. We used these directions to compute a correlation matrix between colour characters in order to explore the potential integration and modularity of the melanisation variation. We tested for melani-

sation integration in organisms by comparing correlations between characters. If melanisation is an integrated process over the entire body, correlation between "light" variables of the different body parts should be mostly positive.

Comparison of colour, genetic and geographic diversity

In order to test the congruence of colour pattern variation with genetic and geographic variation, the distances between colour patterns were compared to the geographic, climatic and genetic distances using pairwise vectors correlation (RV) tests on principal coordinates. The RV coefficient is a multivariate equivalent of the correlation coefficient and addresses the relationship between two sets of variables drawn from a same sample [47], [48]. Its significance was tested using an approximation of a permutation distribution with the library "FactoMineR" in R [49], [50]. The different distances were simultaneously available for 84 specimens from eight populations because the genetic sampling comprised fragmented specimens or specimens from the queen caste. Both individual and population distances were analysed. Geographic distances were computed as geodesic distances between GPS coordinates of the different sample localities using the library "oce" [51]. Genetic distances were computed as allele-shared distances and haplotypic distances. Climatic distances were computed on scores of a Principal Component Analysis of eight climatic variables used in a previous niche modelling of the species [52]: Annual mean temperature, Temperature seasonality, Maximum and Minimum annual temperatures, Annual precipitation, Maximum and Minimum monthly precipitations and Precipitation seasonality. These variables were extracted from the BIOCLIM database as five arc-minutes grids (http://www.worldclim.org/; [53]) on the basis of GPS coordinates, and scaled by their standard deviation. Colour pattern variation was described from scores of the CA. Principal coordinates requiring Euclidean distances, genetic and geodesic distances were transformed following Lingoes [54]. Disparity between the distance

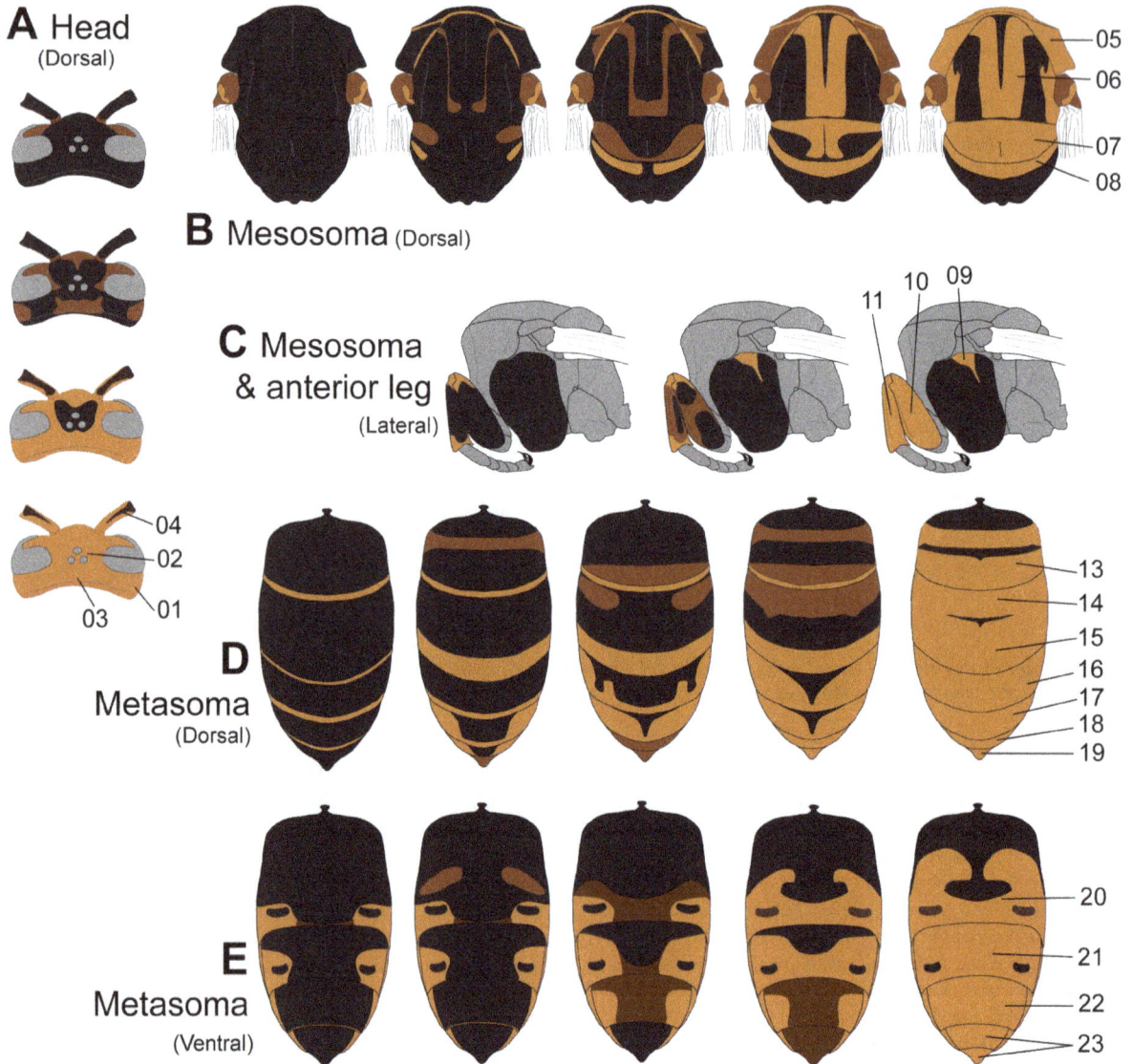

Figure 4. Modalities of colour characters. Each colour character was depicted independently from the others. The ensemble of colour patterns gathered in a given illustration is thus not necessarily reflecting an actual coloration found in wild organisms. **A**. Variation of the four head characters in dorsal view. **B**. Variation of the four dorsal mesosomal characters. **C**. Variation of the lateral mesosoma character and the two anterior leg characters. The hind leg with a spot at the apex of the hind-tibia was not depicted. **D**. Variation of the seven dorsal metasomal characters. **E**. Variation of the four ventral metasomal characters.

matrices was visualised using neighbour joining on the dissimilarity matrix with one minus the squared RV coefficient as a dissimilarity index. We illustrated details of the dissimilarity between haplotypic and phenotypic data by plotting the correlation matrix of distances per individuals between the two datasets and their corresponding hierarchical clustering with the "gplots" package [55]. Under the hypothesis of a strong similarity, the structure of the two clustering trees should be equivalent and the correlation of distances for each individual should approach one. Consequently, the correlation matrix of distances should present clear blocks of high and low correlations related to the corresponding clusters.

The low intra-population diversity produced a similar structure in every distance matrix computed on individual distances. In order to test the correlation of phenotypic, geographic and genetic distances among populations, we also computed distances between populations' colour, climatic, geodesic and genetic averages. We used respectively population mean colour and climatic scores, mean GPS coordinates and mean DAS and haplotype distances.

Previous studies used F_{st} and Q_{st} estimates to compare phenotypic and neutral genetic differentiation between populations and to test for geographic clines (*e.g.* [56], [27]). However, these estimates involve the intra-population variation [57] which can hardly be estimated for lowly sampled populations such as some of ours. Furthermore, Q_{st} computation requires an assessment of the heritability and of the additive genetic variance of the phenotype [27], [58]. These values were unknown for the wing shape of social wasps and a sensitivity analysis showed that they critically influenced our Q_{st} estimates. Therefore, we chose not to use these estimates in our study.

Results

Haplotype diversity

We found 25 different haplotypes in our mitochondrial DNA sampling of *V. velutina*. Populations were separated into two main clusters: a cluster from the Indonesian archipelago and a cluster from mainland populations (Fig. 5A). More than 15 mutations occurred between these two clusters, while the maximum of divergence within each cluster reaches 11 and 12 mutations for mainland and Indonesian haplotypes respectively (Fig. 5B, C).

In the ML tree Indonesian haplotypes were grouped in the cluster that was the sister group of all other Asian haplotypes. They were more similar to Nepalese haplotypes than to haplotypes from other geographically closer populations. Indonesian haplotypes exhibited high genetic diversity within and between populations without shared haplotypes between islands.

Haplotypes from mainland Asia were split in two groups not entirely congruent with the geography of populations. Nepalese haplotypes were basal or grouped with haplotypes from the southern part of Vietnam but this group was not supported by bootstrap values. Thai and Yunnan haplotypes were related to Zhejiang haplotypes in a relatively well supported group.

Genotype diversity

Allele-shared distances were congruent with distance computed with CO1 in distinguishing mainland from Indonesian specimens

in the NJ tree. Overall, mainland specimens were split between a western cluster with the specimens from Nepal, Thailand and Vietnam, and an eastern cluster with specimens from Zhejiang. Specimens from Yunnan formed three different groups spread over these two clusters.

Nepal and Zhejiang populations displayed on average similar genetic diversity ($\theta = 2.29$ and 2.55 respectively) with similar number of localities sampled, but the highest genetic diversity was found in the Yunnan population ($\theta = 2.64$).

We found a significant departure from Hardy-Weinberg Equilibrium for many loci of the Yunnan population and for the L2015 and the R4-114 loci of the Nepal population. We thus excluded them from the Bayesian clustering.

Bayesian clustering distinguished Indonesian and continental populations from K = 2 (Fig. 6). According to Evanno's method, K = 4 would be the best estimate number of clusters in the sampling [59], [60]. At K = 4, continental specimens were split across three clusters, one with Nepal and Thai specimens, a second with Zhejiang specimens and a third with those from Vietnam. With increasing number of K, new clusters occurred mostly within the Zhejiang and Nepal clusters. These relationships may be obscured by the higher allelic diversity found in the Zhejiang and Nepal samples. With K greater than 7, new clusters were not congruent among iterations of simulations. The Indonesian cluster remained as a whole in most iteration, but Sulawesi specimens

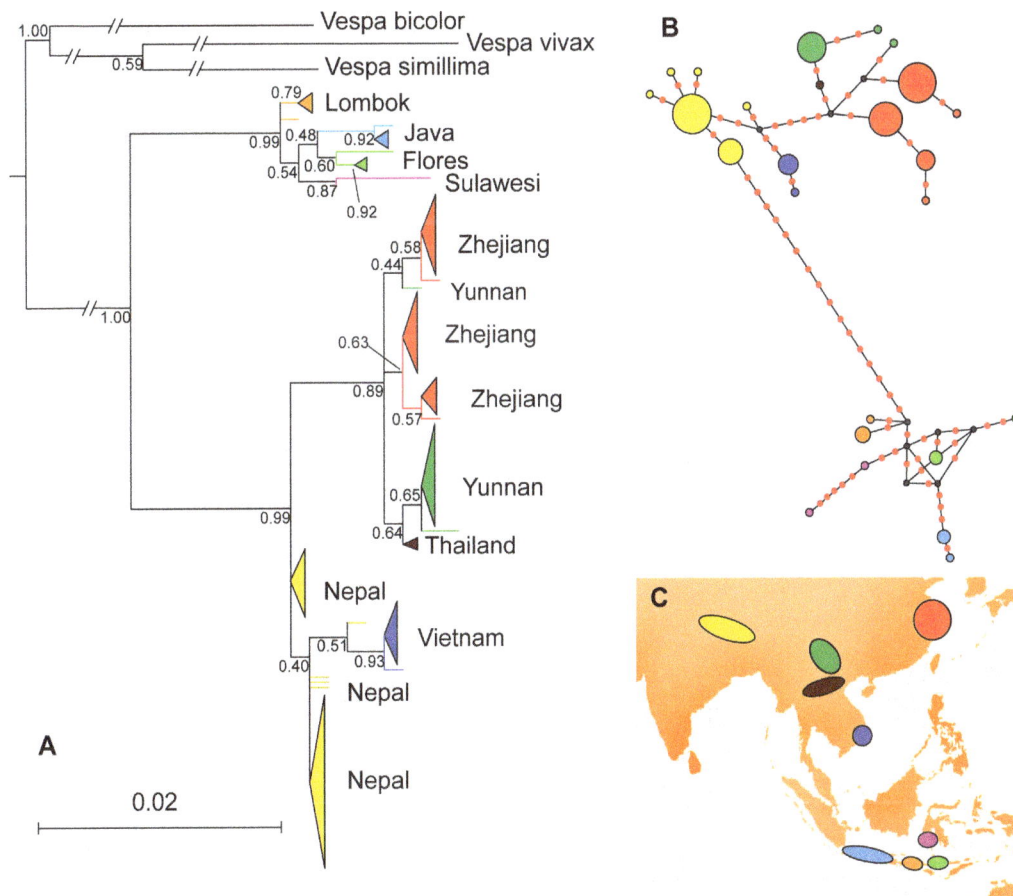

Figure 5. ML tree and haplotype network of CO1 variability of the populations of *V. velutina*. **A** - ML tree computed on CO1 sequences. Scale bar represents the expected mutation per site, node values are bootstrap supports. **B** - Haplotype network. White diamonds are the inferred mutations. **C** - Populations sampled. Size of triangles (A) and circles (C) are proportional to the related number of specimens having these haplotypes.

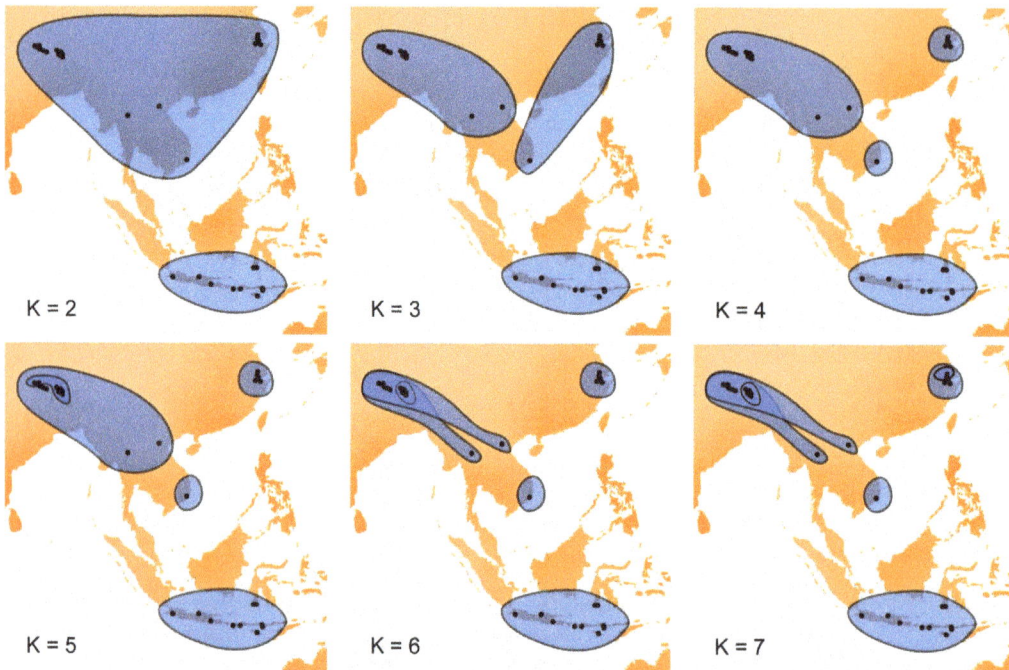

Figure 6. Bayesian clustering of *Vespa velutina* specimens with microsatellite data. Most recurrent results of Bayesian clustering on microsatellite data with increasing number of clusters K. These results were the clusters found in more than 60% of the analyses.

were found separated from other Indonesian populations in few iterations from K = 5.

Colour diversity

The CA returned a colour space with 23 dimensions of which the first dimension gathered 40.64% of the total variation. This axis described a variation from dark to light characters states (Fig. 7). All the vectors of the lightest modalities were on the positive part of the axis. The second dimension encompassed 13.05% of the variation and essentially opposed the coloration of the metanotum and anterior metasomal dorsum to that of the head, legs and pronotum. The third component, accounting for 9.23% of the variation, opposed the coloration of antennal scape and mesepisternum to the coloration of ventral and apico-dorsal metasomal surfaces and the presence of a spot on hind-tibia. The fourth to seventh components mostly described the variation of two characters in relation to all the others: the spot at the apex of the hind-tibia and the spot on the mesepisternum.

Cross-validation confirmed that the colour space separated most of the colour morphs: more than 92.05% of specimens were correctly attributed to their respective colour morph. Of the 31 specimens misidentified from their colour coding, six corresponded to bad discrimination between the *auraria* and *pruthii* morphs, for which the main divergence concerns leg colours not taken into account in the analysis.

The MCA resulted in a similar distribution of the specimens with the first axis of global melanisation encompassing 20.8% of the variation. This analysis returned a colour-space with higher dimensionality allowing only for a slightly more accurate attribution of the specimens to their respective colour morphs with cross-validation (93.08%). Results of the MCA will thus not be further discussed.

The correlation matrix computed between the vectors of lighter coloration of the different body parts in the colour space defined

by the CA showed 96.84% of positive correlations (Fig. 8). The negative correlations opposed primarily the presence of a colour spot at the apex of hind tibia and the darkening of the fourth metasomal tergum to the darkening of the scape, the metanotum, the first three metasomal terga and the fourth metasomal sternum. Correlation between colour patterns revealed two complexes of correlated characters: one including the three head characters, the anterior leg, pronotum, mesoscutum and scutellum and the other the colour patterns of the metanotum and the three first metasomal terga.

Comparison of colour, genetic and geographic diversity

Comparisons using RV test on the individual distances resulted in a significant relatedness between every dataset. The RV coefficient was the highest between climatic and geodesic distances (RV = 0.680) and the lowest between mitochondrial and colour distances (RV = 0.163). Correlation matrix and hierarchical clustering showed low correlations and clear differences in the structure of these haplotypic and phenotypic distances (Fig. 9). Correlation between the two types of distances was low for most of the specimens. Overall the correlation matrix was not highly structured. Clustering using mitochondrial distances grouped individuals per populations except for Sulawesi, Flores and one specimen from Yunnan, and distinctly separated continental and Indonesian populations. On the contrary, phenotypic distances split most populations and mixed distant specimens. The two main colour clusters separated specimens from Zhejiang, Java, Sulawesi and some specimens from Nepal in one group characterized by a dark head and mesosoma, and specimens from Yunnan, Zhejiang, Vietnam, Flores and Lombok populations presenting a lighter head and mesosoma, in a second group.

Population distances returned fewer significant relationships between the different datasets. Mitochondrial, geodesic and climatic distances were still significantly related but these datasets

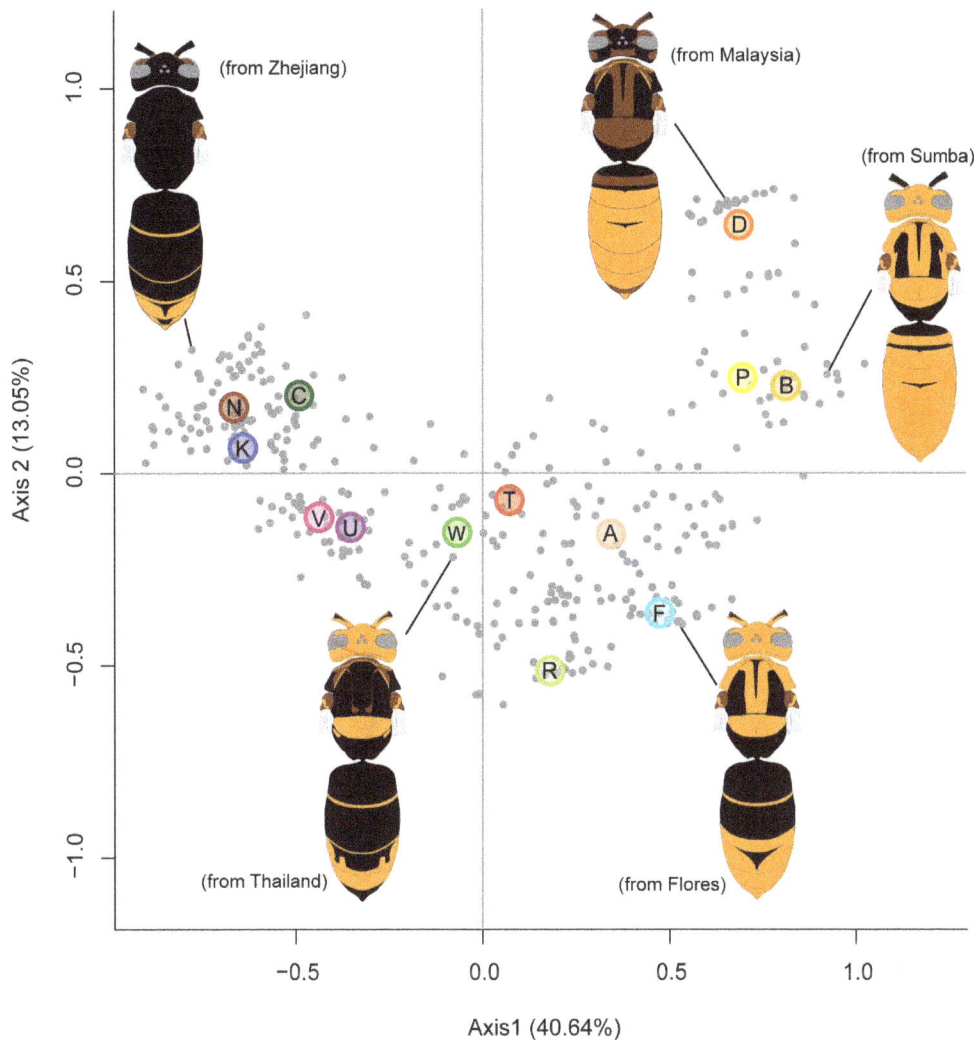

Figure 7. Colour space of *Vespa velutina* **specimens.** Two first dimensions of the colour space resulting from a correspondence analysis on melanisation described by two binary variables of the extreme light and dark coloration for each body part. Intermediate modalities were integrated using a fuzzy coding. Coloured spots described the mean values of each colour morphs (see Fig. 1). Colour forms: N = *nigrithorax*. K = *karnyi*. C = *celebensis*. V = *velutina*. U = *flavitarsus*. W = *variana*. T = *timorensis*. R = *ardens*. A = *auraria*. F = *floresiana*. D = *divergens*. P = *pruthii*. B = *sumbana*.

were no more related to the colour distances and to distances based on genotype data (Table 3).

Discussion

Evolution of *Vespa velutina*

Our results highlighted two main genetic groups of populations: one from continental Asia and the other from the Indonesian islands. Genetic variation for CO1 within the Indonesian cluster was equivalent to the variation observed within the continental group. However, the structure of the Indonesian variation was not congruent with geography. The low sampling of these islands limited a more detailed interpretation of this structure and may explain the absence of resolution found with the Bayesian clustering on microsatellite markers. Within the continental cluster, mitochondrial and nuclear markers returned distinct results. Sequences of CO1 clustered populations from Nepal and Vietnam while Thailand, Yunnan and Zhejiang specimens were in a different clade (Fig. 5). The microsatellite markers supported a different partition of populations with Nepal and Thailand as a

group, Zhejiang and Vietnam being independent populations (Fig. 6). These differences could be explained by the different transmission of mitochondrial and nuclear DNA combined with the unequal sample sizes of populations. Bayesian clustering returned subgroups within the two populations with the largest samples, Nepal and Zhejiang (Fig. 4), while the discrepancies in results between markers involved Vietnam and Thailand populations, both having very low sample diversity. Subgroups found within the continental group should therefore not be considered as relevant biological groups without further genetic data. Both markers confirmed nonetheless that Yunnan sample was genetically heterogeneous, potentially influenced by high migration from the surrounding populations.

The strong genetic difference between continental and Indonesian populations suggested an ancient divergence of these groups of populations. It also raised the question of a potential speciation of southern populations. Two populations from Sumatra and Peninsular Malaysia are located between these continental and southern populations. The absence of molecular data from these two intermediate populations limits our understanding of the

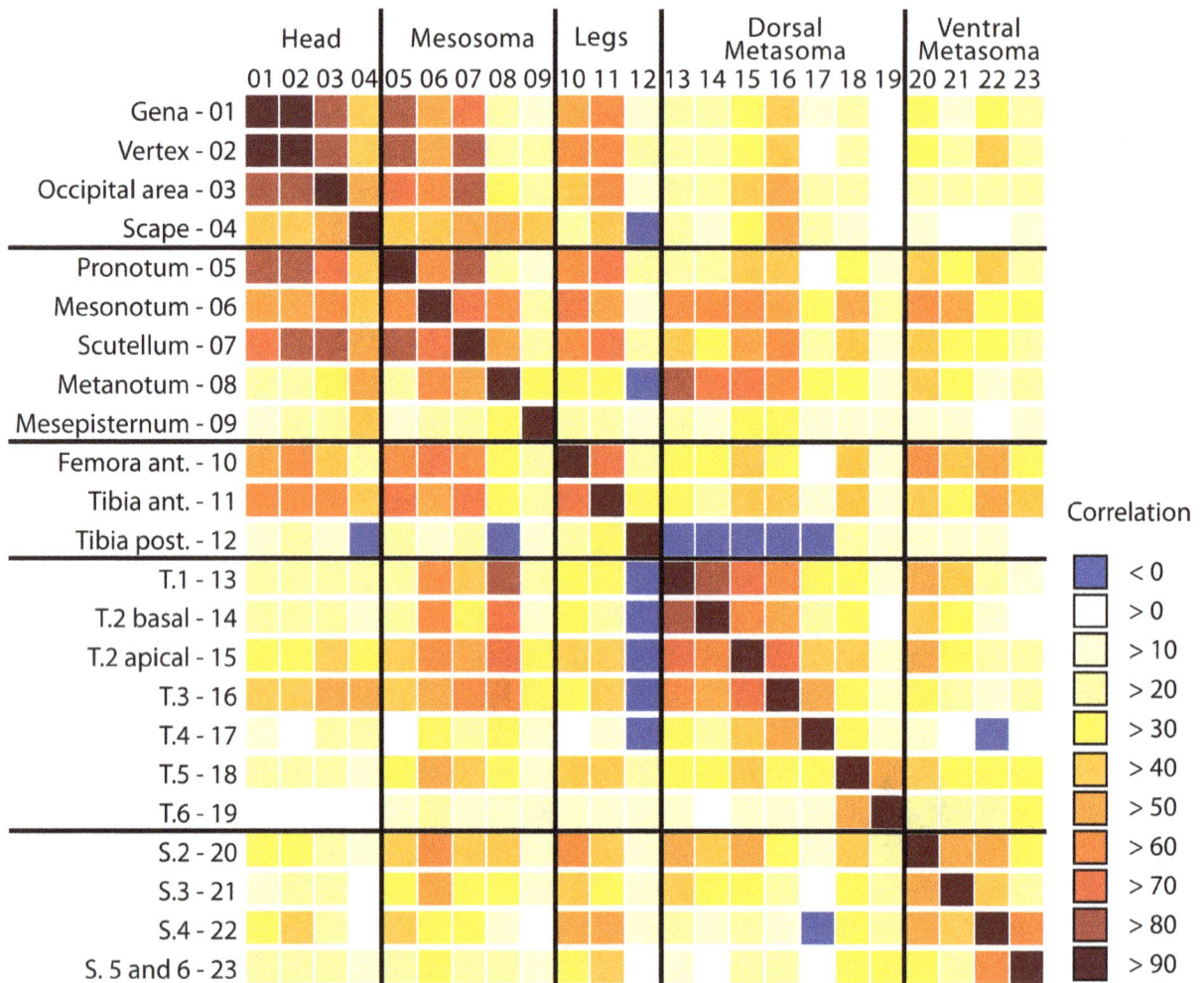

Figure 8. Correlation between the colour variation of body parts. Correlation matrix of the vectors of variation of light colour characters in the colour space (see Fig. 7). Blue marks indicate negative correlation among the two characters.

isolation of the Indonesian and continental populations. Furthermore, the lack of morphological characters discriminating the two main groups did not support the hypothesis of a long isolation leading to a speciation event.

The presence of a continental cluster confirmed that populations of the *auraria* colour form belong to the same species as populations of the *nigrithorax* and *variana* forms [31]. Populations of the *auraria* colour form were long considered as a different species on the basis of sympatry with the colour form *nigrithorax* in northeastern India, Nepal, Burma, Assam and western China [29]. In fact, these two colour forms may interbreed: they present intermediate coloration forms in a single locality, as in our Nepal sample presenting typical *auraria* specimens and darker specimens with a coloration somewhat similar to the *nigrithorax* form as observed in northern Vietnam [31]. Considering the close relationships between Nepalese *nigrithorax* and *auraria* specimens, the different populations of *nigrithorax*, and possibly *auraria*, observed in Figure 1 are likely convergent evolution in different populations.

Patterns of melanisation

We quantified the variation of melanism across *V. velutina* distribution by decomposing the global melanisation pattern into a

suite of discrete variations of the different body parts. Some of the colour variation between the populations of *V. velutina* could not be included in this analysis focusing on defined melanic patterns. For example, the differences in light colour that are clearly yellow or reddish in the different populations were not coded and some variation in leg melanisation was too labile or too rare to be taken into account without overweighting these characters. However, the high rate of correct identification of specimens to their colour morph confirmed that a characterisation of melanin patterns of the different sclerites is a good estimate of the colour variation across the species. Furthermore, the use of a fuzzy coding forcing each character variation to follow a linear direction in the colour space minimally affected the results, when compared to a more complex colour space of each variation considered as a discrete binary variable.

This quantification of melanism highlighted that the melanisation of numerous sclerites varies differently in *V. velutina* (Table 2). Each sclerite tended to have a well defined set of particular and complex melanin patterns across the distribution (Fig. 4). These patterns involved mostly an antero-posterior extension of the black stripes, on the metasoma (Fig. 4.D and E). Other segments like mesoscutum, metanotum or mesepisternum presented well-defined marks of light colours which size may vary (Fig. 4.B).

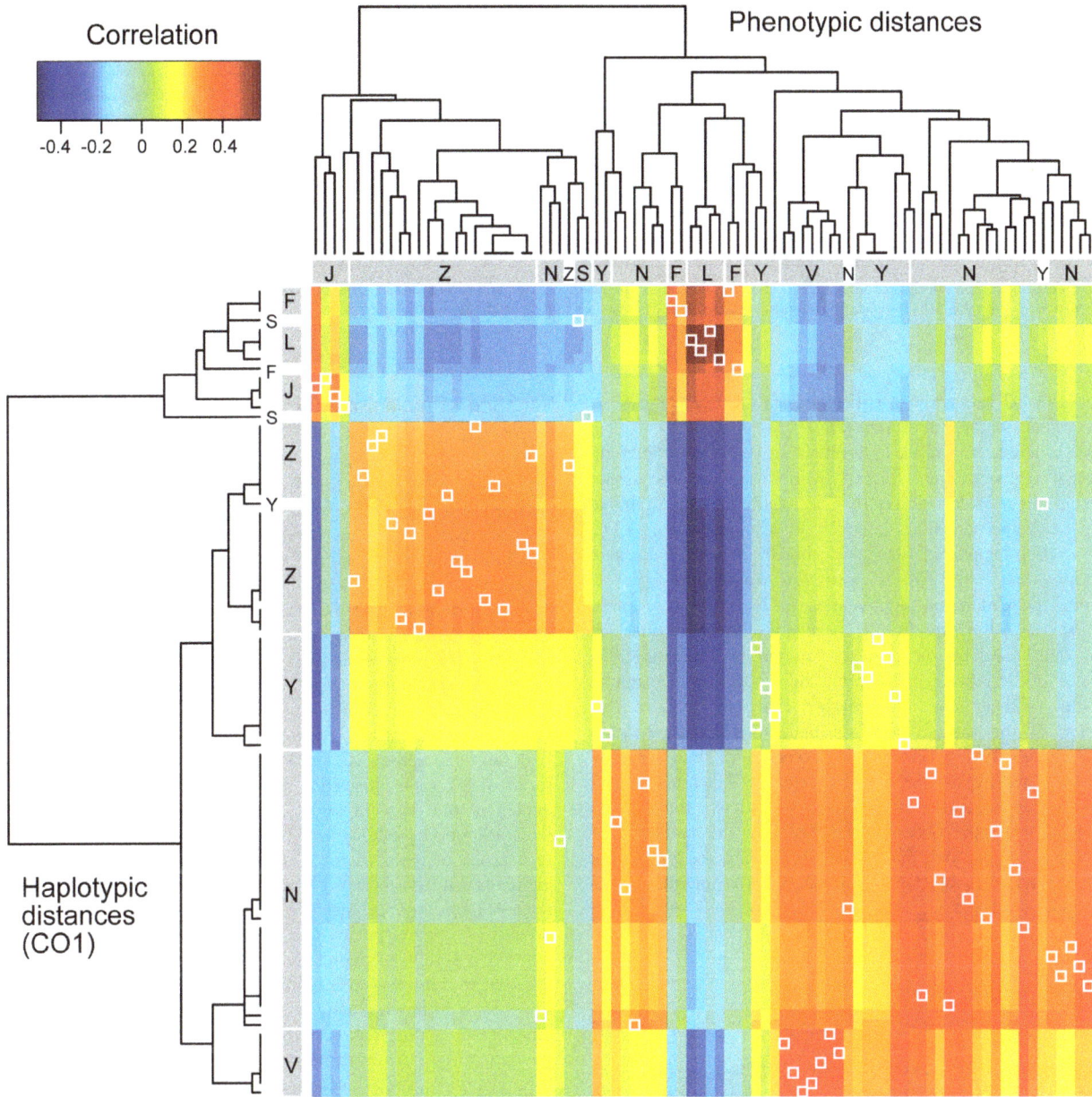

Figure 9. Detailed dissimilarity of haplotypic and colour distances between individuals. Correlation matrix between haplotypic distances (rows) and phenotypic distances (columns) with associated dendrograms. Correlation between distances of a same individual are marked with white squares. Under the hypothesis of similarity between the data, the trees should have the same structure and the individual correlations should approach one. Furthermore, high correlations should be organised in well delimited blocks corresponding to clusters. Haplotypic distances were computed on CO1 sequences and phenotypic distances were computed as the Euclidean distance between individuals in the colour space. Correlation coefficients ranged from −0.6 to 0.6. Dendrograms resulted from complete-linkage clusterings and should not be interpreted as evolutionary trees. Order of specimens differs in rows and columns. Populations: L= Lombok; F= Flores; J= Java; N= Nepal; S= Sulawesi; V= Vietnam; Y= Yunnan; Z= Zhejiang.

Finally, these patterns seemed restricted to a given sclerite, not extending to different segments.

Genetic bases of melanin production and patterning genes

The genetic control of pigmentation has been widely studied, notably in fruit flies and butterfly wings [61]–[64]. This genetic control occurs at different scales in melanin synthesis: directly on the genes coding for proteins of the synthesis chain, like the *yellow* gene [61], [62]; indirectly on genes coding for proteins altering melanin precursors, like *ebony* [65]; or on genes modifying the expression of these two previous groups of genes, like *engrailed* or *Abdomen-B* [66]. Patterning expression of these genes has been related to the diffusion of small compounds called "morphogens" from source cells through the tissues during the development [67]–[69]. Morphogen diffusion is tightly related to temperature, but also to body topology: for example, wing veins act as barriers in signal diffusion across the wing [70], [71]. This may explain why some light or dark spots in hornets do not extend to their

Table 3. Results of pairwise RV tests between populations.

	Haplotype (CO1)	Genotype (Das)	Geodesic distances	Climatic dissimilarity	Coloration dissimilarity
Haplotype (CO1)	-	0.53090	0.00882*	0.00456*	0.63067
Genotype (Das)	0.390	-	0.61353	0.51280	0.62455
Geodesic distances	**0.824**	0.387	-	0.00586*	0.69145
Climatic dissimilarity	**0.900**	0.479	**0.827**	-	0.58485
Colour dissimilarity	0.192	0.533	0.181	0.267	-

RV tests were applied on principal coordinates of genotypic, haplotypic, geodesic, climatic and coloration distances between populations. RV coefficients are in the lower triangle of the table with significant relationship in bold, P of the tests are in italic in the upper triangle.
Significance: "*" <0.05.

neighbouring sclerites through cuticular sutures, such as those on the metanotum or mesepisternum (Fig. 4).

In our analysis, the black and yellow stripes of the different metasomal segments were formed by an apical yellow stripe extending more or less anteriorly and often interrupted by a median band and sometimes lateral spots of melanin (Fig. 4.D). This variation appeared strikingly similar to the patterned activities of three genes Hox, *decapentaplegic*, *wingless* and *engrailed*, regulating the abdominal pigmentation in *Drosophila melanogaster* [72]–[74] (Fig. 10). In the fruitfly, *engrailed* was found responsible for the light band on the posterior margin by repressing the expression of the *yellow* gene, blocking the production of melanin [74]. On another hand, *decapentaplegic* seems to enhance the formation of the median band of melanin [72]. Interaction between these three genes and the variation of their level of expression could explain the complex patterns of melanisation of the metasomal segments of social wasps [75]. These three genes are probably not the only genes involved in the variation of melanin patterns, but they may be good candidates for identifying genes involved in the formation of patterns in metasoma melanisation.

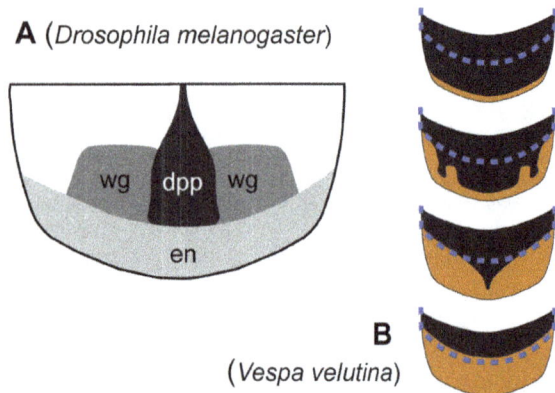

Figure 10. Expression patterns of regulatory genes and variation of melanisation in metasomal segments. A. Expression patterns of *engrailed* (en), *wingless* (wg) and *decapentaplegic* (dpp) on an abdominal tergum of *Drosophila* (modified from [75], after [72], [74]). **B.** Variation of melanisation of the third metasomal tergum of *V. velutina*. The blue dotted lines represent the part of the segment covered by the second tergum.

Variation of melanisation between the different segments

The first axis of variation in the colour space described a global variation in melanisation and appeared positively correlated to the 23 variables of light coloration (Fig. 7). Furthermore these variables were positively correlated in 96.84% of the pairwise comparisons. The correlation between vectors of melanisation of each body part was not homogeneous and suggested two modules: one combining the melanisation of the cephalic capsule together with the pronotum and scutellum, and another the melanisation of the three first metasomal terga with metanotum and in a lesser extent with mesoscutum (Fig. 8). Together, these results suggested a partial integration of melanisation variation over the body, with potential regionalisation of this phenotype.

Because the aposematic signal is located over the entire body in social wasps, such integration and modularity may enhance the quality of the warning signal of the different variants. Colour integration over the body may result from variation of the general regulation of coloration. Two regulation processes of the pigmentation are hormonal levels, melanin production being related to ecdysone titre [75], [76], and genetic variations. Sensitivity of genes involved in melanin synthesis to morphogens may differ from one population to another: the influence of the different morphogens in inhibiting or stimulating a gene expression results from the ability of their transcriptional regulators to bind to a regulatory region of the DNA next to the coding part of the gene [77]. For the *yellow* gene, a *cis*-regulatory promoter has been identified as responsible for pigmentation difference between species due to evolutionary changes altering the number of binding sites of the regulators [78], [66]. Variation in these *cis*-regulatory sequences returned different pigmentations of several body parts, either the whole abdomen, sometime with the thorax, or only the two last abdominal segments. The integration of melanism over the body and its structure in modules may thus be linked to variation in hormonal production, but also to a difference in regulatory genes' sensitivity to morphogens' signals that evolutionary changes alter from one population to another.

Colour variation between populations

Our results of comparisons of distance matrices showed that colour pattern diversity did not match the evolutionary history described by our analysis of molecular markers, nor the geographic distances or the climatic similarity between populations (Figs. 9). Climate, geographic and mitochondrial based distances presented similar structure, but the data based on genotypic and phenotypic distances were differently structured. The absence of congruence between genotypic and CO1 sequences or geography may indicate an influence of the restricted population sampling on the genotypic distances. Another hypothesis would be an unbalanced dispersal of males and females in the species, with males dispersing more than females; however this is not congruent with our current knowledge of this species [79], [52].

Non congruence of colour-based distances with the other dataset was explained by the presence of several populations having very dark colour patterns both in Indonesia and on the continent, while a southern Indonesian population has the lightest colour morph, similar to the Malaysian population and, in a lesser extent, to the continental eastern populations (Fig. 1). On the other hand, Indonesian and continental populations are two divergent lineages discriminated by both mitochondrial and nuclear markers. Most of the similar patterns of melanisation found between these two lineages are thus convergences with different evolutionary origins.

In theory, such convergence could be either induced by a high phenotypic plasticity or by convergent evolutions of genes regulating the melanisation process. Phenotypic plasticity of the coloration has been shown in butterflies through day-length influence on the hormonal production [80] and in paper wasps through variation in rearing temperature [81] and diet [82]. Tibetts highlighted that diet influenced only limited melanin patterns involved in social signal such as the clypeus markings [82]. MacLean and her collaborators showed that colour variation in paper wasps induced by a rearing temperature change did not reach the inter-population variation [16]. Furthermore, diet of a single colony of hornet can change through the season in temperate areas [79] without inducing a colour variation in individuals (pers. obs.). Finally, in contrast to the paper wasps studied by Green et al. [81], hornet larvae develop in enclosed nests with buffered temperature and humidity, lowering the influence of external physical factors [83], [84]. Dark specimens of V. velutina from semi-tropical China have been accidentally introduced to temperate areas of France and Korea several years ago [52], [85] and these invasive populations did not change in coloration [85], [86]. It is therefore unlikely that the observed convergences in coloration of several V. velutina populations are due to phenotypic plasticity induced by similar local factors.

A more probable hypothesis is a parallelism in the evolution of the regulatory and patterning genes either by similar or dissimilar mutations [87]. This evolution may also have been influenced by extrinsic factors through selection: melanism, with a genetic basis, can have both positive and negative impacts on the organism's fitness [19], [20]. The different factors that may induce a selection on melanism are related to the climate and associated environments, but also to the local communities of predators and of Müllerian mimics.

Climatic pressures

Climate is known to have an influence on melanism in insects. Melanism has been related to thermoregulation [88] and dessication resistance [89], [90]. Higher melanic insects were found to warm up faster under the sun, to reach higher temperature, but also to lose less water and resist desiccation better than lighter morphs [88], [89].

However, studies do not agree on the distribution of melanic forms. Some authors suggest that melanism occurs mostly in dry environments [89], while other argues that they are found in cool and wet habitats [16], [88] or in tropical areas [91]. Our results concur only partially with this last study. Two of the least melanised populations occur respectively in Nepal and in the driest island of the distribution of V. velutina in Indonesia while geographically intermediate populations are mostly darker, with the exception of the population from Malaysia. However, we found only low structured climatic differences between the localities of light and dark coloured populations and no cline is visible (Fig. 1). Melanism does not appear to be influenced by altitude, average or extreme temperatures or humidity of the locality of sampling. Especially, the difference between the geographically close Malaysian and Sumatran colour forms, respectively the lightest and the darkest, rejects the hypothesis of climate pressures as the major evolutionary force driving colour polymorphism in V. velutina.

Predator pressures

Besides physical elements of the environment, hornets are likely to be selected for the efficiency of their warning colours [4]. Like other social wasps, hornets have a painful sting and are distasteful due to their venom gland, making them avoided by predators [8] [92], [93]. They benefit from being recognized by potential predators that may attack them or disturb their nests. Furthermore, predator avoidance may be critical in one part of the life-cycle of V. velutina: like in many other social wasps, each colony is annual and founded by one solitary queen [79]. At this time, selective pressures within the species are high, as each queen has to survive for weeks before being able to produce the next generation. As such, warning colours play an important role in social wasps by protecting queens during their solitary phase.

The notion of warning colours depends on the perception of colours by predators [94], [95]. While dark coloration may be advantageous for crypsis [19], the yellow coloration seems to be a warning colour naturally avoided by bird predators [9]. Vidal-Cordero and his collaborators also showed a direct link between the lightness of the metasoma pigmentation in a paper wasp and the size of its venom gland [96]. Light coloration may therefore intervene in predator avoidance through its intensity.

On another hand, black and yellow stripes patterns help in prey recognition both by vertebrate [97], [9] and invertebrate predators [8]. This may explain the presence of black and yellow metasomal stripes in most of the V. velutina populations and the potential patterns observed in mesosoma: the pronotum, the scutellum and sometimes the metanotum can become lightly coloured while the mesoscutum and the propodeum always present black markings. This coloration creates over the mesosoma an alternating pattern of light and black colours extending the striped pattern of the metasoma (Fig. 4.B). It is therefore unsurprising to find yellow marks even on the darkest morphs of V. velutina as well as black segments and markings on the lightest populations (Figs. 1 & 4).

Müllerian mimicry

The influence of warning signal on predator behaviour depends on other harmful species encountered by the predators through the process of Müllerian mimicry, different harmful species sharing a similar aposematic signal, thus mimicking each other. This mimicry reinforces the impact of warning coloration by augmenting the probability of association between the bad experience of the predator and the signal displayed [4]. This phenomenon has been widely studied in butterflies [98], [12], but is also known among bumblebee species [91], [99], [100] and wasps [101]. Both being stinging Hymenoptera, bumblebees, bees and wasps are likely to be recognized as similar prey items by predators and may share the Müllerian effect of their coloration. For example, Hines and Williams pointed out drastic variation of colour in bumblebees from Malaysia, for which specimens are fully orange [100], matching the singular orange morph divergens of V. velutina found in the same region. Another example is the mimicry between the nigrithorax form of V. velutina and queens of two yellow jacket species

Vespula koreensis (Radoszkowski, 1887) and *Vespula orbata* (du Buysson, 1902) found in southern China.

Presence of Müllerian mimicry is also often accompanied by strong phenotypic variation structured in mosaic across the distribution of the species [28]. Polymorphism between populations of *V. velutina* may thus be the result of different selective pressures induced both by variation in the local stinging Hymenoptera communities and by the pressures on melanism-related traits: crypsis, thermoregulation, desiccation and pathogen resistance. While hornets may not be highly sensitive to some of these pressures, co-mimic species with open nests or solitary habits may be, thus leading to an indirect effect of these factors on hornet coloration. Furthermore, the critical phase in colony development associated with the potential founder effects in island colonisation during *Vespa velutina* evolutionary history may explain the high variety of coloration found across islands while colour variation is more progressive and colour forms more widespread in continental Asia.

In order to test for this hypothesis, further studies could focus on characterising the melanism of the different species of stinging Hymenoptera of similar sizes in different regions. This work has already been started on bumblebees [91], [99], [100] but should be extended over the different families of Hymenoptera, as mimetism is not restrained taxonomically.

Conclusion

Our study described the changes in patterns of melanisation over the distribution of the hornet species *Vespa velutina*. The main axis of pattern variation in the resulting colour space described a global melanisation of the body indicating that the melanisation is a partially integrated process across the distribution. Correlations among the melanisation of the different body parts revealed a structured variation with two apparent modules: one composed of the head capsule, anterior part of mesosoma and anterior legs and the other connecting melanisation of the anterior metasomal dorsum and metanotum.

Based on mitochondrial sequences and microsatellite markers, we identified two highly divergent lineages within the *V. velutina* species: a continental lineage and a lineage restricted to southern Indonesian islands. The existence of a continental lineage confirmed the synonymy of *V. auraria* under *V. velutina*. The evolutionary history of the species could not explain the observed colour variations: the variation in melanisation clearly included convergence in different populations. Comparison of the climatic, genetic, geographic and colour diversities showed that melanism was unlikely driven by abiotic factors such as climate variation or latitudinal clines. This variation may result instead of mutations

selected by high local constrains on aposematism and Müllerian mimicry with potential influence of the founder effects in islands.

Together, these results confirmed that colour patterns should not be regarded as reliable criteria for population relationships or species distinction in social wasps. They also suggest that colour variation in aposematic species is not tightly related to abiotic conditions. Further studies could use the quantification of colour patterns to track correlated changes between species within communities of aposematic species and test for the influence of Müllerian mimicry, predators and habitat types on the geographic variation of aposematic coloration.

Supporting Information

Appendix S1   List of light colour characters with their modalities. All modalities are depicted in Figure 4. The coding for the different modalities is mentioned in brackets for the 23 light coloration characters. The coding for the 23 dark characters is equivalent to one minus the corresponding light coloration character.

Acknowledgments

We thank Adrien Quiles and Claire Capdevielle Dulac for their help with microsatellite analysis, and Nelly Gidaszewski, Mathieu Joron, Alexandre Roulin and the two anonymous reviewers for their helpful comments on the subject and the manuscript. We are grateful to Alain Roques (INRA, Orléans), Agnès Rortais (CNRS, Gif-sur-Yvette), but also the LIPI authorities from Indonesia, especially our counterparts Yayuk Suhardjono and Oscar Effendy, NARC from Nepal, Truong Quang Tam (Institute of Tropical Biology, Vietnam) and the Pr. Xue-xin Chen (University of Zhejiang, China) for helping us to collect specimens in Asia. We are also grateful to Josef Gusenleitner, Jean Haxaire, Jean-Luc Renesson and Pierre Tripotin that shared their private collections, and Rémi Perrard for his help with posterior GPS coordinates of ancient sampling localities. Special thanks are due to curators Alain Drumont, Yvonnick Gérard, Kees van Achterberg, Fritz Gusenleitner and Gavin Broad who allowed the study in their respective museum collections. This study benefited of other public collections: the American Museum of Natural History (New York, USA), the University of Zhejiang (Hangzhou, China), the Museum Zoologicum Bogoriense (Bogor, Indonesia) and the Muséum National d'Histoire Naturelle (Paris, France).

Author Contributions

Conceived and designed the experiments: AP MA JFS CV. Performed the experiments: AP MA QR FM JT SB HN CV. Analyzed the data: AP MA RB MB CV. Contributed reagents/materials/analysis tools: AP MA QR FM JT SB HN JFS JMC CV. Wrote the paper: AP MA MB JFS JMC CV.

References

1. Poulton EB (1890) The colours of animals: their meaning and use, especially considered in the case of insects. D. Appleton. 360 p.
2. Gehara M, Summers K, Brown JL (2013) Population expansion, isolation and selection: novel insights on the evolution of color diversity in the strawberry poison frog. Evol Ecol 27: 797–824.
3. Lehtonen PK, Laaksonen T, Artemyev AV, Belskii E, Both C, et al. (2009) Geographic patterns of genetic differentiation and plumage colour variation are different in the pied flycatcher (*Ficedula hypoleuca*). Mol Ecol 18: 4463–4476.
4. Mallet J, Joron M (1999) Evolution of diversity in warning color and mimicry: polymorphisms, shifting balance, and speciation. Annu Rev Ecol Syst 30: 201–233.
5. Ruxton GD, Sherratt TN, Speed MP (2004) Avoiding attack: the evolutionary ecology of crypsis, warning signals, and mimicry. Oxford University Press Oxford. 249 p.
6. Stevens M, Ruxton GD (2012) Linking the evolution and form of warning coloration in nature. Proc R Soc Lond B Biol Sci 279: 417–426.
7. Wallace AR (1878) Tropical Nature and Other Essays. Kessinger Publishing, LLC. 356 p.
8. Kauppinen J, Mappes J (2003) Why are wasps so intimidating: field experiments on hunting dragonflies (Odonata: *Aeshna grandis*). Anim Behav 66: 505–511.
9. Haglund K, Hagen SB, Lampe HM (2006) Responses of domestic chicks (*Gallus gallus domesticus*) to multimodal aposematic signals. Behav Ecol 17: 392–398.
10. Yachi S, Higashi M (1998) The evolution of warning signals. Nature 394: 882–884.
11. Speed MP, Ruxton GD (2005) Aposematism: what should our starting point be? Proc R Soc Lond B Biol Sci 272: 431–438.
12. Joron M, Jiggins CD, Papanicolaou A, McMillan WO (2006) *Heliconius* wing patterns: an evo-devo model for understanding phenotypic diversity. Heredity 97: 157–167.
13. Wang IJ, Shaffer HB (2008) Rapid color evolution in an aposematic species: a phylogenetic analysis of color variation in the strikingly polymorphic strawberry Poison-Dart Frog. Evolution: 2742–2759.
14. Archer ME (1991) The number of species that can be recognized within the genus *Vespa* (Hym., Vespinae). Entomol Month Mag 127: 161–164.

15. van der Vecht J (1957) The Vespinae of the Indo-Malayan and Papuan areas (Hymenoptera, Vespidae). Zool Verhandel 34: 1–82.

16. MacLean B K, Chandler L, MacLean DB (1978) Phenotypic expression in the paper wasp *Polistes fuscatus* (Hymenoptera: Vespidae). Great Lakes Entomol 11: 105–116.

17. Nappi AJ, Vass E (1993) Melanogenesis and the generation of cytotoxic molecules during insect cellular immune reactions. Pigment Cell Res 6: 117–126.

18. Wilson K, Cotter SC, Reeson AF, Pell JK (2001) Melanism and disease resistance in insects. Ecol Lett 4: 637–649.

19. True JR (2003) Insect melanism: the molecules matter. Trends Ecol Evol 18: 640–647.

20. Roff DA, Fairbairn DJ (2013) The costs of being dark: the genetic basis of melanism and its association with fitness-related traits in the sand cricket. J Evol Biol 26: 1406–1416.

21. Lindstedt C, Lindström L, Mappes J (2009) Thermoregulation constrains effective warning signal expression. Evolution 63: 469–478.

22. du Buysson R (1905) Monographie des guêpes ou *Vespa*. Ann Soc Entomol Fr 72: 260–288.

23. Bequaert J (1930) On the generic and subgeneric divisions of the Vespinae (Hymenoptera). Bull Brooklyn Entomol Soc 25: 59–70.

24. Olson EC, Miller RI (1958) Morphological Integration. University of Chicago Press, Chicago, USA. 355 p.

25. Klingenberg CP (2013) Cranial integration and modularity: insights into evolution and development from morphometric data. Hystrix 24: 43–58.

26. Klingenberg CP (2008) Morphological integration and developmental modularity. Annu Rev Ecol Evol Syst 39: 115–132.

27. Antoniazza S, Burri R, Fumagalli L, Goudet J, Roulin A (2010) Local adaptation maintains clinal variation in melanin-based coloration of European Barn Owls (*Tyto alba*). Evolution 64: 1944–1954.

28. Sherratt TN (2008) The evolution of Müllerian mimicry. Naturwissenschaften 95: 681–695.

29. Archer ME (1994) Taxonomy, distribution and nesting biology of the *Vespa bicolor* group (Hym., Vespinae). Entomol Month Mag 130: 149–158.

30. Van der Vecht J (1959) Notes on Oriental Vespinae, including some species from China and Japan (Hymenoptera, Vespidae). Zool Mededel 13: 205–232.

31. Nguyen LTP, Saito F, Kojima J, Carpenter JM (2006) Vespidae of Viet Nam (Insecta: Hymenoptera) 2. Taxonomic Notes on Vespinae. Zoolog Sci 23: 95–104.

32. Perrard A, Pickett K, Villemant C, Kojima J, Carpenter J (2013) Phylogeny of hornets: a total evidence approach (Hymenoptera, Vespidae, Vespinae, *Vespa*). J Hym Res 32: 1–15.

33. Folmer O, Black M, Hoeh W, Lutz R, Vrijenhoek R (1994) DNA primers for amplification of mitochondrial cytochrome c oxidase subunit I from diverse metazoan invertebrates. Mol Mar Biol Biotechnol 3: 294–299.

34. Hajibabaei M, Ivanova NV, Ratnasingham S, Dooh RT, Kirk SL, et al. (2005) Critical factors for assembling a high volume of DNA barcodes. Philos Trans R Soc Lond B Biol Sci 360: 1959–1967.

35. Hall TA (1999) BioEdit: a user-friendly biological sequence alignment editor and analysis program for Windows 95/98/NT. Nucleic Acids Symp Ser 41: 95–98.

36. Arca M, Capdevielle-Dulac C, Villemant C, Mougel F, Arnold G, et al. (2011) Development of microsatellite markers for the yellow-legged Asian hornet, *Vespa velutina*, a major threat for European bees. Conserv Genet Resour 4: 1–4.

37. Daly D, Archer ME, Watts PC, Speed MP, Hughes MR, et al. (2002) Polymorphic microsatellite loci for eusocial wasps (Hymenoptera: Vespidae). Mol Ecol Notes 2: 273–275.

38. Arca M (2012) Caractérisation génétique et étude comportementale d'une espèce envahissante en France: *Vespa velutina* Lepeletier (Hymenoptera, Vespidae) Doctoral Thesis, Paris: Université Pierre et Marie Curie. 199p.

39. Bandelt HJ, Forster P, Röhl A (1999) Median-joining networks for inferring intraspecific phylogenies. Mol Biol Evol 16: 37–48.

40. Hordijk W, Gascuel O (2005) Improving the efficiency of SPR moves in phylogenetic tree search methods based on maximum likelihood. Bioinformatics 21: 4338–4347.

41. Saitou N, Nei M (1987) The neighbor-joining method: a new method for reconstructing phylogenetic trees. Mol Biol Evol 4: 406–425.

42. Chakraborty R, Jin L (1993) A unified approach to study hypervariable polymorphisms: statistical considerations of determining relatedness and population distances. In: Pena S, Jeffreys A, Epplen J, Chakraborty R, editors. DNA fingerprinting, current state of the science. Basel: Birkhauser. pp. 153–175.

43. Paradis E, Claude J, Strimmer K (2004) Ape: Analyses of phylogenetics and evolution in R language. Bioinformatics 20: 289–290.

44. Pritchard JK, Stephens M, Donnelly P (2000) Inference of population structure using multilocus genotype data. Genetics 155: 945–959.

45. Excoffier L, Laval G, Schneider S (2005) Arlequin (version 3.0): an integrated software package for population genetics data analysis. Evol Bioinform Online 1: 47–50.

46. Chevene F., Doléadec S, Chessel D (1994) A fuzzy coding approach for the analysis of long-term ecological data. Freshw Biol 31: 295–309.

47. Escoufier Y (1973) Le traitement des variables vectorielles. Biometrics 29: 751–760.

48. Robert P, Escoufier Y (1976) A unifying tool for linear multivariate statistical methods: The RV Coefficient. J R Stat Soc C Appl Stat 25: 257–265.

49. Josse J, Pagès J, Husson F (2008) Testing the significance of the RV coefficient. Comput Stat Data An 53: 82–91.

50. Husson F, Josse J, Lê S, Mazet J (2013) FactoMineR: Multivariate Exploratory Data Analysis and Data Mining with R, R package version 1.24. CRAN website. Available: http://CRAN.R-project.org/package = FactoMineR. Accessed 2013 September 18.

51. Kelley D (2009) oce: Analysis of Oceanographic data. R package version 0.1–80. CRAN website. Available: http://CRAN.R-project.org/package = oce. Accessed 2013 September 18.

52. Villemant C, Barbet-Massin M, Perrard A, Muller F, Gargominy O, et al. (2011) Predicting the invasion risk by the alien bee-hawking Yellow-legged hornet *Vespa velutina nigrithorax* across Europe and other continents with niche models. Biol Cons 144: 2142–2150.

53. Hijmans RJ, Cameron SE, Parra JL, Jones PG, Jarvis A (2005) Very high resolution interpolated climate surfaces for global land areas. Int J Climatol 25: 1965–1978.

54. Lingoes JC (1971) Some boundary conditions for a monotone analysis of symmetric matrices. Psychometrika 36: 195–203.

55. Warnes GR, Bolker B, Bonebakker L, Gentleman R, Huber W, et al. (2013) gplots: Various R programming tools for plotting data. R package version 2 113. CRAN website. Available: http://CRAN.R-project.org/package = gplots. Accessed 2013 September 18.

56. Chapuis E, Martin G, Goudet J (2008) Multivariate Qst–Fst comparisons: a neutrality test for the evolution of the G matrix in structured populations. Genetics 180: 2135–2149.

57. Leinonen T, McCairns RS, O'Hara RB, Merilä J (2013) Qst–Fst comparisons: evolutionary and ecological insights from genomic heterogeneity. Nat Rev Gen 14: 179–190.

58. Brommer JE (2011) Whither PST? The approximation of Qst by Pst in evolutionary and conservation biology. J Evol Biol 24: 1160–1168.

59. Evanno G, Regnaut S, Goudet J (2005) Detecting the number of clusters of individuals using the software STRUCTURE: a simulation study. Mol Ecol 14: 2611–2620.

60. Waples RS, Gaggiotti O (2006) INVITED REVIEW: What is a population? An empirical evaluation of some genetic methods for identifying the number of gene pools and their degree of connectivity. Mol Ecol 15: 1419–1439.

61. Walter MF, Black BC, Afshar G, Kermabon A-Y, Wright TR, et al. (1991) Temporal and spatial expression of the *yellow* gene in correlation with cuticle formation and DOPA decarboxylase activity in *Drosophila* development. Dev Biol 147: 32–45.

62. Wittkopp PJ, True JR, Carroll SB (2002) Reciprocal functions of the *Drosophila yellow* and *ebony* proteins in the development and evolution of pigment patterns. Development 129: 1849–1858.

63. Wittkopp PJ, Beldade P (2009) Development and evolution of insect pigmentation: genetic mechanisms and the potential consequences of pleiotropy. Semin Cell Dev Biol 20: 65–71.

64. Joron M, Frezal L, Jones RT, Chamberlain NL, Lee SF, et al. (2011) Chromosomal rearrangements maintain a polymorphic supergene controlling butterfly mimicry. Nature 477: 203–206.

65. Koch PB, Lorenz U, Brakefield PM (2000) Butterfly wing pattern mutants: developmental heterochrony and co-ordinately regulated phenotypes. Dev Genes Evol 210: 536–544.

66. Jeong S, Rokas A, Carroll SB (2006) Regulation of body pigmentation by the *Abdominal-B* Hox protein and its gain and loss in *Drosophila* evolution. Cell 125: 1387.

67. French V, Brakefield PM (1992) The development of eyespot patterns on butterfly wings: morphogen sources or sinks? Development 116: 103–109.

68. French V, Brakefield PM (1995) Eyespot development on butterfly wings: the focal signal. Dev Biol 168: 112–123.

69. Murray JD (2003) Mathematical Biology II: Spatial Models and Biomedical Applications. 2003. Springer-Verlag, New York. 814 p.

70. Koch PB, Nijhout HF (2002) The role of wing veins in colour pattern development in the butterfly *Papilio xuthus* (Lepidoptera: Papilionidae). Europ J Entomol 99: 67–72.

71. Reed RD, Gilbert LE (2004) Wing venation and *Distal-less* expression in *Heliconius* butterfly wing pattern development. Dev Genes Evol 214: 628–634.

72. Kopp A, Blackman RK, Duncan I (1999) Wingless, decapentaplegic and EGF receptor signaling pathways interact to specify dorso-ventral pattern in the adult abdomen of *Drosophila*. Development 126: 3495–3507.

73. Kopp A, Duncan I, Carroll SB (2000) Genetic control and evolution of sexually dimorphic characters in *Drosophila*. Nature 408: 553–559.

74. Kopp A, Duncan I (2002) Anteroposterior patterning in adult abdominal segments of *Drosophila*. Dev Biol 242: 15–30.

75. Nijhout HF (2010) Molecular and physiological basis of colour pattern formation. Adv In Insect Phys 38: 219–265.

76. Koch BP, Merk R, Reinhardt R, Weber P (2003) Localization of ecdysone receptor protein during colour pattern formation in wings of the butterfly *Precis coenia* (Lepidoptera: Nymphalidae) and co-expression with *Distal-less* protein. Dev Genes Evol 212: 571–584.

77. Wray GA (2007) The evolutionary significance of *cis*-regulatory mutations. Nat Rev Genet 8: 206–216.

78. Gompel N, Prud'homme B, Wittkopp PJ, Kassner VA, Carroll SB (2005) Chance caught on the wing: *cis*-regulatory evolution and the origin of pigment patterns in *Drosophila*. Nature 433: 481–487.

79. Matsuura M, Yamane S (1990)Biology of the vespine wasps. Springer Verlag. 323 p.

80. Nijhout HF (2003) Development and evolution of adaptive polyphenisms. Evol Dev 5: 9–18.

81. Green JP, Rose C, Field J (2012) The role of climatic factors in the expression of an intrasexual signal in the paper wasp *Polistes dominulus*. Ethology 118: 766–774.

82. Tibbetts EA (2010) The condition dependence and heritability of signaling and nonsignaling color traits in paper wasps. Am Nat 175: 495–503.

83. Riabinin K, Kozhevnikov M, Ishay JS (2004) Ventilating activity at the hornet nest entrance. J Ethol 22: 49–53.

84. Jones JC, Oldroyd BP (2006) Nest thermoregulation in social insects. Adv In Insect Phys 33: 153–191.

85. Villemant C, Haxaire J, Streito JC (2006) Premier bilan de l'invasion de *Vespa velutina* Lepeletier en France (Hymenoptera, Vespidae). Bull Soc Entomol Fr 111: 535–538.

86. Perrard A, Muller F, Rome Q, Villemant C (2011) Observations sur le Frelon asiatique à pattes jaunes, *Vespa velutina* Lepeletier, 1836 (Hymenoptera, Vespidae). Bull Soc Entomol Fr 116: 159–164.

87. Wittkopp PJ, Williams BL, Selegue JE, Carroll SB (2003) *Drosophila* pigmentation evolution: divergent genotypes underlying convergent phenotypes. Proc Natl Acad Sci 100: 1808–1813.

88. Pereboom JJM, Biesmeijer JC (2003) Thermal constraints for stingless bee foragers: the importance of body size and coloration. Oecologia 137: 42–50.

89. Parkash R, Singh S, Ramniwas S (2009) Seasonal changes in humidity level in the tropics impact body color polymorphism and desiccation resistance in *Drosophila jambulina*—Evidence for melanism-desiccation hypothesis. J Insect Physiol 55: 358–368.

90. Parkash R, Chahal J, Sharma V, Dev K (2012) Adaptive associations between total body color dimorphism and climatic stress-related traits in a stenothermal circumtropical *Drosophila* species. Insect Sci 19: 247–262.

91. Williams P (2007) The distribution of bumblebee colour patterns worldwide: possible significance for thermoregulation, crypsis, and warning mimicry. Biol J Linn Soc Lond 92: 97–118.

92. Mostler G (1935) Beobachtungen zur Frage der Wespenmimikry. Z Morphol Oekol Tiere 29: 381–454.

93. Gilbert F (2005) The Evolution of Imperfect Mimicry. In: Fellowes M, Holloway G, Rolff J, editors. Insect evolutionary ecology. Wallingford: CABI Publishing. pp. 231–288.

94. Théry M, Gomez D (2010) Insect colours and visual appearance in the eyes of their predators. Adv In Insect Phys 38: 267–353.

95. Stelzer RJ, Raine NE, Schmitt KD, Chittka L (2010) Effects of aposematic coloration on predation risk in bumblebees? A comparison between differently coloured populations, with consideration of the ultraviolet. J Zool 282: 75–83.

96. Vidal-Cordero JM, Moreno-Rueda G, López-Orta A, Marfil-Daza C, Ros-Santaella JL, et al. (2012) Brighter-colored paper wasps (*Polistes dominula*) have larger poison glands. Front Zool 9: 1–5.

97. Schuler W, Hesse E (1985) On the function of warning coloration: a black and yellow pattern inhibits prey-attack by naive domestic chicks. Behav Ecol Sociobiol 16: 249–255.

98. Joron M, Wynne IR, Lamas G, Mallet J (1999) Variable selection and the coexistence of multiple mimetic forms of the butterfly *Heliconius numata*. Evol Ecol 13: 721–754.

99. Plowright RC, Owen RE (1980) The evolutionary significance of bumble bee color patterns: a mimetic interpretation. Evolution 34: 622–637.

100. Hines HM, Williams PH (2012) Mimetic colour pattern evolution in the highly polymorphic *Bombus trifasciatus* (Hymenoptera: Apidae) species complex and its comimics. Zool J Linn Soc 166: 805–826.

101. O'Donnell S, Joyce F J (1999) Dual mimicry in the dimorphic eusocial wasp *Mischocyttarus mastigophorus* Richards (Hymenoptera: Vespidae). Biol J Linn Soc Lond 66: 501–514.

7

The Evolution of Vp1 Gene in Enterovirus C Species Sub-Group That Contains Types CVA-21, CVA-24, EV-C95, EV-C96 and EV-C99

Teemu Smura[1,2], Soile Blomqvist[1], Tytti Vuorinen[3], Olga Ivanova[4], Elena Samoilovich[5], Haider Al-Hello[1], Carita Savolainen-Kopra[1], Tapani Hovi[1], Merja Roivainen[1]

1 Department of Infectious Disease Surveillance and Control, National Institute for Health and Welfare (THL), Helsinki, Finland, 2 Department of Virology, Haartman Institute, Faculty of Medicine, University of Helsinki, Helsinki, Finland, 3 Department of Virology, University of Turku, Turku, Finland, 4 M.P. Chumakov Institute of Poliomyelitis and Viral Encephalitides, Russian Academy of Medical Sciences, Moscow, Russia, 5 Republican Research and Practical Center for Epidemiology and Microbiology, Minsk, Republic of Belarus

Abstract

Genus *Enterovirus* (Family *Picornaviridae*,) consists of twelve species divided into genetically diverse types by their capsid protein VP1 coding sequences. Each enterovirus type can further be divided into intra-typic sub-clusters (genotypes). The aim of this study was to elucidate what leads to the emergence of novel enterovirus clades (types and genotypes). An evolutionary analysis was conducted for a sub-group of *Enterovirus C* species that contains types Coxsackievirus A21 (CVA-21), CVA-24, Enterovirus C95 (EV-C95), EV-C96 and EV-C99. VP1 gene datasets were collected and analysed to infer the phylogeny, rate of evolution, nucleotide and amino acid substitution patterns and signs of selection. In VP1 coding gene, high intra-typic sequence diversities and robust grouping into distinct genotypes within each type were detected. Within each type the majority of nucleotide substitutions were synonymous and the non-synonymous substitutions tended to cluster in distinct highly polymorphic sites. Signs of positive selection were detected in some of these highly polymorphic sites, while strong negative selection was indicated in most of the codons. Despite robust clustering to intra-typic genotypes, only few genotype-specific 'signature' amino acids were detected. In contrast, when *different* enterovirus types were compared, there was a clear tendency towards fixation of type-specific 'signature' amino acids. The results suggest that permanent fixation of type-specific amino acids is a hallmark associated with evolution of different enterovirus types, whereas neutral evolution and/or (frequency-dependent) positive selection in few highly polymorphic amino acid sites are the dominant forms of evolution when strains *within* an enterovirus type are compared.

Editor: Patrick C. Y. Woo, The University of Hong Kong, Hong Kong

Funding: This study was supported by Academy of Finland. The funders had no role in study design, data collection and analysis, decision to publish, or preparation of the manuscript.

Competing Interests: The authors have declared that no competing interests exist.

* E-mail: Teemu.Smura@thl.fi

Introduction

Enteroviruses (genus *Enterovirus*, family *Picornaviridae*) are small non-enveloped positive strand RNA viruses with icosahedral capsid symmetry. Enteroviruses are classified to twelve species, *Enterovirus A* to *H*, *J* and *Rhinovirus A* to *C* [1]. Seven of the species, *Enterovirus A* to *D* (formerly named *Human enterovirus A* to *D*) and *Rhinovirus A* to *C* (formerly named *Human rhinovirus A* to *C*) are known to infect humans. Enteroviruses are associated with a variety of clinical diseases, such as aseptic meningitis, encephalitis, paralytic disease, respiratory infections, and acute haemorrhagic conjunctivitis (AHC) although most enterovirus infections are sub-clinical [2]. Enteroviruses are transmitted mostly via fecal-oral and respiratory routes.

Enterovirus genome contains approximately 7500 nucleotides. The genome consists of a single open reading frame (ORF) that is flanked by 5′ end and 3′ end untranslated regions (5′UTR and 3′UTR). The ORF is translated to a single polypeptide that is autocatalytically cleaved to P1, P2 and P3 polyproteins. The P1

polyprotein is further cleaved to capsid proteins VP4 to VP1, whereas P2 and P3 are cleaved to non-structural proteins 2A–2C and 3A–3D, respectively.

On the basis of the capsid coding P1 region, enteroviruses form genetically highly diverse types that are equivalent to serotypes defined by antigenic properties [3,4]. Currently, enterovirus types are defined by the sequence divergences in the capsid protein VP1 coding region. Members of the same type have more than 75% nucleotide (nt) and more than 88% amino acid (aa) similarities in the VP1 region [3,5]. Correspondingly, the strains that have less than 70% nucleotide and 85% amino acid similarities are classified to different types. Empirical observations suggest that, in general, the pair wise sequence similarities between enterovirus strains do not form a continuum across the type designations, but rather the members of different types are clearly separated [3]. However, divergent strains that have pairwise nt/aa similarities in the 'grey-zone' of current typing (i.e. 70–75% nt and/or 85–88% aa similarity), such as some CVA-24 and EV-C99 strains, are detected regularly [5].

Although the current typing standard is of great practical value, little is known about the evolutionary processes behind the observed sequence similarity patterns. Likewise, although the implementation of molecular methods for genetic enterovirus typing has led to the discovery of a large number of novel enterovirus types (reviewed in [6]), much less is known about evolutionary processes behind the formation of the enterovirus types.

The aim of this study was to elucidate the evolutionary processes behind the origin and emergence of novel enterovirus clades (types and genotypes). The clearly resolved hierarchical phylogeny of EV-C species [5,7,8] provides an opportunity to study the evolutionary patterns that have led to the emergence of novel enterovirus clades. In this study, an evolutionary analysis was conducted for a sub-group of EV-C species that contains types CVA-21, CVA-24, EV-C95, EV-C96 and EV-C99. Of these enterovirus types, CVA-21 induces mainly mild respiratory infections [9,10], whereas EV-C96, EV-C99 and some of the (prototype-like) CVA-24 strains have been isolated from faecal samples of healthy individuals and patients with paralytic disease [5,8,11–20]. A distinct lineage of CVA-24, so-called variant strains (CVA-24v), induce acute haemorrhagic conjunctivitis (AHC) [21] and are transmitted via direct or indirect contact with eye-secretions [2]. EV-C95 is a recently discovered enterovirus type (Helene Norder unpublished [1]. So far, only two strains of this type have been characterized [17]. Parallel datasets of VP1 sequences were collected and analysed to infer the phylogenetic structure, rate of evolution, nucleotide and amino acid substitution patterns and signs for selection in the VP1 coding region.

Material and Methods

Ethics statement

The virus samples were collected with the consent of The Institutional Review Board of National Institute for Health and Welfare (THL), the Ethics Committee of M.P. Chumakov Institute of Poliomyelitis and Viral Encephalitides of Russian Academy of Medical Sciences and the Ethical Committee of the Minsk Municipality and analyzed anonymously. The sewage samples were obtained from Viikki wastewater treatment plant in Helsinki, Finland. No specific permission was required for the surveillance for enteroviruses from sewage. The field studies did not involve endangered or protected species.

Viruses

The virus strains isolated in this study are listed in Table 1. Virus strains were isolated from sewage samples collected during environmental surveillance for polioviruses in Finland using a two-phase concentration method [22]. In addition, clinical enterovirus isolates sent to the national enterovirus reference laboratory (National Institute for Health and Welfare, THL) from other Finnish laboratories were included in the study [23]. The rest of the virus strains were received as untypeable nonpolio enteroviruses (NPEV) from a number of National Polio Laboratories of the WHO Polio Laboratory Network supporting the Global Poliovirus Eradication Initiative. Human rhabdomyosarcoma (RD), human colorectal adenocarcinoma (CaCo-2), human cervical carcinoma (HeLa) and green monkey kidney (GMK) cell lines were used for virus isolation.

Partial VP1 RT-PCR and sequencing

Viral RNA was extracted from infected cell cultures with RNeasy Total RNA kit (Qiagen, Hilden, Germany) or E.Z.N.A. Total RNA Kit Omega (Bio-Tek Inc., Doraville, GA, USA)

according to the manufacturer's instructions. RT-PCR was carried out as described previously [24] using primers 292 and 222. PCR amplicons were purified with the QIAquick gel extraction kit (Qiagen). Sequencing reactions with BigDye Terminator cycle sequencing ready reaction kit v3.1 (Life Technologies, Carlsbad, CA, USA) and sequencing with ABI3730 Automatic DNA Sequencer (Life Technologies) were performed by the Institute for Molecular Medicine Finland (FIMM) Sequencing Laboratory. The electropherograms were analysed using Geneious Pro 6.0 software (Biomatters Ltd, Auckland, New Zealand, http://www.geneious.com).

Sequence dataset collection

Two parallel datasets were constructed from the sequences. The first dataset contained partial VP1 sequences characterized in this study (N = 72) (Table 1), the overlapping CVA-21, CVA-24, EV-C95, EV-C96 and EV-C99 sequences retrieved from the GenBank and the overlapping sequences of the prototype strains of other EV-C types. Altogether this dataset consisted of 264 sequences. The consensus alignment of partial VP1 dataset consisted of 343 nucleotide sites. The second dataset contained the strain CVA-24-FIN05-1-7920 and all complete CVA-21, CVA-24, EV-C95, EV-C96 and EV-C99 VP1 sequences found in the GenBank (27.2.2013). The complete VP1 dataset consisted of 97 strains. The consensus alignment of this dataset contained 927 nucleotides. To relieve computational demands and correct data collection bias, only one representative sequence from the clusters that shared more than 99% similarity with each other was included in the analysis.

Sequence analysis

The sequences were aligned using ClustalW algorithm (for codons) implemented in MEGA version 5.05 [25] followed by manual refinement. Phylogenetic trees were constructed using neighbor-joining (NJ) method implemented in MEGA version 5.05 and Bayesian Monte Carlo Markov Chain (MCMC) method implemented in BEAST version 1.7.4 [26]. For NJ-trees, bootstrap resampling with 1000 replicates was conducted. Various substitution models including Tamura-Nei (TN93) [27] and general time reversible (GTR) [28] were utilized.

The rates of evolution and divergence times of the virus lineages were estimated using Bayesian MCMC method implemented in BEAST version 1.7.4 [26]. Different molecular clock and demographic models were compared by calculating the marginal likelihoods of the data [29]. Bayes factors (BF) were calculated for each pair of models with Tracer 1.5 [30]. The uncorrelated relaxed molecular clock approach (log-normal distribution fitted the data significantly better than strict clock or the relaxed molecular clock approach with exponential distribution (log BF>400). The Bayesian skyline demographic model fitted the data significantly better than constant or exponential models (log BF>400). Therefore, the analyses were performed using a relaxed molecular clock model (the uncorrelated log-normal distributed model) [31], GTR model of substitution and Bayesian skyline demographic model. The Bayesian analyses were run for 100 million states and sampled every 1000 states. The analyses were carried out on the Bioportal server, University of Oslo (www.bioportal.uio.no) [32] and in CSC – IT Center for Science Ltd. (Espoo, Finland). The analyses were run in duplicate and the log-files were combined to increase the effective sample size. Posterior probabilities were calculated with a burn-in of 10 million states and checked for convergence using Tracer version 1.5. [30].

Codon-specific synonymous (dS) to non-synonymous (dN) rates were estimated using single likelihood ancestor counting (SLAC),

Table 1. The strains sequenced in this study.

Strain	Collection Date	Country	Sample Type
CVA-21-EST06-E1783-20171_28-Feb-2006	28-Feb-2006	Estonia	Sewage
CVA-21-FIN03-862-36252	2003	Finland	Stool
CVA-21-FIN06-E1906-28163_27-Sep-2006	27-Sep-2006	Finland	Sewage
CVA-21-FIN06-E2006-32088_19-Dec-2006	19-Dec-2006	Finland	Sewage
CVA-21-FIN06-EV06-34A-30796_14-Nov-2006	14-Nov-2006	Finland	Stool
CVA-21-LVA03-756_5-Mar-2003	5-Mar-2003	Latvia	
CVA-21-RUS01-15341_4-Jul-2001	4-Jul-2001	Russia	Stool
CVA-21-SVK05-E1571_Skalica_17-Feb-2005	17-Feb-2005	Slovak Republic	Sewage
CVA-21-SVK07-E2058_ Dunajska Streda_10-Jan-2007	10-Jan-2007	Slovak Republic	Sewage
CVA-24-AUT05-1600_12-Apr-2005	12-Apr-2005	Austria	
CVA-24-FIN02-671	2002	Finland	Sewage
CVA-24-FIN04-EV04-27A-2124	2004	Finland (India)	Stool
CVA-24-FIN04-EV04-34A-3787	2004	Finland (India)	Stool
CVA-24-FIN05-1-7920-2005	2005	Finland (China)	Stool
CVA-24-FIN05-1663-13794_18-Oct-2005	18-Oct-2005	Finland	Stool
CVA-24-FIN05-EV05-17A_31-Oct-2005	31-Oct-2005	Finland (China)	Stool
CVA-24-FIN06-1869-25202_6-Sep-2006	6-Sep-2006	Finland	Stool
CVA-24-FIN06-EV06-32A-30807_24-Oct-2006	24-Oct-2006	Finland	Stool
CVA-24-FIN06-EV07-2A-29392_18-Dec-2006	18-Dec-2006	Finland	Stool
CVA-24-FIN07-EV07-12A-33320_15-May-2007	15-May-2007	Finland	Stool
CVA-24-FIN07-EV07-13A-33425_28-May-2007	28-May-2007	Finland	Stool
CVA-24-FIN07-EV07-25A-37012_22-Aug-2007	22-Aug-2007	Finland	Stool
CVA-24-FIN07-EV07-26A-37474_30-Aug-2007	30-Aug-2007	Finland	Stool
CVA-24-FIN07-EV07-42A-51242_2007	2007	Finland	Stool
CVA-24-FIN07-EV08-1A-51415_18-Dec-2007	18-Dec-2007	Finland	Stool
CVA-24-FIN08-EV08-28A-59927_10-Oct-2008	10-Oct-2008	Finland	Stool
CVA-24-FIN08-EV08-3A-51551_30-Jan-2008	30-Jan-2008	Finland	Stool
CVA-24-FIN09-2996-72114_20-May-2009	20-May-2009	Finland (India)	Stool
CVA-24-FIN09-3137-78084_23-Oct-2009	23-Oct-2009	Finland	Stool
CVA-24-FIN09-3140-78104_22-Oct-2009	22-Oct-2009	Finland	Stool
CVA-24-FIN09-EV09-11B-70968_14-May-2009	14-May-2009	Finland	Stool
CVA-24-FIN09-EV09-12B-70971_5-May-2009	5-May-2009	Finland	Stool
CVA-24-FIN09-EV09-14A-72174_15-Jun-2009	15-Jun-2009	Finland	Stool
CVA-24-FIN09-EV09-7A-70636_27-Feb-2009	27-Feb-2009	Finland	Stool
CVA-24-FIN10-E3319-A341_23-Mar-2010	23-Mar-2010	Finland	Sewage
CVA-24-FIN10-EV10-23B-91082_30-Aug-2010	30-Aug-2010	Finland	Stool
CVA-24-FIN10-EV10-2B_2-Mar-2010	2-Mar-2010	Finland	Stool
CVA-24-LVA03-757_19-Jun-2003	19-Jun-2003	Latvia	
CVA-24-RUS-00-14038_21-Nov-2000	21-Nov-2000	Russia	Stool
CVA-24-RUS-01-14455_14-Feb-2001	14-Feb-2001	Russia	Stool
CVA-24-RUS-TKM01-14868_10-Mar-2001	10-Mar-2001	Russia (Turkmenistan)	Stool
CVA-24-RUS-KGZ01-15071_18-Jun-2001	18-Jun-2001	Russia (Kyrgyzstan)	Stool
CVA-24-RUS-UZB01-15213_20-Jun-2001	20-Jun-2001	Russia (Uzbekistan)	Stool
CVA-24-RUS-TKM01-15327_23-Jun-2001	23-Jun-2001	Russia (Turkmenistan)	Stool
CVA-24-RUS-KGZ01-15876_20-Jul-2001	20-Jul-2001	Russia (Kyrgyzstan)	Stool
CVA-24v-FIN07-EV07-20A-34537_15-Jul-2007	15-Jul-2007	Finland	Conjunctival secretion
CVA-24v-FIN07-EV07-22B_13-Aug-2007	13-Aug-2007	Finland	Stool
EV-C96-FIN08-EV08-10A-55557_14-May-2008	14-May-2008	Finland	Stool
EV-C96-FIN09-2983-70820_28-May-2009	28-May-2009	Finland	Stool

Table 1. Cont.

Strain	Collection Date	Country	Sample Type
EV-C96-FIN09-EV09-13B_5-May-2009	5-May-2009	Finland	Stool
EV-C96-FIN09-EV09-9A-70153_24-Mar-2009	24-Mar-2009	Finland	Stool
EV-C96-FIN10-EV10-3A-84622_27-Apr-2010	27-Apr-2010	Finland	Stool
EV-C96-FIN12-EV12-11B_9-May-2012	9-May-2012	Finland	Stool
EV-C96-FIN06-EV06-31B-30779_6-Nov-2006	6-Nov-2006	Finland	Stool
EV-C99-FIN07-EV07-24A-37056_17-Aug-2007	17-Aug-2007	Finland	Stool
EV-C99-FIN07-EV08-2A-51416_20-Dec-2007	20-Dec-2007	Finland	Stool
EV-C99-FIN08-EV08-19A-56305_11-Aug-2008	11-Aug-2008	Finland	Stool
EV-C99-FIN08-EV08-7A-55125_7-Apr-2008	7-Apr-2008	Finland	Stool
EV-C99-RUS-TKM00-13831_8-Sep-2000	8-Sep-2000	Russia (Turkmenistan)	Stool
EV-C99-SVK03-23-20226_7-Feb-2003	7-Feb-2003	Slovak Republic	
EV-C99-SVK04-E1152-44722_19-May-2004	19-May-2004	Slovak Republic	Sewage
EV-C99-FIN09-EV09-15A-72176_22-Jun-2009	22-Jun-2009	Finland	Stool
EV-C99-FIN11-4266-10449_6-Oct-2011	6-Oct-2011	Finland	Stool
EV-C99-FIN09-2991-70979_2-Jun-2009	2-Jun-2009	Finland (Nigeria)	Stool
EV-C99-BLR00-32864-2000	2000	Republic of Belarus	Stool
EV-C99-BLR00-32878-2000	2000	Republic of Belarus	Stool
EV-C99-BLR00-32881-2000	2000	Republic of Belarus	Stool
EV-C99-BLR00-32887-2000	2000	Republic of Belarus	Stool
EV-C99-BLR00-33291-2000	2000	Republic of Belarus	Stool
EV-C99-BLR00-33305-2000	2000	Republic of Belarus	Stool
EV-C99-BLR00-33405-2000	2000	Republic of Belarus	Stool
EV-C99-BLR00-33483-2000	2000	Republic of Belarus	Stool

fixed effects likelihood (FEL), [33] and FUBAR [34] methods implemented in Datamonkey website (www.datamonkey.org) [35]. For this analysis, the nucleotide substitution model was selected using model test implemented in the Datamonkey website and the phylogenetic trees were reconstructed using the Neighbour-Joining method.

The signatures of directional selection were sought using McDonald–Kreitman test (MK-test) [36] implemented in DnaSP v.5.10 [37] and modified MK-tests [38,39] implemented in The standard and generalized MKT website (http://mkt.uab.es/mkt/) [38] and Adapt-A-Rate v.1.0 software [40]. The MK-test compares the ratio of fixed non-synonymous to synonymous differences between two predefined groups and the ratio of polymorphic non-synonymous to synonymous differences and assumes that under neutral evolution these ratios should be equal. In the modification by Egea et al., [38] the estimated number of mutations instead of the number of sites in each class are counted. The modification by Bhatt et al., [39] utilizes proportional counting algorithm in which proportional fixation, polymorphic, silent and replacement "site scores" are utilized instead of unambiguously assigned numbers of sites in each class. The modified MK-tests were run independently either including or excluding potentially mildly deleterious low-frequency variants (with <0.05 frequency). The modified test by Bhatt et al., [39] was also run using polymorphic (neutral) classes with frequencies 0-0.50 [41] and 0.05-0.75 (i.e. excluding both potentially mildly deleterious low-frequency variants and highly polymorphic sites that are potentially under frequency-dependent antigenic selection).

The locations of amino acid substitutions were estimated with 3-dimensional structure model of Coxsackievirus A21 pentamer (PDB ID: 1Z7Z) [42] using Jmol [43].

GenBank accession numbers

The GenBank accession numbers for the sequenced strains are KF128985 - KF129056.

Results

Phylogeny and sequence diversities of partial vp1 sequences

To gain insight into the evolution and epidemiology of enterovirus VP1 coding sequences, partial VP1 coding regions of six EV-C96 strains, 19 EV-C99 strains, nine CVA-21 strains, 36 CVA-24 strains and two CVA-24v strains that were isolated during poliovirus surveillance were sequenced and subjected to phylogenetic analysis together with overlapping sequences retrieved from the GenBank (search 27.2.2013).

Consistently with the previous studies [5,7,44,45], in the VP1 coding region, all of the EV-C types clustered into three sub-groups, which were designated here as A (CVA-1, CVA-19, CVA-22, EV-C104, EV-C105, EV-109, EV-C116, EV-C117 and EV-C118), B (EV-C95, EV-C96, EV-C99, CVA-21 and CVA-24) and C (CVA-13, CVA-17, CVA-20, EV-C102, PV-1, PV-2 and PV-3) (Fig. 1), with the branching order of sub-group A diverging first from the common ancestor of sub-groups B and C. CVA-11 grouped together with the strains of sub-groups B or C depending on the method of phylogenetic tree construction.

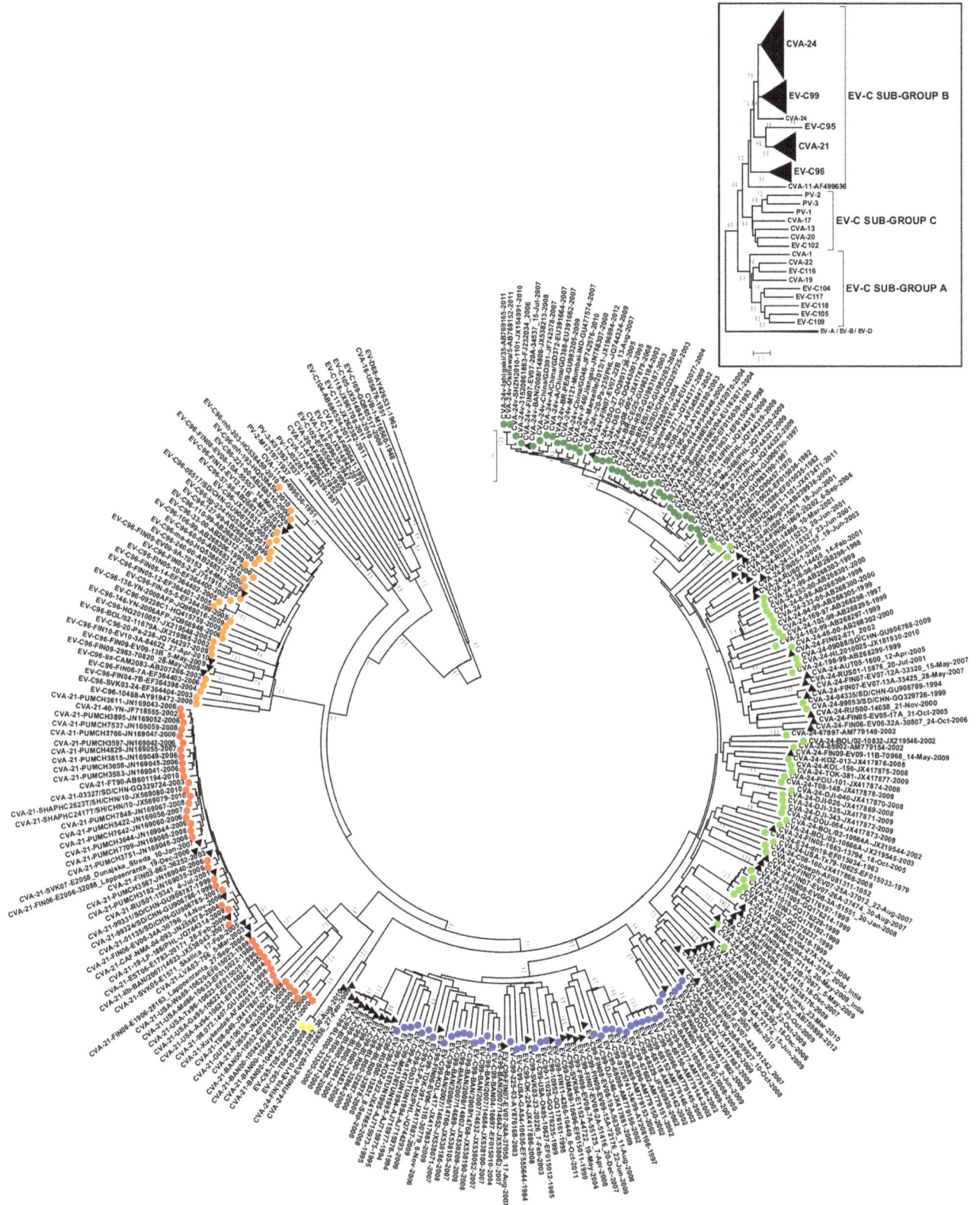

Figure 1. The phylogeny of partial VP1 sequences of EV-C species, including strains isolated in this study. The phylogenetic tree was constructed from partial VP1 coding region (consensus alignment 343 nucleotides) of CVA-21 (red), CVA-24 (green), EV-C95 (yellow), EV-C96 (orange) and EV-C99 (blue) strains and the prototype strains of other EV-C types. The strains sequenced in this study are indicated with black triangles. The tree was constructed using the Neighbour-Joining method and the Tamura-Nei substitution model. The bootstrap support values were calculated for 1000 replicates. The bootstrap support values >70 are shown. The inset represents the same tree with EV-C subgroups A to C identified.

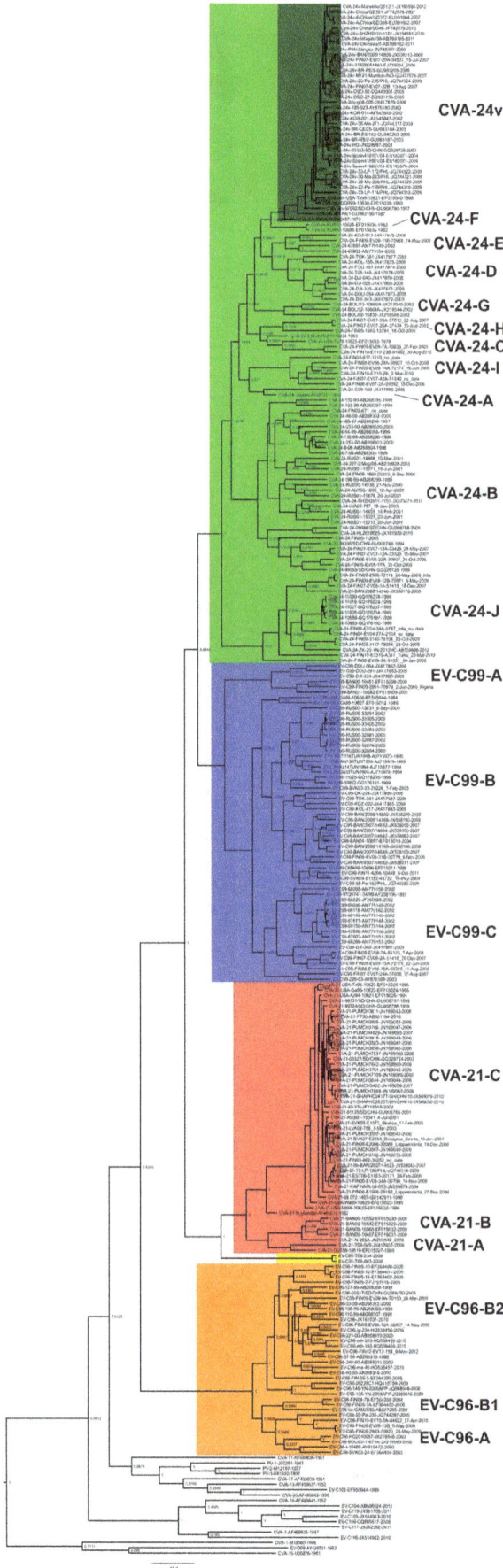

Figure 2. Maximum clade credibility tree of partial VP1 sequences of EV-C species, including strains isolated in this study. The maximum clade credibility tree was constructed from partial VP1 coding region (consensus alignment 343 nucleotides) of CVA-21 (red), CVA-24 (green), EV-C95 (yellow), EV-C96 (orange) and EV-C99 (blue) strains and the prototype strains of other EV-C types. The phylogenetic tree was constructed using Bayesian MCMC method with GTR model of substitution and Bayesian skyline demographic model. Posterior probabilities are shown in each node.

Within EV-C sub-group B, the phylogenetic analysis suggested well-supported hierarchical branching orders. EV-C96 formed an outgroup to the EV-C95/CVA-21/CVA-24/EV-C99 cluster, suggesting that EV-96 has diverged first from a common ancestor of EV-C sub-group B. This divergence has been followed by the branching of the ancestor of EV-C95/CVA-21 from the ancestor of CVA-24/EV-C99 and finally splitting of the ancestors of EV-C95/CVA-21 and CVA-24/EV-C99 into distinct types. Each of the types in EV-C sub-group B were further divided into intra-typic sub-clusters (or genotypes) with robust (i.e. >70) bootstrap support (Fig 1) or with high posterior probability (Fig 2). These genotypes are designated with upper case letters in Figure 2.

Generally similar branching patterns were observed using both NJ (Fig 1.) and the Bayesian MCMC methods (Fig 2). However, there was slight variation in the intra-typic sub-clustering pattern between the methods. In the NJ-tree, the non-AHC-causing CVA-24 strains formed a loose paraphyletic group with low bootstrap support (Fig 1), whereas in MCC tree, all CVA-24 strains formed a monophyletic group with high posterior probability (Fig 2). Many of the intra-typic clusters of EV-C96 and CVA-24 were detected in the MCC-tree (Fig 2) but not in the NJ-tree (Fig. 2). These tentative genotypes were designated as CVA-24-A to J and EV-C96 A to C, respectively (Figure 2). The AHC-causing variant strains of CVA-24 (CVA-24v) formed a monophyletic sub-cluster with both methods. All sub-clusters of CVA-21 were highly supported by both methods. However, a common ancestor for the genotypes CVA-21-A and –B was suggested by the Bayesian method only. For EV-C99 both methods suggested similar branching pattern. In addition, both methods suggested a common ancestor for the previously designated EV-C99 genotypes B and C [17,46].

Representatives of all EV-C96 and EV-C99 intra-typic genotypes and nearly all CVA-24 genotypes (excluding clusters genotypes A, F and G) were detected in this study. For CVA-21, all the strains sequenced in this study clustered to genotype C. Most of the strains sequenced in this study showed no evidence of geographic or temporal clustering. However, a distinct lineage of EV-C99 was detected in December 2007, April 2008, August 2008 and June 2009 in Finland (Fig 3a). Likewise, a distinct lineage of CVA-24 (designated as CVA-24-I in Fig 2) was detected repeatedly between years 2003 and 2010 (and every year between 2006 and 2010) in Finland (Fig 3b). Another CVA-24 lineage was detected in years 2007 and 2009 in Finland (Fig 3c). This lineage was related to a strain isolated in Bangladesh in 2008 [19]. In addition, eight strains of EV-C99 isolated in an orphanage in Minsk, Belarus during an oral poliovirus vaccination study [47] formed a separate cluster in the phylogenetic analysis (Fig 3d).

A wide variation in the nucleotide and amino acid sequences of the VP1-coding region was observed within each of the EV-C sub-group B types (Table 2) and, between some of the strains, the intra-typic similarities exceeded the 75/85% nucleotide/amino acid identity limit for classification into a single type. The inter-typic nucleotide and amino acid similarities between EV-C99 and CVA-24 were 68.0-81.3% and 78.8–94.9%, respectively. Thus, in

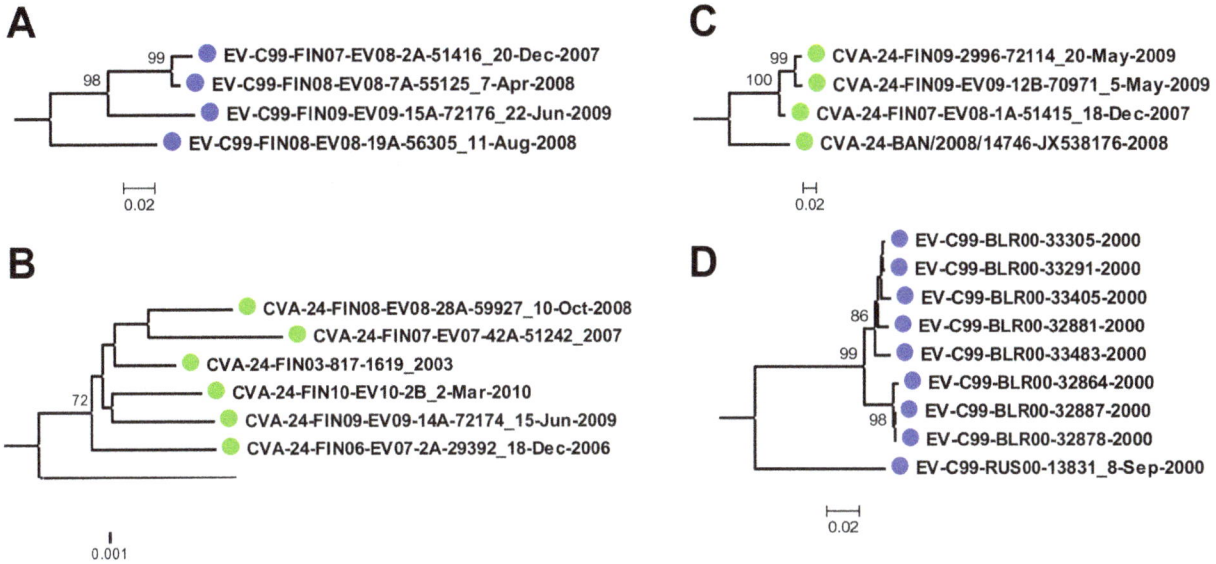

Figure 3. Sub-trees of strains that showed geographic clustering. The phylogenetic trees were constructed from partial VP1 coding region (343 nucleotides) using Neighbour-Joining method and the Tamura-Nei substitution model. Bootstrap values >70 are shown.

agreement with the previous studies [5], although both of these viruses formed clearly separated monophyletic groups, all pair wise comparisons between the strains of these types did not reach the limit of grouping into different types.

Codon-specific substitution patterns and selection

In order to elucidate the evolutionary forces leading to the observed phylogeny, a second dataset that contained the strain CVA-24-FIN05-1-7920 and all complete CVA-21, CVA-24, EV-C95, EV-C96 and EV-C99 VP1 sequences found from the GenBank (search 27.2.2013) was analysed. The major EV-C sub-groups A, B and C, as well as the branching order within EV-C sub-group B, were congruent with those observed in the partial VP1 data (Fig 4). Representatives of all intra-typic genotypes of designated using partial VP1 dataset, excluding CVA-24 geno-types G to J, were included in this dataset. The range of intra and inter-typic nucleotide and amino acid similarities were slightly lower in the complete VP1 dataset compared to those observed in the partial VP1 dataset (Table 2).

The VP1 consensus alignment of EV-C96, EV-C95, CVA-21, CVA-24 and EV-C99 contained 309 codons. In comparison to EV-C96, the VP1 amino-terminal domains of CVA-21, CVA-24, EV-C95 and EV-C99 contained deletions of 9, 4, 9 and 5-6 amino acids corresponding to sites 21-29 in the alignment. EV-C99 showed intra-typic differences in the length of the deletion, since the strains of EV-C99 group A had a deletion of 5 amino acids whereas the other strains of EV-C99 had deletions of 6 amino acids. Furthermore, the amino acid sequences of EV-C99 group A and group B/C were highly divergent in this region. The strains of EV-C99 group A also had another deletion of a single amino acid corresponding to site 152 in the consensus alignment. CVA-21 and EV-C95 had a deletion of two amino acids corresponding to sites 103–104 in the consensus alignment.

Table 2. Nucleotide (lower left) and amino acid (upper-right) similarities (%) between VP1 sequences of EV-C sub-group B serotypes.

		EV-C96	EV-C95	CVA-21	EV-C99	CVA-24
Partial VP1	**EV-C96**	**71.9/86.2**	62.6–73.5	60.4–70.8	70.0–82.8	70.0–82.9
	EV-C95	56.4–64.1	**98.0/99.3**	80.1–86.1	66.4–76.2	67.6–75.7
	CVA-21	53.3–66.3	66.5–71.7	**74.1/87.5**	65.7–76.7	66.4–78.6
	EV-C99	60.5–72.5	58.7–66.3	58.8–69.3	**71.1/84.0**	78.8–94.9
	CVA-24	59.9–73.8	59.5–68.0	61.3–72.5	68.0–81.3	**71.8/84.1**
Complete VP1	**EV-C96**	**75.3/89.3**	70.1–71.8	66.4–70.8	72.9–80.9	71.5–77.4
	EV-C95	61.5–64.5	**98.0/99.3**	82.9–84.2	73.7–77.5	73.8–77.2
	CVA-21	61.0–65.8	68.3–71.5	**76.1/91.9**	72.7–76.8	73.2–78.5
	EV-C99	65.1–71.5	64.2–67.5	64.2–69.9	**73.1/86.1**	80.4–88.7
	CVA-24	63.6–70.5	64.3–68.7	65.1–70.5	69.8–75.9	**75.0/88.2**

Minimum nucleotide/amino acid similarity within each type is shown in the diagonal (bolded).

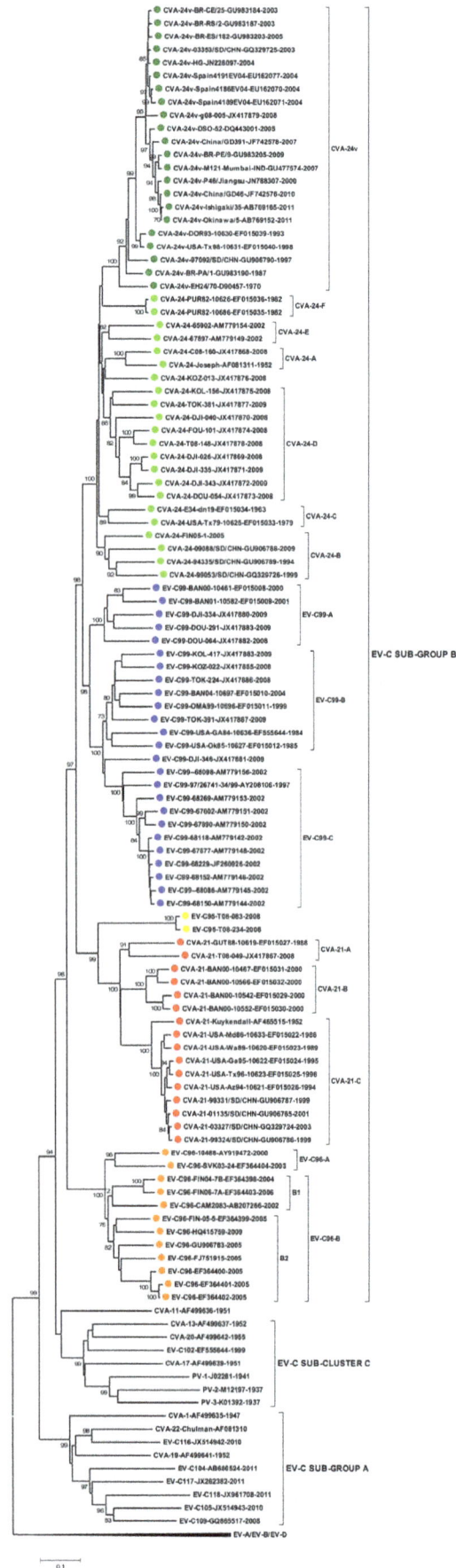

Figure 4. The phylogeny of EV-C species VP1 gene. The phylogenetic tree was constructed from complete VP1 coding region (consensus alignment 927 nucleotides) of those CVA-21 (red), CVA-24 (green), EV-C95 (yellow), EV-C96 (orange) and EV-C99 (blue) strains, of which a complete VP1 sequence was available (GenBank search 27.2.2013), and the prototype strains of other EV-C types. The tree was constructed using the Neighbour-Joining method and the Tamura-Nei substitution model. The bootstrap support values were calculated for 1000 replicates. The bootstrap support values >70 are shown.

Codon-specific non-synonymous to synonymous substitution frequencies were assessed using single likelihood ancestor counting (SLAC), fixed effects likelihood (FEL) and FUBAR methods available at the Datamonkey facility (www.datamonkey.org) [35]. The synonymous and non-synonymous substitution frequencies and the normalized dS-dN estimated using FUBAR method are shown in Fig 5. The analyses suggested that strong negative selection was occurring over most of the codons of the VP1-coding region in all of the types. However, elevated dN to dS ratios were detected for some codons by all of the methods, and statistically significant evidence for positive selection was suggested for codons 104 and 105 in EV-C96, codons 105 and 296 in EV-C99 and codon 105 in CVA-24 by at least one of the methods (Table 3). When the AHC-causing variant strains of CVA-24 (i.e. CVA-24v) and other CVA-24 strains were treated as separate groups, evidence for positive selection was detected in different sites (Fig 6). Elevated dN/dS was detected in codons 25 and 150 for CVA-24v and in codon 105 for non-AHC inducing CVA-24 (Table 3). Notably, CVA-24v strains showed strict conservation of alanine at site 105 (only one CVA-24v strain had aspartic acid in this site) whereas the non-AHC-causing strains of CVA-24 had high amino acid polymorphism (i.e. E, D, V, G, R, S, Q, N or T) in this site. Generally, the codons with signs of positive selection were highly polymorphic at amino acid level within the respective type (Table 3). Parallel evolution was also commonly detected in these sites (i.e. similar amino acid substitution had apparently occurred independently in different genetic lineages).

Within each type, most of the non-synonymous substitutions concentrated in distinct regions (Fig 5) that were most likely (on the basis of CVA-21 structure [42]) located at the structurally disordered N- and C-terminal regions and the loops between beta-sheets (Figure S1). Although the codons with high rate of non-synonymous substitutions overlapped partially between the types, at several regions, there were differences between types both in the location of polymorphic amino acid sites and the frequency of amino acid substitutions in these sites. The disparities in the locations of polymorphic sites between EV-C96, EV-C99 and non-AHC causing CVA-24 strains were subtle, polymorphic sites being often dislocated only by few amino acids, whereas more pronounced differences were detected between CVA-21, CVA-24v and other types. For example, the amino acid sites 223-230 were highly polymorphic among EV-C96, EV-C99 and non-AHC causing CVA-24 strains but conserved completely within CVA-21 (LEGENTDA) and CVA-24v (LKDETVS). Notably, several amino acid sites that were polymorphic among non-AHC causing CVA-24 sites were conserved among CVA-24v strains. Such sites included codons 22 (N, Q, K, S vs. Q), 103 (I, T, N, S, E vs. T), 172 (Q, T, I, R, K vs. R), 174 (A, N, T, K vs. T), 295 (N, D, S, A, G, E vs. D), 296 (S, A, E, L, T vs. S) and 304 (A, E, N, K, T, S, D vs. E).

Signs of directional selection between CVA-21, CVA-24, EV-C95, EV-C96 and EV-C99 and between the intra-typic genotypes within these enterovirus types were sought using McDonald-Kreitman test (MK-test) [36] and modifications of this test [38,39].

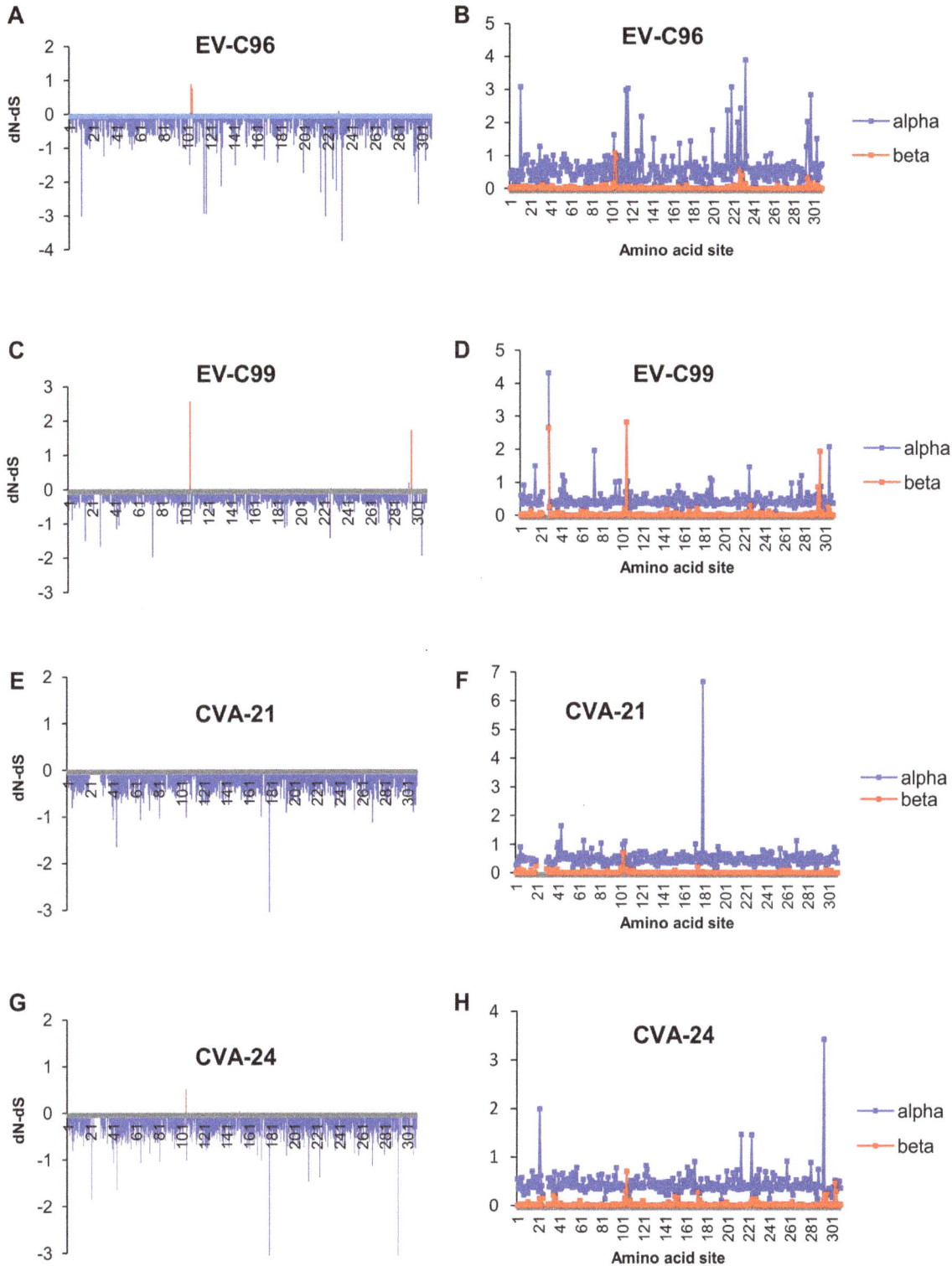

Figure 5. Codon-specific selection. The codon-specific differences between non-synonymous and synonymous rates (normalised dN-dS) (a,c,e,g) and the codon specific posterior distribution means of the synonymous (alpha) and non-synonymous (beta) substitution rates (b,d,f,h) estimated using FUBAR-method. Codons with statistically significant (posterior probability >0.9) evidence for positive selection are shown in red (a, c, e, g).

The results of the MK-tests are summarized in Table 4. The numbers of sites/substitutions in each MK-test class for different EV-types and intra-typic genotypes are shown in Tables S1–S6. The MK-tests suggested significantly increased proportion of non-synonymous mutations among the fixed sites for all inter-typic comparisons. Likewise, the comparison between EV-C99 groups A and B/C suggested statistically significant evidence of higher than expected number of fixed non-synonymous substitutions

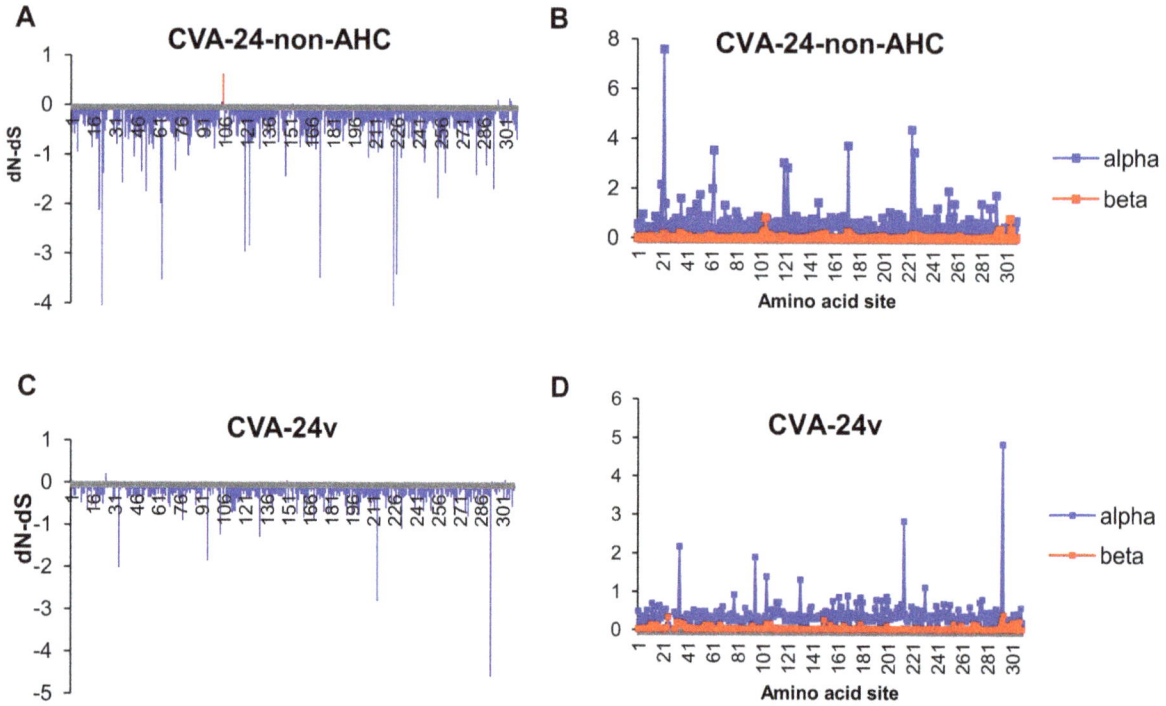

Figure 6. Codon-specific selection in CVA-24v and non-AHC-causing strains of CVA-24. The codon-specific differences between non-synonymous and synonymous rates (normalised dN-dS) (a,c) and the codon specific posterior distribution means of the synonymous (alpha) and non-synonymous (beta) substitution rates (b,d) estimated using FUBAR-method. Codons with statistically significant (posterior probability >0.9) evidence for positive selection are shown in red (a, c).

between these genotypes. In contrast, no consistent deviation from neutrality was observed between intra-typic genotypes of EV-C96, CVA-21 and CVA-24 (Table 4).

Evolutionary rates

The rates of evolution and times of most recent common ancestors (tMRCA) were estimated for both the complete and the partial VP1 datasets using Bayesian MCMC method with relaxed molecular clock (Table 5). The estimated rates of substitution were 1.972×10^{-3} (high-probability distribution [95% HPD] range $1.097–2.903 \times 10^{-3}$) and 2.839×10^{-3} (95% HPD range $2.177–3.509 \times 10^{-3}$) substitutions/site/year for complete and partial VP1

datasets, respectively. The corresponding estimated rates of evolution and tMRCAs for each type in EV-C sub-cluster B are shown in the Table 5.

Discussion

In order to elucidate the evolutionary processes that may lead to emergence of novel enterovirus types or genotypes, we determined evolutionary patterns of EV-C species sub-group B (i.e. strains of CVA-21, CVA-24, EV-C95, EV-C96 or EV-C99) VP1 gene using sequence data collected during entero/poliovirus surveillance together with sequence data retrieved from the GenBank.

Table 3. The VP1 codons that had signs of positive selection.

Type	Codon	Normalized dN-dS			Amino acid composition of the site
		SLAC	FEL	FUBAR	
EV-C96	104	0.89	2.43	**0.89**	S, T, N, A, V, L, M
	105	**1.86**	**3.84**	0.76	E, S, T, A, V, I,
EV-C99	105	0.36	**3.30**	**2.57**	R, K, E, S, T, N, Q, G, P, I, A, V
	296	0.42	**4.57**	**1.73**	S, T, A
CVA-24	105	0.29	**0.57**	**0.52**	R, K, D, E, S, T, N, Q, G, A, V
CVA-24 (non-AHC)	105	**2.63**	**1.04**	0.45	R, K, D, E, S, T, N, Q, G, A, V
CVA-24v	25	1.76	**3.36**	0.21	H, S, P, L
	150	2.03	**1.76**	0.32	S, T, A

The normalized dN-dS values with statistically significant evidence of positive selection are shown in boldface. The significance level (p-value) of 0.1 or posterior probability of 0.9 were used as confidence limits.

Table 4. The summary of the results of MK-tests.

Clusters compared (N of strains in cluster)				MK (McDonald & Kreitman 1991)	Modified MK (Egea et al., 2008)		Modified MK (Bhatt et al., 2010) Neutral site threshold frequency		
						low-frequency (<5%) variants excluded	0-<1	<0.50	0.05-0.75
EV-C96	(12)	CVA-21	(16)	***	***	***	***	***	***
EV-C96	(12)	CVA-24	(42)	***	***	***	***	***	***
EV-C96	(12)	EV-C99	(24)	***	***	***	***	***	***
EV-C96	(12)	EV-C95	(2)	***	***	***	*	*	*
CVA-21	(16)	CVA-24	(42)	***	***	***	***	***	***
CVA-21	(16)	EV-C99	(24)	***	***	***	***	***	***
EV-C95	(2)	CVA-21	(16)	***	***	***	***	***	***
EV-C95	(2)	EV-C99	(24)	***	***	***	***	***	***
EV-C95	(2)	CVA-24	(42)	***	***	***	***	***	***
EV-C99	(24)	CVA-24	(42)	***	***	***	***	***	***
EV-C96-A	(2)	EV-C96-B	(10)	-	-	-	-	-	-
EV-C96-A	(2)	EV-C96-B1	(3)	-	-	-	-	-	-
EV-C96-A	(2)	EV-C96-B2	(7)	-	-	-	-	-	-
EV-C96-B1	(3)	EV-C96-B2	(7)	-	-	-	-	-	-
CVA-21-A	(2)	CVA-21-B	(4)	-	-	-	-	-	-
CVA-21-A	(2)	CVA-21-C	(10)	-	-	-	-	-	-
CVA-21-B	(4)	CVA-21-C	(10)	*	-	-	-	-	-
EV-C99-A	(5)	EV-C99-B/C	(19)	***	***	***	***	**	***
EV-C99-A	(5)	EV-C99-B	(8)	***	***	***	***	***	***
EV-C99-A	(5)	EV-C99-C	(10)	***	**	*	-	-	-
EV-C99-B	(8)	EV-C99-C	(10)	-	-	-	-	-	-
CVA-24	(22)	CVA-24v	(20)	-	**	***	-	-	-
CVA-24-A	(2)	CVA-24-B	(4)	-	-	-	-	-	-
CVA-24-A	(2)	CVA-24-C	(2)	-	-	-	-	-	-
CVA-24-A	(2)	CVA-24-D	(9)	-	-	-	-	-	-
CVA-24-A	(2)	CVA-24-E	(2)	-	-	-	-	-	-
CVA-24-A	(2)	CVA-24-F	(2)	-	-	-	-	-	-
CVA-24-A	(2)	CVA-24v	(20)	-	-	-	-	-	*
CVA-24-B	(4)	CVA-24-C	(2)	-	-	-	-	-	-
CVA-24-B	(4)	CVA-24-D	(9)	-	-	-	*	-	*
CVA-24-B	(4)	CVA-24-E	(2)	-	-	-	-	-	-
CVA-24-B	(4)	CVA-24-F	(2)	-	-	-	-	-	-
CVA-24-B	(4)	CVA-24v	(20)	-	-	-	-	-	*
CVA-24-C	(2)	CVA-24-D	(9)	-	-	-	-	-	*
CVA-24-C	(2)	CVA-24-E	(2)	-	-	-	-	-	-
CVA-24-C	(2)	CVA-24-F	(2)	-	-	-	-	-	-
CVA-24-C	(2)	CVA-24v	(20)	-	-	-	-	-	-
CVA-24-D	(9)	CVA-24-E	(2)	-	*	*	-	-	-
CVA-24-D	(9)	CVA-24-F	(2)	-	-	-	-	-	-
CVA-24-D	(9)	CVA-24v	(20)	*	*	**	-	-	-
CVA-24-E	(2)	CVA-24-F	(2)	-	-	-	-	-	-
CVA-24-E	(2)	CVA-24v	(20)	-	-	-	-	-	*
CVA-24-F	(2)	CVA-24v	(20)	**	-	*	**	-	**
CVA-24-A	(2)	CVA-24-B-F/v	(40)	-	-	**	-	-	-
CVA-24-B/C	(6)	CVA-24-D-F/v	(35)	-	-	-	-	-	*
CVA-24-D-F	(15)	CVA-24v	(20)	-	-	*			

Statistical significance for higher than expected proportion of fixed non-synonymous substitutions (i.e. nF/nP >> sF/sP) between different clusters is shown. P-values were calculated using chi-squared test. (* 0.05>P>0.01; ** 0.01>P>0.001; *** P<0.001; - not significant).

General aspects of molecular evolution and epidemiology of CVA-21, CVA-24, EV-C96 and EV-C99

Altogether 72 strains of EV-C sub-group B were sequenced in this study. Most of the strains were isolated in Finland during routine entero/poliovirus surveillance [23]. Previous reports have suggested low prevalence of EV-C viruses in temperate regions, whereas in tropical and sub-tropical regions EV-C viruses are highly prevalent [17,18,48–51]. In this study, EV-C strains were found regularly in Finland. However, in most cases the possibility of travel-related importation from tropical or sub-tropical regions cannot be excluded. Therefore, on the basis of this study it cannot be stated unambiguously whether the EV-C strains are capable of wide circulation in Finland. The majority of the strains showed no evidence of geographic or temporal clustering, suggesting no evidence for EV-C outbreaks in Finland. Although some lineages of EV-C99 and CVA-24 were detected during several years in Finland, repeated introduction from other countries cannot be excluded.

The mean nucleotide substitution rates of EV-C sub-group B (ranging from 1.170 to 3.625×10^{-3} with combined HPD interval of 1.170 to 3.625 for partial VP1) estimated using Bayesian MCMC method were slightly lower than the corresponding estimates for other enterovirus types (ranging from 4.2 to 8.6×10^{-3} for complete VP1 and 4.8 to 12×10^{-3} for partial VP1) [52–63]. Previously, similar low substitution rate has been estimated for partial VP1 coding region of CVA-24v strains using linear regression [64]. Correspondingly, the estimated tMRCAs of EV-C sub-group B dated earlier than those estimated for most other EV-types. Previously, correlation between the epidemiological fitness and higher mutation, genome replication and nucleotide substitution rates has been found with noroviruses [65]. Further studies should be conducted to assess if such correlation also applies for enteroviruses. However, it should be noted that the molecular clock analysis contains several limitations. For some of the types (e.g. EV-C96), only recently isolated strains are available. The estimated tMRCAs of the more distant ancestors of different EV-types (i.e. the deep nodes in the phylogenetic tree) are therefore highly unreliable since sequence data is available only from the most recent few decades. Likewise, the datasets analysed are likely to be biased due to sparse sampling and (both temporally and geographically) uneven sampling frequencies. Therefore, more sequence data, possibly from archived samples, would be needed to confirm the apparently lower rate of evolution in this EV-C sub-group.

Evolutionary patterns within and between types

Each of the enterovirus types analysed in this study had relatively high intra-typic sequence diversities in VP1 (approaching the <25% nt divergence limit for grouping into the same type). The phylogenetic analysis of nucleotide sequences suggested robust grouping into intra-typic genotypes within each type. Within each type the vast majority of nucleotide substitutions were synonymous and the non-synonymous substitutions (i.e. amino acid substitutions) tended to cluster at distinct highly polymorphic sites. Signs of positive selection were detected in some of these highly polymorphic sites, while strong negative selection was indicated in most of the codons of VP1. Despite robust clustering to intra-typic genotypes, only few genotype-specific 'signature' amino acids were detected between these intra-typic clusters, whereas several amino acid sites showed evidence of parallel evolution (i.e. similar amino acid substitutions had apparently occurred independently in different genetic lineages).

In contrast, when *different* types were compared, the McDonald-Kreitman tests suggested a clear tendency towards fixation of type-specific 'signature' amino acids. Furthermore, several type-specific insertions/deletions were detected and the locations of highly polymorphic or positively selected amino acid sites overlapped only partially between different types. These results suggest that permanent fixation of type-specific amino acids seems to be a hallmark associated with evolution of *different* enterovirus types, whereas neutral evolution and/or (most likely frequency-dependent, see below) positive selection in few highly polymorphic amino acid sites were the dominant forms of evolution when strains *within* a type were compared. An exception to the rarity of permanently fixed signature amino acids in most intra-typic lineages was EV-C99 genotype A that showed similar differences (i.e. fixation of 'signature' amino acids and insertion/deletion sites) to those detected in inter-typic comparisons. The strains of EV-C99 genotype A also had relatively low pairwise similarities (a minimum of 73.1% nucleotide and 86.1% amino acid similarity) with some of the strains of EV-C99 genotypes B and C. Curiously, EV-C99 groups A and B/C may also have antigenic differences since the strains of genotype A show cross-neutralization with the antibodies raised against CVA-24-Joseph whereas the strains of EV-C99 group B do not show such cross-neutralization [5]. Therefore, the EV-C99 genotypes A and B/C may be under a process of divergent evolution that might ultimately lead to two distinct virus types. Further complete genome sequencing is needed to evaluate whether the strains of EV-C99 genotype A are divergent enough to merit classification to a separate type.

The different evolutionary patterns within and between EV types may also have implication on the genetic classification of enteroviruses. In the current classification scheme, EV-strains are classified into the same type, if they have more than 75% nucleotide and more than 85% (or 88%) amino acid similarities in the VP1 region and into different types, if the strains that have less than 70% nucleotide and 85% amino acid similarities [3,5]. However, divergent strains that have pairwise nt/aa similarities in the 'grey-zone' of current typing (i.e. 70–75% nt and/or 85–88% aa similarity) are detected regularly. The hallmarks of inter-typic comparisons (fixation of type-specific amino acids and insertion/deletion sites) could be applied as an additional classification criterion in such cases. In this study, EV-C99 and CVA-24 were clearly separated on the basis of MK-tests despite the 'grey-zone' nucleotide/amino acid similarities between these types. However, further studies on other enterovirus species should be conducted to study the universality of these potential novel classification criteria first.

Possible structural constraints in intra-typic evolution

Within each EV-type analysed, most of the codons in VP1 were detected to be under negative selection. This suggests a strong evolutionary pressure to retain the amino acid sequence and, thus, the structure of VP1. On the basis of the structure of CVA-21 [42], the highly polymorphic amino acid sites are most likely located in the loops between beta-sheets and the structurally disordered amino- and carboxyl-terminal segments of VP1. Such pattern may be explained by frequency-dependent selection (a rare variant has greater fitness than a common variant) posed by the host immune system to amino acids at the virus surface (e.g., antigenic sites). While a mutation at antigenic site could allow the virus to escape from the host immune response, due to the adaptability of the host immune system, the advantage could be short term. At the virus type-level, this changing selective pressure could result in a highly polymorphic site. Alternatively, neutral amino acid substitutions located in the surface loops or amino- and carboxy-terminal segments may have remained in the virus population due to genetic drift.

Table 5. The estimated mean rates of evolution (substitutions/site/year) and tMRCAs (year, BCE) for EV-C sub-cluster B partial VP1 sequence dataset.

Cluster	N (strains)	Mean rate (x 10⁻³)	1st site	2nd site	3rd site	Coefficient of variation	Years isolated (range)	tMRCA (year)
EV-C96	36	3.306 [0.472–5.949]	0.12 [0.16–0.23]	0.096 [0.075–0.12]	2.71 [2.67–2.74]	0.178 [4.642×10^{-5}–0.31]	1998–2012	1887 [1850–1920]
CVA-21	48	3.139 [2.313–4.077]	0.16 [0.12–0.20]	0.056 [0.036–0.077]	2.78 [2.73–2.83]	0.304 [0.026–0.54]	1952–2010	1843 [1784–1896]
EV-C99	60	3.652 [0.930–6.469]	0.25 [0.22–0.28]	0.0975 [0.080–0.12]	2.66 [2.62–2.69]	0.23 [0.068–0.39]	1984–2011	1814 [1753–1871]
CVA-24	118	1.170 [0.993–2.424]	0.21 [0.19–0.24]	0.013 [0.12–0.15]	2.66 [2.63–2.68]	0.446 [0.307–0.589]	1952–2012	1853 [1810–1890]
EV-C sub-group B	264	2.839 [2.177–3.509]	0.32 [0.28–0.35]	0.23 [0.18–0.28]	2.46 [2.40–2.52]	0.404 [0.315–0.493]	1952–2012	1548 [1385–1694]

The analysis was conducted using BEAST program. GTR model of substitution, lognormal relaxed clock and Bayesian Skyline demographic model were used in the analysis. The ranges of high-probability distribution [95% HPD] are given in brackets. EV-C95 was excluded from the table due to small number (2) of known isolates.

Similar results (i.e. high proportion of negatively selected codons combined with neutral evolution and/or evidence of positive selection in few codons) have been detected also for other EV-types including PV [66–68], E-30 [69], EV-71 [56], CVA-4 [58], E-6 [57] and CVB-5 [61]. These results suggest that the intra-typic evolution of the VP1 protein is most likely dominated by (mostly neutral) synonymous mutations combined with frequency-dependent selection and/or neutral evolution at antigenic sites. Intra-typic lineages would therefore have similar structural constraints with variation occurring predominantly at specific sites.

Interestingly, the AHC-causing CVA-24v strains showed less amino acid diversity in VP1 than the other CVA-24 strains. Furthermore, the amino acid polymorphisms were located in different sites in CVA-24v compared to the non-AHC-causing CVA-24 strains. The AHC-causing variants of CVA-24 emerged in Singapore in the year 1970 [21]. Both epidemiological and phylogenetic studies suggest a single origin for CVA-24v strains [5,21]. Therefore, it is likely that CVA-24v strains have undergone a strict population bottle-neck during the colonisation of a new target tissue (i.e. conjunctiva). The high degree of amino acid conservation between CVA-24v strains may be due to this population bottle-neck. Alternatively, CVA-24v strains might be under more strict negative selection than the other CVA-24 lineages. However, the overall dN/dS values were similar for both CVA-24v and the non-AHC-causing CVA-24 genotypes suggesting similar strength of negative selection for both groups. In contrast to the intra-typic evolution, the inter-typic comparisons between EV-C96, CVA-21, EV-99 and CVA-24 suggest that mere frequency-dependent selection in antigenic sites is not sufficient to explain the differences between the EV types. Apparently, the evolution leading to a formation of a new enterovirus type has included permanent fixation and insertion/deletion of distinct amino acids. Furthermore, the polymorphic sites overlap only partially between different EV types. These observations suggest that, in addition to amino acid changes at antigenic sites, also structural changes may have occurred during divergent evolution of EV-types. Hypothetically, a structural change could expose new amino acid sites to selective pressure imposed by the host immune system and, on the other hand, require fixation at sites where polymorphism was previously allowed. Furthermore, an (fitness decreasing) amino acid mutation may require compensatory amino acid fixation(s) elsewhere in the capsid protein.

Alternatively, fixation of neutral amino acid substitutions may have occurred in the ancestral populations of EV-types due to genetic drift during population bottle-necks. However, although the estimated tMRCAs of the more distant ancestors of different EV-types (i.e. the deep nodes in the phylogenetic tree) are highly unreliable, it is likely that the type-specific amino acids have remained conserved for hundreds of years. This suggests strong selection pressures favouring these type-specific amino acids in the given genetic context. Therefore, it is more likely that the permanent fixation of type-specific amino acids is due to dissimilar selection pressures subjected to the ancestor lineages of different enterovirus types. Further studies are needed to elucidate the processes leading to fixation of type-specific amino acid substitutions.

Concluding remarks

The sequence analysis presented in this article suggests different modes of evolution within enterovirus types (resulting in intra-typic lineages) and during evolution, which leads to larger scale (type-specific) differences. In this respect, intra-typic genetic change would be dominated by silent mutations accompanied by amino acid polymorphism occurring dominantly at immunogenic sites.

This genetic change can be observed as a high rate of synonymous mutations, strong negative selection and amino acid polymorphism and/or positive selection at distinct sites in structurally disordered regions. Inter-typic differences, on the other hand, included permanent fixation and insertion/deletion of distinct 'signature'-amino acids that could be a result of larger scale changes in the capsid structure.

Supporting Information

Figure S1 The capsid pentamer of CVA-21; top view (a, d and g) and side view (b, c, e, f, h and i). VP1 is shown in blue, VP2 in green, VP3 in red and VP4 in yellow. The VP1 amino acids that showed type-specific fixation between CVA-21 and EV-C96 (a-c), CVA-21 and EV-C99 (d–f) or CVA-21 and CVA-24 (g-i) are shown in white. The amino acids that showed evidence of positive selection within EV-C96 (a–c) or EV-C99 (d–f) are shown in cyan. The amino acids that showed evidence of positive selection within CVA-24v cluster or in non-AHC-causing strains are shown in purple and orange, respectively (g–i).

Table S1 The numbers of sites in the MK-test classes (s = synonymous; n = non-synonymous; F = fixed; P = polymorphic). The numbers were calculated using standard MacDonald-Kreitman test [36]. P-values were calculated with chi-squared test (* 0.05>P>0.01; ** 0.01>P>0.001; *** P<0.001; NS = not significant).

Table S2 The numbers of sites in the McDonald-Kreitman test classes (s = synonymous; n = non-synonymous; F = fixed; P = polymorphic). The numbers were calculated using modified MacDonald-Kreitman test [38] with Jukes-Cantor substitution model. P-values were calculated with chi-squared test (* 0.05>P>0.01; ** 0.01>P>0.001; *** P<0.001; NS = not significant).

Table S3 The numbers of sites in the McDonald-Kreitman test classes (s = synonymous; n = non-synonymous; F = fixed; P = polymorphic). The numbers were calculated using modified MacDonald-Kreitman test [38] with Jukes-Cantor substitution model. Low-frequency variants (<5%) were excluded from the analysis. P-values were calculated with chi-squared test (* 0.05>P>0.01; ** 0.01>P>0.001; *** P<0.001; NS = not significant).

Table S4 The numbers of sites in the McDonald-Kreitman test classes (s = synonymous; n = non-synonymous; F = fixed; P = polymorphic). The numbers were calculated using modified MacDonald-Kreitman test [39]. Neutral class site frequency thresholds of 0.0–1.0 were used in the analysis. P-values were calculated with chi-squared test (* 0.05>P>0.01; ** 0.01>P>0.001; *** P<0.001; NS = not significant).

Table S5 The numbers of sites in the MK-test classes (s = synonymous; n = non-synonymous; F = fixed; P = polymorphic). The numbers were calculated using modified MacDonald-Kreitman test [39]. Neutral class site frequency thresholds of 0.0–0.5 [40] were used in the analysis. P-values were calculated with chi-squared test (* 0.05>P>0.01; ** 0.01>P>0.001; *** P<0.001; NS = not significant).

Table S6 The numbers of sites in the MK-test classes (s = synonymous; n = non-synonymous; F = fixed; P = polymorphic). The numbers were calculated using modified MacDonald-Kreitman test [39]. Neutral class site frequency thresholds of 0.05–0.75 were used in the analysis. P-values were calculated with chi-squared test (* 0.05>P>0.01; ** 0.01>P>0.001; *** P<0.001; NS = not significant).

Acknowledgments

The authors would like thank Maija Lappalainen (Helsinki University Hospital, Laboratory Services (HUSLAB), Department of Virology and Immunology, Helsinki, Finland), Riina Raud (Healthboard, Tallinn, Estonia), Zdenka Sobotova (National Laboratory for Poliomyelitis and Viral Hepatitis, Public Health Authority of the Slovak Republic, Bratislava, Slovakia) and Prof. Günther Wewalka (Austrian Agency for Health and Food Safety (AGES), Institute for Medical Microbiology and Hygiene, Vienna, Austria) for providing virus isolates. We acknowledge CSC – IT Center for Science Ltd. for the allocation of computational resources.

Author Contributions

Conceived and designed the experiments: TS SB TV OI ES HA CS TH MR. Performed the experiments: TS SB TV OI ES HA CS TH MR. Analyzed the data: TS SB TV OI ES HA CS TH MR. Contributed reagents/materials/analysis tools: TS SB TV OI ES HA CS TH MR. Wrote the paper: TS SB TV OI ES HA CS TH MR.

References

1. Knowles NJ (2013) Picornavirus web page. Available: www.picornaviridae.com. Accessed 2014 Mar 12.
2. Pallansch MA, Roos RP (2007) Enteroviruses: Polioviruses, coxsackieviruses, echoviruses and newer enteroviruses. In: Knipe DM, Howley PM, editors. Fields Virology. Philadelphia, USA: Lippincot Williams & Wilkins. pp. 839–840–894.
3. Oberste MS, Maher K, Kilpatrick DR, Pallansch MA (1999) Molecular evolution of the human enteroviruses: Correlation of serotype with VP1 sequence and application to picornavirus classification. J Virol 73: 1941–1948.
4. Oberste MS, Maher K, Flemister MR, Marchetti G, Kilpatrick DR, et al. (2000) Comparison of classic and molecular approaches for the identification of untypeable enteroviruses. J Clin Microbiol 38: 1170–1174.
5. Brown BA, Maher K, Flemister MR, Naraghi-Arani P, Uddin M, et al. (2009) Resolving ambiguities in genetic typing of human enterovirus species C clinical isolates and identification of enterovirus 96, 99 and 102. J Gen Virol 90: 1713–1723. 10.1099/vir.0.008540-0; 10.1099/vir.0.008540-0.
6. Smura T, Savolainen-Kopra C, Roivainen M (2011) Evolution of newly described enteroviruses. Future Virology 6: 109–131. 10.2217/FVL.10.62.
7. Jiang P, Faase JA, Toyoda H, Paul A, Wimmer E, et al. (2007) Evidence for emergence of diverse polioviruses from C-cluster coxsackie A viruses and implications for global poliovirus eradication. Proc Natl Acad Sci U S A 104: 9457–9462. 10.1073/pnas.0700451104.
8. Smura T, Blomqvist S, Hovi T, Roivainen M (2009) The complete genome sequences for a novel enterovirus type, enterovirus 96, reflect multiple recombinations. Arch Virol 154: 1157–1161. 10.1007/s00705-009-0418-5; 10.1007/s00705-009-0418-5.
9. Lennette EH, Fox VL, Schmidt NJ, Culver JO (1958) The coe virus: An apparently new virus recovered from patients with mild respiratory disease. Am J Hyg 68: 272–287.
10. Xiang Z, Gonzalez R, Wang Z, Ren L, Xiao Y, et al. (2012) Coxsackievirus A21, enterovirus 68, and acute respiratory tract infection, china. Emerg Infect Dis 18: 821–824. 10.3201/eid1805.111376; 10.3201/eid1805.111376.
11. Smura T, Blomqvist S, Paananen A, Vuorinen T, Sobotova Z, et al. (2007) Enterovirus surveillance reveals proposed new serotypes and provides new insight into enterovirus 5′-untranslated region evolution. J Gen Virol 88: 2520–2526. 10.1099/vir.0.82866-0.
12. Xu A, Tao Z, Wang H, Zhang Y, Song L, et al. (2011) The complete genome analysis of two enterovirus 96 strains isolated in china in 2005 and 2009. Virus Genes 42: 323–330. 10.1007/s11262-011-0584-x; 10.1007/s11262-011-0584-x.
13. Bingjun T, Yoshida H, Yan W, Lin L, Tsuji T, et al. (2008) Molecular typing and epidemiology of non-polio enteroviruses isolated from yunnan province, the people's republic of china. J Med Virol 80: 670-679. 10.1002/jmv.21122; 10.1002/jmv.21122.
14. Apostol LN, Suzuki A, Bautista A, Galang H, Paladin FJ, et al. (2012) Detection of non-polio enteroviruses from 17 years of virological surveillance of acute flaccid paralysis in the philippines. J Med Virol 84: 624–631. 10.1002/jmv.23242; 10.1002/jmv.23242.

15. Apostol LN, Imagawa T, Suzuki A, Masago Y, Lupisan S, et al. (2012) Genetic diversity and molecular characterization of enteroviruses from sewage-polluted urban and rural rivers in the philippines. Virus Genes 45: 207–217. 10.1007/s11262-012-0776-z.

16. Bessaud M, Pillet S, Ibrahim W, Joffret ML, Pozzetto B, et al. (2012) Molecular characterization of human enteroviruses in the central african republic: Uncovering wide diversity and identification of a new human enterovirus A71 genogroup. J Clin Microbiol 50: 1650–1658. 10.1128/JCM.06657-11; 10.1128/JCM.06657-11.

17. Sadeuh-Mba SA, Bessaud M, Massenet D, Joffret ML, Endegue MC, et al. (2013) High frequency and diversity of species C enteroviruses in cameroon and neighboring countries. J Clin Microbiol 51: 759–770. 10.1128/JCM.02119-12; 10.1128/JCM.02119-12.

18. Rakoto-Andrianarivelo M, Rousset D, Razafindratsimandresy R, Chevaliez S, Guillot S, et al. (2005) High frequency of human enterovirus species C circulation in madagascar. J Clin Microbiol 43: 242–249. 10.1128/JCM.43.1.242-249.2005.

19. Oberste MS, Feeroz MM, Maher K, Nix WA, Engel GA, et al. (2013) Characterizing the picornavirus landscape among synanthropic nonhuman primates in bangladesh, 2007 to 2008. J Virol 87: 558–571. 10.1128/JVI.00837-12; 10.1128/JVI.00837-12.

20. Cabrerizo M, Rabella N, Tarner N, Castellanos T, Bustillo I, et al. (2012) Molecular identification of an enterovirus 99 strain in spain. Arch Virol 157: 551–554. 10.1007/s00705-011-1207-5; 10.1007/s00705-011-1207-5.

21. Mirkovic RR, Schmidt NJ, Yin-Murphy M, Melnick JL (1974) Enterovirus etiology of the 1970 singapore epidemic of acute conjunctivitis. Intervirology 4: 119–127.

22. World Health Organization (WHO) (2003) Guidelines for environmental surveillance of poliovirus circulation. Dept of Vaccines and Biologicals WHO webpage. Available: http://whqlibdoc.who.int/hq/2003/WHO_V&B_03.03.pdf, Accessed 2014 Mar 12.

23. Blomqvist S, Paananen A, Savolainen-Kopra C, Hovi T, Roivainen M (2008) Eight years of experience with molecular identification of human enteroviruses. J Clin Microbiol 46: 2410–2413. 10.1128/JCM.00313-08; 10.1128/JCM.00313-08.

24. Oberste MS, Nix WA, Maher K, Pallansch MA (2003) Improved molecular identification of enteroviruses by RT-PCR and amplicon sequencing. J Clin Virol 26: 375–377.

25. Tamura K, Peterson D, Peterson N, Stecher G, Nei M, et al. (2011) MEGA5: Molecular evolutionary genetics analysis using maximum likelihood, evolutionary distance, and maximum parsimony methods. Mol Biol Evol 28: 2731–2739. 10.1093/molbev/msr121; 10.1093/molbev/msr121.

26. Drummond AJ, Suchard MA, Xie D, Rambaut A (2012) Bayesian phylogenetics with BEAUti and the BEAST 1.7. Mol Biol Evol 29: 1969–1973. 10.1093/molbev/mss075; 10.1093/molbev/mss075.

27. Tamura K, Nei M (1993) Estimation of the number of nucleotide substitutions in the control region of mitochondrial DNA in humans and chimpanzees. Mol Biol Evol 10: 512–526.

28. Tavare S (1986) Some probabilistic and statistical problems in the analysis of DNA sequences. In: Miura RM, editor. Lectures on Mathematics in the Life Sciences. Providence: Amer. Math. Soc. pp. 57–58–86.

29. Suchard MA, Weiss RE, Sinsheimer JS (2001) Bayesian selection of continuous-time markov chain evolutionary models. Mol Biol Evol 18: 1001–1013.

30. Rambaut A. (2007) Tracer v1.4 available from http://Beast.bio.ed.ac.uk/tracer.

31. Drummond AJ, Ho SY, Phillips MJ, Rambaut A (2006) Relaxed phylogenetics and dating with confidence. PLoS Biol 4: e88. 10.1371/journal.pbio.0040088.

32. Kumar S, Skjaeveland A, Orr RJ, Enger P, Ruden T, et al. (2009) AIR: A batch-oriented web program package for construction of supermatrices ready for phylogenomic analyses. BMC Bioinformatics 10: 357-2105-10-357. 10.1186/1471-2105-10-357; 10.1186/1471-2105-10-357.

33. Kosakovsky Pond SL, Frost SD (2005) Not so different after all: A comparison of methods for detecting amino acid sites under selection. Mol Biol Evol 22: 1208–1222. 10.1093/molbev/msi105.

34. Murrell B, Moola S, Mabona A, Weighill T, Sheward D, et al. (2013) FUBAR: A fast, unconstrained bayesian approximation for inferring selection. Mol Biol Evol 30: 1196–1205. 10.1093/molbev/mst030; 10.1093/molbev/mst030.

35. Delport W, Poon AF, Frost SD, Kosakovsky Pond SL (2010) Datamonkey 2010: A suite of phylogenetic analysis tools for evolutionary biology. Bioinformatics 26: 2455–2457. 10.1093/bioinformatics/btq429; 10.1093/bioinformatics/btq429.

36. McDonald JH, Kreitman M (1991) Adaptive protein evolution at the adh locus in drosophila. Nature 351: 652–654. 10.1038/351652a0.

37. Librado P, Rozas J (2009) DnaSP v5: A software for comprehensive analysis of DNA polymorphism data. Bioinformatics 25: 1451–1452. 10.1093/bioinformatics/btp187; 10.1093/bioinformatics/btp187.

38. Egea R, Casillas S, Barbadilla A (2008) Standard and generalized McDonald-kreitman test: A website to detect selection by comparing different classes of DNA sites. Nucleic Acids Res 36: W157–62. 10.1093/nar/gkn337; 10.1093/nar/gkn337.

39. Bhatt S, Katzourakis A, Pybus OG (2010) Detecting natural selection in RNA virus populations using sequence summary statistics. Infect Genet Evol 10: 421–430. 10.1016/j.meegid.2009.06.001; 10.1016/j.meegid.2009.06.001.

40. Bhatt S, Holmes EC, Pybus OG (2011) The genomic rate of molecular adaptation of the human influenza A virus. Mol Biol Evol 28: 2443–2451. 10.1093/molbev/msr044; 10.1093/molbev/msr044.

41. Williamson S (2003) Adaptation in the env gene of HIV-1 and evolutionary theories of disease progression. Mol Biol Evol 20: 1318–1325. 10.1093/molbev/msg144.

42. Xiao C, Bator-Kelly CM, Rieder E, Chipman PR, Craig A, et al. (2005) The crystal structure of coxsackievirus A21 and its interaction with ICAM-1. Structure 13: 1019–1033. 10.1016/j.str.2005.04.011.

43. [Anonymous] Jmol: An open-source java viewer for chemical structures in 3D. http://Www.jmol.org.

44. Brown B, Oberste MS, Maher K, Pallansch MA (2003) Complete genomic sequencing shows that polioviruses and members of human enterovirus species C are closely related in the noncapsid coding region. J Virol 77: 8973–8984.

45. Lukashev AN, Drexler JF, Kotova VO, Amjaga EN, Reznik VI, et al. (2012) Novel serotypes 105 and 116 are members of distinct subgroups of human enterovirus C. J Gen Virol 93: 2357–2362. 10.1099/vir.0.043216-0; 10.1099/vir.0.043216-0.

46. Bessaud M, Joffret ML, Holmblat B, Razafindratsimandresy R, Delpeyroux F (2011) Genetic relationship between cocirculating human enteroviruses species C. PLoS One 6: e24823. 10.1371/journal.pone.0024823; 10.1371/journal.pone.0024823.

47. Samoilovich E, Roivainen M, Titov LP, Hovi T (2003) Serotype-specific mucosal immune response and subsequent poliovirus replication in vaccinated children. J Med Virol 71: 274–280. 10.1002/jmv.10480.

48. Khetsuriani N, Lamonte-Fowlkes A, Oberst S, Pallansch MA, Centers for Disease Control and Prevention (2006) Enterovirus surveillance—united states, 1970-2005. MMWR Surveill Summ 55: 1–20.

49. Rakoto-Andrianarivelo M, Guillot S, Iber J, Balanant J, Blondel B, et al. (2007) Co-circulation and evolution of polioviruses and species C enteroviruses in a district of madagascar. PLoS Pathog 3: e191. 10.1371/journal.ppat.0030191.

50. Arita M, Zhu SL, Yoshida H, Yoneyama T, Miyamura T, et al. (2005) A sabin 3-derived poliovirus recombinant contained a sequence homologous with indigenous human enterovirus species C in the viral polymerase coding region. J Virol 79: 12650–12657. 10.1128/JVI.79.20.12650-12657.2005.

51. Jegouic S, Joffret ML, Blanchard C, Riquet FB, Perret C, et al. (2009) Recombination between polioviruses and co-circulating coxsackie A viruses: Role in the emergence of pathogenic vaccine-derived polioviruses. PLoS Pathog 5: e1000412. 10.1371/journal.ppat.1000412; 10.1371/journal.ppat.1000412.

52. McWilliam Leitch EC, Bendig J, Cabrerizo M, Cardosa J, Hyypia T, et al. (2009) Transmission networks and population turnover of echovirus 30. J Virol 83: 2109–2118. 10.1128/JVI.02109-08; 10.1128/JVI.02109-08.

53. McWilliam Leitch EC, Cabrerizo M, Cardosa J, Harvala H, Ivanova OE, et al. (2010) Evolutionary dynamics and temporal/geographical correlates of recombination in the human enterovirus echovirus types 9, 11, and 30. J Virol 84: 9292–9300. 10.1128/JVI.00783-10; 10.1128/JVI.00783-10.

54. McWilliam Leitch EC, Cabrerizo M, Cardosa J, Harvala H, Ivanova OE, et al. (2012) The association of recombination events in the founding and emergence of subgenogroup evolutionary lineages of human enterovirus 71. J Virol 86: 2676–2685. 10.1128/JVI.06065-11; 10.1128/JVI.06065-11.

55. Gullberg M, Tolf C, Jonsson N, Mulders MN, Savolainen-Kopra C, et al. (2010) Characterization of a putative ancestor of coxsackievirus B5. J Virol 84: 9695–9708. 10.1128/JVI.00071-10; 10.1128/JVI.00071-10.

56. Tee KK, Lam TT, Chan YF, Bible JM, Kamarulzaman A, et al. (2010) Evolutionary genetics of human enterovirus 71: Origin, population dynamics, natural selection, and seasonal periodicity of the VP1 gene. J Virol 84: 3339–3350. 10.1128/JVI.01019-09; 10.1128/JVI.01019-09.

57. Smura T, Kakkola L, Blomqvist S, Klemola P, Parsons A, et al. (2013) Molecular evolution and epidemiology of echovirus 6 in finland. Infect Genet Evol 16: 234–247. 10.1016/j.meegid.2013.02.011; 10.1016/j.meegid.2013.02.011.

58. Chu PY, Lu PL, Tsai YL, Hsi E, Yao CY, et al. (2011) Spatiotemporal phylogenetic analysis and molecular characterization of coxsackievirus A4. Infect Genet Evol 11: 1426–1435. 10.1016/j.meegid.2011.05.010; 10.1016/j.meegid.2011.05.010.

59. Chu PY, Ke GM, Chen YS, Lu PL, Chen HL, et al. (2010) Molecular epidemiology of coxsackievirus B3. Infect Genet Evol 10: 777–784. 10.1016/j.meegid.2010.04.004; 10.1016/j.meegid.2010.04.004.

60. Savolainen-Kopra C, Paananen A, Blomqvist S, Klemola P, Simonen ML, et al. (2011) A large finnish echovirus 30 outbreak was preceded by silent circulation of the same genotype. Virus Genes 42: 28–36. 10.1007/s11262-010-0536-x; 10.1007/s11262-010-0536-x.

61. Henquell C, Mirand A, Richter J, Schuffenecker I, Bottiger B, et al. (2013) Phylogenetic patterns of human coxsackievirus B5 arise from population dynamics between two genogroups and reveal evolutionary factors of molecular adaptation and transmission. J Virol. 10.1128/JVI.02075-13.

62. Bailly JL, Mirand A, Henquell C, Archimbaud C, Chambon M, et al. (2011) Repeated genomic transfers from echovirus 30 to echovirus 6 lineages indicate co-divergence between co-circulating populations of the two human enterovirus serotypes. Infect Genet Evol 11: 276–289. 10.1016/j.meegid.2010.06.019; 10.1016/j.meegid.2010.06.019.

63. Cabrerizo M, Trallero G, Simmonds P (2013) Recombination and evolutionary dynamics of human echovirus 6. J Med Virol. 10.1002/jmv.23741; 10.1002/jmv.23741.

64. Chu PY, Ke GM, Chang CH, Lin JC, Sun CY, et al. (2009) Molecular epidemiology of coxsackie A type 24 variant in taiwan, 2000–2007. J Clin Virol 45: 285–291. 10.1016/j.jcv.2009.04.013; 10.1016/j.jcv.2009.04.013.

65. Bull RA, Eden JS, Rawlinson WD, White PA (2010) Rapid evolution of pandemic noroviruses of the GII.4 lineage. PLoS Pathog 6: e1000831. 10.1371/journal.ppat.1000831; 10.1371/journal.ppat.1000831.

66. Jorba J, Campagnoli R, De L, Kew O (2008) Calibration of multiple poliovirus molecular clocks covering an extended evolutionary range. J Virol 82: 4429–4440. 10.1128/JVI.02354-07; 10.1128/JVI.02354-07.

67. Hovi T, Savolainen-Kopra C, Smura T, Blomqvist S, Al-Hello H, et al. (2013) Evolution of type 2 vaccine derived poliovirus lineages. evidence for codon-specific positive selection at three distinct locations on capsid wall. PLoS One 8: e66836. 10.1371/journal.pone.0066836; 10.1371/journal.pone.0066836.

68. Hogle JM, Chow M, Filman DJ (1985) Three-dimensional structure of poliovirus at 2.9 A resolution. Science 229: 1358–1365.

69. Bailly JL, Mirand A, Henquell C, Archimbaud C, Chambon M, et al. (2009) Phylogeography of circulating populations of human echovirus 30 over 50 years: Nucleotide polymorphism and signature of purifying selection in the VP1 capsid protein gene. Infect Genet Evol 9: 699–708. 10.1016/j.meegid.2008.04.009; 10.1016/j.meegid.2008.04.009.

8

Emergence of Polymorphic Mating Strategies in Robot Colonies

Stefan Elfwing*, Kenji Doya*

Neural Computation Unit, Okinawa Institute of Science and Technology Graduate University, Okinawa, Japan

Abstract

Polymorphism has fascinated evolutionary biologists since the time of Darwin. Biologists have observed discrete alternative mating strategies in many different species. In this study, we demonstrate that polymorphic mating strategies can emerge in a colony of hermaphrodite robots. We used a survival and reproduction task where the robots maintained their energy levels by capturing energy sources and physically exchanged genotypes for the reproduction of offspring. The reproductive success was dependent on the individuals' energy levels, which created a natural trade-off between the time invested in maintaining a high energy level and the time invested in attracting mating partners. We performed experiments in environments with different density of energy sources and observed a variety in the mating behavior when a robot could see both an energy source and a potential mating partner. The individuals could be classified into two phenotypes: 1) *forager*, who always chooses to capture energy sources, and 2) *tracker*, who keeps track of potential mating partners if its energy level is above a threshold. In four out of the seven highest fitness populations in different environments, we found subpopulations with distinct differences in genotype and in behavioral phenotype. We analyzed the fitnesses of the foragers and the trackers by sampling them from each subpopulation and mixing with different ratios in a population. The fitness curves for the two subpopulations crossed at about 25% of foragers in the population, showing the evolutionary stability of the polymorphism. In one of those polymorphic populations, the trackers were further split into two subpopulations: (*strong trackers*) and (*weak trackers*). Our analyses show that the population consisting of three phenotypes also constituted several stable polymorphic evolutionarily stable states. To our knowledge, our study is the first to demonstrate the emergence of polymorphic evolutionarily stable strategies within a robot evolution framework.

Editor: Stephen R. Proulx, UC Santa Barbara, United States of America

Funding: These authors have no support or funding to report.

Competing Interests: The authors have declared that no competing interests exist.

* E-mail: elfwing@oist.jp (SE); doya@oist.jp (KD)

Introduction

If you come to any more conclusions about polymorphism, I should be very glad to hear the result: it is delightful to have many points fermenting in one's brain, and your letters and conclusions always give one plenty of this same fermentation.

- Charles Darwin, letter to Joseph Hooker, 1846 [1]

Polymorphism has fascinated evolutionary biologists since the time of Darwin [2,3]. Polymorphism is defined as that there exist more than one distinct phenotype of a species occupying the same habitat at the same time [4,5]. Polymorphism does not include continuous variations, but only discrete variations or in the case of continuous traits, such as body size and color, strongly bimodal or multimodal phenotype variation distributions. The existence of more than one distinct phenotype of a species demands an explanation, because the theory of natural selection predicts that the fittest phenotype should drive the other, lesser fit phenotypes to extinction. In general, polymorphism is maintained if the "fitness curves" of the polymorphic phenotypes intersect, where the crossover-point is an evolutionarily stable state, realizing a polymorphic evolutionarily stable strategy (ESS) [6,7]. Common features of the evolution and the maintenance of behavioral polymorphism include: 1) that time or resources can be invested in more than one activity that contributes to the fitness; 2) that the

individuals have rules about how to allocate time and resources among the alternative activities; and 3) that there is a trade-off between the activities that contribute to the fitness, i.e., the allocation of time and resources invested in one activity could be invested in another [8]. Frequency-dependent selection [7,9,10] is considered the most important explanation for the maintenance of polymorphism in a population. Frequency-dependent selection occurs when the fitness of the phenotypes depends on their frequencies in the population, and the fitness curves intersect at a crossover frequency where the phenotypes are equally successful. Alternative mating strategies (or alternative reproductive behaviors) [11–13] is the area of biological research most closely related to this study. Different mating tactics has been observed in a wide variety of species, both in males (e.g., [14–16]) and in females (e.g., [17–19]). However, there are relatively few cases where the differences in mating behavior have been confirmed to have a genetic basis [20–26], and even fewer studies that have suggested equal average reproductive success, i.e., shown crossing of the fitness curves, of alternative phenotypes [21,22,27].

The use of robot evolution experiments to study biological phenomena has gained traction in recent years [28], as a complementary approach to biological studies and theoretical models. In comparison to biological studies, robot evolution has the advantage that the evolution of hundreds of generations of robot controllers can be completed within hours or days. The experiments can easily be repeated for different parameter settings and environmental conditions, which allows for quantitative testing and analysis of robustness and stability. In comparison with theoretical and numerical models, robot models can capture the often complex physical interactions between the agent and the environment, including other agents. Floreano and Keller with different co-authors have used robot evolution experiments to investigate the emergence and reliability of communication [29–32], to quantitatively test Hamilton's rule for the evolution of altruism [33], and to test the influence of genetic architecture and mating frequency on the division of labor in social insect societies [34].

A distinctive feature of our earlier proposed embodied evolution framework [35] is that there is no explicit fitness function or algorithm for selecting individuals for recombination and mutations. Instead, offspring can only be created by the physical exchange of genotypes between two mating robots. In general, the choice of selection method requires careful consideration when using artificial evolution to study ESSs. A strong theoretical assumption underlying ESS analysis is that the population is infinitely large. Fogel *et al.* [36–38] demonstrated in simulation experiments, using the Hawk-Dove game, that for finite populations the results differed, at best, significantly from the theoretical ESS values and, at worst, bore no resemblance to the ESS. They, therefore, questioned the usefulness of ESSs to explain real biological phenomena in populations with limited population sizes. In response, Ficici and Pollock [39] showed that the difference between the theoretical ESS predictions and the observed simulation results could be accounted by the two selection methods used by Fogel *et al.*.

The purpose of this study is to demonstrate that evolutionary stable alternative mating strategies can emerge in a small robot colony without any predefined mating preferences as a result of the trade-off between the resources spent on energy conservation and the resources spent on courtship of mating partners. As alternative mating strategies is a natural precursor for the evolution of sexual dimorphism, this line of research has the potential of increasing our understanding of the emergence of different sexes.

To investigate the ecological conditions for evolution of alternative mating strategies, we performed artificial evolution experiments, in simulation, with a small colony of Cyber Rodent robots [40] using our proposed embodied evolution framework [35]. We performed the experiments in simulation because of the infeasibility of running hundreds of generations of evolution in hardware. In previous work [35], we have shown that learned and evolved behaviors in simulation have similar performance and behavior when transferred to the hardware setting. Each individual interacted in small groups of four robots and during its lifetime of 288 seconds an individual experienced three periods of group interactions, where the participants in each group were randomly selected. We placed the four robots in an arena (2.5×2.5 m) with 4 to 16 energy sources. The robots were equipped with two wheels, an infrared port for the exchange of genotypes, and a camera that could detect energy sources, and tail-lamps and faces of other robots (Figure 1). The robots could execute three basic behaviors, foraging, waiting (for a potential mating partner), and mating, which were learned by reinforcement learning [41]. The mating strategy, i.e., the selection of basic behaviors, was controlled by a linear neural network and the (five) neural network weights were adapted by the evolutionary process (Figure 2). From a biological point of view, the population consisted of simultaneous hermaphrodites, who could reproduce offspring by mating (i.e., an exchange of genotypes with a mating partner). For each of the individuals involved in a mating event, the probability of reproducing offspring was linearly dependent on

Figure 1. Two physical robots with six energy sources. The Cyber Rodent robots used in the experiments were equipped infrared communication for the exchange of genotypes and cameras for visual detection of energy sources (blue), tail-lamps of other robots (green), and faces of other robots (red).

the individual's internal energy level (see Methods and [35]). This created a trade-off, where, in relative terms, an individual could maximize its fitness by maximizing either the frequency of mating events or the energy level at the mating events.

Results

After the evolutionary process converged after 1,000 generations of experiments, we frequently observed a variety in the mating behavior when a robot could see both an energy source and the tail-lamp of a potential mating partner. We classified their mating strategies into two types: 1) *Forager strategy* in which an individual never waited for a potential mating partner and only tried to mate if it saw the face of another robot, and 2) *Tracker strategy* in which an individual waited for potential mating partner to turn around and where the threshold for waiting depended on its current energy level, the distance to closest energy source, and the distance to tail-lamp of the closest potential mating partner (in our preliminary report [42], we called them *roamer* and *stayer*, respectively, borrowing the terminology by Sandell and Liberg [43]). Among the high fitness populations (5 highest fitness populations in each of the 7 environments with different levels of energy source density), 31% (11/35) consisted predominantly of foragers and 49% (17/35) consisted predominantly of trackers. In 20% (7/35) of the populations, and remarkable four of the seven highest fitness populations, there emerged a polymorphic population of foragers and trackers with distinct differences in genotype, phenotype, and behavior. Our analyses show that the polymorphic population could constitute an ESS with an evolutionarily stable state of approximately 25% foragers in the population. In one instance, the trackers were split into two subpopulations: trackers who almost always waited for potential mating partners (strong trackers), and trackers who only waited if the energy level was high and an energy source was close (weak trackers). The analyses show that a population consisting of three phenotypes (foragers, strong trackers, and weak trackers) also could constitute a globally stable polymorphic ESS with several attractors.

The basic behaviors (foraging, waiting, and mating) of each individual were learned from scratch in each generation by reinforcement learning and accelerated by the evolution of the shaping rewards. For a detailed description and analysis of the evolution of shaping rewards and meta-parameters in our

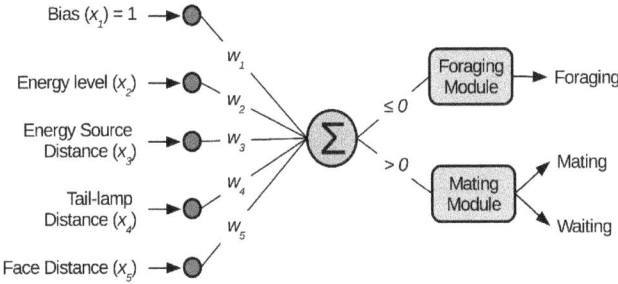

Figure 2. The neural network controller. The control architecture consisted of a linear artificial neural network. The output of the network was the weighted sum ($\sum_i w_i x_i$) of the five network inputs (x_i) and the five evolutionarily tuned neural network weights (w_i). In each time step, if the output was less or equal to zero then the foraging module was selected, otherwise the mating module was selected. The basic behaviors were learned from by reinforcement learning with the aid of evolutionarily tuned additional reward signals and meta-parameters. The foraging module learned a foraging behavior for capturing energy sources. The mating module learned both a mating behavior for the exchange of genotypes, when a face of another robot was visible, and a waiting behavior, when no face was visible.

embodied evolution framework, see [35]. The fitness measured by the number of reproduced offspring correlated significantly with the learning performance of the mating behavior ($R = -0.84$ and $p < 0.0001$; Figure 3). The efficiency of the mating behavior depended strongly on the evolved shaping reward functions. On average, one time step (240 ms) reduction in the time to execute a successful mating lead to an increase in fitness by approximately one offspring.

The purpose of performing experiments with different number of energy sources in the environment was to investigate the effect of energy source density on the emergence of different mating strategies. In general, higher energy source density resulted in higher energy levels at the mating events, which increased the probability of reproducing offspring and, therefore, the fitness (see Figure 3). However, we could not find an effect of energy source density on the evolution of different mating strategies.

Analysis of Mating Strategies

Analyses of the behaviors in the final 20 generations of the evolutionary process show that almost all individuals in all

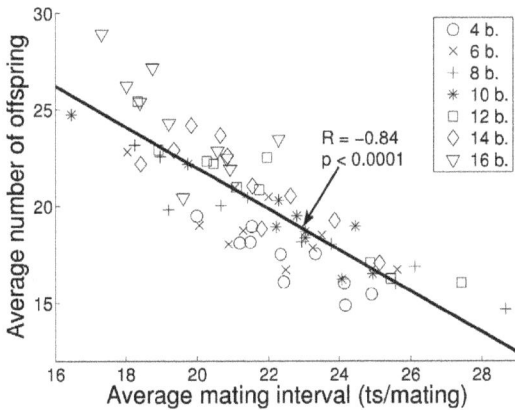

Figure 3. The correlation between the average estimated learning performance (i.e., the average mating interval) and the fitness (i.e., average number of offspring) in the final 20 generations in all experiments. The learning performance was estimated as the number of time steps the mating behavior was selected divided with number of mating events. The seven types of markers indicate the number of energy sources in the environment for each simulation.

experiments executed an opportunistic behavior when either only an energy source was visible or only a face of another robot was visible. In the former situation all individuals always executed the foraging behavior and in the latter situation they always executed the mating behavior. An interesting case happened when a robot could see both an energy source and the tail-lamp of another robot from its behind. While some robots (i.e., foragers) chose to take the foraging behavior of approaching the energy source, others (i.e., trackers) took the mating behavior of keeping track of the tail-lamp and approaching when the face became visible, as illustrated in Figure 4. The choice in general depended on the robot's stored energy level and the distances to the energy source and the tail-lamp.

In order to characterize these behavioral phenotypes, we took the average energy level threshold $\bar{E}_m$ for the mating behavior, by computing the mean value over 676 equidistant visual inputs of the energy source and the tail-lamp (see Methods). With this criterion, the individual mating strategies emerged in the experiments could be classified: *Forager strategy*, in which an individual always took foraging behavior with $\bar{E}_m$ equal or very close to the maximum threshold (1), and 2) *Tracker strategy*, in which an individual waited for the potential mating partners to turn around when its energy level is above a certain threshold, with $\bar{E}_m$ smaller than 1.

We identified the three different types of populations by computing the median and the standard deviation of the average threshold $\bar{E}_m$ for all 1600 individuals in the last 20 generations in each of the 70 experiments. The populations with median values larger than 0.98 were considered forager populations. The remaining populations were identified by their standard deviations. Those with standard deviations smaller than 0.19 were considered tracker populations and those with standard deviations larger than 0.19 were considered polymorphic populations, all of which had multimodal $\bar{E}_m$ distributions.

We found 33, 28, and 9 populations of forager, tracker, and polymorphic strategies, respectively in the 70 populations. If we focus on the higher fitness populations (5 highest fitness populations in each of the 7 environments with different levels of energy source density), 31% (11/35) were foragers, 49% (17/35) were trackers, and 20% (7/35) were polymorphic. Remarkably, four of the seven highest fitness populations were polymorphic.

Analysis of Genotypes, Phenotypes, and Behaviors

The best example of an emerged polymorphic mating strategy is shown in Figures 5–7, which emerged in the experiment with the

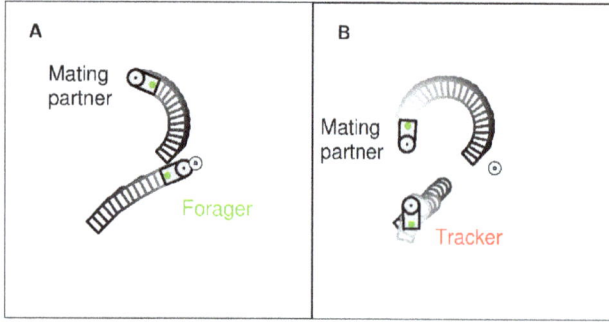

Figure 4. Example trajectories of the learned behaviors for the forager strategy and the tracker strategy. A) The forager ignores the tail-lamp of the mating partner and executes the learned foraging behavior to capture the energy source. B) The tracker executes the learned waiting behavior and adjusts its position according to the trajectory of the mating partner.

highest average fitness in the environment with four energy sources. Figure 5 shows the distribution of the values of w_1, the bias for the mating behavior, and w_5, the weight for the distance to another robot's face, of all 1600 individuals in the final 20 generation. We classified the population into the two subpopulations by k-means clustering in the w_1 and w_5 weight space and show the forager subpopulation in black and the tracker subpopulation in red in this and the following figures. Figure 6 shows the histogram of average energy thresholds $\bar{E}_m$ of selection of the mating behavior. The genetically classified foragers and trackers formed clearly distinct distributions. Figure 7 shows the average percentage of the lifetimes executing the three basic behaviors of foraging, waiting (choice of mating despite the face is not visible), and mating.

The result clearly shows a polymorphic population with distinct bimodal distributions in genotype, phenotype, and behavior for a forager subpopulation and a tracker subpopulation. For the foragers, the median of average energy threshold $\bar{E}_m$ was equal the maximum threshold of 1. They spent on average 77.5% of their lifetimes executing the foraging behavior and almost no time (0.6%) executing the waiting behavior. In comparison, for the trackers the median of threshold $\bar{E}_m$ was 0.40. They spent on

average 57.6% of their lifetimes executing the foraging behavior and 21.5% executing the waiting behavior.

Evolutionary Stability of Polymorphism

To investigate the evolutionarily stability of the emerged polymorphic population of foragers and trackers, we performed additional experiments in which we fixed the proportion of the two phenotypes in the polymorphic population shown in Figures 5–7. Individuals from the two genotypes were selected randomly from the final (1000th) generation of the evolutionary experiment, which consisted of 10 foragers and 70 trackers. The experiment was repeated 100 times for each proportion of the two phenotypes. Figure 8 shows the average number of mating events, Figure 9 shows the average normalized energy level at the mating events, and Figure 10 shows the average number of offspring (i.e., the fitness) for the two phenotypes, as functions of the tracker proportion in the population. The number of mating events increased, both for the population as a whole and for the two phenotypes, as the number of trackers in the population increased. The increase in the number of mating events was much larger for the foragers (black line in Figure 8), from 10 to almost 13 mating events, as the tracker proportion increased from 0% to 87.5%. In comparison, the number of mating events for the trackers (red line in Figure 8) increased with only approximately 1 mating event,

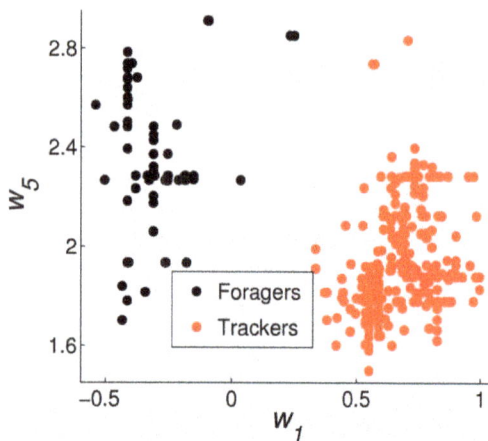

Figure 5. Difference in genotype between the forager and the tracker subpopulations. The distribution of values of the bias weights (w_1) and the face distance weights (w_5) for all 1600 individuals in the final 20 generations.

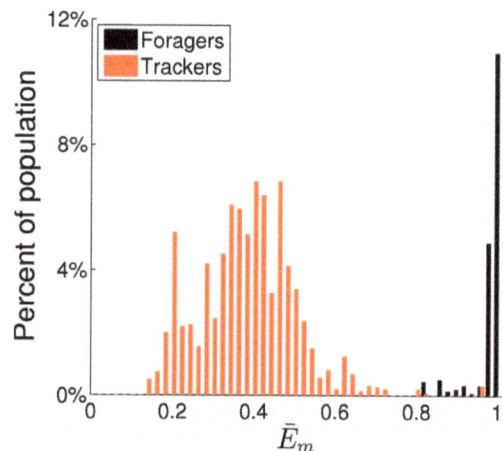

Figure 6. Difference in phenotype between the forager and the tracker subpopulations. The histograms of average waiting threshold values, $\bar{E}_m$, for all 1600 individuals in the final 20 generations.

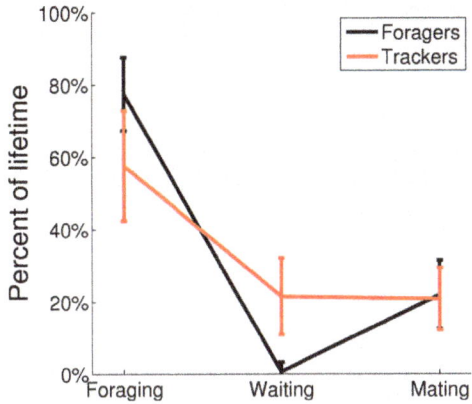

Figure 7. Difference in behavior between the forager and the tracker subpopulations. The mean percentages of the lifetimes, with standard deviation, the individuals spent executing the three basic behaviors for all 1600 individuals in the final 20 generations.

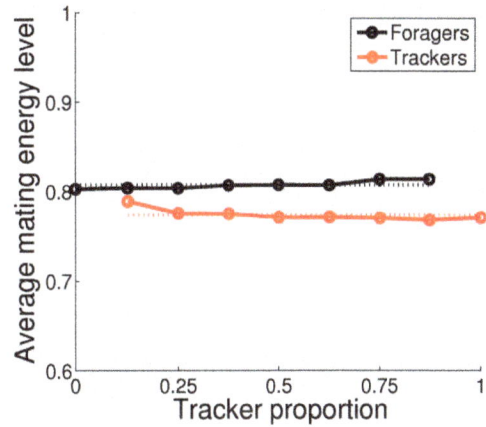

Figure 9. Average energy level at the mating events as functions of the tracker proportion in the population. The dotted lines show the constant approximations as the average values over all phenotype proportions.

from 12 to 13, as the tracker proportion increased from 12.5% to 100%. The average energy levels were approximately constant over all phenotype proportions for both foragers and trackers (dotted lines in Figure 9) with a mean value of 0.81 for the foragers and 0.77 for the trackers. The experimental result clearly show (Figure 10) that the emerged population of foragers and trackers constitute a polymorphic ESS with an evolutionarily stable state of around 25% foragers in the population. The stable state corresponded to group-interactions with only one forager in the environment where the three trackers would all wait and adjust their trajectories according to the forager's behavior.

The foragers demonstrated negative frequency-dependent selection, i.e., the fitness increased as they became rarer. The foragers could be considered as a parasitic phenotype, because they relied on a high proportion of trackers to achieve a large number of mating events and, thereby, high fitness. In relative terms, the foragers tried to maximize the second term of the fitness function (Equation 1) by maximizing their own energy at the mating events, $\bar{E}_{own}$. The trackers demonstrated positive frequency-dependent selection, i.e., the fitness increased as they became

more common. However, the proportion of trackers in population had much less effect on the trackers' fitness. The average fitness increased only about 8% as the tracker proportion increased from 25% to 100%, compared to the foragers that increased their average fitness 21% as the tracker proportion increased from 0% to 75%. In relative terms, the trackers maximized the first term of the fitness function, the mating frequency $\bar{M}$, by using the waiting behavior to attract potential mating partners. The result also shows that the evolved mating strategy could have a significant impact on the average population fitness. There was an almost 4 offspring increase in fitness between a population with only foragers (fitness of about 16 offspring) and a population with only trackers or an evolutionarily stable polymorphic population (fitness of about 20).

Mating Dynamics of Foragers and Trackers

Now let us investigate the dynamic interactions of the two subpopulations behind the observed fitness curves in Figure 10.

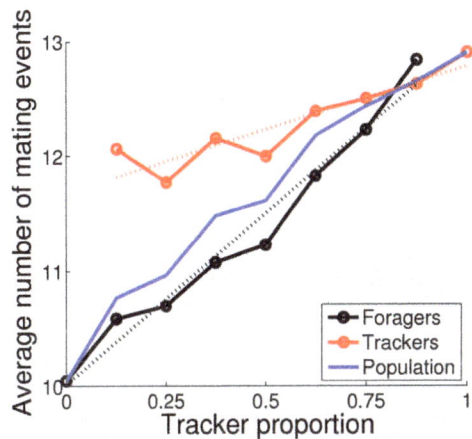

Figure 8. Average number of number of mating events as functions of the tracker proportion in the population. The dotted lines show the best linear fit for the two subpopulations and the blue line shows average values for the population as a whole.

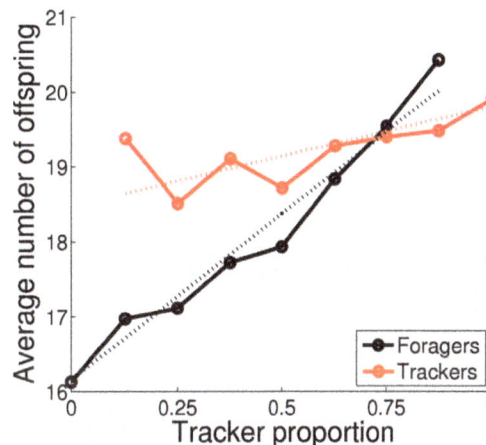

Figure 10. Average number of offspring (i.e., the fitness) as functions of the tracker proportion in the population. The dotted lines show the estimated fitness values using Equations 7 and 9.

The average fitness $\bar{F}$, i.e., the number of reproduced offspring, is a function of the average number of mating events, $\bar{M}$, the average normalized energy level at the mating occasions (for offspring created by the individual), $\bar{E}_{own}$, and the mating partners' average normalized energy level (for offspring created by the mating partners), $\bar{E}_{other}$:

$$\bar{F} = \bar{M}(\bar{E}_{own} + \bar{E}_{other}). \qquad (1)$$

In order to model the average fitness, we take the following hypotheses: H1) The trackers achieved almost perfect mixing of mating partners. H2) The foragers mated more frequently with trackers than the proportion x_t of trackers in the population. H3) The mating frequencies of the foragers, $\bar{M}_f$, and the trackers, $\bar{M}_t$, were functions of the tracker proportion x_t. Under these conditions, we predict the fitness curves of the two subpopulations.

Let us denote the average numbers of mating events in a lifetime of a forager with foragers by $\bar{M}_{f \to f}$, with trackers by $\bar{M}_{f \to t}$, and with either by $\bar{M}_f = \bar{M}_{f \to f} + \bar{M}_{f \to t}$. We also denote the average numbers of mating events of a tracker $\bar{M}_{t \to f}$, $\bar{M}_{t \to t}$, and $\bar{M}_t$ in the same convention.

H1 can be represented as

$$\frac{\bar{M}_{t \to t}}{\bar{M}_t} \approx x_t. \qquad (2)$$

This was true for all tested phenotype proportions, as shown by the red line in Figure 11.

H2 means that

$$\frac{\bar{M}_{f \to t}}{\bar{M}_f} = x_t + \epsilon \qquad (3)$$

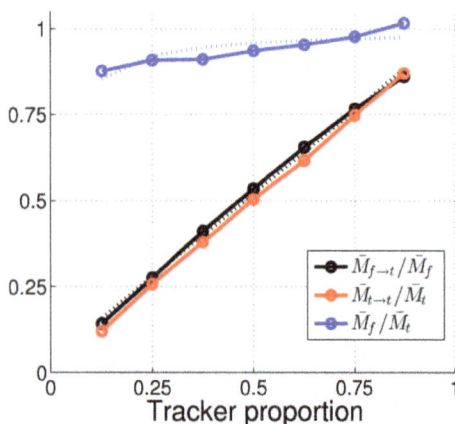

Figure 11. Average proportion of mating events with tracker mating partners as functions of the tracker proportion in the population, for the forager (black solid lines with circles) and the tracker (red solid lines with circles) subpopulations, and the ratio of average number of forager mating events to average number of tracker mating events (blue solid lines with circles). The black and red dotted lines show the best linear fit for the two subpopulations. The blue dotted line shows the predicted ratio of forager mating events to tracker mating events from Equation 5 with $\epsilon = 0.0214$.

with a positive ϵ. This was the case in our experiment with $\epsilon \approx 0.0214$, as shown by the black line in Figure 11.

From H1 (Equation 2) and the symmetry condition $x_f \bar{M}_{f \to t} \equiv x_t \bar{M}_{t \to f}$, we have $\bar{M}_{f \to t} = \frac{x_t}{x_f} \bar{M}_{t \to f} = x_t \bar{M}_t$. From this and H2 (Equation 3), the ratio of the average numbers of mating events by foragers and trackers ($\bar{M}_f / \bar{M}_t$) can then be derived as

$$\frac{\bar{M}_{f \to t}}{\bar{M}_f} = \frac{x_t \bar{M}_t}{\bar{M}_f} = x_t + \epsilon, \qquad (4)$$

$$\frac{\bar{M}_f}{\bar{M}_t} = \frac{x_t}{x_t + \epsilon}. \qquad (5)$$

In our experiment, the ratio $\bar{M}_f / \bar{M}_t$ of forager mating events to tracker mating events increased from 0.88 to 1.02 when the tracker proportion x_t increased from 0.125 to 0.875, as shown by the blue line in Figure 11. This fits well with the model prediction in Equation 5 with $\epsilon = 0.0214$, shown by the dotted blue line in Figure 11.

From H3, the average fitness (Equation 1) of the foragers, $\bar{F}_f$ and the trackers, $\bar{F}_t$, can be rewritten as

$$\bar{F}_f(x_t) = \bar{M}_f(x_t)\left[\bar{E}_f + \left(\frac{\bar{M}_{f \to f}(x_t)}{\bar{M}_f(x_t)}\bar{E}_f + \frac{\bar{M}_{f \to t}(x_t)}{\bar{M}_f(x_t)}\bar{E}_t\right)\right] \qquad (6)$$

$$\approx \bar{M}_f(x_t)\left[2\bar{E}_f + (\bar{E}_t - \bar{E}_f)(x_t + \epsilon)\right], \qquad (7)$$

$$\bar{F}_t(x_t) = \bar{M}_t(x_t)\left[\bar{E}_t + \left(\frac{\bar{M}_{t \to f}(x_t)}{\bar{M}_t(x_t)}\bar{E}_f + \frac{\bar{M}_{t \to t}(x_t)}{\bar{M}_t(x_t)}\bar{E}_t\right)\right] \qquad (8)$$

$$\approx \bar{M}_t(x_t)\left[\bar{E}_t + \bar{E}_f + (\bar{E}_t - \bar{E}_f)x_t\right], \qquad (9)$$

The dotted lines in Figure 10 shows the estimation of the average fitness of the foragers and the trackers, using Equations 7 and 9, respectively, with constant $\bar{E}_i$-values, and with the best linear fit of $\bar{M}_i$ from the data (see dotted lines in Figures 9 and 8, respectively).

Population with Three Mating Strategies

As shown in Figure 6, the distribution of the average energy threshold for waiting, $\bar{E}_m$, of all trackers in the final 20 generations was very broad. Interestingly, in the final generation, the trackers' phenotype distribution was bimodal with two distinct peaks (Figure 12): a larger subpopulation (46 individuals) of *strong trackers* with a smaller median $\bar{E}_m$-value of approximately 0.22 and a smaller subpopulation (24 individuals) of *weak trackers* with a larger median $\bar{E}_m$-value of approximately 0.48. The separation in genotype between the strong and weak trackers was not as distinct, but it is clearly visible in w_1 and w_5 weight space (Figure 13).

To investigate the evolutionarily stability of a population consisting of foragers, weak trackers, and strong trackers, we conducted additional experiments in which we fixed the proportions of the three phenotypes and measured their fitnesses. As in the earlier experiments, the genotypes of the three subpopulations were randomly selected from the genotypes in the final generation of the evolutionary experiments and repeated 100 times for each proportion of the three phenotypes. The result of the experiments

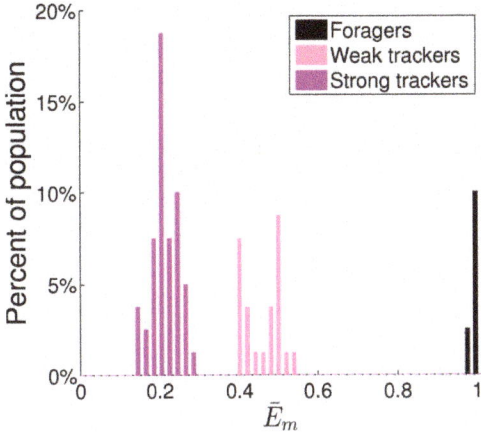

Figure 12. Difference in phenotype between the forager (black), the weak tracker (pink), and the strong tracker (purple) subpopulations. The histograms of average waiting threshold values, $\bar{E}_m$, in the final generation.

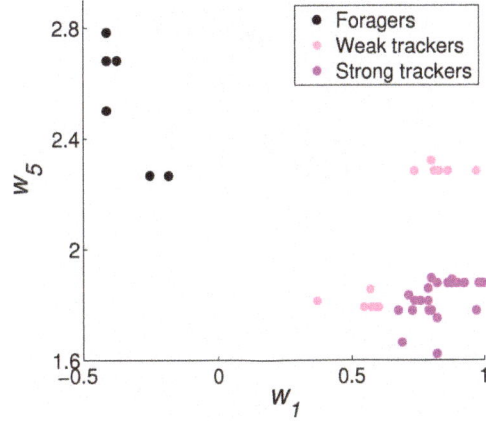

Figure 13. Difference in genotype between the forager (black), the weak tracker (pink), and the strong tracker (purple) subpopulations. The distribution of values of the bias weights (w_1) and the face distance weights (w_5) in the final generation.

is summarized in the DiFinetti diagram in Figure 14. The proportion of each phenotype increases from the side of the triangle to the opposite vertex. The black circles represent the tested phenotype proportions. In general, the average population fitness (shown by the background coloring of the diagram) increased with the number of strong trackers in the population. The average population fitness was 16.2 offspring in population with only foragers, 18.2 offspring in population with only weak trackers, and 19.7 offspring in a population with only strong trackers, which was close to the maximum average fitness of 19.9 for a population consisting of 87.5% strong trackers and 12.5% foragers.

Assuming a fixed population size, the proportion of phenotype i in the next generation, x_i', can be calculated by the discrete replicator dynamic equation:

$$x_i' = \frac{x_i F_i(x)}{\sum_j x_j F_j(x)}, \quad (10)$$

where, F_i is the average fitness of phenotype i with proportion x_i of the population in the current generation. The black arrows in Figure 14 show the average directions and magnitudes of the changes in phenotype proportions. Three populations consisting of only two of the phenotypes (as seen along the three sides of the triangle) were evolutionarily stable with populations proportions close to: 1) 31.25% foragers and 68.75% strong trackers (bottom side); 2) 87.5% weak trackers and 12.5% strong trackers (right side); and 3) 12.5% foragers and 87.5% weak trackers (left side). Populations consisting of all three phenotypes were globally stable, because all arrows in the inner triangle, representing populations consisting of at least 12.5% of each of the three phenotypes, point inside the triangle. The diagram suggests that there were two evolutionarily stable attractors where the magnitude of the change in the proportion of the phenotypes was zero, $|\mathbf{x}'| = 0$, with centers close to populations consisting of: 4) 25% foragers, 50% weak trackers, and 25% strong trackers ($|\mathbf{x}'| \approx 0.16$) and 5) 25% foragers, 25% weak trackers, and 50% strong trackers ($|\mathbf{x}'| \approx 0.09$). For populations with 12.5% to 25% foragers, the average magnitudes of the change in phenotype proportions were small, i.e., low selection pressure, which meant that phenotype ratios could move relatively easily between the attractors by random genetic drift.

For example, in the final 20 generations of the evolutionary experiment (illustrated by the gray line ending in the gray triangle), the phenotype ratios were first close to the fourth attractor (fitness of 19.1), and ended up close to the fifth attractor (fitness of 19.3).

Discussion

In this study, we demonstrated that polymorphic mating strategies can emerge in a small robot colony under homogeneous evolutionary conditions, without a selection scheme or an explicit fitness function that promoted a certain outcome. Our study is, to our knowledge, the first to demonstrate the emergence of polymorphic ESSs within a robot evolution framework. This gives further evidence that artificial robot evolution (for an overview see [28]) can be a feasible and a valuable approach for investigating hypotheses of biological phenomena.

The importance of specific details of the genetic algorithm and the structure of the genotype were illustrated in this study. A condition for the evolution of the polymorphic ESSs consisting of foragers and trackers was the small proportion of genotype controlling the mating strategy, in combination with the relatively low crossover rate. The mating strategy was controlled by only 5 out of the 51 genes, located at the beginning of the genotype, and with a crossover rate of 0.1 there was only a 0.8% probability that an offspring would have a mating strategy controlled be a mixture of genes from both parents. A much more frequent mixing of the mating strategy genes would have made it more difficult or even impossible to evolve and maintain separate genetic traits in the same population. An assumption underlying evolutionary game theory [7], is that the payoffs that agents are assumed to be without noise. It is therefore very encouraging that evolutionarily stable polymorphic ESSs could emerge in our experiment with a small population size and with large variances in the performance of similar, and even identical, individuals. The lifetime learning of the basic behaviors by reinforcement learning introduced additional stochasticity. Even in the last part of the evolutionary process, a few individuals failed to capture any batteries or engage in any mating activity. This was usually caused by that the individual got trapped in a corner of the environment and failed to learn how to navigate out of it.

The forager strategy in the evolved polymorphic populations can be seen as a cheater strategy. To achieve high fitness, a forager

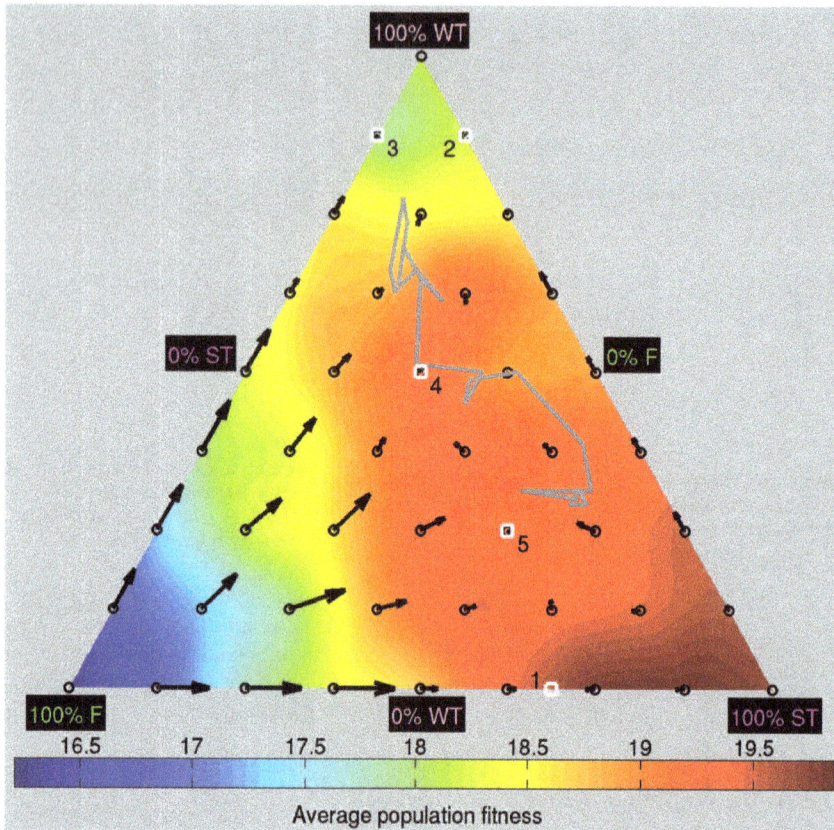

Figure 14. DiFinetti diagram of the directions and magnitudes of the changes in subpopulation proportions for the three phenotypes: foragers (F), weak trackers (WT), strong trackers (ST). The small black circles indicate the tested phenotype proportions. The black arrows show the average direction and magnitude of the change in phenotype proportions (magnified by a factor of five for visualization purposes). The white numbered squares indicate the approximate population ratios of the five evolutionary stable states, where the ratios of the phenotypes (F, WT, ST) were 1: (31.25%, 0%, 68.75%), 2: (0%, 87.5%, 12.5%), 3: (12.5%, 87.5%, 0%), 4: (25%, 50%, 25%), and 5: (25%, 25%, 50%). The gray line, ending in the gray triangle, shows the phenotype proportions in final 20 generations of the evolutionary experiments. The background coloring visualizes the average population fitness.

relies on that all the other individuals in the environment (i.e., trackers) will adjust their behaviors according to the trajectory of the forager. The forager, therefore, avoids the cost of searching for mating opportunities. There exists a rich literature on the potential of cheating in hermaphrodite mating systems (for an overview see [44]). Usually, cheating refers to the attempt of individuals to take on the male role over the female role in mating encounters to avoid the cost offspring reproduction.

The most similar study to ours was conducted by Rold *et al.* [45]. They co-evolved a population of predefined male and female robots. The robots, as in our experiments, remained alive by capturing energy sources and reproduced by physical mating, consisting of touching a robot of the opposite sex. The only difference between males and females was that the males remained reproductive throughout their lives, while the females became non-reproductive for a fixed period of time after an reproductive mating event. In their experiment, the reproduced offspring were not the result of a genetic exchange between mating robots. Instead, the males and females were evolved separately with the number of reproductive mating events used as fitness objective. The evolved behaviors of the males and females had distinct differences and their behaviors corresponded to observed behaviors of males and females in biological studies. Interestingly, the evolved behaviors of the males and females also matched the behavior of the foragers and trackers, respectively, in our study.

The males opportunistically ate all the food they could find while looking for reproductive females. The reproductive females were less active and adopted a mating strategy of waiting for males to mate with them. This give some support to a hypothesis that polymorphic mating strategies, emerged due to basic trade-off between the resources spent on energy conservation and the resources spent on courtship of mating partners, is a precursor of sexual dimorphism. In our experiments, polymorphism could arise because the foragers and trackers optimized, in relative terms, different parts of the fitness function (Equation 1). The foragers maximized their own energy level, $\bar{E}_{own}$, by spending all their lives foraging for energy sources except for when a a potential mating partner was directly visible, while the trackers maximized the mating frequency, $\bar{M}$, by spending considerable amount of their lives waiting for potential mating partners. The evolution of "proto-sexes" is a research venue we plan to explore in future work. In the current experimental setup, both the sender and receiver can reproduce offspring at the mating events and the cost of mating is equal, whether offspring are reproduced or not (see *Methods*). A more biological plausible setup would be that only one of the agents took on the female role, e.g., the receiver, and also bore the main cost of reproducing offspring. The goal would then be to investigate if, and in such case under which conditions, a breeding system with distinct male and female roles could evolve from an initial population of hermaphrodites without any predefined mating preferences, and maybe even more exotic breeding systems such as *androdioecy* (males and hermaphrodites) and *gynodioecy* (females and hermaphrodites).

Methods

Four Cyber Rodent mobile robots [40] were placed in a 2.5×2.5 m arena, together with 4, 6, 8, 10, 12, 14, or 16 energy sources (Figure 1). The task of the robots were to survive by maintaining their internal energy level through foraging of energy sources and by reproduction of offspring through physical exchange of genotypes by infrared communication. We performed the experiments in a simulation environment, developed to mimic the features of the real Cyber Rodent hardware platform. The robots were equipped with a camera system with color blob detection, used to extract the distances and relative angles to nearest energy source (blue), the nearest tail-lamp of another robot (green), and the nearest face of another robot (red). Mimicking the real robotic hardware, the field of view of the simulated vision system set to $[-75^o; 75^o]$. Within an angle range of $[-45^o; 45^o]$, the robots could detect energy sources up to 2 m, tail-lamps up to 1.5 m, and faces up to 1 m. Outside this range, the detection capability decreased linearly down to 0.2 m for the maximum angles.

We performed 1000 generations of evolution and for each energy source density, we ran 10 evolution experiments. To be able to conduct robot evolution experiments with only only four robots, we utilized time-sharing in subpopulations of 20 individuals within each robot. Each individual in a subpopulation took control, in random order, of the robot for three time-sharings of 400 time steps, i.e., the total lifetime was 1200 time steps. An individual had a maximum internal energy level (E_{max}) of 500 energy units. Each time step, the energy level decreased by 1 unit and a capture of an energy source increased the energy level by 100 units. At birth, an individual had an internal energy level of 400 units. If an individual's energy was depleted, then the individual died and was removed from its subpopulation. When a robot captured an energy source it disappeared from its current position and reappeared in new, randomly selected, position.

We did not apply an explicit fitness function or a centralized selection process, instead offspring were created by a mating. The individuals controlling the robots could create offspring by a physical exchange of genotype through infrared communication. The infrared communication ports were located slightly to the right of center in the front of the Cyber Rodent robots, directed straight forward. In the simulation environment, the maximum range of the communication was set to 1 m and the angle range was set to $[-30°; 30°]$. An individual could initiate the infrared communication by executing a predefined action selected by the mating behavior. For a mating event to be successful, both robots had to be within each others mating range before and after the individuals controlling the robots executed the actions of their currently selected reinforcement learning modules. The probability, for each of the two individuals involved in a mating event, of reproducing offspring was linearly depended on the individual's energy level (E/E_{max}). A reproductive mating event created two offspring in the individual's subpopulation by applying one-point crossover with a probability of 0.1. The genes of the two newly created genotypes were then mutated with probability of 0.1, by adding value from a Gaussian distribution with zero mean and a standard deviation of 0.1. After all individuals in a subpopulation had survived for a full lifetime or died a premature death, a new subpopulation was created by randomly selecting a fixed number,

i.e., 20 in our experiments, of the offspring reproduced during the last generation.

The genotype consisted of 51 real-valued genes: 1) 5 genes controlling the mating strategy by encoding the weights of the top-level neural network that selected basic behaviors (Figure 2); 2) 42 genes determining the parameters of the additional reward signals for the basic behaviors in the form of potential-based shaping rewards [46]; and 3) 4 genes determining the meta-parameters of the reinforcement learning algorithm. The five-dimensional input to the neural network consisted of a constant bias of 1 (x_1), the individual's internal energy (x_2), and the inverse distances to the nearest energy source (x_3), tail-lamp (x_4), and face (x_5). The sensory inputs were linearly scaled to a range of $[0; 1]$. If a visual target was not visible, the corresponding input value was set to -1. In each sensory-motor cycle (time step), the output of the neural network ($\sum_i w_i x_i$) determined which of two reinforcement learning modules that was selected. If the output was greater than zero the mating module was selected, otherwise the foraging module was selected. After a successful mating event, whether it reproduced offspring or not, an individual could not select the mating module again until it had captured an energy source or until 20 time steps had passed. During this time, the tail-lamp was turned off.

In the case when only an energy source and a tail-lamp were visible, the energy thresholds for the selection of the mating module, E_m, was given by

$$E_m(x_3, x_4) = -[w_1 + w_3 x_3 + w_4 x_4 - w_5]/w_2, \qquad (11)$$

which depended on the distance to the closest energy source (x_3) and the distance to the closest tail-lamp (x_4). In order to derive the average energy threshold $\bar{E}_m$, we computed the mean of E_m over 676 values of x_3 and x_4 (26 equidistant values between 0 and 1 for each of the two sensory inputs).

The reinforcement learning modules learned their behaviors from scratch in each generation with the aid of evolutionarily tuned potential-based shaping rewards and meta-parameters. The foraging module executed a foraging behavior using the relative angle and the distance to the nearest energy source as state variables. The mating module executed either a mating behavior or a waiting behavior, depending on the current sensory inputs. If a face of another robot was visible, the mating behavior was executed using the relative angle and the distance to the nearest face as state variables. Otherwise, the waiting behavior was executed using the relative angle and the distance to the nearest tail-lamp as state variables. The behaviors were learned by the Sarsa reinforcement learning algorithm [47,48] with tile coding [48] and potential-based shaping rewards [46]. The global reward for the reinforcement learning modules was set to +1 for a successful mating event and +1 for a capture of an energy source, otherwise the reward was set to 0.

The additional experiments, conducted to investigate the evolutionarily stability of the emerged polymorphic ESSs, were performed in a similar manner as the evolution experiments. The only difference was that, in each generation, the subpopulations were created by randomly selecting the genotypes of the different phenotypes from the final generation of the evolution experiment according to the predefined phenotypes ratios.

For a detailed description of our embodied evolution framework and algorithm specifics, see [35].

Author Contributions

Conceived and designed the experiments: SE KD. Performed the experiments: SE. Analyzed the data: SE. Wrote the paper: SE KD.

References

1. Darwin C (1987) The Correspondence of Charles Darwin: 1844–1846. Cambridge University Press.
2. Darwin C (1859) On the Origin of Species by Means of Natural Selection. London: Murray.
3. Darwin C (1874) The Descent of Man, and Selection in Relation to Sex (2nd ed.). London: Murray.
4. Ford EB (1975) Ecological Genetics (4th ed.). London: Chapman & Hall.
5. Smith JM (1998) Evolutionary Genetics (2nd ed.). Oxford University Press.
6. Smith JM, Price GR (1973) The logic of animal conflict. Nature 246: 15–18.
7. Smith JM (1982) Evolution and the Theory of Games. Cambridge University Press.
8. Brockmann HJ (2001) The evolution of alternative strategies and tactics. Advances in the Study of Behavior 30: 1–51.
9. Anderson WW (1969) Polymorphism resulting from the mating advantage of rare male genotypes. Proceedings of the National Academy of Sciences of the United States of America 64: 190–197.
10. Gadgil M (1972) Male dimorphism as a consequence of sexual selection. The American Naturalist 106: 574–580.
11. Rubenstein DI (1980) On the evolution of alternative mating strategies. In: Staddon JER, editor, Limits to Action: The Allocation of Individual Behavior, Academic Press, New York. pp. 65–100.
12. Gross MR (1996) Alternative reproductive strategies and tactics: Diversity within sexes. Trends in Ecology and Evolution 11: 92–98.
13. Brockmann HJ, Taborsky M(2008) Alternative reproductive tactics and the evolution of alternative phenotypes. In: Oliveira R, Taborsky M, Brockmann HJ, editors, Alternative Reproductive Tactics: An integrative Approach, Cambridge University Press. pp. 25–51.
14. Bisazza A (1993) Male competition, female mate choice and sexual size dimorphism in poeciliid fishes. Marine Behaviour and Physiology 23: 257–286.
15. Brockmann HJ, Penn D (1992) Male mating tactics in the horseshoe crab, limulus polyphemus. Animal Behaviour 44: 653–665.
16. Utami SS, Goossens B, Bruford MW, de Ruiter JR, van Hooff JARAM (2002) Male bimaturism and reproductive success in sumatran orang-utans. Behavioral Ecology 13: 643–652.
17. Roulin A, Ducret B, Ravussin PA, Altwegg R (2003) Female colour polymorphism covaries with reproductive strategies in the tawny owl strix aluco. Journal of Avian Biology 34: 393–401.
18. Mappes T, Koivula M, Koskela E, Oksanen TA, Savolainen T, et al. (2008) Frequency and densitydependent selection on life-history strategies a field experiment. Journal of Avian Biology 3: e1687.
19. Svensson E, Abbott JK, Gosden TP, Coreau A (2009) Female polymorphisms, sexual conflict and limits to speciation processes in animals. Evolutionary Ecology 23: 93–108.
20. Zimmerer EJ, Kallman KD (1989) Genetic basis for alternative reproductive tactics in the pygmy swordtail, xiphophorus nigrensis. Evolution 43: 1298–1307.
21. Shuster SM,Wade MJ (1991) Equal mating success among male reproductive strategies in a marine isopod. Nature 350: 608–610.
22. Ryan MJ, Pease CM, Morris MR (1992) A genetic polymorphism in the swordtail xiphophorus nigrensis: testing the prediction of equal fitnesses. American Naturalist 139: 21–31.
23. Thompson CW, Moore IT, Moore MC (1993) Social, environmental and genetic factors in the ontogeny of phenotypic differentiation in a lizard with alternative male reproductive strategies. Behavioral Ecology and Sociobiology 33: 137–146.
24. Radwan J (1995) Male morph determination in two species of acarid mites. Heredity 74: 669–673.
25. Lank DB, Smith CM, Hanotte O, Burke T, Cooke F (1995) Genetic polymorphism for alternative mating behavior in lekking male ruff philomachus pugnax. Nature 378: 59–62.
26. Sinervo B, Lively CM (1996) The rock-paper-scissors game and the evolution of alternative male strategies. Nature 380: 240–243.
27. Sinervo B, Zamudio KR (2001) The evolution of alternative reproductive strategies: fitness differential, heritability, and genetic correlation between the sexes. Journal of Heredity 92: 198–205.
28. Floreano D, Keller L (2010) Evolution of adaptive behaviour in robots by means of darwinian selection. PLOS Biology 8.
29. Floreano D, Mitri S, Magnenat S, Keller L (2007) Evolutionary conditions for the emergence of communication in robots. Current Biology 17: 514–519.
30. Mitri S, Floreano D, Keller L (2009) The evolution of information suppression in communicating robots with conflicting interests. PNAS 106.
31. Mitri S, Floreano D, Keller L (2011) Relatedness influences signal reliability in evolving robots. Proceedings of the Royal Society B 278: 378–383.
32. Wischmann S, Floreano D, Keller L (2012) Historical contingency affects signaling strategies and competitive abilities in evolving populations of simulated robots. PNAS Early Edition.
33. Waibel M, Floreano D, Keller L (2011) A quantitative test of hamilton's rule for the evolution of altruism. PLOS Biology 9.
34. Tarapore D, Floreano D, Keller L (2010) Task-dependent influence of genetic architecture and mating frequency on division of labour in social insect societies. Behavioral Ecology and Sociobiology 64: 675–684.
35. Elfwing S, Uchibe U, Doya K, Christensen HI (2011) Darwinian embodied evolution of the learning ability for survival. Adaptive Behavior 19: 101–120.
36. Fogel DB, Fogel GB (1995) Evolutionary stable strategies are not always stable under evolutionary dynamics. In: Evolutionary Programming. pp. 565–577.
37. Fogel DB, Fogel GB, Andrews PC (1997) On the instability of evolutionary stable strategies. Biosystems 44: 135–152.
38. Fogel GB, Andrews PC, Fogel DB (1998) On the instability of evolutionary stable strategies in small populations. Ecological Modelling 109: 283–294.
39. Ficici S, Pollack JB (2000) Effects of finite populations on evolutionary stable strategies. In: Proceedings of the Genetic and Evolutionary Computation Conference (GECCO2000). pp. 927–934.
40. Doya K, Uchibe E (2005) The cyber rodent project: Exploration of adaptive mechanisms for self-preservation and self-reproduction. Adaptive Behavior 13: 149–160.
41. Sutton RS, Barto A (1998) Reinforcement Learning: An Introduction. MIT Press.
42. Elfwing S, Uchibe E, Doya K (2009) Emergence of different mating strategies in artificial embodied evolution. In: Proceedings of International Conference Neural Information Processing (ICONIP2009). volume 5864, pp. 638–647.
43. Sandell M, Liberg O (1992) Roamers and stayers: A model on male mating tactics and mating systems. The American Naturalist 139: 177–189.
44. Leonard JL (2006) Sexual selection: lessons from hermaphrodite mating systems. Integrative and Comparative Biology 46: 349–367.
45. Da Rold F, Petrosino G, Parisi D (2011) Male and female robots. Adaptive Behavior 19: 317–334.
46. Ng AY, Harada D, Russell SJ (1999) Policy invariance under reward transformations: theory and application to reward shaping. In: Proceedings of the International Conference on Machine learning (ICML1999). San Francisco, CA: Morgan Kaufmann Publishers Inc., pp. 278–287.
47. Rummery GA, Niranjan M (1994) On-line Q-learning using connectionist systems. Technical Report CUED/F-INFENG/TR 166, Cambridge University Engineering Department.
48. Sutton RS (1996) Generalization in reinforcement learning: Successful examples using sparse coarse coding. In: Proceedings of Advances in Neural Information Processing Systems (NIPS1996). Volume 8, pp. 1038–1044.

9

Phenotypic Resistance and the Dynamics of Bacterial Escape from Phage Control

James J. Bull[1,2,3]*, Christina Skovgaard Vegge[4], Matthew Schmerer[3], Waqas Nasir Chaudhry[5,6], Bruce R. Levin[5]

1 The Institute for Cellular and Molecular Biology, The University of Texas, Austin, Texas, United States of America, 2 Center for Computational Biology and Bioinformatics, The University of Texas, Austin, Texas, United States of America, 3 Department of Integrative Biology, The University of Texas, Austin, Texas, United States of America, 4 Department of Veterinary Disease Biology, University of Copenhagen, Frederiksberg, Denmark, 5 Department of Biology, Emory University, Atlanta, Georgia, United States of America, 6 Atta-ur-Rahman School of Applied Biosciences, National University of Sciences and Technology, Islamabad, Pakistan

Abstract

The canonical view of phage - bacterial interactions in dense, liquid cultures is that the phage will eliminate most of the sensitive cells; genetic resistance will then ascend to restore high bacterial densities. Yet there are various mechanisms by which bacteria may remain sensitive to phages but still attain high densities in their presence – because bacteria enter a transient state of reduced adsorption. Importantly, these mechanisms may be cryptic and inapparent prior to the addition of phage yet result in a rapid rebound of bacterial density after phage are introduced. We describe mathematical models of these processes and suggest how different types of this 'phenotypic' resistance may be elucidated. We offer preliminary *in vitro* studies of a previously characterized *E. coli* model system and *Campylobacter jejuni* illustrating apparent phenotypic resistance. As phenotypic resistance may be specific to the receptors used by phages, awareness of its mechanisms may identify ways of improving the choice of phages for therapy. Phenotypic resistance can also explain several enigmas in the ecology of phage-bacterial dynamics. Phenotypic resistance does not preclude the evolution of genetic resistance and may often be an intermediate step to genetic resistance.

Editor: Daniel E. Rozen, Leiden University, Netherlands

Funding: Supported by NIH GM57756 (JJB) and GM 091875 (BRL), Higher Education Commission Pakistan (WNC). JJB is also supported as the Miescher Regents Professor (U. Texas). The funders had no role in study design, data collection and analysis, decision to publish, or preparation of the manuscript.

Competing Interests: The authors have declared that no competing interests exist.

* E-mail: bull@utexas.edu

Introduction

Bacterial viruses – bacteriophages or 'phages' – were central to the foundations of modern genetics and molecular biology, and their properties have been studied thoroughly. And for well over half a century, the standard model for a lytic phage invading a high-density bacterial population has been one of the evolution of genetic resistance: the phage will kill sensitive bacteria, whereupon either sensitive bacteria will be maintained at low density thereafter, or genetically resistant mutants that are initially rare will ascend to abundance [1–4]. If resistance incurs a fitness cost, the final population will include a mix of sensitive and resistant cells, with phage maintained on the sensitive population and possible co-evolution of phage and bacteria [4–7]. A dominant dynamical feature of this process is an initial and profound depression of bacterial densities when the phage first invades, with a gradual recovery of bacterial densities if resistant cells were initially present.

We describe a process that has two important differences from the standard model. First, resistance is partial or quantitative, rather than absolute. Second, resistance is phenotypic (environmental, in quantitative genetics phraseology) rather than genetic. When both conditions are satisfied, the bacterial population may respond to phage infection with only a moderate initial drop in density; bacteria recover quickly to a state in which both bacteria and phage are maintained at high density. Phenotypic resistance may be characterized by just one or a few discrete states in the population up to a virtually continuous distribution of degrees of partial resistance.

The ramifications of phenotypic resistance are subtle but potentially profound. Partially resistant bacteria may be absent or at low frequency in the initial population, and thus be cryptic, but they can ascend quickly and have a major effect on dynamics. They may thus be a cause of phage therapy failure, and understanding the phenomenon may avoid such failures. Phenotypic resistance may also explain the stable coexistence of phage in bacteria in environments that are otherwise predicted to experience oscillations (undamped or of increasing magnitude, e.g., [8]).

The goal of this paper is to elaborate this alternative model and describe some of its consequences. The first section presents and reviews empirical results from a couple well characterized systems to motivate the model. Subsequently, using mathematical models, we describe methods to detect phenotypic resistance and consider its implications to the population and evolutionary dynamics of bacteria and lytic phage.

Results

1. Empirical contradictions with the standard model

The perspective in this paper is motivated by observations that defy the standard model: in short-term bacterial cultures exposed to phage, an initial decline in bacterial densities caused by phage killing is followed by a bacterial rebound amid an abundance of phage, but the cells are predominantly sensitive. This pattern was reported recently for *Streptococcus thermophilus*, and the anomaly of the rise in sensitive bacteria was noted [9]. We likewise report a similar pattern here for *E. coli* O18:H7:K1 (Fig. 1). The anomaly is that bacterial densities increase after initially being depressed by phage, yet phage density is higher during the bacterial increase than it was during the decrease – constancy of parameters cannot explain this pattern [8,9].

Various mechanisms could underlie the rise of sensitive bacteria in the presence of phage. The most plausible is a decline in the effective adsorption rate of phage to bacteria, as a low adsorption rate constant is known to enable coexistence of high densities of phage and sensitive bacteria [8,10]. Any mechanism that involves loss of phage or irreversible binding of phage to debris would reduce viable phage density rather than leave it intact. The fact that the ascending bacteria are sensitive violates the classic model of a genetic resistance mutation. There are several types of non-genetic, partial resistance (low adsorption rate constant) that could explain it, however, as considered next.

2. Interpreting and modeling phenotypic resistance

2.A. Importance of the adsorption rate in phage-bacterial dynamics. In well mixed cultures, phage-bacterial interactions are modeled as a mass action process [1,2]. The rate at which phage-bacterial collisions occur and result in infection is represented as the product of bacterial density (B, per mL), phage density (P, per mL) and the adsorption rate constant (k, mL per

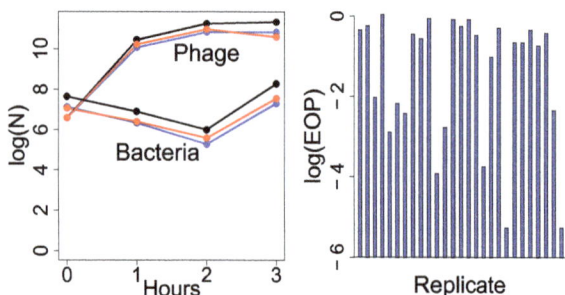

Figure 1. Evidence of phenotypic resistance. (Left) Short-term dynamics of *E. coli* O18:K1:H7 with phage K1-ind(1). This phage does not require the K1 capsule for infection, and although this phage grows well when introduced at low density to a bacterial culture, it and others like it perform poorly in preventing mortality of mice infected with the bacterium [24–27]. The assay here grew bacteria 1 hr at 37° in 10 mL LB with aeration, added phage at an MOI of 0.1 (time 0), then plated phage and bacteria from this culture at times shown by dots. Three replicates were performed, each designated with a different color. Cells from the final, 3 hr culture were plated and 5–10 colonies tested for phage sensitivity; most colonies in each replicate were sensitive to the phage [ratios of sensitive to total were (9+1)/10, (6+1)/7, and (5+2)/9, the second number in parentheses indicating intermediate sensitivity]. As the density of phage exceeded that of cells by 3–4 orders of magnitude at the time of plating, we often observed that sensitive colonies were contaminated by low densities of phage, so these numbers should be considered underestimates of sensitivity. (Right) Efficiency of plating (EOP) values of phage F287 on 27 independent colonies of *Campylobacter jejuni* obtained from a 24 hr culture inoculated from a single colony. The culture has already accumulated considerable variation in phage sensitivity in the absence of phage. EOP is the ratio of number of plaques obtained on the isolate divided by the number of plaques obtained on a standard sensitive strain. In both panels, the vertical axis is $\log_{10}$ of the respective quantity.

min): kPB. The adsorption rate constant is a combined rate at which a bacterium and phage encounter each other by diffusion and also in which the encounter results in an infection. It is typically bounded at the high end by a value between $10^{-8} - 10^{-9}$ [1], but values 2–3 orders of magnitude lower can be observed even when phage are maintained [8]. The fact that the adsorption rate constant is bounded at the high end by such a tiny value is due to physical constraints: the small sizes of bacteria and phage greatly limit their chance encounter from diffusion in a mL of liquid [1]. Thus, when the adsorption rate constant lies near this upper boundary, virtually all collisions that do occur between phage and bacteria result in infection – the probability of infection per contact is near 1. Smaller values of k thus reflect reduced probabilities of infection per contact.

It is well understood that, in a flow-through system (e.g., chemostat) with a single type of phage and bacterium, the adsorption rate has a major effect on equilibrium densities. The equilibrium densities are in fact inversely proportional to the adsorption rate constant [8]. Furthermore, low adsorption rates ($\approx 10^{-11}$ or less) lead to the paradoxical outcome that both phage and sensitive bacteria coexist at high densities. That result is key to understanding the importance of quantitative variation in adsorption rates because the individual bacteria with low adsorption rates differentially survive phage attack and convert the bacterial population into a state of predominantly low adsorption rate.

2.B. Three possible mechanisms of phenotypic variation in resistance. Within the range of attainable values, the adsorption rate constant is a property of both the bacterium and phage, affected by genetic and by non-genetic properties of each. Changes in the number of receptors on the bacterium will affect the adsorption rate constant as will variation in the number of tail fibers on the phage. Bacteria can completely escape phage infection by modification or elimination of essential receptors, but they can also quantitatively reduce the probability of infection by reducing or masking receptors. Indeed, several types of variation in bacterial resistance to a phage can have non-genetic causes. It has long been appreciated that starved or stationary phase cells can have low adsorption rates, perhaps because of reduced cell size and also because of reduced receptor abundance [11,12]. Even with actively growing cells, any factor altering the abundance or availability of receptors can impart quantitative levels of resistance. It is thus plausible that populations of genetically uniform bacteria contain a mix of cells that collectively present a distribution of different adsorption rates (e.g., Fig. 1).

At least three mechanisms of phenotypic resistance can be envisioned. We acknowledge that most of these mechanisms are speculative, being inferred indirectly from other types of evidence on bacterial phenotypes:

a. Induced. Reduced adsorption is due to a change in uninfected bacterial gene expression after exposure to products of phage-lysed bacteria in the environment.

b. Intrinsic. Reduced adsorption is due to a physiological or gene expression state that exists in a subset of the population prior to the introduction of phage. It is a consequence of normal environmental variation or a high mutation rate (Fig. 1).

c. Dynamic. Reduced adsorption results from degradation or blocking of bacterial receptors by phage proteins released by lysed cells [11].

A category whose effects overlap intrinsic resistance is phase variation. Phase variation is a genetic process of rapid mutation between states of expression and suppression of genes encoding bacterial surface receptors, and the consequences of phase variation to evolution of resistance to phages are well appreciated [13], although the implications to dynamics have not been addressed. A major difference between phase variation and our use of intrinsic variation lies in the degree of resistance of individual cells. With phase variation, cells in the resistant state may be completely resistant (lacking the receptor), whereas with intrinsic variation, the resistant state is partial – cells can be infected. Thus bacteria with intrinsic resistance would manifest clear plaques but phase variation would manifest turbid plaques.

We offer these categories as possibilities. Although there seems to be appreciation among phage biologists of some of these mechanisms, for the most part, they and their consequences have not been systematically explored. Nor is it necessarily obvious how to recognize and distinguish them. We thus work out their consequences below.

3. Bacterial populations prior to phage invasion

We offer models to illustrate several consequences of quantitative, phenotypic resistance. The category of 'intrinsic' variation is distinct from the other two categories in that the reduced adsorption rate exists prior to exposure of the population to phages. Because it exists prior to phage invasion, intrinsic variation can be assessed in phage-free cultures, rendering its properties both subtle but easy to model. Thus more of our treatment is on intrinsic variation than on the other two mechanisms. Although most derivations assume just 2 bacterial types in the population, we expect that phenotypic resistance may involve a broad spectrum of partial resistance levels (e.g., Fig. 1).

3.A. Abundance of different types in a phage-free bacterial population. Consider the bacterial population before phage are introduced. Phenotypic resistance in the absence of phage can be ignored with induced and dynamic mechanisms of phenotypic resistance, but it will be present under the intrinsic mechanism. If there are two bacterial states (1, 2) that have no effect on reproduction and survival in the absence of phage, dynamics of the two cell types obey

$$\dot{B}_1 = \alpha B_1 - s_{12}B_1 + s_{21}B_2 - dB_1$$

$$\dot{B}_2 = \alpha B_2 - s_{21}B_2 + s_{12}B_1 - dB_2, \tag{1}$$

where B_i represents the density of bacterial type i, α is the growth rate, d the death rate, and s_{ij} is the switching rate from state i to j. A superior dot indicates differentiation with respect to time.

Without loss of generality, we assume that the death rate matches the birth rate, which will be true at equilibrium under a diversity of mechanisms. The equilibrium abundance of the two states is thus given by

$$\frac{\hat{B}_1}{\hat{B}_2} = \frac{s_{21}}{s_{12}}, \tag{2}$$

with a hat indicating equilibrium. Thus the higher the relative switching rate away from a state, the lower the state's equilibrium relative abundance. The simplicity of this result is due to our assumption that the two states are the same in all parameters except switching rates.

With more than two states and arbitrary switching, s_{ij}, there is no straightforward analytical solution. Equilibrium satisfies

$$(\mathbf{S}^{\mathbf{T}} - \mathbf{I})\vec{\hat{B}} = \vec{0}, \tag{3}$$

where $\mathbf{S}^{\mathbf{T}}$ is the transpose of matrix $\{s_{ij}\}$, $\mathbf{I}$ is the identity matrix, and $\vec{\hat{B}}$ is the vector of equilibrium bacterial densities. The matrix $\mathbf{S}$ has the property that row elements sum to 1. We have not explored the possibility of an analytical solution to (3) nor can we suggest that a solution is unique. If the s_{ij} are known, however, a numerical solution is straightforward.

3.B. Survival of bacteria exposed to high densities of phage: differences in adsorption rate are profoundly important. Under the intrinsic mechanism, many adsorption rate states may initially be rare in the population. Here we show that the addition of a high concentration of phage to the bacteria results in large and rapid changes in the relative abundances of different cell types. These rapid changes may be useful in detecting the otherwise cryptic variation in adsorption rate.

It is first noteworthy that standard adsorption rate assays will be insensitive to minor bacterial variants in the population. Adsorption rate assays are measured from the decline of free phage density in a culture with bacteria – phage are added at time 0, and the number of attached versus unattached phages are counted at one or more later times. To understand how the measured adsorption rate is affected by variation among bacteria, we need to know how the free phage density declines as a property of the different bacterial states in the population. Let the density of bacterial type i be $B_i = Bq_i$, where B is total bacterial density, q_i is B_i/B, hence $\sum_i q_i = 1$. The rate at which phage are lost to adsorption is

$$\dot{P} = -k_1PB_1 - k_2PB_2 - \ldots$$
$$= -\left(\sum_i k_iq_i\right)PB = -\bar{k}PB, \tag{4}$$

where P is phage density, k_i is the adsorption rate constant for bacterial type i, and $\bar{k} = \sum_i q_ik_i$. These dynamics assume the absence of phage reproduction, as applies in adsorption rate assays. And if bacteria already adsorbed to phage continue to adsorb additional phage, the densities of bacteria adsorbing phage can be assumed to stay constant because the sum of live and previously infected bacteria does not change (provided the time frame is confined to the interval before infected bacteria lyse – burst open). From this result it is apparent that any rare bacterial types – those with low q_i – will have little effect on the measured adsorption rate, which is an average over all types.

Now consider changes in densities of surviving bacteria as a function of their adsorption rates. Let α be the growth rate of the uninfected bacteria. Assuming two types of cells, parameters as above, and no phage reproduction (e.g., a brief interval after phage are added to the culture), the rates of change of surviving bacteria in the presence of phage are

$$\dot{B}_1 = -k_1B_1P + \alpha B_1 \tag{5}$$

$$\dot{B}_2 = -k_2B_2P + \alpha B_2.$$

To get an approximate and easily interpretable sense of dynamics, first assume constant phage density (and again no phage reproduction). The density over time of uninfected bacteria of type i ($B_i(t)$) follows

$$B_i(t) = B_i(0)e^{(\alpha - k_i P)t}, \tag{6}$$

so the ratio of two surviving bacterial types obeys

$$\frac{B_1(t)}{B_2(t)} = \frac{B_1(0)}{B_2(0)} e^{P(k_2 - k_1)t}. \tag{7}$$

When phage density is high (on the order of the inverse of the larger adsorption rate constant), even a ten-fold difference in adsorption rate has a profound effect on relative abundances of surviving bacterial types in as little as 10 minutes.

This result points to a way of detecting standing variation in adsorption rate under the intrinsic model: infect a bacterial culture with a high density of phage and plate to measure bacterial survival before phage reproduce (which is also before bacteria lyse). Different cell types will survive at different rates. A more exact calculation than that in (7) is required because the density of free phage is not constant but declines as phage adsorb; equations from Bull and Regoes [14] can be used to correct for this effect. Returning to eqn (4) but making explicit the constancy of bacterial densities for adsorption,

$$\dot{P} = -P \sum_i k_i B_i(0), \tag{8}$$

with the solution

$$P(t) = P(0)e^{-t \sum_i k_i B_i(0)}. \tag{9}$$

Note that bacterial densities are held constant for the sake of phage adsorption, even though some or many of those bacteria are infected and will be unable to form colonies. We are, however, also interested in bacterial survivors, but we need this result to make that calculation.

The decay in each type of surviving (uninfected) bacteria can now be calculated. Assuming a time interval so short that bacterial reproduction can be neglected, eqns (5) and (9) yield

$$\begin{aligned}\dot{B}_i &= -k_i B_i(t) P(t) \\ &= -k_i B_i(t) P(0) e^{-t \sum_i k_i B_i(0)}\end{aligned} \tag{10}$$

with the solution

$$\ln B_i(t) = \frac{k_i P(0)}{\sum_i k_i B_i(0)} \left[e^{-t \sum_i k_i B_i(0)} - 1 \right] + \ln B_i(0). \tag{11}$$

Bacterial survival curves are shown in Fig. 2; it is assumed that the only effect on phage density is loss through adsorption, hence no release of phage progeny occurs within the plotted time frame. The left panel assumes a homogenous cell population (a single k for all cells) in which free phage initially outnumber bacteria. The green and blue curves are for 100-fold excess of phage; the green curve is forced to be linear, as would be obtained with strictly constant phage density, whereas the blue curve accounts for the declining phage density due to adsorptions; curvature of the blue deviates almost imperceptibly from linear. The gray curve represents a 10-fold excess of phage and violates linearity

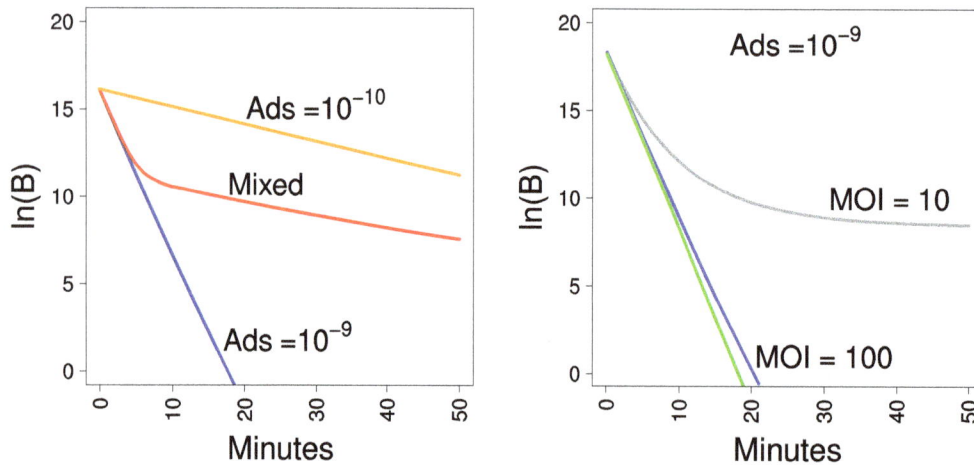

Figure 2. Survival curves of bacteria. Bacteria are exposed to a high initial density of phage, and survivors (uninfected) measured before burst occurs; cells continue adsorbing to phage regardless of how many phage have already adsorbed. The vertical axis represents the natural log of surviving bacterial densities. (Left) Survival curves for homogeneous bacterial populations. The blue curve represents a 100-fold excess of phage to cells (MOI is multiplicity of infection, or the ratio of phage to cells), whereas the gray curve is for only a 10-fold excess; initial phage densities are 10^9/mL, adsorption rate constants are 10^{-9} mL/min. The blue and gray curves account for a decay in phage concentration as phage continue adsorbing to bacteria. The green curve is strictly linear for comparison to the blue curve. The plateau of the gray curve is due to nearly all phage having adsorbed after 20–30 min. Note that all three curves are adjusted to start at the same bacterial density for visual comparison of slopes, even though the blue curve should start at a 10-fold lower bacterial density than the gray curve. (Right) Survival curves for two homogenous bacterial populations differing in adsorption rate constants (orange: 10^{-10} mL/min; blue: 10^{-9} mL/min). The red curve is for a mix of 99% cells with a rate constant of 10^{-9} and 1% cells with rate constant 10^{-10}. Thus, despite the small fraction of cells with low adsorption rate in the mixed population, there is a profound effect on the survival curve after just a few minutes of exposure. Initial phage density is 10^9/mL, and cells are at 1.1×10^7/mL for each homogenous population and 10^7/mL and 10^5/mL for the two types in the mixed population.

profoundly; flattening of the curve is due to the loss of nearly all free phage from adsorption after 20–30 min.

The right panel compares two curves for pure cultures with one curve for a culture containing a mix of bacteria with two resistance levels. Phage start with a 100-fold excess over cells; the blue curve assumes that all cells have an adsorption rate of 10^{-9}, the orange curve assumes that all cells have an adsorption rate of 10^{-10}; the red curve assumes that 99% of the cells have an adsorption rate of 10^{-9} and the other 1% have an adsorption rate of 10^{-10}. Thus a small fraction of cells with reduced adsorption can have a major effect on the survival curve when there is a large excess of phage. It is thus evident that this method offers a means of detecting standing variation in adsorption rate.

This survival curve will be sensitive to the nature of phenotypic resistance and to the properties of the phage suspension added to the cells. The foregoing calculations assumed that resistance is pre-existing (intrinsic) and that no induction of resistance or dynamic resistance occurs. Depending on the time course of the bacterial response, qualitatively similar shapes – curve flattening – would be obtained if the bacteria had no pre-existing variation in resistance but exhibited dynamic or induced resistance in response to the addition of phage. Use of a purified phage preparation (e.g., from a cesium gradient) would presumably avoid both induced and dynamic resistance prior to lysis. Conversely, pre-exposure of the bacteria to a lysate in which all the phage were killed (e.g., by heat or radiation) might cause a marked shift in the adsorption rate before viable phage were added and thus reveal dynamic or induced resistance.

Empirical determination of the survival curve is not trivial. To use these equations, plating of surviving bacteria must be done before burst – because we do not allow phage reproduction or reduction in cells adsorbing phage – but infected bacteria will burst after plating. Once the density of surviving bacteria has fallen several logs, larger volumes must be plated and the plates will have a large excess of phage that may kill otherwise surviving microcolonies; use of phage anti-sera or specific detergents may reduce the effect. In addition, survival rates may be biased upward because of anomalies such as clumps and debris protecting bacteria from phage. Empirical determination of the survival curve will be least problematic when the variation in adsorption rate is evident over only a few logs drop in viable cell density, while volumes plated are still small.

4. Phage growth

4.A. Phage invasion is determined by the average adsorption rate, so phenotypic resistance is cryptic by this measure. As a first step toward understanding the impact of phenotypic resistance on phage dynamics, we consider phage invasion – is phenotypic resistance readily detectable when a phage first invades? Phage growth is easily quantified, and if this measure of invasion reflects long term dynamics, it would provide a convenient short-cut to more elaborate assays. Invasion of a phage into the bacterial population is modeled as if viable bacteria are not limiting and maintain a constant density, in which case exponential phage growth applies soon after the introduction. Following Campbell [2], the rate of change in phage density (P) in a population with a single cell type (single adsorption rate) is given by

$$\dot{P} = -kPB + bkP_L B, \qquad (12)$$

where L is the lysis time, b is burst size, and P_L is the density of phage L time units in the past. Bacterial density B is assumed

constant in time, and we ignore phage loss from death and washout (which is easily included).

The model is readily expanded to include multiple bacterial types that have different effects on phage life history. As above, let B denote total cell density and q_i be the fraction of bacteria with adsorption rate k_i, with lysis time L_i and burst size b_i. With constancy of total bacterial density as well as of the q_i, equation (12) becomes

$$\dot{P} = -P(k_1 B_1 + k_2 B_2 + \ldots) + (b_1 P_{L_1} k_1 B_1 + b_2 P_{L_2} k_2 B_2 + \ldots)$$
$$= -PB\bar{k} + \sum_i b_i P_{L_i} k_i B_i. \qquad (13)$$

When exponential growth has been attained (requiring a stable age-of-infection distribution, [15]), the phage growth rate r obeys

$$r = B\left(\sum_i q_i k_i b_i e^{-rL_i} - \bar{k}\right),$$

or if all infections have the same lysis time and burst size,

$$r = \bar{k}B(be^{-rL} - 1) \qquad (14)$$

Thus if adsorption rate is the only variable, invasion proceeds according to the average adsorption rate in the population, a result that might be inferred from (4) above. Bacteria whose abundance is at most a few per cent have little or no perceptible impact on how rapidly the phage population expands at low density, so their ultimate impact on dynamics is cryptic at invasion. The model is general to all three mechanisms of phenotypic resistance, in that phage are rare enough during invasion that the induced and dynamic mechanisms will not operate.

4.B. Full dynamics with intrinsic resistance. To evaluate the dynamics after invasion, we use a differential equation model similar to that in Levin et al. [8]. There are 6 variables: sensitive bacteria (B_1), semi-resistant bacteria (B_2), infected bacteria of each type (I_1, I_2), phage (P), and resource level in the environment (R). The two types of bacteria accommodate our definition of intrinsic resistance, although the model does not distinguish genetic and non-genetic bases for the two types, provided the switching rates apply. Loss and gain terms for bacteria are straightforward; infected bacteria are included because they continue to adsorb phage and use resources up to lysis. Gains to phage density come solely from the lysing of infected bacteria, whereas phage losses include adsorption to infected and uninfected bacteria as well as death/washout (which diminishes all variables at the same per capita rate). Also, the bacterial growth rate declines as bacterial abundance increases; we use a standard (if detailed) model with a Monod function of metabolism to approximate biological reality [8] but suggest that many simpler functions would behave similarly.

Our equations follow:

$$\dot{B}_1 = v\psi B_1 - k_1 B_1 P - s_{1,2} B_1 + s_{2,1} B_2 - wB_1 \qquad (15)$$

$$\dot{B}_2 = v\psi B_2 - k_2 B_2 P + s_{1,2} B_1 - s_{2,1} B_2 - wB_2$$

$$\dot{P} = bP_L e^{-wL}\left[k_1 B_{1_L} + k_2 B_{2_L}\right] - P(w + k_1 I_1 + k_2 I_2)$$

Table 1. Variables and parameters.

Notation	Description	Values
Variables		
B_1	density of sensitive bacteria	
B_2	density of partially resistant bacteria	
I_1	density of infected bacteria of type 1	
I_2	density of infected bacteria of type 2	
P	phage density	
R	resource level	
S	density of substance affecting adsorption in dynamic model	
Functions of variables		
ψ	Resource scaling level ($=R/(R+M)$)	
$k(S)$	Dynamic adsorption rate ($=[A+(100-A)e^{-\sigma S}]10^{-11}$)	
Parameters		
w	washout/death rate	0.1
b	burst size	200
L	lysis time	25
$s_{1,2}$	transition rate from sensitivity to partial resistance	10^{-4}
$s_{2,1}$	transition rate from partial resistance to sensitivity	5×10^{-3}
k_1	adsorption rate constant for B_1	1×10^{-9}
k_2	adsorption rate constant for B_2	varies
v	maximum bacterial growth rate	0.35
e	conversion efficiency	5×10^{-7}
C	input resource concentration	100
M	Monod constant	0.25
A	minimum adsorption rate determinant in $k(S)$	3, 10
σ	efficacy parameter in $k(S)$	$10^{-6}, 10^{-7}$

$$\dot{I}_1 = -k_1 P_L B_{1_L} e^{-wL} + k_1 P B_1 - w I_1$$

$$\dot{I}_2 = -k_2 P_L B_{2_L} e^{-wL} + k_2 P B_2 - w I_2$$

$$\dot{R} = w(C-R) - \psi ev[(B_1+I_1)+(B_2+I_2)].$$

Parameters and variables are defined in Table 1. A subscript L indicates the value of the variable L minutes in the past – the time from infection to lysis of the bacterium.

To obtain a sense of model behavior, four runs of different conditions are illustrated (Fig. 3). The upper left panel shows the behavior for a baseline set of parameters and initial conditions, and the other three panels change one of those values or conditions. In the absence of phage, semi-resistant bacteria (red curve) would be maintained at 2% of the population. With phage present, those bacteria are invariably maintained at a higher density than the sensitive cells. The various results indicate the sensitivity of outcomes to small changes in parameter values.

Oscillations are typical in this type of model [8], but even with oscillations present, semi-resistant bacteria often remain at high density, usually one to two orders of magnitude above sensitive cell density. Thus phage profoundly change the relative abundances of different resistance levels, and phage are less effective at controlling semi-resistant cells.

4.C. Full dynamics with induced or dynamic resistance. Our model of dynamic resistance differs from that of intrinsic resistance in three ways: (i) only a single bacterial type is present, (ii) the variable S is introduced for the concentration of a substance released by lysing cells, and (iii) adsorption rate is a declining function of S, $k(S)$. Equations are much the same as in the previous model except for the effect of S and the omission of a second bacterial type:

$$\dot{P} = bk(S_L)P_L B_L e^{-Lw} - k(S)PB - k(S)IB - wP \quad (16)$$

$$\dot{B} = v\psi B - k(S)PB - wB$$

$$\dot{S} = k(S_L)P_L B_L e^{-Lw} - wS$$

$$k(S) = [A+(100-A)e^{-\sigma S}]10^{-11}$$

$$\dot{I} = -k(S_L)P_L B_L e^{-Lw} + k(S)PB - wI$$

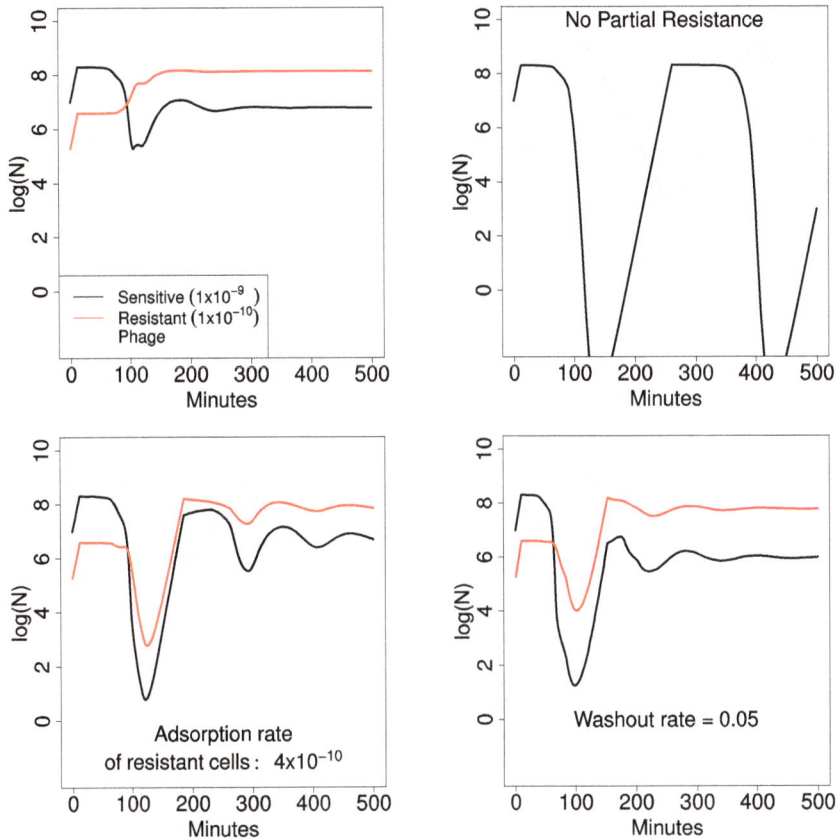

Figure 3. Iterations of the 'intrinsic' model, as given by eqn(15). The upper left panel represents a run with baseline parameter values (Table 1; the key gives adsorption rate constants), and only those properties differing from this baseline are noted in the other panels. The three other trials differ in one respect from the upper left: partially resistant bacteria are absent (upper right), the adsorption rate constant of the partially resistant bacteria is increased 4-fold (lower left), or the washout rate is halved (lower right). The red curve represents density of the partially resistant cell type, black is of the sensitive type, and light gray is of phage. In the absence of phage, density of the partially resistant bacteria would be 2% of the total, but partially resistant bacteria always comprise the majority when phage are present (the upper right panel is a control run in which partially resistant bacteria are absent, illustrating the typical oscillations). Initial densities were 10^7 for B_1, 2×10^5 for B_2, 10^5 for P, 100 for R, and 0 for both types of infected cells. The vertical axis is $\log_{10}$ of the respective density.

$$\dot{R} = w(C - R) - \psi ev(B + I).$$

Table 1 again provides definitions and values, except that subscripts for B and I are omitted in this model. The adsorption rate function $k(S)$ ranges from a maximum of 1×10^{-9} when S is 0 to a lower bound of $A \times 10^{-11}$ $(0 < A \leq 100)$. The impact of S between these extremes depends exponentially on the product σS, so σ can be regarded as an efficacy parameter (baseline $\sigma = 10^{-6}$). These equations can represent either dynamic resistance (S interferes with adsorption) or induced resistance (S causes bacteria to reduce receptor levels), although the effect of induced resistance might be more accurately modeled as a delay function.

Fig. 4 shows that dynamic/induced resistance can stabilize oscillations. As expected, the stabilizing effect depends on how low the adsorption rate constant becomes through the impact of S. By comparison to the first two panels, the third panel shows that the dynamics are highly sensitive to the 'efficacy' parameter σ.

Discussion

Bacterial genetics of the 1960s and 1970s was dominated by the model of a one-to-one relationship between genotype and phenotype. Bacteria could or could not ferment specific sugars, could or could not synthesize amino acids or vitamins, and were susceptible or resistant to killing by lytic bacteriophages. In recent years it has become clear that there is qualitative as well as quantitative phenotypic variation among genetically identical members of the same bacterial population. Prominent, extreme examples are the genetically identical states of competence and cannibalism in *Bacillus subtilis* and the phenomenon of 'persistence' and its associated phenotypic resistance to antibiotics [16]. More broadly, there is now a wide appreciation of cell-cell variation in metabolism within populations [17–19], with relevance to drug resistance [20]. Phase variation in bacteria (e.g., [13]) is an intermediate category that lies between the standard model of genetic resistance and that of phenotypic resistance – its rapid switching between genetic states renders it similar to our use of phenotypic resistance, but the complete resistance and modification of the DNA sequence renders it 'genetic.'

The existence and relevance of bacterial phenotypic variation to phage dynamics has received little attention. Weitz and Dushoff ([12]) developed a model in which the adsorption rate constant declined with increasing cell density, reflecting the common observation that cells at or near stationary phase do not lead to productive infections [21]. That model has multiple stable dynamical states, with phage unable to invade at high bacterial

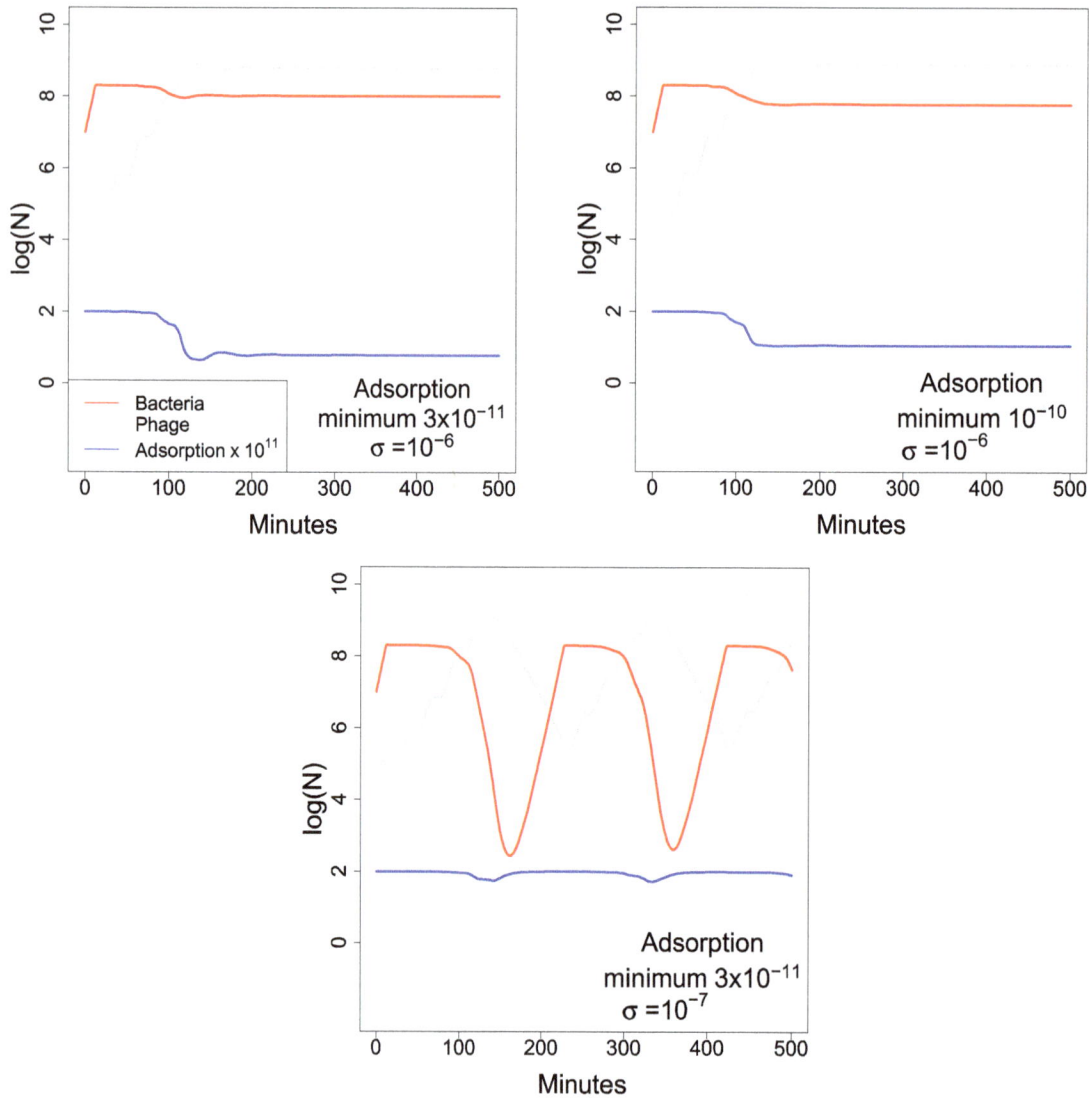

Figure 4. Iterations of the 'dynamic' or 'induced' model of phenotypic resistance, as given by eqn(16). Only one bacterial type is present. The maximum adsorption rate is 10^{-9} in all three panels, but the minimum adsorption rate is $A \times 10^{-11}$: $A = 3$ in the upper left and lower panels, $A = 10$ in the upper right panel. Collectively these figures illustrate that the magnitude of oscillations is moderately robust to differences in the minimum absorption rate but sensitive to the adsorption rate function (as determined by σ). Red (light gray) curves are for bacteria (phage), the blue curve is the adsorption rate multiplied by 10^{11}, which varies with S. Parameter values are as given in Table 1 except where indicated. Initial densities were 10^7 for B, 10^5 for P, 100 for R, 100 for S, and 0 for infected cells. The vertical axis is $\log_{10}$ of the respective density.

density but maintained at lower cell density, for example. Closer to the theme of our study, a recent article presented evidence for phenotypic variation in the susceptibility of bacteria to lytic phage [9]: when infected with the phage 2972, a subset of a genetically susceptible population of *Streptococcus thermophilus* not only survives in the presence of high phage densities, it replicates. Here we presented qualitatively similar results for an *E. coli* strain and its lytic phage, reported within-culture variation in efficiency of plating for a *Campylobacter* strain and its phage, and also addressed the dynamical implications of phenotypic resistance.

The rapid replication of bacteria phenotypically resistant to phage distinguishes the phenomenon from bacterial persisters that resist antibiotics – whose cells do not replicate. The replication of phenotypically resistant bacteria enables them to rebound to potentially high densities in the presence of phages, leading to the

enigma that initially effective phages cease to control bacteria even though the bacteria technically remain sensitive.

We offered three possible types of phenotypic resistance: induced, intrinsic, and dynamic. Of the three, only intrinsic resistance exists prior to phage introduction. The genetic and molecular mechanisms behind the different types remain largely unexplored, however. Furthermore, mechanisms of intrinsic resistance may span the spectrum from formal genetic mutations to epigenetics to stochastic variation in protein numbers following cell division. Indeed, phase variation is a mechanism of intrinsic resistance that generates rapid switching of resistance states in ways that might appear to be epigenetic or non-genetic. Considerable empirical effort may thus be required to distinguish the alternative mechanisms.

The implications of phenotypic resistance to phages have both practical and basic dimensions. In phage therapy – the use of phages to control medical and industrial bacterial parasites and pests – phenotypic resistance can lead to a mismatch between the short term growth of a phage and its long term ability to control bacterial populations. Thus a phage that grows well when first introduced may transform the bacterial population into a state that is recalcitrant to phage control even though the bacteria remain sensitive. This mismatch can complicate the wise choice of phages suitable for therapy, and an appreciation of how phenotypic resistance arises may aid the choice of phages suited to therapy (see below).

From the broader perspective of phage-bacterial dynamics, phenotypic resistance may explain a long standing enigma – the failure to observe wild and undamped oscillations in chemostats and other experimental populations. The theory for bacterial-phage growth in chemostats has been worked out for over 3 decades [8]. There are straightforward equilibria [2,8], but simulations reveal a strong tendency for oscillations around those equilibria, and the oscillations do not dampen (unpublished and [8]). In stunning contrast, empirical studies of phages in chemostats have consistently failed to observe such behavior.

Schrag and Mittler [22] addressed this problem and suggested that wall populations in chemostats acted as refuges that stabilized the dynamics. It is now also appreciated that biofilms provide refuges for bacteria while also enabling phage to 'graze' upon the planktonic migrants from the biofilm [23]. All three systems – wall growth, biofilms, and phenotypic resistance – give rise to the paradox of high levels of phage maintained on sensitive bacterial populations. Phenotypic resistance is unique in providing a mechanism of escape in purely planktonic bacterial populations and thus may be relevant to phage-bacterial dynamics many contexts.

Phenotypic resistance to phages has not been explored in depth, so little is known about its mechanisms. We can nonetheless suggest how its properties might differ from those of genetic resistance in its implications for the coexistence of phage and bacteria. One is that mechanisms of phenotypic resistance may be more diverse and abundant than mechanisms of genetic resistance because they can be affected by subtle changes in gene expression and by molecules in the external environment. Second, and for similar reasons, phenotypic resistance may have less severe fitness consequences than does genetic resistance. Last, as noted above, phenotypic resistance will often be cryptic in the sense that the initial, phage-free population gives little evidence of it, yet it rapidly dominates the population when phage are added. The properties of phenotypic resistance may be specific to the receptors used by a particular phage and thus differ for different phages; genetic resistance also exhibits dependence on phage receptors, of course.

The perspective here may solve a puzzle in the classic phage therapy work of Smith and Huggins [24]. They showed that, when treating an artificial *E. coli* O18:H7:K1 infection of mice, phages requiring the K1 capsule for infection (K1-dep) rescued at nearly 100%, but phages whose infection was independent of the capsule (K1-ind) were much less successful. Although Smith and Huggins commented that K1-ind phages grew poorly and often failed to clear a culture, Bull et al. [25–27] found that the growth rate of K1-ind phages was high both in culture and in mice, certainly more than enough to overwhelm the bacterial population if extrapolated across a few hours. It now appears relevant that the growth rate assays of Bull et al. [25,26] were conducted at low MOI and thus would not be affected by phenotypic resistance (recall eqn(14)), whereas the alternative outcomes of lysis versus non-lysis of an entire culture would be affected by phenotypic resistance. By extension to other systems, phenotypic resistance

could be a cause of phage therapy failure even though the phages are able to form clear plaques and grow well when first introduced to an infection or culture.

If phenotypic resistance explains the failure of K1-ind phages to cure an infection or clear a culture, is there a plausible reason that phenotypic resistance is not a problem for K1-dep phages in the Smith-Huggins system? Bull et al. [26] noted that K1-dependent phages from different taxonomic groups all encoded endosialidase tailspikes. In infections of mice, free enzyme produced as unattached tailspikes will contribute to infection clearance by stripping capsules and thereby augmenting the immune system (e.g., treatment with enzyme alone cures the infection, [28]). This mechanism should not contribute to clearing an *in vitro* culture however, as there is no immune system to remove cells with reduced capsules. Instead, enhanced clearing of a culture might be attributable to limited phenotypic resistance, if complete loss of the capsule through phenotypic means is rare. At present, this model is conjecture but is readily testable.

If phenotypic resistance is sufficiently prevalent as to commonly hamper therapeutic success of arbitrarily chosen phages, a fruitful avenue of research will be to look for generalities in the foundations of phenotypic resistance. Some types of phage receptors may be less prone to the problem than others, and if the rules of phenotypic resistance can be discovered, it may be possible to screen phages and choose ones least likely to fail in the application. The work with *E. coli* O18:H7:K1 is one possible example of the feasibility of this approach, although in that system, there may be a dual benefit *in vivo* of phages that target the bacterial capsule.

Phenotypic resistance does not prevent the evolution of classical genetic resistance. Genetic resistance may still be favored in the long term because it reduces the death rate from phage, although phenotypic resistance should ascend first and then slow the ascent of genetic resistance, and observations restricted to the long term will not easily detect phenotypic resistance after it has been replaced by genetic resistance. However, the dynamical behavior of the phage and bacterial populations with a mix of genetic and phenotypic resistance may nonetheless exhibit signatures characteristic of phenotypic resistance (in preparation).

Materials and Methods

Bacteria and phage

E. coli O18:H7:K1 strain RS218 from frozen aliquots was added to 10 mL LB broth (10 g NaCl, 10 g Bacto tryptone, 5 g Bacto yeast extract per liter) in 125 mL flasks and grown at $37°C$ with agitation for 1 hr to a density of $10^7/mL$. Phage K1-ind(1) [26] were added to a tenth the bacterial concentration and the culture grown without dilution for another 3 hr. Phage and bacteria were sampled at hourly intervals. Bacteria were plated in soft agar (7 g Bacto agar/L in LB) overlaid on agar plates (15 g Bacto agar/L in LB) and incubated overnight. Plating bacteria in soft agar maintains small colony size and thus reduces the contact rate with phage also present in the sample.

Bacterial sensitivity to phage was tested on colony isolates by first placing a streak of phage (~50 μL of a suspension of at least 10^9 phage/mL) across an agar plate and allowing it to dry. Bacterial colonies were individually suspended in 100 μL of LB, vortexed, and ~10 μL streaked across and perpendicular to the phage streak. Streaks were scored after overnight growth; sensitive colonies showed dense growth up to the phage but not beyond.

Campylobacter jejuni NCTC1168 [29] was routinely cultivated on blood agar base II (Oxoid) supplemented with 5% bovine blood and incubated at $37°C$ under microaerobic conditions (6% O_2,

6% CO2, 4% H2, 84% N2). A single colony was restreaked twice and applied as inoculum for a 40 ml culture in brain heart infusion broth supplemented with 1 mM CaCl2 and 10 mM MgCl2. Cell density was measured by plating at inoculation (4.4×10^4 CFU/ mL) and at 24 h (2.5×10^9 CFU/mL), with plates grown at 37°C under microaerobic conditions as described above. To investigate variation of phage susceptibility within the culture, individual colonies plated at time 0 h and 24 h were restreaked and their distinct phage susceptibility tested by plaque assays with phage F287 [30] essentially as described [31]. Efficiency of plating was calculated relative to the phage titer of the inoculum and measured across a series of dilutions, with 2×10^6 pfu at the highest concentration.

Simulations and graphics

Differential equations were evaluated numerically in the program Berkeley Madonna (v. 9.0.112 beta) with a step size of 10^{-4} and method Runge-Kutta 4; models will be made available on request. The numerical output was transferred to R for presentation. Figures were drawn in R [32].

Acknowledgments

M. Holst Sorensen is greatly acknowledged for providing *C. jejuni* phage F287 and for discussions of the *C. jejuni* phage infection experiment. We thank E. Miller, D. Rozen and two anonymous reviewers for comments and pointing out some errors in equations. I. J. Molineux and A. Campbell both independently suggested the possibility of our dynamic resistance mechanism.

Author Contributions

Conceived and designed the experiments: JJB BRL WNC MS CSV. Performed the experiments: WNC MS CSV. Analyzed the data: JJB BRL. Contributed reagents/materials/analysis tools: JJB BRL CSV. Wrote the paper: JJB BRL. Conceived of idea: JJB BRL. Mathematical derivations: JJB. Generated and analyzed simulation model: JJB BRL.

References

1. Adams MH (1959) Bacteriophages. New York: Interscience Publishers.
2. Campbell A (1961) Conditions for the existence of bacteriophage. Evolution 15: 153–165.
3. Horne MT (1970) Coevolution of *Escherichia coli* and bacteriophages in chemostat culture. Science 168: 992–993.
4. Bohannan BJ, Lenski RE (2000) Linking genetic change to community evolution: insights from studies of bacteria and bacteriophage. Ecology Letters 3: 362–377.
5. Chao L, Levin BR, Stewart FM (1977) A complex community in a simple habitat: an experimental study with bacteria and phage. Ecology 58: 369–378.
6. Spanakis E, Horne MT (1987) Co-adaptation of *Escherichia coli* and coliphage λvir in continuous culture. J Gen Microbiol 133: 353–360.
7. Wei Y, Ocampo P, Levin BR (2010) An experimental study of the population and evolutionary dynamics of *Vibrio cholera* O1 and the bacteriophage JSF4. Proc Biol Sci 277: 3247–3254.
8. Levin BR, Stewart FM, Chao L (1977) Resource-limited growth, competition, and predation – a model and experimental studies with bacteria and bacteriophage. American Naturalist 111: 3–24.
9. Levin BR, Moineau S, Bushman M, Barrangou R (2013) The population and evolutionary dynamics of phage and bacteria with CRISPR-mediated immunity. PLoS genetics 9: e1003312.
10. Levin BR, Bull JJ (2004) Population and evolutionary dynamics of phage therapy. Nat Rev Microbiol 2: 166–73.
11. Koibong L, Barksdale L, Garmise L (1961) Phenotypic alterations associated with the bacterio-phage carrier state of *Shigella dysenteriae*. J Gen Microbiol 24: 355–367.
12. Weitz JS, Dushoff J (2008) Alternative stable states in hostphage dynamics. Theoretical Ecology 1: 13–19.
13. Bikard D, Marraffini LA (2012) Innate and adaptive immunity in bacteria: mechanisms of programmed genetic variation to fight bacteriophages. Curr Opin Immunol 24: 15–20.
14. Bull JJ, Regoes RR (2006) Pharmacodynamics of non-replicating viruses, bacteriocins and lysins. Proc Biol Sci 273: 2703–12.
15. Bull JJ (2006) Optimality models of phage life history and parallels in disease evolution. J Theor Biol 241: 928–38.
16. Balaban NQ, Merrin J, Chait R, Kowalik L, Leibler S (2004) Bacterial persistence as a phenotypic switch. Science 305: 1622–1625.
17. Elowitz MB, Levine AJ, Siggia ED, Swain PS (2002) Stochastic gene expression in a single cell. Science 297: 1183–1186.
18. Veening JW, Smits WK, Kuipers OP (2008) Bistability, epigenetics, and bethedging in bacteria. Annu Rev Microbiol 62: 193–210.
19. Davidson CJ, Surette MG (2008) Individuality in bacteria. Annu Rev Genet 42: 253–268.
20. Levin BR, Rozen DE (2006) Non-inherited antibiotic resistance. Nature Reviews Microb 4: 556–562.
21. Gallet R, Lenormand T, Wang IN (2012) Phenotypic stochasticity protects lytic bacteriophage populations from extinction during the bacterial stationary phase. Evolution 66: 3485–3494.
22. Schrag SJ, Mittler JE (1996) Host-parasite coexistence: the role of spatial refuges in stabilizing bacteria-phage interactions. Amer Natur 148: 348–377.
23. Cornelissen A, Ceyssens P, T'Syen J, Van Praet H, Noben J, et al. (2011) The t7-related *Pseudomonas putida* phage 15 displays virion-associated biofilm degradation properties. PLoS One 6: e18597.
24. Smith HW, Huggins MB (1982) Successful treatment of experimental *Escherichia coli* infections in mice using phage: its general superiority over antibiotics. J Gen Microbiol 128: 307–18.
25. Bull JJ, Levin BR, DeRouin T, Walker N, Bloch CA (2002) Dynamics of success and failure in phage and antibiotic therapy in experimental infections. BMC Microbiol 2: 35.
26. Bull JJ, Vimr ER, Molineux IJ (2010) A tale of tails: Sialidase is key to success in a model of phage therapy against k1-capsulated *Escherichia coli*. Virology 398: 79–86.
27. Bull JJ, Otto G, Molineux IJ (2012) *In vivo* growth rates are poorly correlated with phage therapy success in a mouse infection model. Antimicrob Agents Chemother 56: 949–954.
28. Mushtaq N, Redpath MB, Luzio JP, Taylor PW (2004) Prevention and cure of systemic *Escherichia coli* K1 infection by modification of the bacterial phenotype. Antimicrob Agents Chemother 48: 1503–8.
29. Parkhill J, Wren BW, Mungall K, Ketley JM, Churcher C, et al. (2000) The genome sequence of the food-borne pathogen *Campylobacter jejuni* reveals hypervariable sequences. Nature 403: 665–668.
30. Hansen VM, Rosenquist H, Baggesen DL, Brown S, Christensen BB (20078)Characterization of *Campylobacter* phages including analysis of host range by selected *Campylobacter* penner serotypes. BMC Microbiology 7: 90.
31. Holst Soerensen MC, van Alphen LB, Fodor C, Crowley SM, Christensen BB, et al. (2012) Phase variable expression of capsular polysaccharide modifications allows *Campylobacter jejuni* to avoid bacteriophage infection in chickens. Frontiers in Cellular and Infection Microbiology 2: 11.
32. R Development Core Team (2012) R: A Language and Environment for Statistical Computing. R Foundation for Statistical Computing, Vienna, Austria. Available: http://www.R-project.org/. ISBN 3-900051-07-0.

10

Species-Specific Expansion and Molecular Evolution of the 3-hydroxy-3-methylglutaryl Coenzyme A Reductase (HMGR) Gene Family in Plants

Wei Li[1], Wei Liu[1], Hengling Wei[2], Qiuling He[1], Jinhong Chen[1], Baohong Zhang[3], Shuijin Zhu[1]*

1 Department of Agronomy, Zhejiang University, Hangzhou, Zhejiang, China, **2** State Key Laboratory of Cotton Biology, Cotton Research Institute, Chinese Academy of Agricultural Sciences, Anyang, Henan, China, **3** Department of Biology, East Carolina University, Greenville, North Carolina, United States of America

Abstract

The terpene compounds represent the largest and most diverse class of plant secondary metabolites which are important in plant growth and development. The 3-hydroxy-3-methylglutaryl coenzyme A reductase (HMGR; EC 1.1.1.34) is one of the key enzymes contributed to terpene biosynthesis. To better understand the basic characteristics and evolutionary history of the HMGR gene family in plants, a genome-wide analysis of HMGR genes from 20 representative species was carried out. A total of 56 HMGR genes in the 14 land plant genomes were identified, but no genes were found in all 6 algal genomes. The gene structure and protein architecture of all plant HMGR genes were highly conserved. The phylogenetic analysis revealed that the plant HMGRs were derived from one ancestor gene and finally developed into four distinct groups, two in the monocot plants and two in dicot plants. Species-specific gene duplications, caused mainly by segmental duplication, led to the limited expansion of HMGR genes in *Zea mays*, *Gossypium raimondii*, *Populus trichocarpa* and *Glycine max* after the species diverged. The analysis of Ka/Ks ratios and expression profiles indicated that functional divergence after the gene duplications was restricted. The results suggested that the function and evolution of HMGR gene family were dramatically conserved throughout the plant kingdom.

Editor: Tianzhen Zhang, Nanjing Agricultural University, China

Funding: The works are support by the National Basic Research Program (973 program, No: 2010CB126006) and the National High Technology Research and Development Program of China (2011AA10A102, 2013AA102601). The funders had no role in study design, data collection and analysis, decision to publish, or preparation of the manuscript.

Competing Interests: The authors have declared that no competing interests exist.

* E-mail: shjzhu@zju.edu.cn

Introduction

Plants produce thousands of secondary metabolites that play important roles in numerous biological processes. Structurally and functionally different terpenes represent the largest and most diverse class of secondary metabolites [1]. In addition to their physiological functions in photosynthesis, respiration, and growth and development, many specialized terpenes also have ecological roles in mediating plant interactions with various biotic and abiotic factors [2]. For example, terpenes can serve as phytoalexins in defense against phytopathogens and herbivores [3,4,5,6], and low-molecular-weight terpene compounds may released as odors that attract pollinators or induce defense responses in neighboring plants [7,8].

In plant cells, two distinct pathways are responsible for the biosynthesis of terpene compounds, the cytosolic mevalonate pathway (MVA pathway) and the plastidial 2-C-methyl-D-erythritol-4-phosphate pathway (MEP pathway) [9]. The reaction catalyzed by the enzyme 3-hydroxy-3-methylglutaryl coenzyme A reductase (HMGR) is the first committed step in the MVA pathway [10]. The gene encoding HMGR has been found that widely present in eukaryotes and prokaryotes, and well studied in mammals due to its critical role in mediating cholesterol biosynthesis [11]. In plants, the HMGR genes also have been extensively cloned and characterized from a number of species

including *Arabidopsis thaliana* [12,13], *Oryza sativa* [14], *Triticum aestivum* [15], *Gossypium hirsutum* [16], *Solanum tuberosum* [17], *Cucumis melo* [18], *Hevea brasiliensis* [19] and some medicinal plants [20,21,22]. Additionally, HMGR is considered as a key enzyme for biotechnological purposes and can be utilized to increase terpenes content in plants. As a result, up-regulation of HMGR genes could improve terpenes productivities in the transgenic plants [23,24,25,26,27]. Furthermore, it was reported that transgenic tomato plants that constitutively expressed a heterologous HMGR gene from melon showed a significant increase in fruit size [28].

HMGR protein is comprised of three domains, i.e., the transmembrane domains with changeable number in N-terminal region, the highly divergent linker domain, and the long and conserved catalytic domain in C-terminus. Within the catalytic domain, subdomains have been defined as the small helical N-terminal domain, the large and central L-domain harboring two HMG-CoA binding motifs and an NADP(H) binding motif, and the smallest S-domain harboring an NADP(H) binding motif. The two HMG-CoA binding motifs, EMPVGYVQIP and TTEGCLVA, and two NADP(H) binding motifs, DAMGMNM and GTVGGGT, are functionally important and thus highly conserved in all HMGR proteins [29,30].

Although HMGR genes have been systematically analyzed in Gramineae plants [30], little is known about their features in the genome-wide level in plant kingdom. Thus, a comprehensive investigation about the basic characteristics and evolutionary history of this gene family is necessary in plants. Fortunately, the recent availability of whole genome sequences of various plant species in public databases offers an opportunity to identify the complete set of HMGR genes in many species. In present study, an extensive survey of HMGR families in 20 species ranging from unicellular algae to higher plants was conducted. Subsequently, the distribution, protein architecture, exon/intron organization, phylogenetic relationship and expansion pattern of this gene family were assessed, and the expression profiles of HMGR genes in *Zea mays* and *Glycine max* using published transcriptome data were analyzed as well.

Materials and Methods

Identification and verification of HMGRs in different plant genomes

The amino acid sequences of *Arabidopsis thaliana* HMGR genes [2] were retrieved from The Arabidopsis Information Resource (TAIR, http://www.arabidopsis.org) and used as queries to search against other plant genome databases with BlastP and tBlastN programs (default parameters). The 20 completely sequenced genomes of species from unicellular green algae to multicellular higher plants were used in this study (Table 1 and Table S1). Subsequently, all hits considered as candidate sequences were submitted to Pfam database (http://pfam.sanger.ac.uk/) to confirm the presence of the conserved domain (PF00368).

Protein motif and gene structure analysis

The conserved motifs encoded by each HMGR gene were identified using the program of Multiple Em for Motif Elicitation (MEME; version 4.9.0) [31] at the website (http://meme.nbcr. net/meme/cgi-bin/meme.cgi). The analysis was performed with a set of parameters as follows: number of repetitions, any; minimum width for each motif, 6; maximum width for each motif, 100; and maximum number of motifs to be found, 5. All obtained motifs were searched in the InterPro database with InterProScan [32]. The exon/intron structures of HMGR genes were obtained by comparing the genomic sequences and their predicted coding sequences (CDS) using GSDS (http://gsds.cbi.pku.edu.cn/) [33].

Multiple sequence alignment and phylogenetic reconstruction

Multiple sequence alignments of the full-length protein sequences were performed by Clustal X version 2.0 program [34] with default parameters. The neighbor-joining (NJ) phylogenetic tree was constructed using MEGA 5.2 [35] with pairwise deletion option. The reliability of obtained phylogenetic tree was tested using bootstrapping with 1000 replicates.

Chromosomal mapping and gene duplications

The locations of HMGR genes in *Zea mays*, *Gossypium raimondii*, *Populus trichocarpa* and *Glycine max* were collected from the genome annotation data of the corresponding organism, respectively. The chromosomal distribution images of these HMGR genes were generated by MapInspect software according to their starting positions in chromosomes [36]. Gene duplication events of HMGR genes in *Zea mays*, *Gossypium raimondii*, *Populus trichocarpa* and *Glycine max* were also investigated. Gene duplication was

Table 1. Species and number of HMGR genes used in this study.

Lineage	Species	Abbreviation	Number of HMGR genes
Green algae	Ostreococcus lucimarinus	Ol	0
	Ostreococcus tauri	Ot	0
	Micromonas pusilla	Mp	0
	Chlamydomonas reinhardtii	Cr	0
	Chlorella variabilis	Cv	0
	Volvox carteri	Vc	0
Mosses	Physcomitrella patens	Pp	3
Lycophytes	Selaginella moellendorffii	Sm	1
monocots	Brachypodium distachyon	Bd	3
	Oryza sativa	Os	3
	Zea mays	Zm	7
	Sorghum bicolor	Sb	3
Eudicots	Vitis vinifera	Vv	3
	Citrus sinensis	Cs	2
	Gossypium raimondii	Gr	9
	Carica papaya	Cp	3
	Brassica rapa	Br	3
	Arabidopsis thaliana	At	2
	Glycine max	Gm	8
	Populus trichocarpa	Pt	6
Total			56

defined according to (1) the length of aligned sequence cover > 80% of the longer gene; and (2) the identity of the aligned regions >80% [37,38]. With the chromosomal locations of HMGR genes, two types of gene duplications were recognized, i.e., tandem duplication and segmental duplication.

Estimating the divergence time for duplicated gene pairs

The pairwise alignment of HMGR duplicated gene pairs from four plants was performed using Clustal X version 2.0 program [34]. Then pairwise synonymous (Ks) and non-synonymous (Ka) numbers of substitutions corrected for multiple hits were calculated using the DnaSP v5.0 software (DNA polymorphism analysis) [39]. Finally, the selection pressure for these duplicate HMGR gene pairs was calculated as Ka/Ks ratio. Based on the synonymous substitutions per year (λ) of 6.5×10^{-9} for *Zea mays* [40], 9.1×10^{-9} for *Populus trichocarpa* [41], and 6.1×10^{-9} for *Glycine max* [42], by substituting the calculated Ks values, the approximate age of duplicated events of the duplicate HMGR gene pairs was estimated ($T = Ks/2\lambda \times 10^{-6}$ Mya).

Gene expression analysis

The expression profiles of ZmHMGRs and GrHMGRs were clustered using the Cluster 3.0 [43], respectively. The public expression data for various tissues and developmental stages in *Zea mays* were obtained from the Plant Expression Database (PLEXdb, http://www.plexdb.org/) [44,45] according to the identified ZmHMGR ID, and the transcriptome sequencing datasets of *Glycine max* were downloaded from SoyBase (http://soybase.org/soyseq/) [46] based on the GmHMGR ID.

Results and Discussion

Genomic identification of HMGR genes in plants

In order to identify HMGR genes in Viridiplantae, the blast searches among the 20 completely sequenced genomes (Table S1) were carried out. These genomes represent major evolutionary lineages of the plant kingdom such as algae, mosses, lycophytes, monocots, and eudicots. After removing partial or redundant sequences, and the predicted alternative splice variants, a total of 56 genes encoding HMGR proteins were retrieved in the 14 land plant genomes, and no HMGR genes were detected in algae (Table 1). Because there is no standard naming system for HMGR genes (not including Arabidopsis), the newly identified HMGR genes were assigned according to the species and the gene orders on the chromosomes (Table S2). The Arabidopsis HMGR genes were named following the TAIR website (http://www.arabidopsis.org/).

The Viridiplantae are comprised of two major lineages that split early, i.e., the Chlorophyta (chlorophyte algae) and the Streptophyta (charophyte algae and land plants) [47,48]. Interestingly, it was observed that no HMGR genes were found in genomes of all 6 algal species, which belong to the division Chlorophyta of green algae, suggesting that the HMGR gene family might be lost in the chlorophyte algae during evolution. The HMGR protein is a major rate-limiting enzyme in the MVA pathways [10], and the MVA pathway is considered as an ancestral metabolic route for the biosynthesis of terpene compounds in all the three domains of life (bacteria, archaea, and eukaryotes) [49]. So it could be speculated that the Chlorophyta had abandoned the MVA pathway. These observations were compatible with previous findings that the chlorophyte algae synthesized their terpenes exclusively via the MEP pathway and might develop efficient mechanisms of exporting MEP-pathway-derived terpene intermediates from the plastid for the biosynthesis of cytosolic terpenes

[50,51]. The analysis also provided further proof for that the genes involved in the MVA pathway in the chlorophyte algae were not silenced but really absent.

Although genome sequence data are currently not available for charophyte algae, which is believed to be the closest relatives of land plants [47]. But it has been experimentally substantiated that the charophyte algae *Mesostigma viride* contained the HMGR gene [52]. Moreover, it was also found that the HMGR genes were widespread in land plants. So it could be figured out that the MVA pathway was still operating in the Streptophyta, especially in land plants. The MEP pathway has been identified to be present in plastid-bearing eukaryotes [53]. Therefore, unlike the chlorophyte algae, the land plants simultaneously retained the MVA and MEP pathways. In this case, the MVA pathway was active within the cytoplasm at the same time as the MEP pathway functioned in the plastids. The utilization of both the MVA and MEP pathways could enable plants to separate and optimize the biosynthesis of a wide range of complex terpene-derived specialized metabolites. Obviously, by retaining and compartmentalizing the two pathways, the land plants could gain a selective advantage in interactions with their surrounding environments to overcome sessile-lifestyle constraints.

Additionally, among the land plants, there were 1 to 9 HMGR genes in each species, and most species (10 of 14 species) only had 3 or less HMGR genes. The non-vascular *Physcomitrella patens*, a species of mosses which is a basal lineage of land plants, contained 3 HMGR genes. The *Selaginella moellendorffii*, the oldest extant vascular plant belonging to lycophytes, was the fewest HMGR gene family species among the land plants in our survey, which was expected as it has one of the smallest plant genomes known [54]. In the flowering plants, the number of HMGR genes was varied greatly. Here, *Citrus sinensis* and *Arabidopsis thaliana* contained only two HMGR genes, which was the fewest two species in HMGR gene number among flowering plants we investigated. The other six plants, *Brachypodium distachyon*, *Oryza sativa*, *Sorghum bicolor*, *Vitis vinifera*, *Carica papaya*, and *Brassica rapa*, had three HMGR gene members. While the remaining four species had a relative higher number of HMGR genes, 6 HMGRs in *Populus trichocarpa*, 7 HMGRs in *Zea mays*, 8 HMGRs in *Glycine max*, and 9 HMGRs in *Gossypium raimondii*. The variable size of the HMGR gene family suggested that the gene family underwent species-specific expansion in flowering plants. Furthermore, among these species which underwent the gene expansion, *Zea mays* belongs to the monocots, and the other three belong to the eudicots, which indicated that the gene expansion occurred both in monocot plants and dicot plants.

Conserved protein motifs and exon/intron structure of HMGR genes

The MEME motif search tool was employed to identify the conserved motifs presented in 56 plant HMGR proteins, and 5 conserved motifs were uncovered (Figure S1). After searching in the Interpro database, all motifs corresponded to known domains. The motif 5 was a region including two transmembrane helices, and the others were located in the catalytic domain of HMGR genes (Figure 1).

A common feature of plant HMGRs is the presence of a transmemebrane region consisting of two separate transmemebrane domains that are linked to the cytoplasmic domain bearing the catalytic center [55,56]. Of the 56 HMGRs, only VvHMGR3, GrHMGR8 and GmHMGR3 missed the motif 5, suggesting that most plant HMGR genes have two transmembrane helices in the N-terminus. The result was verified by prediction of transmembrane helices in HMGR proteins using the TMHMM Server v.

(A)

(B)

(C)

(D)

(E)

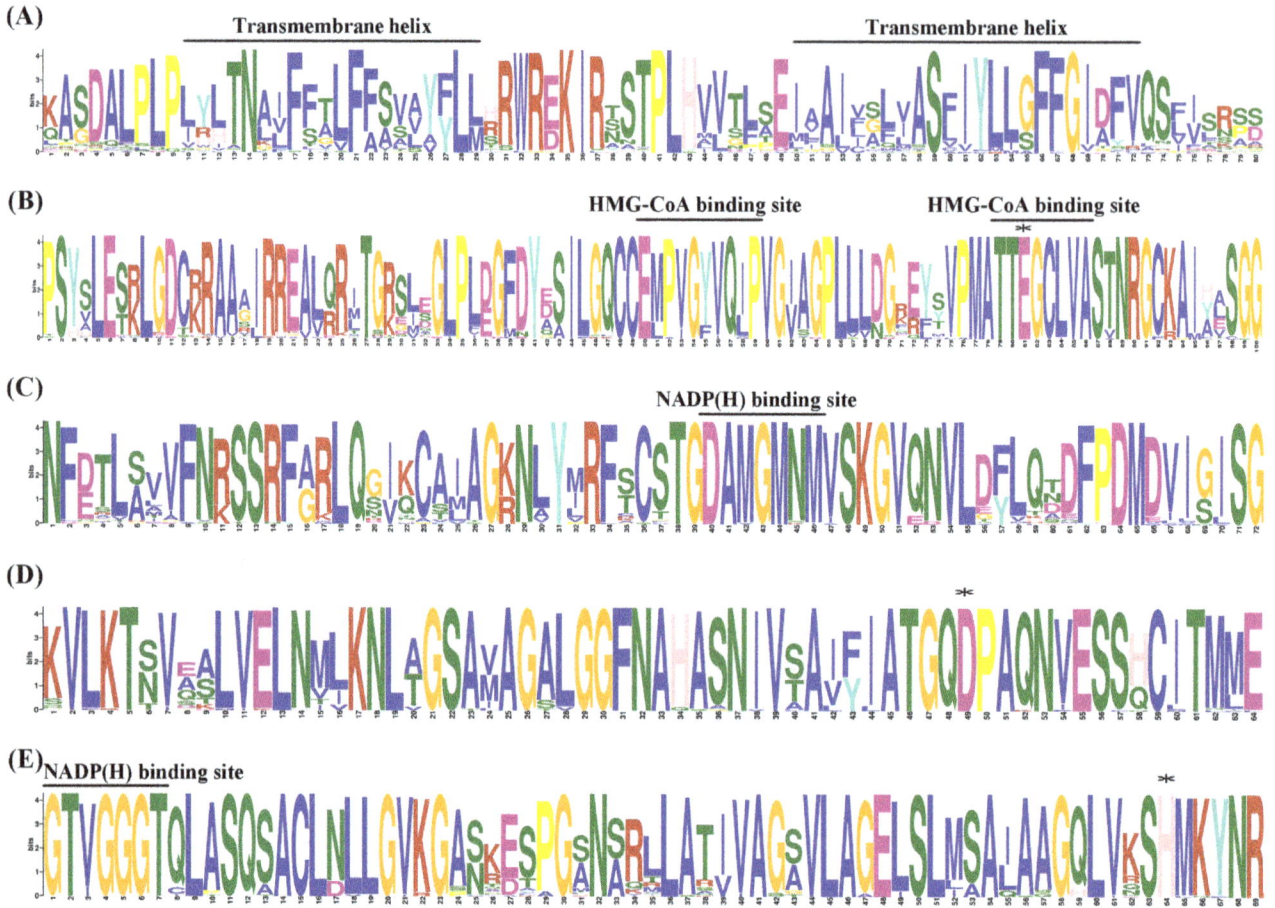

Figure 1. Sequence logos of the five motifs identified using the MEME search tool. (A), (B), (C), (D) and (E) represent the motif 5, 3, 2, 1 and 4, respectively. The height of letter designating the amino acid residue at each position represents the degree of conservation. The numbers on the x-axis represent the residue positions in the motifs. The y-axis represents the information content measured in bits. The two transmembrane helices, two HMG-CoA binding sites (EMPVGYVQIP and TTEGCLVA) and two NADP(H) binding sites (DAMGMNM and GTVGGGT) are represented on the top of the corresponding locations in motifs. Asterisks (*) indicate the conserved residues in the catalytic domain of plant HMGR genes.

2.0 (Table S2). Unexpectedly, existence of the motif 5 and the prediction of two transmembrane helices by TMHMM Server v. 2.0 did not correspond precisely in five HMGR genes. Further analysis found that there were no transmembrane helices that could be predicted by TMHMM Server v. 2.0 due to the diversity of several amino acid sites in the motif 5 in these genes. Overall, the plant HMGRs usually had two transmemebrane domains in the N-terminal region of proteins, and several genes appeared to start sequence variations, or even lost the domain.

In the protein sequences of plant HMGR, there were four conserved motifs including EMPVGYVQIP, TTEGCLVA, DAMGMNM, and GTVGGGT in the catalytic domain. In this study, the motif 3 represented the two HMG-CoA binding sites (EMPVGYVQIP and TTEGCLVA), the motif 2 represented one of the NADP(H) binding sites (DAMGMNM), and the motif 4 represented the other NADP(H) binding sites (GTVGGGT), respectively (Figure 1). Among them, the motif 3 in the N-terminus was lost in GrHMGR8 and GmHMGR3, the motif 2 in the middle was lost in OsHMGR1 and ZmHMGR1, and the motif 4 in the C-terminus was lost in OsHMGR1, ZmHMGR1, ZmHMGR3, ZmHMGR5 and GmHMGR3 (Figure S1). Apparently, the C-terminus of the catalytic domain was more variable than the N-terminus in the evolutionary process of plant HMGRs.

Interestingly, 3 out of 7 ZmHMGRs missed the motif 4 that binds to NADP(H), and the ZmHMGR1 also missed the other NADP(H) binding sites (motif 2). It could be guessed that this might be the need for functional differentiation of HMGR genes after gene expansions in maize.

In order to validate the conservation of residues in the catalytic domain, the sequence logos of the motif 3, motif 2, motif 1 and motif 4 were investigated (Figure 1). It was observed that the high homology region appeared to be centered around the HMG-CoA and NADP(H) binding sites. The amino acids composed of the second HMG-CoA binding site and the two NADP(H) binding sites were almost the same in all analyzed plant HMGRs. While the amino acid residues in the first HMG-CoA binding site were diverse among plant HMGRs, which might contribute to the substrate selectivity. Additionally, the position and orientation of four key catalytic residues (Glu, Lys, Asp and His) [29], which are functionally significant, were highly conserved in HMGR proteins. Among them, three residues except Lys residue were shown in Figure 1.

Analysis of HMGR gene structure for exon/intron organization revealed that the number of introns per gene varied from 1 to 14 (Figure S2). OsHMGR1, ZmHMGR1, and ZmHMGR5 possessed a minimum of one intron each, whereas ZmHMGR3

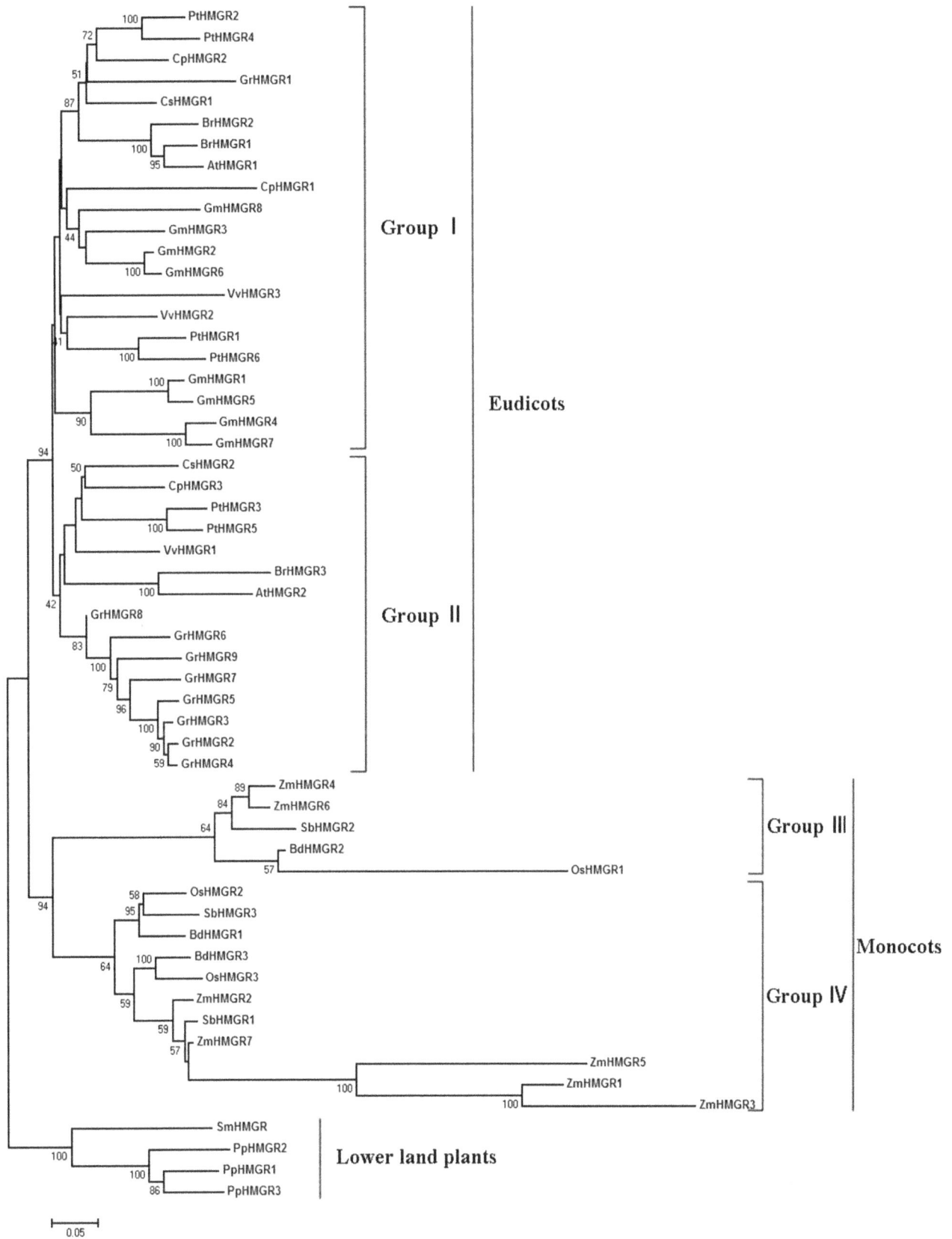

Figure 2. Neighbor-joining phylogenetic tree of plant HMGR proteins. The tree was constructed based on a complete protein sequence alignment of HMGR genes using neighbor-joining method, and the PpHMGRs and SmHMGR were designated as outgroups. Numbers at the nodes represent bootstrap support (1000 replicates). Bootstrap support higher than 40% is indicated at respective nodes.

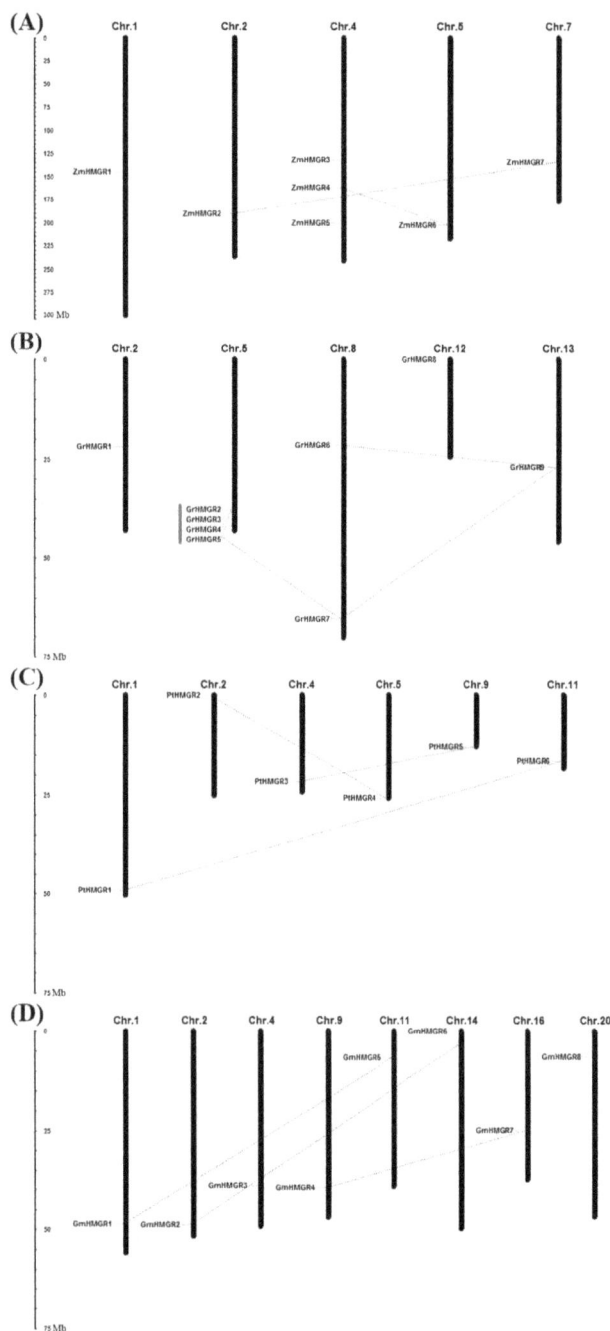

Figure 3. Chromosomal localization of HMGR genes in four selected plant species. (A): *Zea mays*; (B): *Gossypium raimoddii*; (C): *Populus trichocarpa*; and (D): *Glycine max*. The chromosome number is indicated at the top of each chromosome representation. The scale represents megabases (Mb). The segmental duplicated genes are indicated by dotted lines and the tandemly duplicated genes by red vertical lines.

possessed a maximum of fourteen introns. Among all the analyzed genes, the great majority (43 out 56), including all HMGRs from the lower land plants, possessed three introns. These results indicated that the common ancestor of plant HMGRs had three introns. The intron/exon structure of HMGR genes was highly coincided in plant evolution.

Regardless of the less variability, the protein architectures and gene structures of plant HMGRs were remarkably conserved, indicating that the molecular characteristics and biological functions of each plant HMGR were highly conserved during evolution. This was not the same as other gene families involved in plant secondary metabolism, which were very much lineage-dependent and varied tremendously among plant taxa [57]. The results also suggested that the plant HMGR genes might be monophyletic and were descendants of an ancestor.

Phylogenetic analysis of HMGR gene families

To examine the evolutionary relationships of the HMGR genes in Viridiplantae, a neighbor-joining phylogenetic tree (Figure 2) was constructed based on the alignments of full-length HMGR protein sequences. Because mosses and lycophytes are the basal lineages of land plants, PpHMGRs and SmHMGR were designated as outgroups. Firstly, HMGR genes from the flowering plants were divided into two monophyletic clades, the monocots and eudicots. No genes from the two lineages tended to cluster together in the phylogenetic tree, suggesting that the plant HMGRs were derived from one ancestor gene and developed into different branches after the lineages diverged. Within the monocots or eudicots, the HMGR genes fell into two major distinctive groups each with high bootstrap values. As shown in the Figure 2, these groups were named group I, II, III and IV.

In eudicots, there were 36 HMGR genes from 8 plant species. Group I contained 21 members, and group II contained 15 members. Statistically, 7 out 8 dicot plants contained all two groups of HMGR genes, except for *Glycine max*, which only contained group I HMGR genes. Obviously, the significant expansion of group I genes contributed to the increase of HMGR genes in *Glycine max* which had up to 8 members. In addition, those HMGR genes from the same species were not equally classified into the two groups. Most of the 8 plant species had more genes in group I than in group II, except for *Gossypium raimondii* which contained 8 group II genes, but only a single gene in group I. It was indicated that, unlike *Glycine max*, the increase in HMGR genes in *Gossypium raimondii* attributed to the remarkable expansion of group II genes.

In monocots, all of 4 monocot plants contained the two group genes, group III and IV. As the HMGR genes from dicot plants, the 16 HMGR genes from 4 monocot plants were not evenly distributed in the two groups neither. All investigated monocot plants contained more members in group IV than in group III. Such as *Zea mays* that had the largest HMGR gene family in analyzed monocots had five genes in group IV, but two genes in group III.

In general, there were no paralogs of HMGR genes from plants which contained less than or equal to 3 HMGRs, indicating that no gene expansion occurred within the HMGR gene family after the divergence of these plant species. In contrast, four other species, *Zea mays*, *Gossypium raimondii*, *Populus trichocarpa*, and *Glycine max*, underwent considerably more frequent gene duplications, which gave rise to an increase in more members of the HMGR gene family. As we know, the occurrence of most of secondary metabolites and their respective biosynthetic pathways is restricted to specific plants or plant lineages [57]. The evolution of these pathways definitely requires new enzymes and regulatory elements in specific plants. So it can be deduced that the four plant species may need to produce more or wider variety of terpene compounds in their respective development. *Gossypium raimondii*, which contained the largest HMGR gene family (9 genes) in our survey, was known to accumulate a unique group of terpenes included

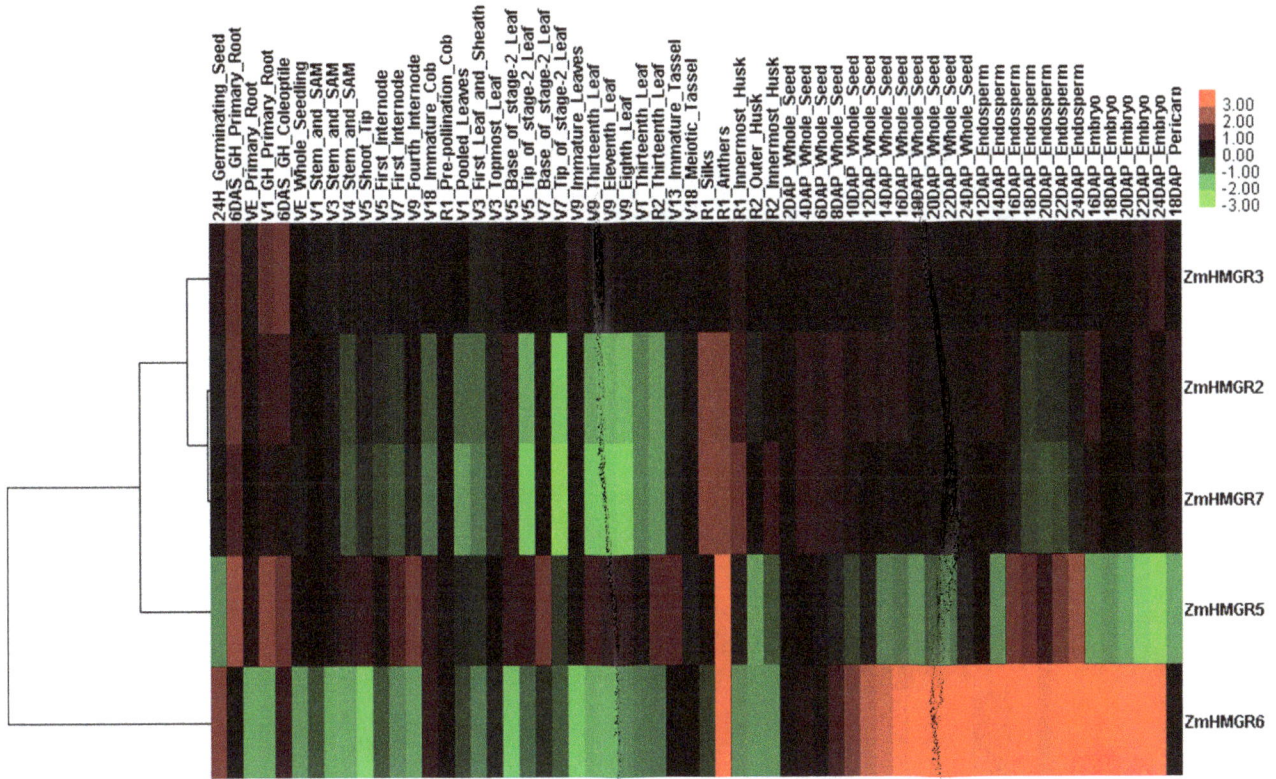

Figure 4. Expression profiles of HMGR genes in *Zea mays* across different tissues and developmental stages. The color scale represents the relative signal intensity values. DAP: Days After Pollination; DAS: Days After Sowing.

desoxyhemigossypol, hemigossypol, gossypol, hemigossypolone, and the heliocides [58], which fit well with this hypothesis.

Chromosomal localization and duplication of HMGR genes in four selected plant species

Based on the coordinate of each HMGR gene on the chromosomes, the chromosomal distribution images of these HMGR genes in four plant species which underwent species-specific gene expansions were generated (Figure 3). The 7 HMGR genes of *Zea mays* were distributed unevenly on 5 chromosomes, and chromosome 4 contained three HMGR genes. In *Gossypium raimondii*, 9 HMGR genes were located on 5 chromosomes, and one gene cluster contained 4 genes was detected on chromosome 5. While in *Populus trichocarpa* and *Glycine max*, the HMGR genes were distributed uniformly, one gene on each chromosome.

A total of fourteen duplication events which contributed to HMGR gene family expansion in the four plant species were found. Among them, two events as segmental duplications were in *Zea mays*, and there were three segmental duplication events in respective genome of *Populus trichocarpa* and *Glycine max*, strongly indicating that the segmental duplication was the main cause of the species-specific expansion of HMGR genes in the three species. Particularly, there were up to six duplication events in *Gossypium raimondii* which had the largest HMGR gene family number in our survey. With three segmental duplications and three tandem duplications, it was suggested that a relatively large extent of species-specific gene expansion was caused by both segmental and tandem duplications after the divergence of *Gossypium raimondii*.

During the process of evolution, the duplicated genes might have undergone divergent fates such as nonfunctionalization (loss of original functions), neofunctionalization (acquisition of novel functions), or subfunctionalization (partition of original functions) [41]. To explore whether Darwinian positive selection was involved in HMGR gene divergence after duplication, the Ka/Ks ratio for each pair of duplicated HMGR genes were calculated (Table S3). Generally, a Ka/Ks ratio >1 indicates accelerated evolution with positive selection, a ratio = 1 indicates that the genes are pseudogenes with neutral selection, while a ratio <1 indicates the functional constraint with negative or purifying selection of the genes. In this study, the Ka/Ks ratios for fourteen duplicated HMGR gene pairs were less than 0.3, suggesting that the HMGR genes from the four plants have mainly experienced strong purifying selection pressure with limited functional divergence after the species-specific duplications. These results suggested that functions of the duplicated HMGR genes did not diverge much during subsequent evolution. The approximate age of segmentally duplicated HMGR gene pairs from *Zea mays*, *Populus trichocarpa* and *Glycine max* were estimated using the Ks as the proxy for time (Table S3). The Ks values of these duplicated HMGR gene pairs were 0.152–0.289, indicating that the duplications might have occurred 12.25–16.05 million years ago (Mya).

Expression profiling of HMGR genes in Zea mays and Glycine max

To understand the temporal and spatial expression patterns of HMGR genes, their expression profiles during *Zea mays* and *Glycine max* development were analyzed using the public expression data.

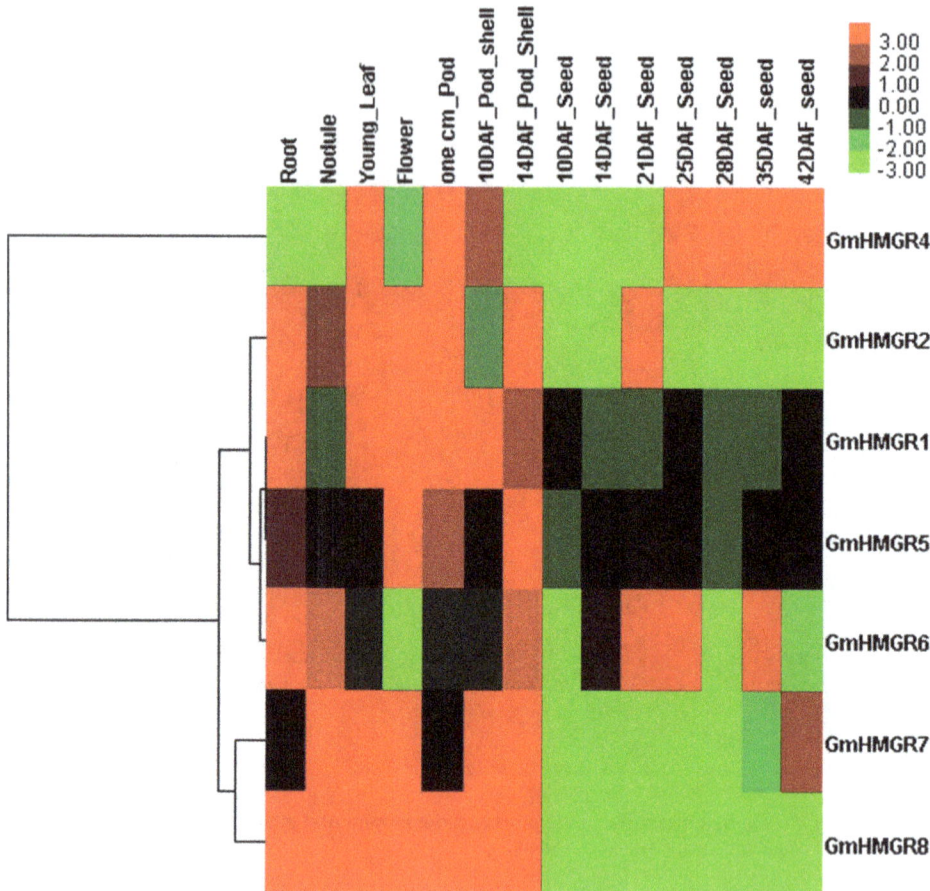

Figure 5. Expression profiles of HMGR genes in *Glycine max* across different tissues and developmental stages. The color scale represents the relative signal intensity values. DAF: Days After Flowering.

The published microarray data of 60 different tissues and developmental stages of *Zea mays* [44] were collected and investigated (Figure 4). Five of the seven ZmHMGRs showed wide expressions in the examined tissues. The ZmHMG6 showed higher expression in seeds (whole seed, endosperm and embryo) than in other organs, which indicated that it may play important roles in seed development or secondary metabolites accumulation in maize seed. By contrast, the ZmHMGR5 just expressed specificity in the endosperm, but not the whole seed and embryo. Additionally, the ZmHMGR5 showed relatively high expression in vegetative tissues (root, internode and leaf). The ZmHMGR2, ZmHMGR3 and ZmHMGR5 also showed remarkable expression in roots. In the anthers, there were four ZmHMGRs (ZmHMGR2, ZmHMGR5, ZmHMGR6 and ZmHMGR7) expressed highly, which might be due to the large demand for terpene compounds in the pollen development [59]. However, two ZmHMGRs (ZmHMGR1 and ZmHMGR4) were not found to have corresponding probes in this dataset, so there were no expression data to be investgated. As the ZmHMGR4 had the typical gene structure (3 introns) and was just diversified in the transmemebrane region of the N-terminus in protein sequence, and it also was involved in a duplicated event with ZmHMGR6 that was expressed widely in maize development. It was deduced that the ZmHMGR4 might undergo the process of subfunctiona-lization after gene duplication and might be expressed similarly with ZmHMGR6. The other ZmHMGR duplicated gene pair,

ZmHMGR2 and ZmHMGR7, which had the same gene structure and protein architecture shared similar expression patterns in nearly all of the organs and developmental conditions analyzed.

The expression profiles of HMGR genes in *Glycine max* [46] were also analyzed (Figure 5). Similar to ZmHMGRs, the GmHMGRs also exhibited abundant transcript across multiple tissues and organs. Moreover, there was also a HMGR gene (GmHMGR4) expressed at relatively high level in the seeds. Especially, it had higher expression in later developmental stages but lower expression in early developmental stages of seed, which suggested that it might contribute to the accumulation of terpene compounds in soybean seed. Additionally, the expression patterns of all identified soybean duplicated gene pairs which had the same gene structure and protein architecture were also investigated in this research. Most gene pairs such as GmHMGR1/GmHMGR5 and GmHMGR2/GmMGR6 were similar. However, it was not the case for GmHMGR4/GmHMGR7. The expression of the duplicated gene pairs were strongly divergent, which might be caused by the significant variation in gene regulations after the duplication events. Overall, the highly similar expression patterns of most duplicated gene pairs in *Zea mays* and *Glycine max* implied that functional divergence after the gene duplications was restricted. These results indicated that the HMGR gene family was dramatically conserved during plant genome evolution.

Conclusions

The HMGR gene family might be lost in the chlorophyte algae during evolution, but still widespread in land plants. The plant HMGR genes might be derived from one ancestor gene and finally developed into two distinct groups within the monocots and eudicots, respectively. The gene structure and protein architecture of all plant HMGRs were considerable conserved. The HMGR family in four flowering plants underwent species-specific expansions after the species diverged due to the large production for terpene compounds in their respective development. Segmental duplication appeared to be the dominant mechanism for the gene duplication events in three species, whereas segmental duplication and tandem duplication played similar roles in the expansion of the HMGR gene family of *Gossypium raimondii*. The functional divergence after the gene duplications was restricted. The findings implied that the HMGR gene family was dramatically conserved during plant evolution, and the HMGR was the committed enzyme for terpene biosynthesis that had essential roles in regulating plant development.

Supporting Information

Figure S1   Distribution of conserved motifs in plant HMGR proteins identified using the MEME search tool. Different motifs are indicated by different colors numbered 1-5. The names of all members of HMGR genes and combined *p*-values are shown on the left side of the figure, and the positions and sizes of motifs are indicated at the bottom of the figure. The motif 5 is a region including two transmembrane helices. The motif 3, 2, 1 and 4 are located in the catalytic domain of HMGR proteins. Moreover, the N-terminus of the motif 3 is at the start position of catalytic domain in each HMGR protein.

Figure S2   Exon/intron organization of plant HMGR genes. Exons are represented by green boxes and introns by black lines. Names of all plant HMGR genes from different lineages are shown on the left side of the figure.

Table S1   Data sources for the genome sequences used to mine for HMGR genes.

Table S2   Summary of 56 HMGR genes identified in Viridiplantae.

Table S3   Ka/Ks analysis and estimated divergence time for the HMGR duplicated genes from *Zea mays*, *Gossypium raimondii*, *Populus trichocarpa* and *Glycine max*.

Acknowledgments

We are grateful to Lei Mei, Ning Zhu, Cheng Li, and Yue Chen (Zhejiang University, China) for their support in this study. We thank Joseph Ryan Polli (East Carolina University, USA) for helpful suggestions and comments on how to improve the manuscript.

Author Contributions

Conceived and designed the experiments: SZ W. Li BZ. Performed the experiments: W. Li W. Liu HW JC QH. Analyzed the data: W. Li HW. Contributed reagents/materials/analysis tools: HW. Wrote the paper: W. Li SZ BZ.

References

1. Degenhardt J, Kollner TG, Gershenzon J (2009) Monoterpene and sesquiterpene synthases and the origin of terpene skeletal diversity in plants. Phytochemistry 70: 1621–1637.
2. Tholl D, Lee S (2011) Terpene Specialized Metabolism in *Arabidopsis thaliana*. Arabidopsis Book 9: e143.
3. Keeling CI, Bohlmann J (2006) Diterpene resin acids in conifers. Phytochemistry 67: 2415–2423.
4. Heiling S, Schuman MC, Schoettner M, Mukerjee P, Berger B, et al. (2010) Jasmonate and ppHsystemin regulate key Malonylation steps in the biosynthesis of 17-Hydroxygeranyllinalool Diterpene Glycosides, an abundant and effective direct defense against herbivores in *Nicotiana attenuata*. Plant Cell 22: 273–292.
5. Hasegawa M, Mitsuhara I, Seo S, Imai T, Koga J, et al. (2010) Phytoalexin accumulation in the interaction between rice and the blast fungus. Mol Plant Microbe Interact 23: 1000–1011.
6. Balkema-Boomstra AG, Zijlstra S, Verstappen FW, Inggamer H, Mercke PE, et al. (2003) Role of cucurbitacin C in resistance to spider mite (*Tetranychus urticae*) in cucumber (*Cucumis sativus* L.). J Chem Ecol 29: 225–235.
7. Arimura G, Ozawa R, Shimoda T, Nishioka T, Boland W, et al. (2000) Herbivory-induced volatiles elicit defence genes in lima bean leaves. Nature 406: 512–515.
8. Pichersky E, Gershenzon J (2002) The formation and function of plant volatiles: perfumes for pollinator attraction and defense. Curr Opin Plant Biol 5: 237–243.
9. Bick JA, Lange BM (2003) Metabolic cross talk between cytosolic and plastidial pathways of isoprenoid biosynthesis: unidirectional transport of intermediates across the chloroplast envelope membrane. Arch Biochem Biophys 415: 146–154.
10. Bach TJ (1995) Some new aspects of isoprenoid biosynthesis in plants—a review. Lipids 30: 191–202.
11. Friesen JA, Rodwell VW (2004) The 3-hydroxy-3-methylglutaryl coenzyme-A (HMG-CoA) reductases. Genome Biol 5: 248.
12. Learned RM, Fink GR (1989) 3-Hydroxy-3-methylglutaryl-coenzyme A reductase from *Arabidopsis thaliana* is structurally distinct from the yeast and animal enzymes. Proc Natl Acad Sci U S A 86: 2779–2783.
13. Enjuto M, Balcells L, Campos N, Caelles C, Arro M, et al. (1994) *Arabidopsis thaliana* contains two differentially expressed 3-hydroxy-3-methylglutaryl-CoA reductase genes, which encode microsomal forms of the enzyme. Proc Natl Acad Sci U S A 91: 927–931.
14. Ha SH, Lee SW, Kim YM, Hwang YS (2001) Molecular characterization of Hmg2 gene encoding a 3-hydroxy-methylglutaryl-CoA reductase in rice. Mol Cells 11: 295–302.
15. Aoyagi K, Beyou A, Moon K, Fang L, Ulrich T (1993) Isolation and characterization of cDNAs encoding wheat 3-hydroxy-3-methylglutaryl coenzyme A reductase. Plant physiology 102: 623–628.
16. Loguercio LL, Scott HC, Trolinder NL, Wilkins TA (1999) Hmg-coA reductase gene family in cotton (*Gossypium hirsutum* L.): unique structural features and differential expression of hmg2 potentially associated with synthesis of specific isoprenoids in developing embryos. Plant Cell Physiol 40: 750–761.
17. Korth KL, Stermer BA, Bhattacharyya MK, Dixon RA (1997) HMG-CoA reductase gene families that differentially accumulate transcripts in potato tubers are developmentally expressed in floral tissues. Plant Mol Biol 33: 545–551.
18. Kato-Emori S, Higashi K, Hosoya K, Kobayashi T, Ezura H (2001) Cloning and characterization of the gene encoding 3-hydroxy-3-methylglutaryl coenzyme A reductase in melon (*Cucumis melo* L. *reticulatus*). Mol Genet Genomics 265: 135–142.
19. Venkatachalam P, Priya P, Jayashree R, Rekha K, Thulaseedharan A (2009) Molecular cloning and characterization of a 3-hydroxy-3-methylglutaryl-coenzyme A reductase 1 (*hmgr1*) gene from rubber tree (*Hevea brasiliensis* Muell. Arg.): A key gene involved in isoprenoid biosynthesis. Physiol Mol Biol Plants 15: 133–143.
20. Dai Z, Cui G, Zhou SF, Zhang X, Huang L (2011) Cloning and characterization of a novel 3-hydroxy-3-methylglutaryl coenzyme A reductase gene from *Salvia miltiorrhiza* involved in diterpenoid tanshinone accumulation. J Plant Physiol 168: 148–157.
21. Shen G, Pang Y, Wu W, Liao Z, Zhao L, et al. (2006) Cloning and characterization of a root-specific expressing gene encoding 3-hydroxy-3-methylglutaryl coenzyme A reductase from *Ginkgo biloba*. Mol Biol Rep 33: 117–127.
22. Akhtar N, Gupta P, Sangwan NS, Sangwan RS, Trivedi PK (2013) Cloning and functional characterization of 3-hydroxy-3-methylglutaryl coenzyme A reductase gene from *Withania somnifera*: an important medicinal plant. Protoplasma 250: 613–622.
23. Chappell J, Wolf F, Proulx J, Cuellar R, Saunders C (1995) Is the Reaction Catalyzed by 3-Hydroxy-3-Methylglutaryl Coenzyme A Reductase a Rate-

Limiting Step for Isoprenoid Biosynthesis in Plants? Plant Physiol 109: 1337–1343.

24. Schaller H, Grausem B, Benveniste P, Chye M, Tan C, et al. (1995) Expression of the *Hevea brasiliensis* (H.B.K.) Mull. Arg. 3-hydroxy-3-methylglutaryl-coenzyme A reductase 1 in tobacco results in sterol overproduction. Plant physiology 109: 761–770.

25. Hey SJ, Powers SJ, Beale MH, Hawkins ND, Ward JL, et al. (2006) Enhanced seed phytosterol accumulation through expression of a modified HMG-CoA reductase. Plant Biotechnol J 4: 219–229.

26. Harker M, Holmberg N, Clayton JC, Gibbard CL, Wallace AD, et al. (2003) Enhancement of seed phytosterol levels by expression of an N-terminal truncated *Hevea brasiliensis* (rubber tree) 3-hydroxy-3-methylglutaryl-CoA reductase. Plant Biotechnol J 1: 113–121.

27. Munoz-Bertomeu J, Sales E, Ros R, Arrillaga I, Segura J (2007) Up-regulation of an N-terminal truncated 3-hydroxy-3-methylglutaryl CoA reductase enhances production of essential oils and sterols in transgenic *Lavandula latifolia*. Plant Biotechnol J 5: 746–758.

28. Omura T, Watanabe S, Iijima Y, Aoki K, Shibata D, et al. (2007) Molecular and genetic characterization of transgenic tomato expressing 3-hydroxy-3-methylglutaryl coenzyme A reductase. Plant biotechnology 24: 107–115.

29. Istvan ES, Deisenhofer J (2000) The structure of the catalytic portion of human HMG-CoA reductase. Biochim Biophys Acta 1529: 9–18.

30. Darabi M, Masoudi-Nejad A, Nemat-Zadeh G (2012) Bioinformatics study of the 3-hydroxy-3-methylglotaryl-coenzyme A reductase (HMGR) gene in Gramineae. Mol Biol Rep 39: 8925–8935.

31. Bailey TL, Elkan C (1994) Fitting a mixture model by expectation maximization to discover motifs in biopolymers. Proc Int Conf Intell Syst Mol Biol 2: 28–36.

32. Mulder NJ, Apweiler R, Attwood TK, Bairoch A, Bateman A, et al. (2005) InterPro, progress and status in 2005. Nucleic Acids Res 33: D201–D205.

33. Guo AY, Zhu QH, Chen X, Luo JC (2007) [GSDS: a gene structure display server]. Yi Chuan 29: 1023–1026.

34. Larkin MA, Blackshields G, Brown NP, Chenna R, McGettigan PA, et al. (2007) Clustal W and Clustal X version 2.0. Bioinformatics 23: 2947–2948.

35. Tamura K, Peterson D, Peterson N, Stecher G, Nei M, et al. (2011) MEGA5: molecular evolutionary genetics analysis using maximum likelihood, evolutionary distance, and maximum parsimony methods. Mol Biol Evol 28: 2731–2739.

36. Zhao Y, Zhou Y, Jiang H, Li X, Gan D, et al. (2011) Systematic analysis of sequences and expression patterns of drought-responsive members of the HD-Zip gene family in maize. PLoS One 6: e28488.

37. Wei H, Li W, Sun X, Zhu S, Zhu J (2013) Systematic analysis and comparison of nucleotide-binding site disease resistance genes in a diploid cotton *Gossypium raimondii*. PLoS One 8: e68435.

38. Jiang H, Wu Q, Jin J, Sheng L, Yan H, et al. (2013) Genome-wide identification and expression profiling of ankyrin-repeat gene family in maize. Dev Genes Evol 223: 303–318.

39. Rozas J, Sanchez-DelBarrio JC, Messeguer X, Rozas R (2003) DnaSP, DNA polymorphism analyses by the coalescent and other methods. Bioinformatics 19: 2496–2497.

40. Gaut BS, Morton BR, McCaig BC, Clegg MT (1996) Substitution rate comparisons between grasses and palms: synonymous rate differences at the nuclear gene *Adh* parallel rate differences at the plastid gene *rbcL*. Proc Natl Acad Sci U S A 93: 10274–10279.

41. Lynch M, Conery JS (2000) The evolutionary fate and consequences of duplicate genes. Science 290: 1151–1155.

42. Schlueter JA, Dixon P, Granger C, Grant D, Clark L, et al. (2004) Mining EST databases to resolve evolutionary events in major crop species. Genome 47: 868–876.

43. de Hoon MJ, Imoto S, Nolan J, Miyano S (2004) Open source clustering software. Bioinformatics 20: 1453–1454.

44. Sekhon RS, Lin H, Childs KL, Hansey CN, Buell CR, et al. (2011) Genome-wide atlas of transcription during maize development. Plant J 66: 553–563.

45. Dash S, Van Hemert J, Hong L, Wise RP, Dickerson JA (2012) PLEXdb: gene expression resources for plants and plant pathogens. Nucleic Acids Res 40: D1194–D1201.

46. Severin AJ, Woody JL, Bolon YT, Joseph B, Diers BW, et al. (2010) RNA-Seq Atlas of *Glycine max*: a guide to the soybean transcriptome. BMC Plant Biol 10: 160.

47. Graham LE, Cook ME, Busse JS (2000) The origin of plants: body plan changes contributing to a major evolutionary radiation. Proc Natl Acad Sci U S A 97: 4535–4540.

48. Lewis LA, McCourt RM (2004) Green algae and the origin of land plants. Am J Bot 91: 1535–1556.

49. Lombard J, Moreira D (2011) Origins and early evolution of the mevalonate pathway of isoprenoid biosynthesis in the three domains of life. Mol Biol Evol 28: 87–99.

50. Lohr M, Schwender J, Polle JE (2012) Isoprenoid biosynthesis in eukaryotic phototrophs: a spotlight on algae. Plant Sci 185–186: 9–22.

51. Vranova E, Coman D, Gruissem W (2013) Network analysis of the MVA and MEP pathways for isoprenoid synthesis. Annu Rev Plant Biol 64: 665–700.

52. Grauvogel C, Petersen J (2007) Isoprenoid biosynthesis authenticates the classification of the green alga *Mesostigma viride* as an ancient streptophyte. Gene 396: 125–133.

53. Lange BM, Rujan T, Martin W, Croteau R (2000) Isoprenoid biosynthesis: the evolution of two ancient and distinct pathways across genomes. Proc Natl Acad Sci U S A 97: 13172–13177.

54. Banks JA (2009) Selaginella and 400 million years of separation. Annu Rev Plant Biol 60: 223–238.

55. Campos N, Boronat A (1995) Targeting and topology in the membrane of plant 3-hydroxy-3-methylglutaryl coenzyme A reductase. Plant Cell 7: 2163–2174.

56. Leivar P, González VM, Castel S, Trelease RN, López-Iglesias C, et al. (2005) Subcellular localization of Arabidopsis 3-hydroxy-3-methylglutaryl-coenzyme A reductase. Plant physiology 137: 57–69.

57. Ober D (2010) Gene duplications and the time thereafter - examples from plant secondary metabolism. Plant Biol (Stuttg) 12: 570–577.

58. Wang K, Wang Z, Li F, Ye W, Wang J, et al. (2012) The draft genome of a diploid cotton *Gossypium raimondii*. Nat Genet 44: 1098–1103.

59. Enjuto M, Lumbreras V, Marin C, Boronat A (1995) Expression of the Arabidopsis HMG2 gene, encoding 3-hydroxy-3-methylglutaryl coenzyme A reductase, is restricted to meristematic and floral tissues. Plant Cell 7: 517–527.

11

Independent Evolutionary Origin of *fem* Paralogous Genes and Complementary Sex Determination in Hymenopteran Insects

Vasco Koch, Inga Nissen, Björn D. Schmitt, Martin Beye*

Institute of Evolutionary Genetics, Heinrich Heine University Duesseldorf, Duesseldorf, Germany

Abstract

The primary signal of sex determination in the honeybee, the *complementary sex determiner* (*csd*) gene, evolved from a gene duplication event from an ancestral copy of the *fem* gene. Recently, other paralogs of the *fem* gene have been identified in several ant and bumblebee genomes. This discovery and the close phylogenetic relationship of the paralogous gene sequences led to the hypothesis of a single ancestry of the *csd* genetic system of complementary sex determination in the Hymenopteran insects, in which the *fem* and *csd* gene copies evolved as a unit in concert with the mutual transfers of sequences (concerted evolution). Here, we show that the paralogous gene copies evolved repeatedly through independent gene duplication events in the honeybee, bumblebee, and ant lineage. We detected no sequence tracts that would indicate a DNA transfer between the *fem* and the *fem1*/*csd* genes between different ant and bee species. Instead, we found tracts of duplication events in other genomic locations, suggesting that gene duplication was a frequent event in the evolution of these genes. These and other evidences suggest that the *fem1*/*csd* gene originated repeatedly through gene duplications in the bumblebee, honeybee, and ant lineages in the last 100 million years. Signatures of concerted evolution were not detectable, implicating that the gene tree based on neutral synonymous sites represents the phylogenetic relationships and origins of the *fem* and *fem1*/*csd* genes. Our results further imply that the *fem1* and *csd* gene in bumblebees, honeybees, and ants are not orthologs, because they originated independently from the *fem* gene. Hence, the widely shared and conserved complementary sex determination mechanism in Hymenopteran insects is controlled by different genes and molecular processes. These findings highlight the limits of comparative genomics and emphasize the requirement to study gene functions in different species and major hymenopteran lineages.

Editor: William Hughes, University of Sussex, United Kingdom

Funding: This work was supported by grants from the Deutsche Forschungsgemeinschaft, DFG. The funders had no role in study design, data collection and analysis, decision to publish, or preparation of the manuscript.

Competing Interests: The authors have declared that no competing interests exist.

* E-mail: martin.beye@uni-duesseldorf.de

Introduction

Complementary sex determination, in which the heterozygous genotype at a certain locus determines femaleness, is widely shared in hymenopteran insects and has a deep ancestry [1,2]. Thus far, the underlying gene *complementary sex determiner* (*csd*) has been identified in the western honeybee (*Apis mellifera*) by positional cloning and knockdown studies [3,4]. The *csd* gene encodes an SR-type protein. Csd proteins derived from the heterozygous *csd* genotype induce the female sex pathway by directing the female splicing of the primary transcripts of the *fem* gene [4,5]. The resulting female mRNAs subsequently encode the functional Fem proteins. Csd proteins derived from the hemizygous or homozygous genotypes are not required for sex determination. The male splicing of the *fem* transcripts results by default. The male-specific exons contain a translational stop codon to prematurely stop translation. The absence of functional Fem proteins leads to the development of maleness [5]. More than 14 *csd* alleles have been identified in local honeybee populations, which show an average of 3% pairwise difference in their entire amino acid encoding sequence [6,7].

The low divergence of the honeybee *csd* and *fem* genes at synonymous sites compared to bumblebee and stingless bee sequences suggests that the *csd* gene was derived from a gene duplication event of an ancestral copy of the *fem* gene in the honeybee lineage [4]. The *csd* gene was shaped by positive selection shortly after it originated [4,8]. *fem* is the putative ortholog of the *transformer* (*tra*) gene [4], a key sex-determining gene in *Drosophila melanogaster*. However, unambiguous homology relies on identities in a 30-amino-acid motif deduced from another dipteran ortholog of the *tra* gene from *Ceratitis capitata* [9,10].

A recent study found repeated duplicates of the *fem* gene in four ant and two bumblebee genomes [11]. The transcripts of these *fem* genes are sex-specifically spliced, suggesting a conserved sex-determining role of this gene. The function of the duplicated copies are thus far unknown [11]. The wasp *Nasonia vitripennis*, however, lacks a sister copy of the *fem*/*tra* gene [12]. In this study, we named the other copies of the *fem* gene *fem1*. This is because we have no functional information as to whether these genes control the complementary sex determination process as in honeybees.

The phylogenetic relationships deduced from coding nucleotide sequences [11] showed that the paralogous gene pairs of *fem* and *csd/fem1* are more closely related in four ant species, the bumblebee and, as previously shown, the honeybee lineage. Figure S1 shows the sequence relationship of the genes for the neutral synonymous sites. Two recent studies [11,13] have proposed that in contrast to a model of independent gene duplications, the most parsimonious explanation of the close relationship between the *fem* and *fem1/csd* sequences is that concerted evolution (either due to repeated unequal crossing-over or gene conversion) homogenized the duplicated copies in the different lineages (Fig. 1). One or a few ancestral duplication events gave rise to the *csd* gene and complementary sex determination observed in the Hymenoptera order [11,13]. The process of concerted evolution between the *fem* and *fem1/csd* genes repeatedly homogenized the two loci, producing the low divergence in the gene pair that we find today.

Here, we readdress the question of whether the *fem1/csd* copies repeatedly evolved through gene duplication (Fig. 1a) or whether the *fem* and *fem1/csd* gene pairs evolved through concerted evolution (Fig. 1b). The clarification of this question will provide fruitful insight into the evolution of paralogous genes and the evolution of a complementary sex determination system.

The arguments given below prompted us to further investigate this question.

1) Studies at the genome-wide scale showed that concerted evolution only affects 2% of the paralogous gene pairs [14], suggesting that this process rarely acts as a homogenizing force between paralogs.

2) The rate for the rise of new paralogous gene copies is 0.01 per gene per million years [15,16]. This suggests that new duplicates of the *fem* gene can repeatedly originate in different hymenopteran lineages, which have an evolutionary history of more than 120 million years [17–20].

3) The evidence for concerted evolution between the paralogs provided thus far, namely, (i) the alternative tree topologies of the *fem* and *fem1/csd* nucleotide sequences [13] and (ii) the putative gene conversion tracts in the nucleotide sequence [11,13], could also result from a heterogeneity in the sequence

divergence, a recombination event between the *csd* alleles, methodological problems or homoplasic (convergent) nucleotide changes [4,6,8,21].

Here, we present evidence suggesting that the paralogous gene copies *fem1/csd* in ants, bumblebees and honeybees evolved independently and repeatedly through a series of gene duplication events (Fig. 1a).

Results

Amino acid changes in the MRCA ancestral sequences of bees and ants are shared between the Fem and Csd/Fem1 proteins

To find further support for either the repeated gene duplication model or the concerted evolution model, we followed the evolutionary trajectory of substitutions that led to amino acid changes in the ancestral sequences of the most recent common ancestor (MRCA) of bees and ants (Fig. 2a). This evolutionary window predates the timing of the different gene duplication events under the repeated duplication model and can therefore provide unique information about the evolutionary history of the sister copies. Under the concerted evolution model, we would expect to find unique substitutions in the ancestral sequences of the MRCA of ants and bees, which are confined to the *fem* or the *csd/fem1* gene (Fig. 2a). This pattern would arise because the two sister copies originated only once in Hymenoptera [11] and accumulated substitutions separately due to their separate evolutionary history, which predates the MRCA of bees and ants [11]. Concerted evolution, the exchange of sequences between evolutionary old paralogous genes, would partly homogenize the sister copy genes, which would thus appear as to have more recent common ancestry in the phylogenetic tree (Fig. S1, S2). Under the repeated gene duplication model, the ancestral sequence in the evolutionary time window that predates the different duplication events should be the same for the *fem* and *csd/fem1* genes because at this time point, only a single copy of the gene existed (Fig. 2a).

We generated separate phylogenetic trees using the amino acid sequence of the Fem and Fem1/Csd proteins, which allowed us to

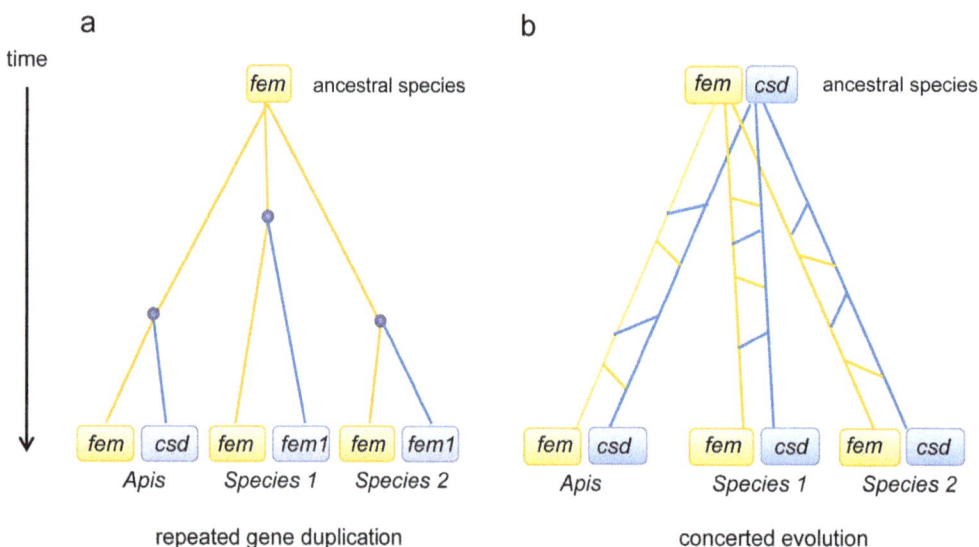

Figure 1. Two models for the evolutionary history of *fem* paralogous genes in ants and bees: (a) repeated gene duplication and (b) concerted evolution. Points in (**a**) denote gene duplication events giving rise to two gene copies. Connecting lines in (**b**) between branches indicate concerted evolution events resulting from unequal crossing over and/or gene conversion.

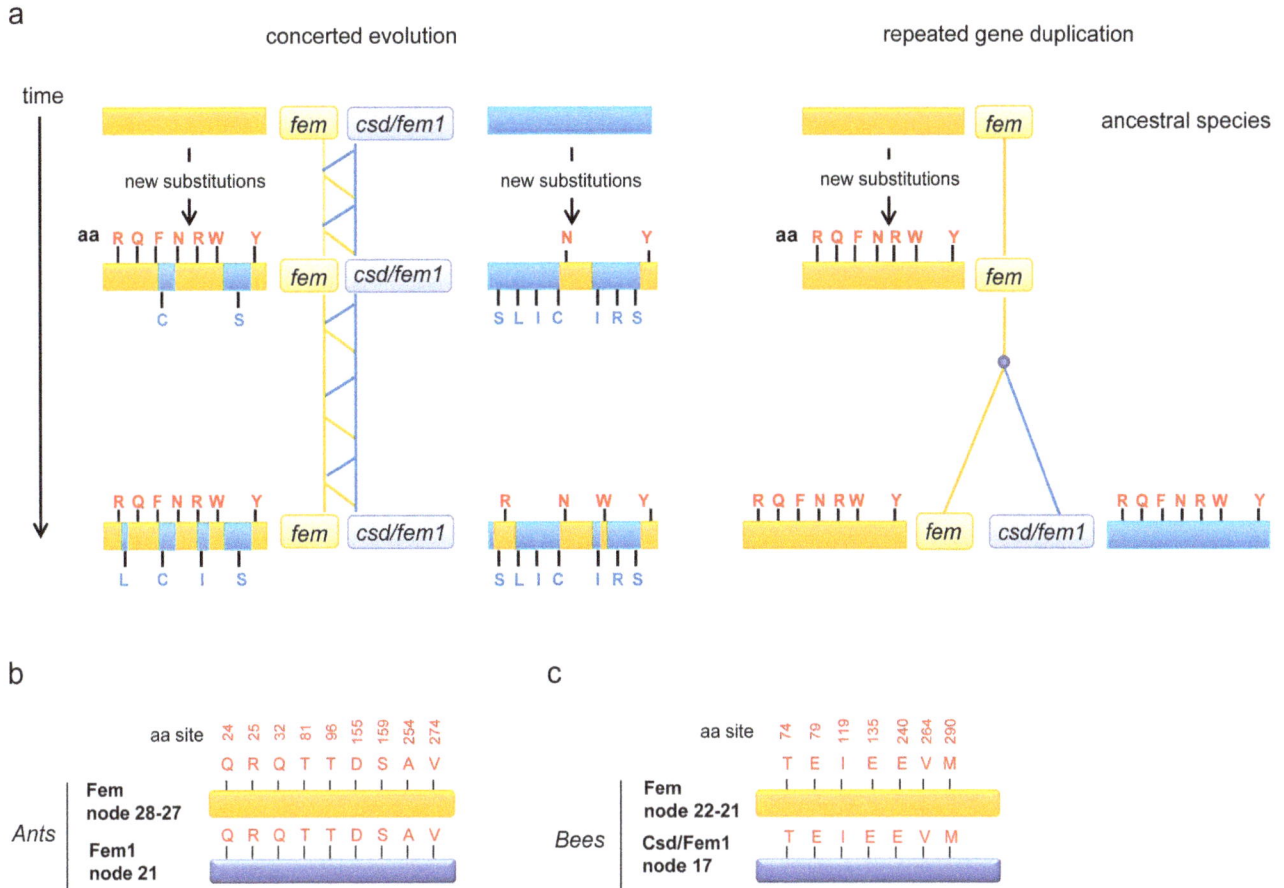

Figure 2. The evolutionary fate of *fem* gene substitutions in an evolutionary window predating the putative repeated gene duplications. (a) The expected evolutionary fate of *fem* substitutions in the paralogous genes *fem* and *csd/fem1* under the model of concerted evolution and repeated gene duplication. **(b, c)** The letters above the yellow boxes show the inferred amino acid changes in the Fem protein tree that evolved during the evolutionary window of the MRCA of ants and bees and the MRCAs of ants **(b)** and of bees **(c)**. Letters above the blue boxes indicate the amino acid residues that are found at the homologous sites in the ancestral Csd/Fem1 protein sequence of the MRCA of ants **(b)** and of bees **(c)**. Numbers above the letters designate the homologous sites in the Fem amino acid sequence alignment. Numbers before the boxes indicate the nodes (Fig. S2) used to infer the ancestral sequence information. aa denotes amino acid.

trace the putative separate evolutionary history of these sister copy genes. We inferred the ancestral amino acid sequences of the MRCAs of bees and of ants (Fig. 2, S2) using the maximum likelihood method [22]. These nodes had high statistical support and represented evolutionary time windows before the putative repeated gene duplication events. For the large evolutionary distances between ants and bees, we analyzed amino acid changes instead of synonymous substitutions, which were saturated, at least for the less degenerate sites. We identified changes in the Fem protein of the MRCA amino acid sequence of bees and ants by comparing the MRCA sequences of bees and of ants. We found 7 changes in the MRCA sequence of bees and 9 in the MRCA sequence of ants. In ants, we found the same 9 amino acid changes in the sister copy sequence of the paralogous Fem1 protein (Fig. 2b, Fig. S3). In bees, we found the same 7 amino acid changes in the MRCA sequence of the paralogous Csd/Fem1 protein (Fig. 2c, Fig. S4). These informative changes in the ant and bee sequences are found in different parts of the protein, suggesting that we obtained information that covered the entire protein. Our study found no amino acid changes that were confined to only one of the sister copies, which would indicate a deeper ancestry of the gene duplication that predated the MRCA of bees and of ants.

Therefore, the sequences harbor no information that can provide evidence for a separate history of the *fem* and *csd/fem1* genes that predates those of the ant and bee lineages, or for a single gene duplication event in Hymenoptera.

No evidence of concerted evolution is found in bumblebee sequences

We studied whether we can detect sequence tracts that would indicate a transfer of sequences between paralogous genes (concerted evolution) in the bumblebee lineage. The phylogenetic clustering of the *fem* and *fem1* sequences in the bumblebee lineage (Fig. S1) suggests that concerted evolutionary events should also occur in the bumblebee lineage. We used seven methods (RDP, GENECONV, BootScan, MaxChi, Chimaera, SiScan and 3Seq) designed to detect tracts of recombination events in the nucleotide sequences, which are included in the RDP 3.44 software package [23]. We applied this method to a single sequence alignment, which included the *fem* and *fem1* sequences of the two *Bombus* species and the *fem* sequence from *A. mellifera* as an outgroup reference. We classified the detected transfer events as either (i) concerted evolution events if they occurred between paralogous genes, (ii) recombination events if they occurred between the same gene, and (iii) falsely discovered events (FDE) if these events are biologically implausible. Such biologically implausible events are events in which sequence transfer between orthologous genes should give rise to a paralogous recombinant sequence or are events in which the sequence of the outgroup reference species is involved. The results of the analysis are shown in Figure 3. We found no sequence tracts that were transferred between paralogous genes, suggesting that concerted evolution played no role in the evolution of the paralogous genes of the *Bombus* lineage.

We next studied whether we can identify signatures of concerted evolution by confining our analysis to single amino acid substitutions. We tested whether some *fem* or *fem1* substitutions that newly evolved in each bumblebee species were transferred to its paralogous sister gene (Fig. 4a). Such shared evolved states between the two paralogous genes within each species would indicate a transfer of the corresponding nucleotide sequence by concerted evolution. We determined the ancestral sequences of the Fem and Fem1 protein of the MRCA of *Bombus terrestris* and *Bombus impatiens*, identified the evolutionary changes and studied whether these changes were also present in the paralogous sister copy (Fig. 4a). We found 4 newly evolved amino acid changes in the Fem protein of *B. impatiens* and 2 in that of *B. terrestris*. All of these newly evolved changes were not present in the sister Fem1 proteins (Fig. 4b, Fig. S5). We detected 8 newly evolved amino acid changes in the Fem1 protein of *B. impatiens* and 11 in *B. terrestris*, and these newly evolved changes were not present in the sister Fem protein (Fig. 4b, Fig. S6). Taken together, we found that all 12 newly evolved amino acid changes in *B. impatiens* and all 13 newly evolved amino acid changes in *B. terrestris* were absent in their corresponding sister copies. This survey covered different parts of the approximately 400 amino acid (aa)-long protein, providing evidence that concerted evolution events were absent in the *Bombus fem* and *fem1* sequences.

No evidence of concerted evolution in sequences of the honeybee lineage

Next we applied the RDP, GENECONV, BootScan, MaxChi, Chimaera, SiScan and 3Seq methods to identify sequence tracts of concerted evolution between the paralogous *fem* and *csd* nucleotide sequences of the honeybee. We included the same coding nucleotide sequences as in a previous study [13], comprised of

36 *csd* and 1 *fem A. mellifera*, 16 *csd* and 1 *fem Apis cerana csd*, 19 *csd* and 1 *fem Apis dorsata* sequence and a *fem B. terrestris* sequence as an outgroup reference.

In this study, we applied the methods to a single sequence alignment that included all *fem* and *csd* sequences. In the previous study from Privman et al., 100 alignments were used, each consisting of the *fem* sequences and a randomly chosen *csd* allele sequence from each *Apis* species and a *fem* sequence from *B. terrestris*. The rationale behind our altered experimental design was that the detection methods used in the RDP 3.44 software program are designed for large datasets, to identify the recombinant sequence and the two sequences from which the recombinant sequence was derived [23]. We removed a sequence (GenBank accession #: AY352276) from the analysis because it was a chimeric sequence of the *csd* and *fem* gene. This sequence resulted from a misassembly of cDNA sequences derived from the *fem* and *csd* gene at a point in time when we had no knowledge about a second gene in the genome and the nature of allelic variation [3]. This sequence entry has now been removed from GenBank. We also updated the sequence (GenBank accession #: AY350616), which is a *fem* and not a *csd* sequence. The results of this sequence analysis are shown in (Fig. 5).

We found no tracts of gene conversion (Fig. 5a), suggesting that a DNA transfer between the *fem* and *csd* gene did not occur in the honeybee lineage. We observed 3 events in which recombination between the *Apis csd* ortholog sequences gave rise to a paralogous *Apis fem* sequence, suggesting falsely detected events (FDE) that were identified by the program. We confirmed this falsely detection by demonstrating that the putatively transferred fragment is indeed *fem* derived which we showed by the clustering into the *fem* gene cluster in the phylogenetic tree analysis. However, we detected multiple recombination events between the *csd* sequences (alleles) derived from the same and different *Apis* species (Fig. 5b). This suggests that recombination is a regular process between alleles of the *csd* gene, consistent with previous reports [7,21].

We next evaluated how the number of sequences affects the false detection of events. We generated 20 sequence alignments, each consisting of two randomly chosen *csd* sequences from a single *Apis* species and one *fem* sequence from *B. terrestris*, which served as the outgroup reference sequence for the alignment above. For these alignments, the GENECONV method detected

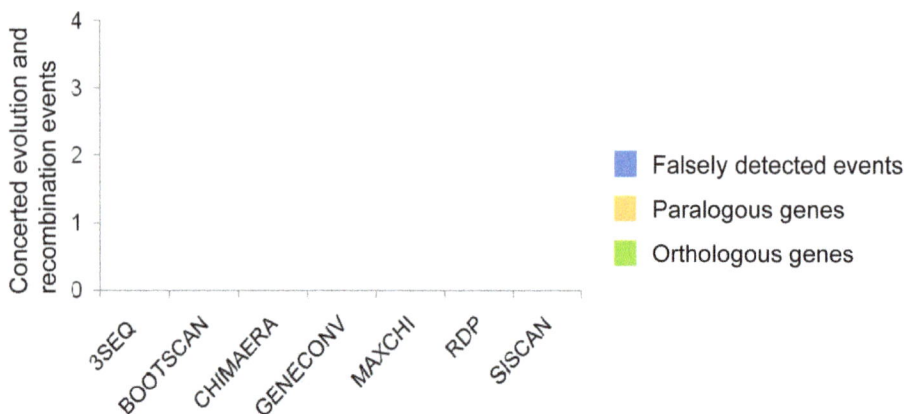

Legend:
- Falsely detected events (blue)
- Paralogous genes (yellow)
- Orthologous genes (green)

Figure 3. Number of gene conversion and recombination events in *B. terrestris* and *B. impatiens* sequences. Tracts of putatively recombined sequence were detected by the 7 methods as shown on the x-axis and using RDP 3.44 software program. The analysis was run on a single alignment of the *fem* and *fem1* sequences of the *B. terrestris, B. impatiens* and *A. mellifera fem* sequence. Gene conversion events refer to DNA transfers between the paralogous genes *fem* and *fem*1. Recombination events indicate transfer events between sequences of the same gene. Falsely detected events (FDE) refer to biologically implausible events (see Materials and Methods).

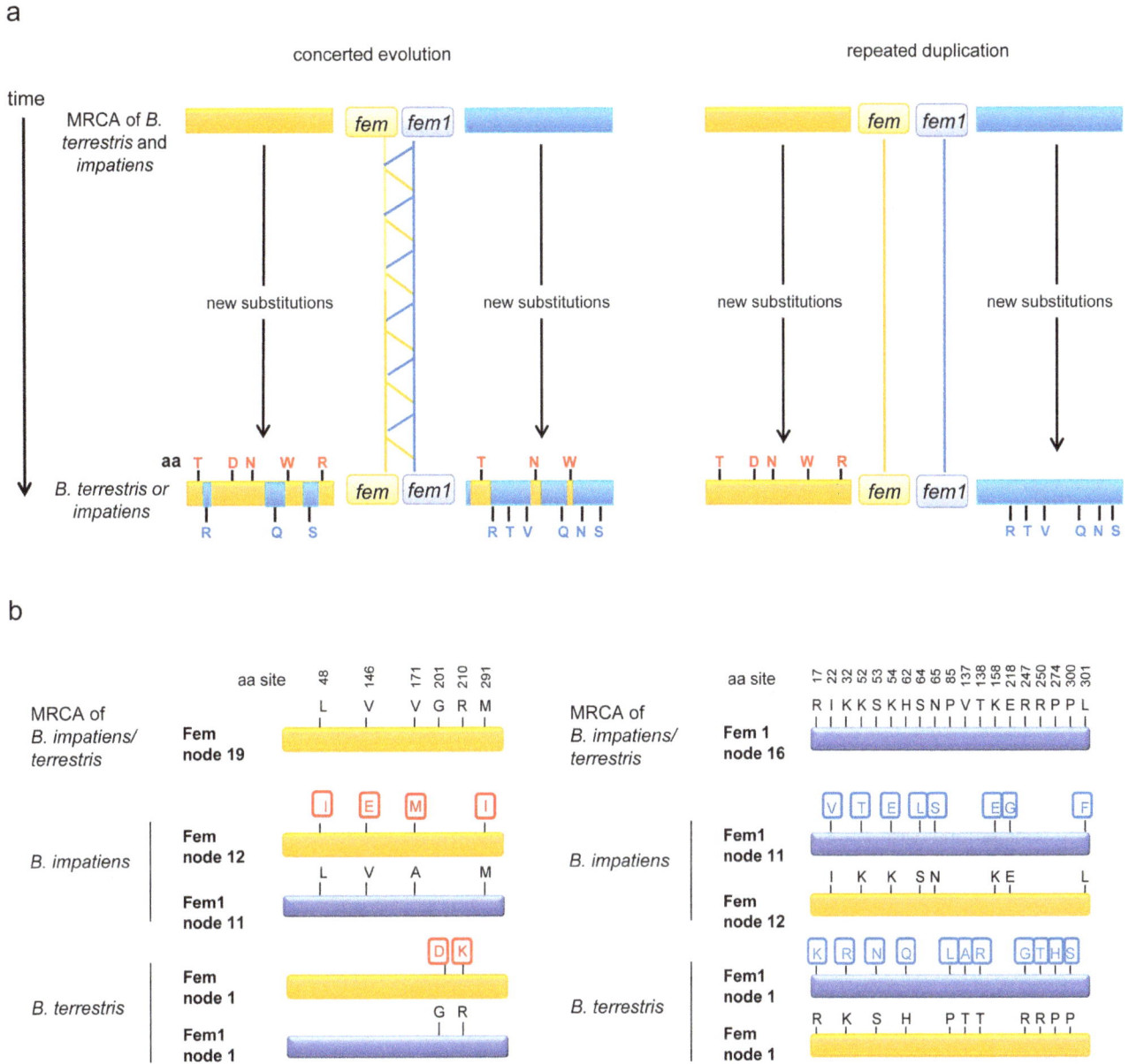

Figure 4. The evolutionary trajectory of *fem* gene substitutions in the evolutionary window that follows the putative gene duplication event in the *Bombus* lineage. (a) The expected evolutionary fate of *fem* substitutions in the paralogous genes *fem* and *fem1* under the models of concerted evolution and repeated gene duplication. **(b)** Deduced amino acid changes. The yellow box indicates the Fem protein, and the blue box indicates the Fem1 protein. Black letters above the boxes indicate the ancestral state of the amino acid residues found in the MRCA of *B. terrestris* and *B. impatiens*. Red letters in the red frame indicate the amino acid residues that evolved since the MRCA of *B. terrestris* and *B. impatiens* in the Fem protein. Blue letters in the blue frame indicate the amino acid residues that evolved since the MRCA of *B. terrestris* and *B. impatiens* in the Fem1 protein. Numbers before the boxes indicate the nodes (Fig. S2) used to infer the ancestral sequence information. aa denotes amino acid.

8, the MaxChi method detected 12, and the Chimaera method detected 9 tracts of sequences in which DNA was putatively transferred between the *csd* sequences of the *Apis* and *fem* gene of the *B. terrestris* sequences (Fig. S7). However, these events are biologically implausible, because the nucleotide differences we observe today in the *csd* alleles evolved after the split into different *Apis* species [7]. Hence, the *Bombus* sequence cannot have contributed through concerted evolutionary events to the *csd* polymorphism. We confirmed this falsely detection in a sample of detected events by demonstrating that the putatively transferred fragment is not derived from the *Bombus* sequence, which we

showed by the clustering into the *csd* gene cluster in the phylogenetic tree analysis. We detected no such transfer between the *Apis csd* alleles and the *B. terrestris fem* sequences if all sequences are included in a single alignment (Fig. 5) suggesting that having fewer sequences in an alignment increases the rate of falsely detecting DNA transfer between paralogs.

We did not perform single amino acid substitution analysis in the honeybee as we did for *Bombus* because we have not robustly identified enough newly evolved sites.

a

Concerted evolution between *fem*
and *csd* sequences

b

Recombination events between *fem* or *csd*
sequences

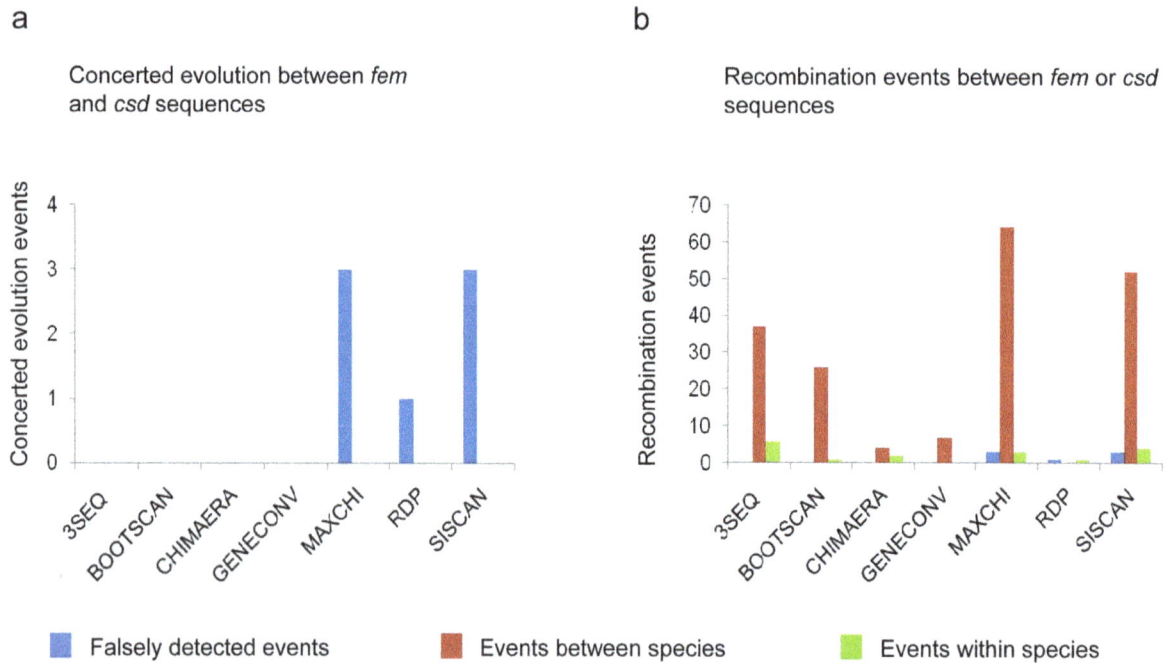

Falsely detected events   Events between species   Events within species

Figure 5. Events of gene conversion and recombination in the *fem* and *csd* sequences of *A. mellifera*, *A. dorsata* and *A. cerana*. Tracts of putatively recombined sequences were detected by seven different methods as indicated on the x-axis using the RDP 3.44 software program. The analysis was run on a single alignment of 71 *csd* and 4 *fem* sequences. (a) The number of concerted evolution events that refer to the DNA transfer between paralogous genes *fem* and *csd*. (b) The number of recombination events that identify events between sequences of the same gene. Falsely detected events (FDE) in (a) and (b) refer to biologically implausible events. The outgroup reference sequence, *B. terrestris fem*, was never involved in one of the detected events.

d_N/d_S ratio differences between paralogous genes suggest a directional DNA transfer process

We consistently observed that the d_N/d_S ratios (nonsynonymous (d_N) to synonymous (d_S) per site substitutions) of the *fem1/csd* sequence pairs were higher compared to those of the *fem* sequence pairs ((Table 1); χ^2-test,$P<0.05$ for all comparisons), suggesting that selection operates differently on the *fem* and the *fem1/csd* genes. The ratio of the differences is most pronounced for the *Apis* sequences ($d_N/d_{Sfem} = 0.1$–0.2 versus $d_N/d_{Scsd} = 0.8$–1), consistent with previous findings [4,8], and for the bumblebee sequences ($d_N/d_{Sfem} = 0.16$ versus $d_N/d_{Scsd} = 0.56$). The difference in the d_N/d_S ratios is less pronounced in the ants.

We further evaluated how differences in the d_N/d_S ratios are compatible with the mutual transfer of DNA and concerted evolution. The mutual transfer between the paralogous genes would also transfer the differences in the d_N/d_S ratios between paralogous genes. We assume that the entire sequences are in equilibrium of homogenization through concerted evolution and divergence. This is consistent with the model that concerted evolution is a random mutational and ongoing process that occurs through hymenopteran phylogeny. At this equilibrium, the gene-wide d_N/d_S values are good approximations for DNA fragments that are, on average, transferred between paralogs.

First, we showed that there are not different rates at synonymous sites in the honeybee and the bumblebee clade (Tajima's relative rate test, P>0.05). This result suggests that the differences in the d_N/d_S ratios between the genes reflect substitution rate differences at nonsynonymous sites (d_N).

DNA fragments transferred from the *fem1/csd* can only reach the lower d_N/d_S ratios in the *fem* gene as evolutionary time progresses if new mutations occur at the neutral synonymous sites

along with purifying selection at the nonsynonymous sites. We approximated the mean number of neutral pairwise substitutions per site (d_S) that is required to reach the lower d_N/d_S ratios. The *csd* sequences in honeybees show an average ratio of $d_N/d_{S\ csd} = 0.9$ (Table 1), suggesting that during the separation time of the two paralogous genes (in terms of $d_{Scsd/fem} = 0.18$, Table S1), approximately $d_{N\ csd} = 0.16$ pairwise substitutions in the *csd* gene have accumulated. We next estimated d_S, which has accumulated so that a transferred *csd* fragment (that has on average a ratio d_N/d_S $_{csd} = 0.9$) can reach the observed d_N/d_S ratio of the *fem* gene (in which $d_N/d_{S\ fem} = 0.17$). We assume the most conservative model, in which all newly arising nonsynonymous mutations were removed by purifying selection and only new synonymous mutations became fixed. A *csd* fragment can only adjust for the *fem*'s d_N/d_S ratio if $d_{S\ x} = 0.77$ additional synonymous substitutions have accumulated. This result suggests that a transferred *csd* DNA requires on average $d_S = 0.77$ synonymous substitutions to observe the low d_N/d_S ratio of the *fem* gene.

Similarly, we approximated $d_{S\ x}$ for bumblebee sequences. *fem1*-derived sequences in the *fem* sequence ($d_{S\ fem1/fem} = 0.22$, d_N/d_S $_{fem1} = 0.56$, $d_{N\ fem1} = 0.12$, $d_N/d_{S\ fem} = 0.16$) would require $d_{S\ x} = 0.53$ pairwise synonymous differences to accumulate in order for the sequence to reach the same $d_N/d_{S\ fem}$ ratio.

$d_{S\ x} = 0.77$ in honeybees and $d_{S\ x} = 0.53$ in bumblebees, as required for the transferred *csd* DNA to reach the observed d_N/d_S ratios of the *fem* gene are largely exceeding $d_S = 0.39$ that have accumulated between the bumblebee and honeybee species. This suggests that these transfers should predate the MRCA of bees, which is inconsistent with our previous result that there would be an absence of such transfers in this evolutionary window (Fig. 2). In addition, $d_{S\ x}$ largely exceeds the divergence between paralogous genes in the bumblebee ($d_{S\ fem1/fem} = 0.22$) and the

Table 1. The d_N and d_S values and ratios for the interspecies comparisons of the *fem* and *fem1/csd* genes.

Clade	Species		*fem* gene			*fem1/csd* gene			d_N/d_{Sfem}
			dN (SE)	dS (SE)	dN/dS	dN (SE)	dS (SE)	dN/dS	$<d_N/d_{Sfem\ 1/csd}$ *
Apis	Ador	Amel	0.02 (0.004)	0.09 (0.02)	0.22	0.13 (0.01)	0.13 (0.02)	1.00	P<0.0001
	Amel	Acer	0.01 (0.003)	0.08 (0.02)	0.13	0.12 (0.01)	0.15 (0.02)	0.8	P<0.0001
Bombus	Bter	Bimp	0.008 (0.003)	0.051 (0.013)	0.16	0.05 (0.007)	0.09 (0.02)	0.56	P<0.01
	Cflo	Hsal	0.26 (0.01)	0.51 (0.029)	0.51	0.35 (0.02)	0.5 (0.03)	0.7	P<0.01
	Cflo	Pbar	0.19 (0.01)	0.44 (0.03)	0.43	0.3 (0.01)	0.43 (0.03)	0.7	P<0.01
Ants	Hsal	Pbar	0.24 (0.01)	0.53 (0.03)	0.45	0.26 (0.01)	0.46 (0.03)	0.57	P<0.02
	Acep	Pbar	0.11 (0.01)	0.37 (0.03)	0.30	0.16 (0.01)	0.41 (0.03)	0.39	P<0.05
	Cflo	Acep	0.21 (0.01)	0.46 (0.03)	0.46	0.32 (0.01)	0.49 (0.03)	0.65	P<0.01
	Acep	Hsal	0.23 (0.01)	0.5 (0.03)	0.46	0.28 (0.01)	0.44 (0.03)	0.64	P<0.01

Species names: Amel, A. mellifera; Ador, A. dorsata; Acer, A. cerana; Bimp, B. impatiens; Bter, B. terrestris; Hsal, H. saltator; Pbar, P. barbatus; Acep, Atta cephalotes; Cflo, Camponotus floridanus. SE: standard error.
*A one-tailed χ^2 - test was conducted using the absolute number of synonymous and nonsynonymous differences.

honeybee ($d_{S\ fem/csd} = 0.18$), suggesting that the divergence between paralogs is too low to be compatible with such transfers and ratios. The results of this simple transfer model imply that the transfer processes cannot be bidirectional between paralogous genes, as predicted under a model of concerted evolution in which sequences are mutual exchanged and can become fixed through positive selection and genetic drift. Only a directional transfer from *fem* to *csd/fem1* would be compatible with data suggesting that relaxed or positive selection could substantially increase the d_N/d_S ratio. This directional transfer process is consistent with the gene duplication model, in which a new gene copy becomes neofunctionalized [4].

Additional sequence copies are repeatedly found at other genomic loci in bees and ants

To find further evidence for the repeated duplication model, we searched for other gene duplication tracts of the *fem* and *fem1/csd* genes. Using BlastN searches, we found genomic sequences with a high similarity to the coding nucleotide sequences of the *fem* or *csd* genes in the western honeybee, two bumblebee species and two ant species at other genomic loci (Fig. 6a). These duplicate genes exist as inactive genes with no complete open reading frame (ORF), suggesting that they are pseudogenes (*ps*). For the *ps1 csd* gene of *A. mellifera*, we confirmed by RT-PCR that this gene is transcriptionally inactive. We estimated the synonymous pairwise divergence (d_S) of the pseudogenes and of the *fem* or *fem1/csd* genes within each species (Fig. 6b), which we related to the MRCA events by providing the *fem* d_S values of different species. For *A. mellifera*, we found two pseudogenes, one derived from the *csd* gene and one derived from the *fem* gene (Fig. 6). The latter of these gave rise to a new female intron sequence in the *fem* gene [5]. The d_S divergence between pseudogenes and the *fem* or *csd* gene is smaller compared to the d_S between the *csd* and *fem* and the d_S between the *fem* of *A. mellifera* and *A. dorsata* (Fig. 6b), suggesting that both sequences were duplicated recently in the *A. mellifera* lineage. The d_S divergence in the bumblebee genomes suggests that the origin of these duplications predates the split between the *B. terrestris* and *B. impatiens* species (Fig. 6b) but that it originated after the split of the current functional gene copies of *fem* and *fem1*. In the ant *Harpegnathos saltator*, we observed one pseudogene that originated after and one that arose during the split between the functional *fem* and *fem1/csd* gene copies. However, in the ant *Pogonomyrmex*

barbatus, we also found a pseudogene with a much deeper ancestry (*Pbarps1 fem*; $d_S = 0.27$) than the functional gene pairs. Our results demonstrate that other gene duplication events occurred throughout the phylogeny and even within the *A. mellifera* lineage.

Discussion

Our study presents several lines of evidence that support the repeated gene duplication model, but reject the concerted evolution model in which the low divergence of paralogs resulted from homogenization. We studied *fem* and *csd/fem1* paralogous genes in several bee and ant species, representing 120 million years of evolution [17–20]. We first showed that there were no unique changes in the Fem or Csd/Fem1 proteins at a point in time that would indicate a separate history of the two gene copies, predating the MRCAs of bees and of ants. We detected no sequence tracts that would indicate a DNA transfer between the paralogs in two bumblebee and three honeybee species by using different methods. We also identified other tracts of duplicated copies of the *fem* and *fem1/csd* gene at other genomic loci in different ant and bee species, suggesting that repeated gene duplication is a frequent process in the evolution of these genes. Finally, we showed that the major differences in the d_N/d_S ratio between the *fem* and *fem1/csd* genes in bees exclude a mutual transfer of sequences but suggest a directional transfer from *fem* to the *fem1/csd* gene, which is consistent with gene duplication and a neofunctionalization model [4] and not with a mutual exchange of sequences under concerted evolution. We conclude from these results that the *fem1/csd* genes repeatedly originated through gene duplication in the bumblebee, honeybee and ant lineages. Concerted evolution played no detectable role in the evolution of these genes, suggesting that the phylogenetic relationship of the paralogs is represented by a gene tree based on neutral synonymous sites (Fig. S1).

Our finding is consistent with frequency estimates of gene duplication and concerted evolution events. Previous studies estimated that a gene will, on average, duplicate every 100 million years [15,16], which is consistent with our finding of repeated gene duplications of the *fem* gene in the phylogeny of ants and bees which split approximately 120 million years ago [17–20]. Another study showed that gene conversion is a rare event, detectable in only 2% of duplicated genes, and that this process requires physical distances smaller than 9 kb [14]. Contrary to the latter

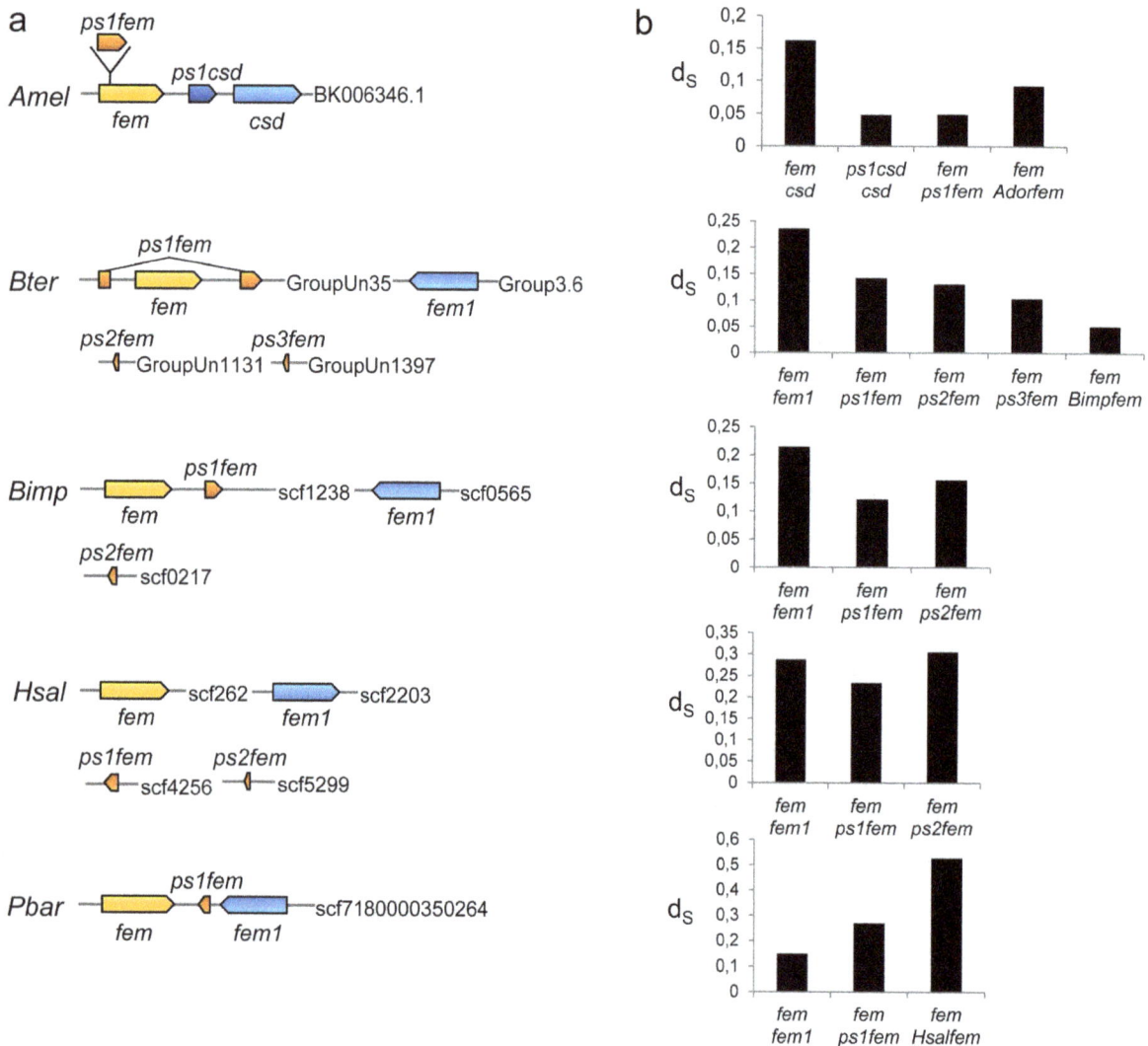

Figure 6. Pseudogenes (ps) of the fem and csd genes in the ant, bumblebee and western honeybee genomes. (a) The orientation and location of the pseudogenes (psfem, pscsd). Boxes denote the genes or pseudogenes. The box length of pseudogenes indicates the relative degree of homology to the coding nucleotide sequences of the fem or csd genes. The phylogenetic relationship assignments are based on the lowest d_S estimates or the ancestral state. Numbers behind the bars indicate the genomic scaffold, linkage group or the GenBank accession number. **(b)** Evolutionary distance between duplicated fem and fem1/csd gene copies are presented in terms of pairwise synonymous divergence per synonymous site (d_S). Abbreviations: Amel, A. mellifera; Bimp, B. impatiens; Bter, B. terrestris; Hsal, H. saltator; Pbar, P. barbatus.

requirement, fem and csd gene are separated by more than 12 kb in the honeybee (A. mellifera). Studies of genes that have multiple copies in the genome have demonstrated that new copies constantly originate by gene duplication [24–35]. Some of the duplicated copies are maintained in the genome for an extended period of time, while other copies were deleted or became nonfunctional through the accumulation of deleterious mutations.

Previous studies [11,13] proposed that concerted evolution produced the low divergence of the fem and fem1/csd genes. The authors suggested that (i) alternative tree topologies of the nucleotide sequences [13] and (ii) putative gene conversion tracts [11,13] are evidence for concerted evolution. Schmieder et al. identified gene conversion tracts in the genomic sequences of the paralogs. Our results showed that de novo duplications of the fem and fem1/csd genes generated such tracts. Privman et al. counted more than 100 recombination events in the honeybee fem and csd sequences, which they take as evidence for concerted evolution. In

our reanalysis of the same sequences using the same methods (Fig. 5), we distinguished whether the transfer of DNA occurred between paralogous genes (concerted evolution) or between alleles of the same (orthologous) gene (recombination events). We found no tracts of gene conversion events between the fem and csd sequences (Fig. 5a), suggesting the absence of concerted evolution in honeybees. However, we found repeated transfers between alleles of the same (csd) gene (Fig. 5b), a finding which has been repeatedly reported [7,21].

In our reanalysis, we used the entire 75-sequence data set in a single alignment, in contrast to the Privman et al. study, which used 100 alignments of 7 fem and csd sequences with different sets of csd alleles chosen from each species. We inspected some of the results using the Privman et al. alignments and repeatedly found tracts that suggested recombination events between the csd alleles. We also demonstrated in this study that, as in the work of Privman et al., small alignments of only a few sequences (Fig. S7) can generate an increase of the number of falsely detected events.

Privman et al. also proposed that the differences in the phylogenetic relationships of recombinant and non-recombinant regions provide further evidence for concerted evolution. Because these falsely detected "recombinant" sequence tracts are sequences from the same gene (csd) (see Fig. 5) these inconsistencies in the phylogenetic relationships are no further evidence of concerted evolution events. These "inconsistencies" have been previously reported for csd alleles. The combined forces of meiotic recombination and balancing selection generate a heterogeneity of divergence across the csd gene [6,7,21]. Recombination redistributes a small subset of variants of the 5′ region with multiple, highly diverged 3′ variants, which generates inconsistencies in the resulting phylogenetic relationships as previously shown [7,21,36].

Privman et al. found fem-specific substitutions for csd allele AY352276. However, this sequence is actually a chimeric sequence that was generated by the misassembly of the fem and csd cDNA sequences. This sequence was generated at a point in time when we searched for a third csd allele and had no knowledge of a second gene or the nature of the allelic diversity [3]. This sequence entry has been removed from GenBank. Privman et al. also suggested that alternative tree topologies of fem and fem1 sequences in the ants indicate gene conversion events [13]. We argue that the divergence of these sequences is too high ($d_S = 0.5$ for most species pairs) to exclude the possibility that ambiguous trees (and split phylogenetic networks) resulted from homoplasic (convergent) substitutions in the sequences.

Our results imply that the fem1 and csd genes in the ant, bumblebee and honeybee species are not orthologs because they originated independently through gene duplications (Fig. S1). The csd gene originated in the honeybee lineage [4,5]. Hence, complementary sex determination in bumblebee and ant species [1,2] is regulated by other genes and not by the orthologs of the csd gene. Consistent with this conclusion, the csd alleles of honeybees share a hypervariable region of asparagine- and tyrosine-enriched repeats [7] that are consistently absent in the fem1 genes of ant and bumblebee species.

Our results show that a new gene for complementary sex determination originated in honeybees [4], while the phylogenetic distribution of complementary sex determination indicates a deep ancestry in Hymenopteran insects [1,2]. One explanation for the replacement of a complementary sex determination gene is that ancestral complementary sex determiner genes degenerate over evolutionary time [4,10]. This is because meiotic recombination is suppressed at the sex determiner gene locus [21], allowing more deleterious mutations to accumulate over time [37–39]. This process could generate an adaptive advantage for evolving new sex determination genes that would eventually replace the older, malfunctioning, complementary sex determination gene. Such a degeneration process has been proposed for sex chromosomal systems [40–45] and may also explain the rapid evolution of the complementary sex determination system.

Characterizing the gene functions of the fem1 genes and the molecular basis of complementary sex determination in ants and bumblebees would provide interesting insights into the evolution of this sex determination system. Our results imply that the conserved phenotype (complementary sex determination) is only loosely evolutionary associated with the controlling molecular process. These findings highlight the limits of comparative genomics and emphasize the requirement to study gene functions in different species and major hymenopteran lineages.

Materials and Methods

The fem paralogous sequences defined in this study as fem1 (in ants and bumblebees) or csd (in honeybees) were taken from Schmieder et al. [11] and were provided by Schmieder, S. and Poirie, M. The genomic locus of the fem/fem1 gene of Solenopsis invicta was not accessible in public libraries. The honeybee fem and csd sequences that we used to detect tracts of DNA transfers were kindly provided by Privman, E. The coding sequences were aligned based on the deduced amino acids, assuming a standard genetic code table. We used either the Clustal or the Muscle program that was implemented in the MEGA5 program suite [46] to align the coding nucleotide sequences according to the deduced amino acid sequences. The alignments were edited manually. Nucleotide and amino acid substitution analyses were conducted using the MEGA5 program. Maximum likelihood fits of the substitution models with the lowest Bayesian information criterion (BIC) score were used to choose a substitution model for amino acids and nucleotides when possible. Pairwise gaps in the alignment were deleted.

Ancestral sequences

Ancestral amino acid sequences were inferred using the maximum likelihood method [22] under the Jones-Taylor-Thornton (JTT) matrix-based model [47]. The rates between sites were treated as a gamma distribution. The ancestral sequences were inferred from separate fem and fem1/csd sequence alignments, and Figure S2 shows the gene tree used. All informative changes used had a probability $P > 0.5$. We had to exclude some sites where the homology between sites of the Fem protein and the Csd/Fem1 alignment were ambiguous. In three cases of the bumblebee analysis, we changed the ancestral state we obtained from the MEGA analysis, because it contradicted the parsimony evolution of the sequence in the Fem and Fem1 tree (Fig. S5 and Fig. S6).

Evolutionary divergence between sequences

The d_N/d_S ratios of the interspecies comparisons between the ants, bumblebees and honeybees were inferred from sequence alignments that included either the bee or ant sequences, which greatly improved the number of identical positions in the alignments. The evolutionary distances between the ants and bees suggested that the less degenerated synonymous sites were saturated, making the d_S estimates between the bee and ant sequences less reliable. Analyses were conducted using the Nei-Gojobori model to estimate either the nonsynonymous and synonymous substitutions per site or the absolute numbers. A χ^2-square test was used to test the ratio differences in terms of absolute numbers. We tested the equality of the evolutionary rate at the most degenerate third codon position (which is presumably largely synonymous) by using Tajima's relative rate [48], which we performed using MEGA5 software [46]. We tested equality among the Apis csd and fem sequences by using the Bombus fem sequence as the outgroup. Similarly, we tested equality among Bombus fem1 and fem sequences by using the Apis fem sequence as an outgroup.

RT-PCR analysis

We performed RT-PCR experiments on embryonic and larval RNA [5] to identify possible splice products of mRNA in regions of the identified pseudogenes. PCR fragments were sequenced and compared with the genomic data.

Phylogenetic relationships

The phylogenetic relationship of the *fem* and *fem1/csd* sequences was determined based on presumably neutral synonymous differences by excluding the *tra* sequences of *Ceratitis capitata* and *Nasonia vitripennis* (because d_S could not be estimated for these species) using the neighbor-joining method. The confidence probability (multiplied by 100) that the interior branch length is greater than 0 was estimated using the bootstrap test with 10,000 replicates. The evolutionary distances were computed using the Pamilo-Bianchi-Li method and are displayed in the units of the number of synonymous substitutions per synonymous site. For each sequence pair, all ambiguous positions were removed.

Sequence tracts of DNA transfers

Tracts of DNA transfers in the honeybee *fem* and *csd* sequences were identified using the RDP 3.44 software program [23]. We used the following tests: RDP [49] with internal and external references, GENECONV [50], BootScan [51], MaxChi [52], Chimaera [53], SiScan [54] and 3Seq [55]. The methods implemented in the RDP 3.44 software program relied on the identification of recombinant sequences and the parental sequences from which these recombinant sequences were derived, which is facilitated by having a large set of sequences [23]. We thus generated a single alignment. Our alignment included 36 *A. mellifera*, 16 *A. cerana* and 19 *A. dorsata csd* coding nucleotide sequences, the single *fem* nucleotide sequence from each *Apis* species, and the *fem* nucleotide sequence of *B. terrestris* as an outgroup reference sequence. We used the same program settings as described in a previous study [13].We removed sequence GenBank accession #: AY352276 from the analysis, as this is a chimeric sequence of the *fem* and the *csd* gene (the entry has now been deleted). We also updated the sequence (GenBank accession #: AY350616) because it is not a *csd* but a *fem*-derived sequence. To evaluate the effect of the number of sequences in the alignment, we generated 20 sequence alignments consisting of three sequences. Each alignment included two randomly chosen *csd* sequences from a single *Apis* species as well as the *fem* sequence from *B. terrestris*. We classified the detected transfer events as (i) concerted evolution events if they occurred between paralogous genes, (ii) recombination events if they occurred between the same gene and (iii) falsely discovered events (FDE) if they were biologically implausible. Such biologically implausible events are events in which the sequence transfer between orthologous genes should give rise to a paralogous recombinant sequence or events in which the sequence of the outgroup reference species was involved. We confirmed falsely detection by inferring the clustering of the sequence tracts into the gene cluster in phylogenetic tree analyses.

Supporting Information

Figure S1 Gene tree of the *fem* and *fem1/csd* sister copies in ants and bees, which were inferred from synonymous differences. The evolutionary history was inferred using the neighbor-joining method. The confidence probability (multiplied by 100) that the interior branch length is greater than 0 was estimated using the bootstrap test (10000 replicates are shown next to the branches). The tree is drawn to scale, with branch lengths in the same units as those of the evolutionary distances used to infer the phylogenetic tree. The evolutionary distances were computed using the Pamilo-Bianchi-Li method [1] and are in the units of the number of synonymous

substitutions per synonymous site. All ambiguous positions were removed for each sequence pair. There were a total of 575 positions in the final dataset. Evolutionary analyses were conducted in MEGA 5 [2]. The sequences of *Nasonia* and *Ceratitis* were excludeto estimate d_S. Abbreviations: *Acep, Atta cephalotes; Acer, Apis cerana; Aech, Acromyrmex echinatior; Ador, Apis dorsata; Amel, Apis mellifera; Bimp, Bombus impatiens; Bter, Bombus terrestris; Cflo, Camponotus floridanus; Hsal, Harpegnathos saltator; Mcom, Melipona compressipes; Pbar, Pogonomyrmex barbatus; Sinv, Solenopsis invicta.*

Figure S2 The initial tree of the Fem (a) and Fem1/Csd (b) proteins that were used to infer the ancestral amino acid sequences. The evolutionary history was inferred by using the maximum likelihood method based on the JFF model [1]. Numbers in the tree assign the different nodes for which the ancestral sequence were obtained. The tree is drawn to scale, with branch lengths measured in the number of substitutions per site. Abbreviations: *Acep, Atta cephalotes; Acer, Apis cerana; Aech, Acromyrmex echinatior; Ador, Apis dorsata; Aflo, Apis florea; Amel, Apis mellifera; Bimp, Bombus impatiens; Bter, Bombus terrestris; Ccap, Ceratitis capitata; Cflo, Camponotus floridanus; Hsal, Harpegnathos saltator; Mcom, Melipona compressipes; Nvit, Nasonia vitripennis; Pbar, Pogonomyrmex barbatus; Sinv, Solenopsis invicta.*

Figure S3 The informative substitutions found in the ant lineage that were used in Figure 2. The identity of the different species and nodes of the Fem and Fem1 protein tree is shown. Site number (#) indicates the positions in the Fem and in the Csd/Fem1 protein sequence alignment.

Figure S4 The informative substitutions found in the bee lineage that were used in Figure 2. The identity of the different species and nodes of the Fem and Csd/Fem1 protein tree is shown. Site number (#) indicates the positions in the Fem and in the Csd/Fem1 protein sequence alignment.

Figure S5 The informative substitutions found in the *Bombus* lineage (Fem tree) that were used in Figure 3. The identity of the different species and nodes of the Fem and Fem1 protein tree is shown. Site number (#) indicates the position in the alignment of the Fem and of the Csd/Fem1 protein sequences.

Figure S6 The informative substitutions found in the *Bombus* lineage (Fem1 tree) that were used in Figure 3. The identity of the different species and nodes of the Fem and Fem1 protein tree is shown. Site number (#) indicate the position in the alignment of the Fem and of the Csd/Fem1 protein sequences.

Figure S7 The number of falsely detected events (FDE) using the methods as indicated on the X axis. These programs were implemented in the RDP 3.44 software program. The methods were run on each of the 20 alignments which consisted of two randomly chosen *csd* sequences from a single *Apis* species and one *fem* sequence from *B. terrestris*. These events are falsely detected as this transfer involve the outgroup *Bombus* sequence and the polymorphism between csd alleles which newly evolved in the different *Apis* species.

Table S1   The $d_{S\,i}$ values for the paralogous gene pairs *fem* and *csd/fem*1 within each species (*Ador*, *Apis dorsata*; *Amel*, *Apis mellifera*; *Bimp*, *Bombus impatiens*; *Bter*, *Bombus terrestris*).

Author Contributions

Conceived and designed the experiments: VK IN BS MB. Analyzed the data: VK IN BS MB. Contributed reagents/materials/analysis tools: VK IN BS MB. Wrote the paper: MB.

References

1. Heimpel GE, de Boer JG (2008) Sex determination in the hymenoptera. Annu Rev Entomol 53: 209–230.
2. Cook JM (1993) Sex determination in the hymenoptera: A review of models and evidence. Heredity (Edinburgh) 71: 421–435.
3. Beye M, Hasselmann M, Fondrk MK, Page RE, Omholt SW (2003) The gene *csd* is the primary signal for sexual development in the honeybee and encodes an SR-type protein. Cell 114: 419–429.
4. Hasselmann M, Gempe T, Schiott M, Nunes-Silva CG, Otte M, et al. (2008) Evidence for the evolutionary nascence of a novel sex determination pathway in honeybees. Nature 454: 519–522.
5. Gempe T, Hasselmann M, Schiott M, Hause G, Otte M, et al. (2009) Sex determination in honeybees: two separate mechanisms induce and maintain the female pathway. PLoS Biol 7: e1000222.
6. Hasselmann M, Beye M (2004) Signatures of selection among sex-determining alleles of the honey bee. Proc Natl Acad Sci U S A 101: 4888–4893.
7. Hasselmann M, Vekemans X, Pflugfelder J, Koeniger N, Koeniger G, et al. (2008) Evidence for convergent nucleotide evolution and high allelic turnover rates at the *complementary sex determiner* gene of Western and Asian honeybees. Mol Biol Evol 25: 696–708.
8. Hasselmann M, Lechner S, Schulte C, Beye M (2010) Origin of a function by tandem gene duplication limits the evolutionary capability of its sister copy. Proc Natl Acad Sci USA 107: 13378–13383.
9. Pane A, Salvemini M, Bovi PD, Polito C, Saccone G (2002) The transformer gene in *Ceratitis capitata* provides a genetic basis for selecting and remembering the sexual fate. Development 129: 3715–3725.
10. Gempe T, Beye M (2011) Function and evolution of sex determination mechanisms, genes and pathways in insects. Bioessays 33: 52–60.
11. Schmieder S, Colinet D, Poirie M (2012) Tracing back the nascence of a new sex-determination pathway to the ancestor of bees and ants. Nat Commun 3: 895.
12. Verhulst EC, Beukeboom LW, van de Zand L (2010) Maternal control of haplodiploid sex determination in the wasp *Nasonia*. Science 328: 620–623.
13. Privman E, Wurm Y, Keller L (2013) Duplication and concerted evolution in a master sex determiner under balancing selection. Proc Biol Sci 280: 20122968.
14. Semple C, Wolfe KH (1999) Gene duplication and gene conversion in the *Caenorhabditis elegans* genome. J Mol Evol 48: 555–564.
15. Lipinski KJ, Farslow JC, Fitzpatrick KA, Lynch M, Katju V, et al. (2011) High spontaneous rate of gene duplication in *Caenorhabditis elegans*. Curr Biol 21: 306–310.
16. Lynch M, Conery JS (2000) The evolutionary fate and consequences of duplicate genes. Science 290: 1151–1155.
17. Grimaldi D, Engel MS (2005) Evolution of the Insects. Cambridge: University Press Cambridge, UK.
18. Ramirez SR, Nieh JC, Quental TB, Roubik DW, Imperatriz-Fonseca VL, et al. (2010) A molecular phylogeny of the stingless bee genus *Melipona* (Hymenoptera: Apidae). Mol Phylogenet Evol 56: 519–525.
19. Wilson EO, Holldobler B (2005) The rise of the ants: a phylogenetic and ecological explanation. Proc Natl Acad Sci U S A 102: 7411–7414.
20. Brady SG, Schultz TR, Fisher BL, Ward PS (2006) Evaluating alternative hypotheses for the early evolution and diversification of ants. Proc Natl Acad Sci U S A 103: 18172–18177.
21. Hasselmann M, Beye M (2006) Pronounced differences of recombination activity at the sex determination locus (SDL) of the honey bee, a locus under strong balancing selection. Genetics 174: 1469–1480.
22. Nei M, Kumar S (2000) Molecular Evolution and Phylogenetics. New York: Oxford University Press.
23. Martin DP, Lemey P, Lott M, Moulton V, Posada D, et al. (2010) RDP3: a flexible and fast computer program for analyzing recombination. Bioinformatics 26: 2462–2463.
24. Su C, Nei M (2001) Evolutionary dynamics of the T-cell receptor VB gene family as inferred from the human and mouse genomic sequences. Mol Biol Evol 18: 503–513.
25. Rooney AP, Piontkivska H, Nei M (2002) Molecular evolution of the nontandemly repeated genes of the histone 3 multigene family. Mol Biol Evol 19: 68–75.
26. Piontkivska H, Rooney AP, Nei M (2002) Purifying selection and birth-and-death evolution in the histone H4 gene family. Mol Biol Evol 19: 689–697.
27. Ota T, Nei M (1995) Evolution of immunoglobulin VH pseudogenes in chickens. Mol Biol Evol 12: 94–102.
28. Ota T, Nei M (1994) Divergent evolution and evolution by the birth-and-death process in the immunoglobulin VH gene family. Mol Biol Evol 11: 469–482.
29. Niimura Y, Nei M (2005) Evolutionary changes of the number of olfactory receptor genes in the human and mouse lineages. Gene 346: 23–28.
30. Niimura Y, Nei M (2003) Evolution of olfactory receptor genes in the human genome. Proc Natl Acad Sci U S A 100: 12235–12240.
31. Eirin-Lopez JM, Gonzalez-Tizon AM, Martinez A, Mendez J (2004) Birth-and-death evolution with strong purifying selection in the histone H1 multigene family and the origin of orphon H1 genes. Mol Biol Evol 21: 1992–2003.
32. Robertson HM, Gadau J, Wanner KW (2010) The insect chemoreceptor superfamily of the parasitoid jewel wasp *Nasonia vitripennis*. Insect Mol Biol 19 Suppl 1: 121–136.
33. Robertson HM, Wanner KW (2006) The chemoreceptor superfamily in the honey bee, *Apis mellifera*: expansion of the odorant, but not gustatory, receptor family. Genome Res 16: 1395–1403.
34. Nei M, Gu X, Sitnikova T (1997) Evolution by the birth-and-death process in multigene families of the vertebrate immune system. Proc Natl Acad Sci U S A 94: 7799–7806.
35. Nei M, Rooney AP (2005) Concerted and birth-and-death evolution of multigene families. Annu Rev Genet 39: 121–152.
36. Beye M, Seelmann C, Gempe T, Hasselmann M, Vekemans X, et al. (2013) Gradual molecular evolution of a sex determination switch through incomplete penetrance of femaleness. Curr Biol 23: 2559–2564.
37. Graves JA (2006) Sex chromosome specialization and degeneration in mammals. Cell 124: 901–914.
38. Charlesworth B, Charlesworth D (1978) A model for the evolution of dioecy and gynodioecy. Am Nat 112: 975–997.
39. Charlesworth D, Charlesworth B, Marais G (2005) Steps in the evolution of heteromorphic sex chromosomes. Heredity 95: 118–128.
40. Marshall Graves JA (2008) Weird animal genomes and the evolution of vertebrate sex and sex chromosomes. Annu Rev Genet 42:565–86.: 565–586.
41. Mank JE, Avise JC (2009) Evolutionary diversity and turn-over in sex determination in teleost fishes. Sex Dev 3: 60–67.
42. Bull JJ (1983) Evolution of Sex Determining Mechanisms. Menlo Park,Calif.: Benjamin/Cummings Publishing Company.
43. Sanchez L (2008) Sex-determining mechanisms in insects. Int J Dev Biol 52: 837–856.
44. Schutt C, Nothiger R (2000) Structure, function and evolution of sex-determining systems in Dipteran insects. Development 127: 667–677.
45. Marin I, Baker BS (1998) The evolutionary dynamics of sex determination. Science 281: 1990–1994.
46. Tamura K, Peterson D, Peterson N, Stecher G, Nei M, et al. (2011) MEGA5: molecular evolutionary genetics analysis using maximum likelihood, evolutionary distance, and maximum parsimony methods. Mol Biol Evol 28: 2731–2739.
47. Jones DT, Taylor WR, Thornton JM (1992) The rapid generation of mutation data matrices from protein sequences. Comput Appl Biosci 8: 275–282.
48. Tajima F (1993) Simple methods for testing the molecular evolutionary clock hypothesis. Genetics 135: 599–607.
49. Martin D, Rybicki E (2000) RDP: detection of recombination amongst aligned sequences. Bioinformatics 16: 562–563.
50. Padidam M, Sawyer S, Fauquet CM (1999) Possible emergence of new geminiviruses by frequent recombination. Virology 265: 218–225.
51. Martin DP, Posada D, Crandall KA, Williamson C (2005) A modified bootscan algorithm for automated identification of recombinant sequences and recombination breakpoints. AIDS Res Hum Retroviruses 21: 98–102.
52. Smith JM (1992) Analyzing the mosaic structure of genes. J Mol Evol 34: 126–129.
53. Posada D, Crandall KA (2001) Evaluation of methods for detecting recombination from DNA sequences: computer simulations. Proc Natl Acad Sci U S A 98: 13757–13762.
54. Gibbs MJ, Armstrong JS, Gibbs AJ (2000) Sister-scanning: a Monte Carlo procedure for assessing signals in recombinant sequences. Bioinformatics 16: 573–582.
55. Boni MF, Posada D, Feldman MW (2007) An exact nonparametric method for inferring mosaic structure in sequence triplets. Genetics 176: 1035–1047.

12

Human Migration Patterns in Yemen and Implications for Reconstructing Prehistoric Population Movements

Aida T. Miró-Herrans[1,2,3*¤], **Ali Al-Meeri**[4], **Connie J. Mulligan**[2,3]

1 Genetics and Genomics Graduate Program, University of Florida, Gainesville, Florida, United States of America, **2** University of Florida Genetics Institute, University of Florida, Gainesville, Florida, United States of America, **3** Department of Anthropology, University of Florida, Gainesville, Florida, United States of America, **4** Clinical Biochemistry Department, Faculty of Medicine and Health Sciences, Sana'a University, Sana'a, Republic of Yemen

Abstract

Population migration has played an important role in human evolutionary history and in the patterning of human genetic variation. A deeper and empirically-based understanding of human migration dynamics is needed in order to interpret genetic and archaeological evidence and to accurately reconstruct the prehistoric processes that comprise human evolutionary history. Current empirical estimates of migration include either short time frames (i.e. within one generation) or partial knowledge about migration, such as proportion of migrants or distance of migration. An analysis of migration that includes both proportion of migrants and distance, and direction over multiple generations would better inform prehistoric reconstructions. To evaluate human migration, we use GPS coordinates from the place of residence of the Yemeni individuals sampled in our study, their birthplaces and their parents' and grandparents' birthplaces to calculate the proportion of migrants, as well as the distance and direction of migration events between each generation. We test for differences in these values between the generations and identify factors that influence the probability of migration. Our results show that the proportion and distance of migration between females and males is similar within generations. In contrast, the proportion and distance of migration is significantly lower in the grandparents' generation, most likely reflecting the decreasing effect of technology. Based on our results, we calculate the proportion of migration events (0.102) and mean and median distances of migration (96 km and 26 km) for the grandparent's generation to represent early times in human evolution. These estimates can serve to set parameter values of demographic models in model-based methods of prehistoric reconstruction, such as approximate Bayesian computation. Our study provides the first empirically-based estimates of human migration over multiple generations in a developing country and these estimates are intended to enable more precise reconstruction of the demographic processes that characterized human evolution.

Editor: David Caramelli, University of Florence, Italy

Funding: This research was supported by National Science Foundation (NSF) grant #BCS-0518530 to CJM and an NSF Graduate Research Fellowship to ATMH. The funders had no role in study design, data collection and analysis, decision to publish, or preparation of the manuscript.

Competing Interests: The authors have declared that no competing interests exist.

* E-mail: amiroherrans@utexas.edu

¤ Current address: Department of Anthropology, University of Texas at Austin, Austin, Texas, United States of America

Introduction

Humans' facility for dispersal has played a large role in our evolutionary history, yet our understanding of how and why humans have moved throughout history is unclear. Most data on human movement come from ethnographic and archaeological studies, comparisons of birthplaces from birth certificates, and census data. While ethnographic studies offer insight into social and environmental factors that influence human movement, they generally involve seasonal or temporary movements, as in the case of migrant workers [1] or hunter- gatherers [2,3]. In order to understand how migration has influenced our evolutionary history, it is necessary to address migration as the movement to a new location for permanent settlement. Although archaeological studies can provide information about movement over longer periods of time, they are often limited by the availability of data [4] and restricted to specific regions and time periods. Birth certificate and census data allow us to trace movement across longer periods of time as well, but studies using these data generally focus either on the proportion of migrants or the distance moved, do not usually use multi-generational families, and can typically only be studied in developed countries [5–10]. A deeper understanding of migration over multiple generations in a developing country offers the possibility of describing more general patterns of human migration and of identifying factors that may have influenced migration throughout human evolution.

Since human migration has had the largest effect on genetic variation over human evolution [11], a better understanding of human migration patterns would allow more accurate reconstructions of demographic processes. Comparisons of empirical genetic data to simulated genetic variation generated from models that realistically represent the demographic process under study offer the possibility of reconstructing prehistoric demographic processes [12]. Values for migration parameters estimated from human migration patterns, such as the proportion of the population that is moving, could define some model parameters in order to generate more realistic demographic scenarios. The ability to include empirically-informed values to fix or set ranges on migration parameters increases the probability of identifying the best model to explain the data.

Yemen is a developing country [13] that has a heterogeneous landscape with coastal plains on the west and south, mountain ranges in the west and desert in the north, thus providing a fertile setting in which to investigate environmental factors that may have influenced prehistoric population movements. Yemen has a patrilocal and patrilineal society with a primarily shared language and religion [14], which are social factors that could play a role in migration, as well. Migration within a population of mostly agriculturalists and pastoralists should provide more realistic values of distance and proportion of migration for prehistoric movements since the advent of agriculture. The values should also provide informative lower limits for describing the migration of prehistoric hunter-gatherers, who typically exhibit more movement than agriculturalists [4].

In this study, we use GPS coordinates from birthplaces and places of residence across four generations in Yemen to calculate the proportion, the distance and the direction of migration between each generation. We test for differences in these values between the generations, we identify factors that influence migration patterns, and we discuss possible effects of the migration patterns on genetic variation. Based on our results, we provide estimates for the proportion and distance of migration in a developing country, which can define parameter values for evolutionary models used to reconstruct prehistoric demographic processes. Our use of empirical data on population movements over four generations in Yemen provides knowledge that will allow for more accurate reconstruction of prehistoric processes of migration.

Methods

Ethics Statement

This study has been approved by the Western Institutional Review Board Olympia, Washington (WIRB project #20070219). Samples were collected with verbal informed consent approved by WIRB. This modified inform consent was used because a majority of the population is illiterate. Only individuals who gave consent provided both saliva samples and information for the sample collection sheet, and were entered into the database of study participants.

Samples and Data

In 2007, saliva samples were collected throughout mainland Yemen for genetic analysis. Data were also collected from each study participant on current place of residence, place of birth, parents' place of birth and grandparents' place of birth. Since all sampled individuals were adults, their current residence was used as a proxy for the location of the next generation, i.e. their offspring, therefore providing data on residence patterns for four generations in the study. For the purposes of this study, the individuals in each generation were considered independent samples. Location names for all birthplaces (and place of residence) were translated from Arabic and GPS coordinates were obtained using Geonames.org. In instances where a town name was not identifiable in the Geonames database, but the larger district could be identified, a GPS coordinate was obtained for the centroid of the district. Samples for which town or district locations could not be determined were removed. Ultimately, the resulting dataset contained GPS coordinates for the sampled individual's place of residence and place of birth, mother's and father's places of birth, and maternal-grandmother's, maternal-grandfather's, paternal-grandmother's, and paternal-grandfather's places of birth for 351 sampled individuals (2,457 total sample locations).

Estimation and Analysis of Migration

The occurrence of migration was determined by the difference in birthplace or residence location between generations. The current place of residence, considered a proxy for the "offspring" generation (G0) of the sampled individuals, was used to identify migration in the sampled individual's generation (G1). Thus, a migration event occurred in the sampled individual's generation (G1) if the place of residence was different from the birthplace. A migration in the parental generation (G2) occurred if the parent's offspring was born in a different location than the parent's birthplace (i.e. if the sampled individual's birthplace was different from their mother's or father's birthplace). Similarly, a migration event in the grandparental generation (G3) occurred if the parent's birthplace was different from the grandparent's birthplace. Migration events were determined for eight different groups: female sampled individuals (G1$_{fem}$), male sampled individuals (G1$_{male}$), mothers (G2$_{fem}$), fathers (G2$_{male}$), maternal-grandmothers (G3$_{mfem}$), maternal-grandfathers (G3$_{mmale}$), paternal-grandmothers (G3$_{pfem}$), paternal-grandfathers (G3$_{pmale}$). The frequency of migration events was calculated for each of the eight groups (sample sizes were 70 in G1$_{fem}$, 281 in G1$_{male}$, and 351 in each group in G2 and in G3. The observed frequencies were compared through goodness-of-fit tests.

The age of the sampled individuals ranged from 18 to 69, which meant that each generation group (G1, G2, G3) essentially included two generation time periods. To account for the possibility of migration events occurring over different generation time periods within each generation group, the eight groups were further divided into two age groups with a 25 year generation time between them, based on the ages of the sampled individuals (under and over 40 years old). Only 10% of the samples in any generation were in the over 40 years old sub-group, suggesting that any difference in migration event frequencies could be due instead, to the unbalanced sample size; thus no further analyses were performed with the groups partitioned by age over and under 40 years.

Migration distance was calculated from the geographic distance between birthplaces/residences in two different generations using the GPS coordinates. G1 migration distances were calculated as the geographic distance between the sampled individual's birthplace and place of residence. G2 migration distances were calculated from the parent's birthplace and the sampled individual's birthplace. Migration distances were calculated for G3 from the difference in grandparent's birthplace and parent's birthplace. The migration distances were compared between sex in each generation and between generations using Wilcoxon Rank tests and Kruskal-Wallis analyses of variance.

Different models including generation group, sex, birthplace location (latitude and longitude), and residence location (latitude and longitude) were tested in logistic regressions to see which model (and parameters) best explained migration. AIC (Akaike information criterion) were used to select the best model. Additionally, the migration events were plotted geographically and the mean direction of the migrations was calculated for each collection site (to account for sampling) using ESRI ArcMap10 [15].

Results

The proportion of migrants was calculated from the frequency of migration events for females and males in three generations (G1$_{fem}$ = 0.314, G1$_{male}$ = 0.267, G2$_{fem}$ = 0.376, G2$_{male}$ = 0.311, G3$_{mfem}$ = 0.120, G3$_{mmale}$ = 0.111, G3$_{pfem}$ = 0.097, G3$_{pmale}$ = 0.080) (Figure 1). Within each generation, the proportion of migrants between male and female groups was not significantly different. However, more recent generations G1 and G2 had a significantly larger proportion of migrants than G3 (p = 0.0005). The proportion of migrants for each generation (males and females combined) was G1 = 0.276, G2 = 0.343, G3 = 0.102. We also calculated a multi-generation proportion of migrants for G3 to correct for back migration events by determining the number of migration events in which the grandparents' birthplace was different than the residence location. This produced a multi-generation proportion of migrants for G3 of 0.086.

The distance of migration was also calculated for each of the eight groups. G1 and G2 migration distances were significantly larger than G3 (p<2.2×10^{-16}) (Figure 2). Density plots combining the migration distance (including non-migrants) and the frequency of these distances revealed that $G1_{fem}$ not only had the largest migration distance, but had more migrations at longer distances (>250 km), than the other groups. However, when compared by sex within generations, female distances were not significantly different from male distances. Summary statistics on migration distances were calculated on all individuals and on only migrating individuals (Table 1).

Correlation analyses were performed on marital pairs in G2 and G3 to determine whether marital pairs were moving together. A low correlation coefficient (<0.1) would suggest the marital pair migrations were completely independent from each other and a high correlation coefficient (>0.9) would suggest that the marital pairs were moving together and should be treated as one group (instead of female and male groups). G2 had a significant (p = 2.2×10^{-16}) Spearman's rho correlation coefficient of 0.589. Maternal grandparents ($G3_M$) had a rho coefficient of 0.782 (p = 2.2×10^{-16}) and paternal grandparents ($G3_P$) had a rho coefficient of 0.623 (p = 2.2×10^{-16}). These results showed there was a moderate and significant correlation between all marital pairs. These coefficients suggest that a portion of the marital pairs are moving together, but the correlations are not high enough (>0.9) to consider the marital pairs as a single group. Female and male marital pair distances were plotted and showed that correlated migrations were of the same distance, which is consistent with marital pairs moving to the same place (Figure 3). Out of the 121 migration events in G3, 56% were of marital pairs moving together. These results suggest that many of the individuals may be moving due to post-marital residence dynamics (i.e. husbands and wives moving together).

Logistic regression models, including different combinations of generation, sex, birthplace coordinates and residence location coordinates, were performed to explain presence or absence of migration. The model with the lowest AIC included generation, sex, birth latitude and longitude and residence latitude (Table 2). This best model demonstrated, that relative to $G1_{fem}$ (as the baseline group), the probability of migration decreased in G3, decreased in males (consistent with females moving with their husbands' families) and decreased with a more easterly birthplace. In contrast, the probability of migration increased in G2 and increased with more northern birthplaces and places of residence. However, of these factors, only G3 had a coefficient above one, suggesting that G3 contributes the most to the probability of migration, and specifically, belonging to the G3 generation decreases the probability of migration.

Although birthplace latitude, birth place longitude and residence location latitude had small coefficients, their statistically significant contribution to the migration probability suggests that there could be factors "pushing" individuals away from a place (leave one's birthplace) or "pulling" individuals to a place (move to a new place) [16]. The birthplace and residence coordinates were used to plot the directionality of migration to assess whether or not there was a pattern in directionality that could explain the "pushing" and "pulling" effects (Figure 4). The mean migration direction was calculated from these migration vectors for each sample collection site (to account for the effect of sampling). While the mean migration directions seem to have a southbound tendency, the circular variance (which describes the variation associated with the directional mean, where values close to 0 represent a similar direction for all migration vectors and values close to 1 correspond to vectors in all compass directions) was moderate to high for all collection sites, ranging from 0.675 to 0.867 (Table 3), suggesting movement in all directions.

The mean migration directions were further calculated by collection site for each generation group (Figure S1 and Table S1). Within generation groups G2 and G3, female and male migration directions were similar in many collection sites, supporting the idea that marital pairs moved together. The mean migration lengths were generally larger for G1 and G2 than for G3, reflecting the decreased migration distance in G3. For each collection site, the mean migration directions varied greatly between generation groups, suggesting a level of stochasticity to the migration directions. When the mean migration directions were spatially compared to geographic features (i.e. elevation, land use/land

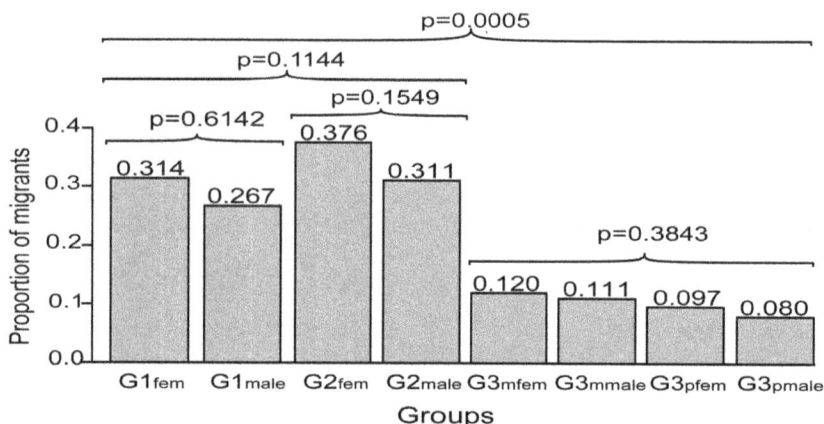

Figure 1. Proportion of migrants by sex for each generation group. P-values are shown for goodness-of-fit tests between groups.

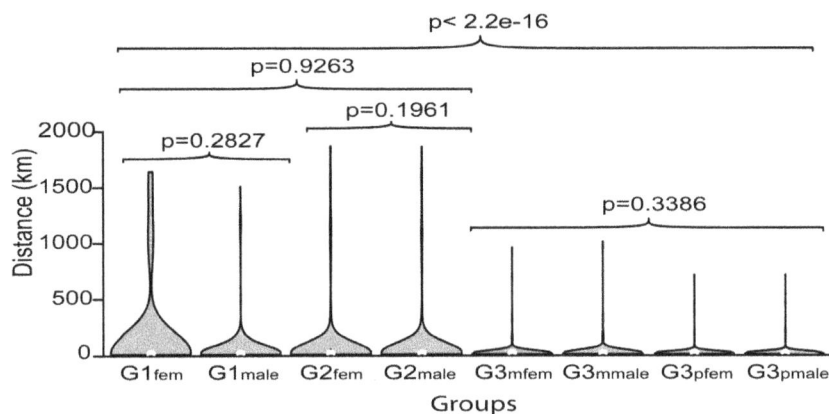

Figure 2. Density plots combining migration distance and frequency of the distance for each group. Wilcoxon Rank tests were performed for G1 and G2 within generation comparisons and Kruskal-Wallis tests were performed for G3 within generation comparison and between generation comparisons. P-values are shown for the respective tests.

cover, and watershed), no pattern arose (data not shown), further supporting stochasticity in the directionality of migrations.

Discussion

Our study helps elucidate human migration patterns using empirical population movement data across multiple generations in Yemen. Our results show that the proportion and distance of migration increased in recent generations. While movement in the recent generations may reflect social and political changes that have occurred in the last 50 years [17], the reduced movement in

the oldest generation most likely reflects a lack of technology and associated mobility [16], suggesting that this generation may be most representative of prehistoric movements. The correlated distance and directionality of migrations within marital pairs illustrate the prevalence of post-marital residence dynamics. The significance of birthplace and residence locations in the probability of migration, but lack of pattern in the direction of migration, suggest a degree of stochasticity in terms of human movements. These cultural factors affecting modern movement have most likely played important roles in prehistoric migrations as well,

Table 1. Summary statistics for migration distances.

All individuals	Mean[a]	±SD	Median[a]	Mode[a]
G1 (351[b])	69	249	0	0
G1Female (70)	156	405	.	.
G1Male (281)	48	186	.	.
G2 (702)	72	265	.	.
G2Female (351)	73	269	.	.
G2Male (351)	72	262	.	.
G3 (1404)	10	66	.	.
G3Female (702)	9	61	.	.
G3Male (702)	10	71	.	.
Migrating individuals	**Mean**	**±SD**	**Median**	**Mode**
G1 (97)	251	424	81	103
G1Female (22)	497	601	103	103
G1Male (75)	179	328	75	103
G2 (241)	211	419	29	103
G2Female (132)	193	411	23	103
G2Male (109)	232	430	44	103
G3 (143)	96	188	26	26
G3Female (76)	82	169	24	17
G3Male (67)	111	208	28	26

Median and mode for "All individuals" was zero for all groups.
[a]Distances in km.
[b]Numbers in parentheses represent sample size.

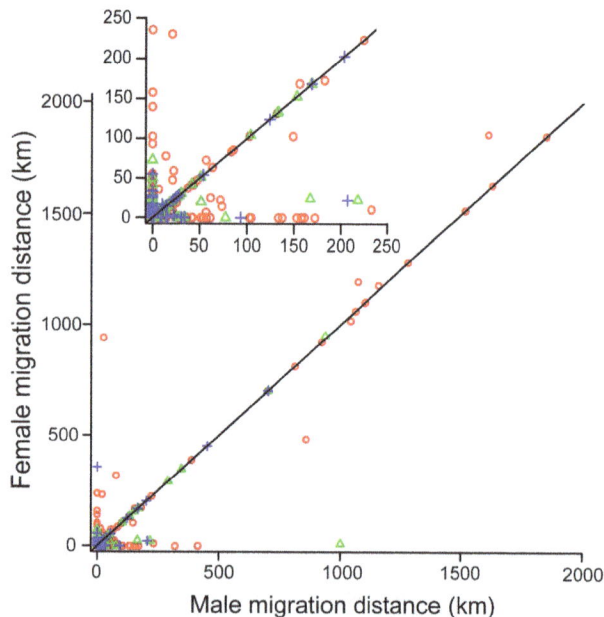

Figure 3. Plot of migration distances for marital pairs. G2 (red circle), G3$_M$ (green triangle), G3$_P$ (blue cross). The solid line shows a theoretical 1:1 relationship, where females and males have the same dispersal distance. The inner box shows a close-up of the relationship for distances less than 250 km.

Table 2. Best model to explain probability of migration.

Factor	Coefficient
Intercept	−1.924
Generation:G2	0.256
Generation:G3	−1.244
Sex:Male	−0.240
Birth Latitude	0.221
Birth Longitude	−0.121
Residence Latitude	0.225

$p < 0.04$ for all factors. The probability of migration decreases in G3, decreases in males, and decreases with a more eastern birthplace, in comparison to G1$_{fem}$. The probability of migration increases in G2 and increases with more northern birthplaces and places of residence.

suggesting that the migration patterns and estimates described in our results provide information to make more accurate prehistoric inferences.

Patrilocality and Genetic Signals

Moderate correlation coefficients for G2 and G3 marital pairs and the plot of migration distances in marital pairs suggests that pairs are moving together and the correlation seems to strengthen with increasing distance (Figure 3). Our best fit model, which shows that females are more likely to move than males when accounting for other contributing variables, suggests that patrilocality (females moving to their husbands' family) may be driving the movement. This is supported by ethnographic accounts that ~90% of the Yemeni population is patrilocal [18]. However, the coefficient of the effect that being male has on the probability of migration is low (−0.240) and within each generation the migration distance is not significantly different between females and males. This suggests that males are only slightly less likely to migrate than females and that males are travelling similar distances compared to females. In a perfect patrilocal post-marital residence dynamic, males move short distances and stay close to their family, while females move longer distances to be near their husbands' family. The similar migration distances between females and males suggest there is not strict patrilocality in Yemen and that other factors are influencing male movement. This interpretation is supported by ethnographic data showing that males may occasionally migrate large distances from their birthplace for socioeconomic or political reasons [14,18]. Our data show that male migration has occurred more often in the last 50 years (as shown by the increase in dispersal in G1 and G2 relative to G3).

The similar migration distances between females and males, and consequent imperfect patrilocality may be the principal contributor to the lack of association observed between geographic and genetic distance in male lineages (i.e. Y chromosome) in Yemen

[19]. Females moving with their husbands may also explain why shared mitochondrial DNA (mtDNA) haplotypes have been found between east and west Yemen, over 750 km apart [20].

Patterns of Migration

Logistic regressions were used to test the effect of birthplace and residence locations on the probability of migration in order to assess whether there were factors "pushing" or "pulling", respectively, individuals to a new location. Birthplace latitude and longitude and residence latitude were significant parameters in explaining the probability of migration. Given this result, birthplace and residence coordinates were used to plot migration directions and determine whether a pattern could be observed that could account for the effects of birthplace and residence locations. Mean migration directions were calculated by collection site (to account for sampling bias) to summarize the overall migration direction patterns (Figures 4 and S1, and Table 3). Although the mean migration directions had a southbound trend, the circular variances were large, suggesting overall dispersal in multiple directions (Figure 4 and Table 3). Additionally, mean migration directions calculated by collection site for each generation showed that the collection sites had different mean directions between generations, further supporting migration in multiple directions (Figure S1 and Table S1). We also spatially compared the migration directions with different geographic features (i.e. elevation, land use/land cover, and watershed) to identify environmental factors that may influence migration direction. We found no pattern associated with the migration directions and the geographic features (data not shown). These results suggest that while there may be factors "pushing" and "pulling" individuals to move, the overall direction of migration has little or no pattern. These results contrast with island migration patterns (e.g., Polynesia) where migration direction has a pattern from larger islands to smaller islands [21–23]. Given that continental migrations are less limited by the carrying capacity of new colonization sites than islands, our results are not surprising.

While island migrations have been well described by ethnographic and archaeological data [21,23], continental migration patterns have been primarily addressed through genetic data. Genetic evidence has suggested that overall continental migrations have a linear pattern, such that increasing distance from Africa is correlated with decreasing genetic diversity [24,25]. Our data suggest that the smaller scale migrations (Figures 4 and S1) that led to this continental pattern may have been less directed. Our results are consistent with the idea that smaller migrations, which

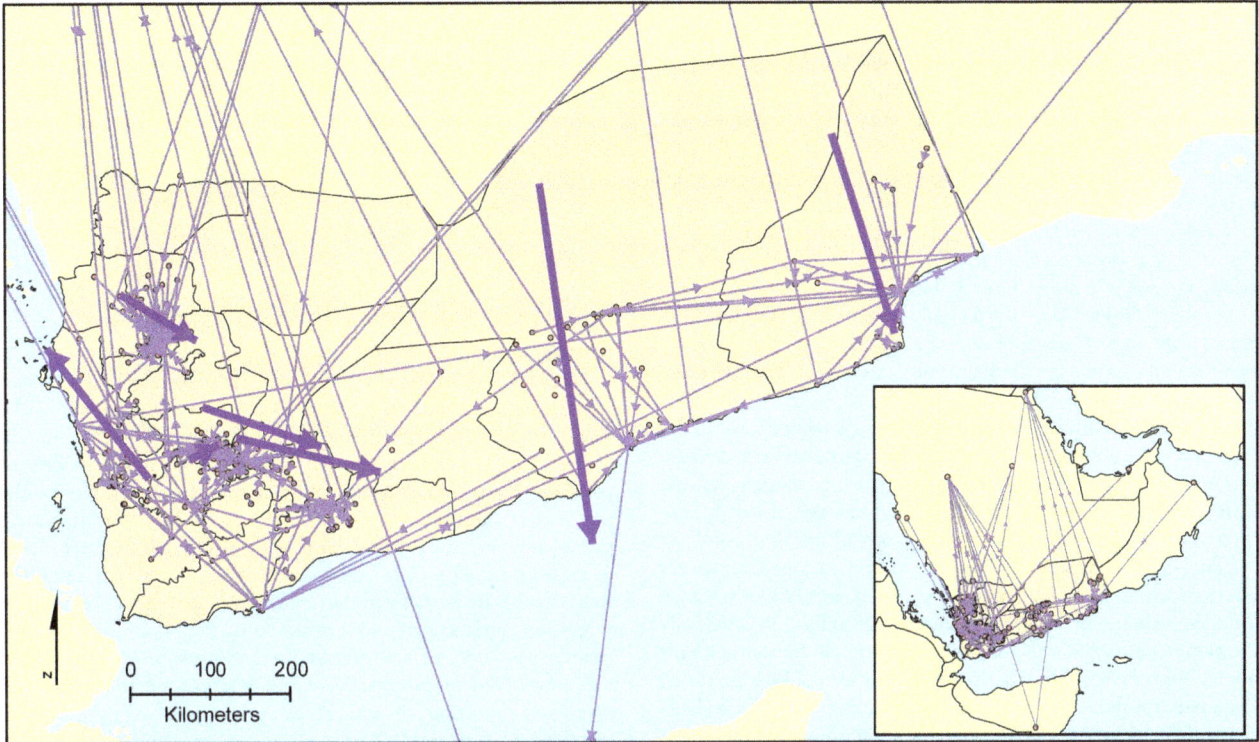

Figure 4. Migration direction vectors and mean migration direction by collection site over all three generations. Small arrows represent migration vectors and large arrows represent mean migration direction.

consider the movement of individuals, tend to be more random, while larger scale movements focused on populations have more directionality associated with them [4,26].

Empirical Estimates of Migration

Comparisons of proportion of migrants and migration distances across four generations showed that migration was significantly lower around fifty years ago (G3). Furthermore, the best fit model to explain the probability of migration shows that G3 has not only the biggest effect, but a negative effect on the probability of migration (i.e. belonging to G3 decreases the probability of a migration event). Spatial patterns of migration in G3 (Figure S1c) show, that although there are some long migration distances, on

average, the distances are short. Yemen's less-developed state and poor transportation infrastructure [17] combined with the significantly reduced migration in G3, suggests that our data from the G3 generation can provide empirically-based estimates of migration frequency and distance that are reflective of prehistoric movements.

We calculated the mean and median migration distances for G3 (Table 1). The mean migration distance for all individuals (i.e. including both individuals who migrated and those who did not) was 10 km. The mean and median distances for migrating individuals only were 96 km and 26 km, respectively. The shorter migration distance values (10 km and 26 km) are within the range of previously reported average migration distances [27–29]. These

Table 3. Estimates for the direction of migration in each collection site across all three generation groups.

Collection site	Mean directional angle[a]	Mean distance[b]	Circular variance[c]
Abyan	103.19	1.659	0.676
Al Bayda	107.87	1.405	0.806
Al Hudaydah	321.06	1.814	0.867
Al Mahra	161.96	2.282	0.833
Amran	120.51	0.995	0.704
Dhamar	62.78	0.370	0.758
Hadramout	171.57	3.953	0.675

[a]Mean directional angle is measured clockwise from due North.
[b]Mean distance is measured in decimal degrees.
[c]Circular variance describes the variation associated with the directional mean, where values close to 0 represent a similar direction for all migration vectors and values close to 1 correspond to vectors in all compass directions.

shorter migration distances potentially demarcate the distances within which post-marital residence patterns (patrilocality in the case of Yemen) have a distinguishable effect on genetic structure [19,29]. In contrast, at distances beyond these values, isolation by distance is probably more predominant, and sex-biased migration is less detectable.

Since most populations before the advent of agriculture (~10 kya) were hunter-gatherers, we wished to identify whether our results provided estimates that may be informative in reconstructing prehistoric processes throughout these different periods in human history. Our shorter migration distance values (10 km and 26 km) are within the range of 10–30 km that Ammerman and Cavalli-Sforza [27] believe is plausible for migration distance in agriculturalist societies. Furthermore, dividing 26 km by a generation time of 25 years results in a migration speed of 1.04 km/year. This value is comparable to the 1 km/year migration speed for the Neolithic transition estimated from archeological data [4,30]. These similarities suggest that the shorter distance values, particularly the median distance, are representative of migration distances of agriculturalist groups.

Hunter-gatherers generally migrate more and longer distances than agriculturalists. Therefore, our mean migration distance estimated using only migrating individuals offers a potentially informative migration value for the more mobile hunter-gatherer populations. Specifically, a migration speed (3.84 km/year) calculated from the mean value for only migrating individuals (96 km) falls within the broad range of hunter-gatherer migration speeds based on archeological evidence. Fort et al [31] estimated the speed of the hunter-gatherers' recolonization of northern Europe after the last glacial maxima between 0.7 and 1.4 km/year. Hamilton and Buchanan [32] estimated a speed of 5–8 km/year for the colonization of North America, while Hazelwood and Steele [4] obtained estimates of 6–10 km/year. Because our value is intermediate to the values of these region-specific studies, it provides a distance that may be more generally applicable to migration processes, particularly *de novo* colonization migration distances by hunter-gatherers. This can be seen when we compare our migration speed estimate with Macaulay et al's [33] inferred migration speed for the colonization of Southeast Asia. Based on founder time estimates from Eurasian and Australasian mtDNAs and the distance between India and Australasia, Macaulay et al infer a migration speed of 4 km/year. Our empirical estimate of 3.84 km/year suggests that their proposed migration process is in fact plausible.

While migration distance has been estimated through different approaches, few studies have estimated the proportion of migrants [5,7,29,34]. We calculated the proportion of migrants for G3 to be 0.102 (or 0.086 when adjusting for back migration in the four generations). These values are smaller than the 0.4 proportion of migrants that can be calculated from Wood et al's [34] dataset on migration between parishes in Papua New Guinea or the 0.366 estimate obtained from the calculation of individuals that were not born in the same parishes as their parents in La Cabrera, Spain [7]. These differences from our estimates seem reasonable as Wood et al's estimates are from a more recent population (and are closer to our G1 and G2 estimates) and Boatinni et al's estimates are from a more developed country. Our estimates are somewhat larger than the 0.032 proportion of migrants into the island of Pingelap in Micronesia presented by Morton et al [5]. However, our adjusted proportion of migrants (0.087) is closer to Morton et al's value. We also calculated the maximum and average number of individuals moving between a pair of locations, for a proportion of migrants of 0.0036 and 0.0011, respectively. These lower values are consistent with findings by Deshpande et al [35], where the

genetic estimates of proportion of migrants (i.e. migration rates) for a world-wide colonization model are less than 0.01. Our values are similar to findings by Miró-Herrans and Mulligan [11], where the most probable proportion of migrants exchanged between African and non-Africans populations was 0.001 and are similar to the migration rate for non-African populations (1.5×10^{-3}) obtained by Cox et al [36]. The similarity of our estimates with those from other migration studies suggests that our values can be used in different scenarios to generate testable models for prehistoric reconstruction.

Application of Migration Estimates in Prehistoric Demographic Modeling

Model-based approaches for inferring prehistoric processes from genetic variation are becoming increasingly popular [37]. These approaches, such as approximate Bayesian computation [12], require the generation of explicit demographic models to compare to empirical data. Including specific values for known parameters and informative ranges of values for unknown parameters increases the probability of identifying the best model to explain the data. The results from our study provide estimates that can be used to fix or set ranges on parameters related to migration, such as gene flow or founding population size, so that other parameters of interest can be addressed in greater depth, e.g., time of a demographic event. For example, the maximum and average proportion of individuals moving between a pair of locations (0.0036 and 0.0011) can be used to define gene flow (or migration rates) between populations stretching from southern Asia to northern Africa to create simulated DNA for models that address the back-migration into Africa. The larger migration values (0.102 or 0.086) can be used to define the founding population sizes for each new population out-of-Africa and back-to-Africa. Defining these parameters would allow for an in-depth exploration of the timing of the back-migration.

Additionally, our results provide estimates to generate more geographically explicit models. Our mean and median migration distances (96 km and 26 km) provide estimates for the distance between populations, particularly for large scale movements, such as the back-migration from southern Asia. The migration distance between each population would define the number of populations to be simulated for the region under study. For example, a distance of 100 km between each population would require ~70 populations between southern Asia and northern Africa (approx. 7,000 km). Understanding the possible distances involved in large scale movements also helps us determine how rapidly a migration could have occurred and how levels of gene flow may have been affected between the populations.

The lack of migration directionality in our results suggests that explicitly including stochasticity or multidirectionality when describing the movement between populations might more accurately reflect the large-scale migration process. For example, the back-migration to Africa probably included movement through established populations, where the migrants settled in some of the established populations, but not in others. Therefore, a lattice stepping-stone migration model, that includes some randomness in terms of when a migration occurs and between which populations, might better reflect this migration process.

Our results show there is over a 58% correlation between female and male movement in marital pairs, in which more pairs move together with increasing distance. Additionally, we show that 56% of migration events in G3 were by marital pairs. This means that at least 50% of the migrants have a 1:1 female to male ratio. Even if the remaining 50% of migrants are only female or male, the ratio is at most 3:1. These results argue for, at most, a 3:1 ratio (for either sex) of sex-biased migration for migrations at short distances, where post-marital residence has a larger effect on population structuring [19,29]. Alternatively, for longer migrations, such as the migration from southern Asia to northern Africa, our results suggest that a female to male ratio closer to 1:1 more accurately models demographically balanced populations that would have been reproductively self-sustaining.

Conclusions

In this study, we analyzed empirical data on migration patterns over four generations of human populations in Yemen in order to gain insight into the factors that influence migration, and specifically may have affected prehistoric movements throughout human evolution. Our approach to trace migration over generations has enabled the study of migration patterns throughout a developing country that would otherwise have been unfeasible. We provide empirical estimates for migration-related parameters that can be used to generate demographic models in model-based methods of prehistoric reconstruction. Our empirical estimates of generation G3 provide values for proportion of migrants, with values ranging from 0.102 or 0.086 proportion of overall migration, to 0.0036 or 0.0011 proportion of migrants between two specific populations. We also provide migration distances (96 km and 26 km, mean and median, respectively) that can be used to define the distance between populations and therefore the number of populations for the area under study. Using our approach, populations employing other modes of subsistence, such as hunter-gatherers, may be studied to further improve our knowledge on human migration.

The findings from this study shed light on human migration patterns and enable more accurate reconstruction of the demographic processes that characterized human evolution. Improved models of human demographic changes and the associated genetic variation can provide a powerful tool to test for selective pressures, as well as to model the evolutionary history of co-evolving organisms. In this way, reconstruction of human demography and evolution may further provide insight into the movement and evolution of human pathogens and other co-evolving organisms.

Supporting Information

Figure S1   Migration direction vectors and mean migration direction for each collection site by generation group. a)G1. b)G2. c)G3. Females: purple, Males: green. Small arrows: migration vectors. Large arrows: mean migration direction.

Acknowledgments

We thank David Reed, Michael Miyamoto, and Steven Brandt for insightful discussion and comments on this manuscript. We thank Tania Saade for her contribution to the translation of the names of geographic locations used in this study.

Author Contributions

Conceived and designed the experiments: ATMH CJM. Performed the experiments: ATMH CJM AA. Analyzed the data: ATMH. Contributed reagents/materials/analysis tools: ATMH CJM AA. Wrote the paper: ATMH CJM AA.

References

1. de Haan A, Rogaly B (2002) Introduction: Migrant workers and their role in rural change. Journal of Development Studies 38: 1–14.
2. Hahn CHL, Vedder H, Fourie L (1966) Native tribes of south west africa: Frank Cass and Company.
3. Marlowe FW (2010) The hadza: Hunter-gatherers of tanzania. Berkeley, CA: University of California Press. 336 p.
4. Hazelwood L, Steele J (2004) Spatial dynamics of human dispersals. Journal of Archaeological Science 31: 669–679.
5. Morton N, Harris D, Yee S, Lew R (1971) Pingelap and mokil atolls: Migration. American Journal of Human Genetics 23: 339–349.
6. Mielke J, Relethford JH, Eriksson A (1994) Temporal trends in migration in the aland islands: Effects of population size and geographic distance. Human Biology 66: 399–410.
7. Boattini A, Blanco Villegas MJ, Pattener D (2007) Genetic structure of la cabrera,spain, from surnames and migration matrices. Human Biology 79: 649–666.
8. Levy M (2010) Scale-free human migration and the geography of social networks. Physica A: Statistical Mechanics and its Applications 389: 4913–4917.
9. Davis KF, D'Odorico P, Laio F, Ridolfi L (2013) Global spatio-temporal patterns in human migration: A complex network perspective. PLoS One 8: e53723.
10. Gray C, Bilsborrow R (2013) Environmental influences on human migration in rural ecuador. Demography.
11. Miró-Herrans AT, Mulligan CJ (2013) Human demographic processes and genetic variation as revealed by mtdna simulations. Molecular Biology and Evolution 30: 244–252.
12. Beaumont MA, Wenyang Z, Balding DJ (2002) Approximate bayesian computation in population genetics. Genetics 162: 2025–2035.
13. Malik K (2013) Human development report 2013. United Nations Development Program.
14. Dresch P (1989) Tribes, government, and history in yemen. New York: Clarendon Press.
15. ESRI (2011) Arcgis desktop: Release 10. Redlands, CA: Environmental Systems Research Institute.
16. Lee E (1966) A theory of migration. Demography 3: 47–57.
17. Federal-Research-Division (2008) Country profile: Yemen. In: Congress Lo, editor. Available: http://lcweb2locgov/frd/cs/profiles/Yemenpdf.
18. Weir B (2007) A tribal order: Politics and law in the mountains of yemen. Austin, TX: The University of Texas Press.
19. Raaum RL, Al-Meeri A, Mulligan CJ (2013) Culture modifies expectations of kinship and sex-biased dispersal patterns: A case study of patrilineality and patrilocality in tribal yemen. American Journal of Physical Anthropology 150: 526–538.
20. Cerny V, Mulligan CJ, Ridl J, Zaloudkova M, Edens CM, et al. (2008) Regional differences in the distribution of the sub-saharan, west eurasian, and south asian mtdna lineages in yemen. American Journal of Physical Anthropoly 136: 128–137.
21. Kirch P (1980) Society polynesian prehistory: Cultural adaptation in island ecosystems: Oceanic islands serve asarchaeological laboratories for studying the complex dialectic between human populations and their environments. American Scientist 68: 39–48.
22. Jobling MA, Hurles ME, Tyler-Smith C (2004) Human evolutionary genetics: Origins, peoples & disease. New York: Garland Science.
23. Clark G, Anderson A, Wright D (2006) Human colonization of the palau islands, western micronesia. The Journal of Island and Coastal Archaeology 1: 215–232.
24. Li JZ, Absher DM, Tang H, Southwick AM, Casto AM, et al. (2008) Worldwide human relationships inferred from genome-wide patterns of variation. Science 319: 1100–1104.
25. Ramachandran S, Deshpande O, Roseman CC, Rosenberg NA, Feldman MW, et al. (2005) Support from the relationship of genetic and geographic distance in human populations for a serial founder effect originating in africa. Proceedings of the National Academy of Sciences 102: 15942–15947.
26. Skellam JG (1951) Random dispersal in theoretical populations. Biometrika 38: 196–218.
27. Ammerman AJ, Cavalli-Sforza LL (1984) The neolithic transition and the genetics of populations in europe. Princeton, New Jersey: Princeton University Press.
28. Wijsman EM, Cavalli-Sforza LL (1984) Migration and genetic popoulation structure with special reference to humans. Annual Review of Ecology and Systematics 15: 279–301.
29. Marks SJ, Levy H, Martinez-Cadenas C, Montinaro F, Capelli C (2012) Migration distance rather than migration rate explains genetic diversity in human patrilocal groups. Molecular Ecology 21: 4958–4969.
30. Pinhasi R, Fort J, Ammerman AJ (2005) Tracing the origin and spread of agriculture in europe. PLoS Biology 3: e410.
31. Fort J, Pujol T, Cavalli-Sforza LL (2004) Palaeolithic populations and waves of advance. Cambridge Archaeological Journal 14: 53–61.

32. Hamilton MJ, Buchanan B (2007) Spatial gradients in clovis-age radiocarbon dates across north america suggest rapid colonization from the north. Proceedings of the National Academy of Sciences USA 104: 15625–15630.

33. Macaulay V, Hill C, Achilli A, Rengo C, Clarke D, et al. (2005) Single, rapid coastal settlement of asia revealed by analysis of complete mitochondrial genomes. Science 308: 1034–1036.

34. Wood JW, Smouse PE, Long JC (1985) Sex-specific dispersal patterns in two human populations of highland new guinea. The American Naturalist 125: 747–768.

35. Deshpande O, Batzoglou S, Feldman M, Cavalli-Sforza L (2009) A serial founder effect model for human settlement out of africa. Proceedings of the Royal Society B: Biological Sciences 276: 291–300.

36. Cox MP, Woerner AE, Wall JD, Hammer MF (2008) Intergenic DNA sequences from the human x chromosome reveal high rates of global gene flow. BMC Genetics 9: 76.

37. Marjoram P, Tavaré S (2006) Modern computational approaches for analysing molecular genetic variation data. Nature Reviews Genetics 7: 759–770.

13

Evolution of the F-Box Gene Family in *Euarchontoglires*: Gene Number Variation and Selection Patterns

Ailan Wang[1,2], Mingchuan Fu[1,2], Xiaoqian Jiang[1,2], Yuanhui Mao[1,2], Xiangchen Li[1,2], Shiheng Tao[1,2]*

1 State Key Laboratory of Crop Stress Biology in Arid Areas and College of Life Sciences, Northwest A & F University, Yangling, Shaanxi, China, **2** Bioinformatics Center, Northwest A&F University, Yangling, Shaanxi, China

Abstract

F-box proteins are substrate adaptors used by the SKP1–CUL1–F-box protein (SCF) complex, a type of E3 ubiquitin ligase complex in the ubiquitin proteasome system (UPS). SCF-mediated ubiquitylation regulates proteolysis of hundreds of cellular proteins involved in key signaling and disease systems. However, our knowledge of the evolution of the F-box gene family in *Euarchontoglires* is limited. In the present study, 559 F-box genes and nine related pseudogenes were identified in eight genomes. Lineage-specific gene gain and loss events occurred during the evolution of *Euarchontoglires*, resulting in varying F-box gene numbers ranging from 66 to 81 among the eight species. Both tandem duplication and retrotransposition were found to have contributed to the increase of F-box gene number, whereas mutation in the F-box domain was the main mechanism responsible for reduction in the number of F-box genes, resulting in a balance of expansion and contraction in the F-box gene family. Thus, the *Euarchontoglire* F-box gene family evolved under a birth-and-death model. Signatures of positive selection were detected in substrate-recognizing domains of multiple F-box proteins, and adaptive changes played a role in evolution of the *Euarchontoglire* F-box gene family. In addition, single nucleotide polymorphism (SNP) distributions were found to be highly non-random among different regions of F-box genes in 1092 human individuals, with domain regions having a significantly lower number of non-synonymous SNPs.

Editor: Marc Robinson-Rechavi, University of Lausanne, Switzerland

Funding: The work was the project from National Natural Science Fund Program of China (31271917) (http://www.nsfc.gov.cn/Portal0/default152.htm). The funders had no role in study design, data collection and analysis, decision to publish, or preparation of the manuscript.

Competing Interests: The authors have declared that no competing interests exist.

* E-mail: shihengt@nwsuaf.edu.cn

Introduction

To maintain homeostasis or to undergo specified developmental decisions, an organism must be able to respond rapidly to a variety of environmental changes. Protein turnover plays a critical role in the control of many signaling pathways. More than 80% of all proteins are estimated to be degraded via the ubiquitin-proteasome system (UPS) [1]. Protein ubiquitination is an enzymatic cascade in which ubiquitin is activated by an E1 enzyme, transferred to an E2 ubiquitin-conjugating enzyme and then transferred to a substrate selected by an E3 ubiquitin ligase [2]. An E3 ubiquitin ligase must rapidly and uniquely bind to target proteins in response to stimuli. One of the best characterized E3s are the S phase kinase-associated protein 1 (SKP1)–cullin 1 (CUL1)–F-box protein (SCF) type ubiquitin ligase complexes [3]. CUL1 serves as a scaffold for assembling the ubiquitin-conjugating machinery. The C-terminus of CUL1 interacts with the RING-box protein 1 (RBX1), whereas its N- terminus binds to SKP1, which, in turn, binds to an F-box protein.

F-box proteins contain an N-terminal 48-amino-acid F-box domain (first identified in Cyclin F), which binds to SKP1 to create a link to CUL1. In addition, F-box proteins generally contain C-terminal variable protein-interaction domains, such as Trp–Asp repeats (also called WD40) and leucine-rich repeats (LRR), as well as unknown motifs, which are responsible for binding specific substrates [4]. As a core component of UPS, F-box proteins are involved in a wide range of cellular processes, from cell cycle control to gene transcription and organism development. Given this critical role, misregulation of F-box protein-mediated ubiquitination has been implicated in many human diseases, such as cancers and viral infections [5,6].

The number of F-box genes varies dramatically even among closely related species [7]. For instance, lineage-specific expansion has been found in annual Arabidopsis but not in the perennial Populus, suggesting an adaptive advantage conferred by F-box genes for particular physiological processes in Arabidopsis [8]. Given the wide involvement of F-box proteins in cellular processes in human cells, pursuing research on the evolution of F-box proteins in humans and other closely related species is very important. However, most previous works in the field have focused on the evolutionary pattern of F-box genes only in plants [9–13], where F-box gene expansion was more frequent. Our knowledge of the evolutionary mechanisms responsible for the emergence, maintenance, and loss of F-box gene duplications in animals is rather limited. In the current study, we investigated the variation in the number of F-box genes, as well as underlying mechanisms for such variation, in eight *Euarchontoglire* species with high-quality genome sequences.

Several studies have demonstrated that in nematodes and plants, F-box genes are under strong positive selection pressure at sites in their substrate-binding domains [12–14]. In order to find out whether advantageous natural selection has driven F-box genes in *Euarchontoglires* undergoing adaptive evolution, we studied the selection patterns of the F-box gene family. Within a protein, different structural or functional domains are likely to be subject to different functional constraints and evolve at different rates [15]. Therefore, we assessed selective pressures acting on orthologous F-box genes at various levels, such as full-length, partial segments, and single amino acid sites. In addition, mutational burden within different regions of F-box genes was assessed in the human population using 1000 Genomes data. Our results provide insights into the evolutionary regime that has continually reshaped the protein-protein interaction domains responsible for broadening or altering the substrate specificity of F-box genes.

Materials and Methods

Genome-wide identification of F-box genes in eight genomes

The hidden Markov model (HMM) profile of the F-box domain (PF00646) was downloaded from Pfam [16]. All sequence data were downloaded from ENSEMBL (version 69, October 2012) [17]. The downloaded HMM profile was used to search the entire set of annotated proteins from eight species, namely, *Callithrix jacchus* (marmoset), *Gorilla gorilla* (gorilla), *Homo sapiens* (human), *Macaca mulatta* (macaque), *Mus musculus* (mouse), *Pan troglodytes* (chimpanzee), *Pongo abelii* (orangutan), and *Rattus norvegicus* (rat) using the hmmsearch program implemented in the HMMER package [18]. We used the default cut-off values to filter the results of these queries. HMMER-predicted proteins were then scanned for F-box and other domains using InterProScan, which is a tool that integrates multiple signature-recognition methods into one resource [19,20].

For each F-box domain-containing protein identified by the hmmsearch and InterProsScan programs, additional PSI-BLAST [21] searches (with an e-value cut-off of 1e-20) were performed against the entire set of annotated proteins to identify additional F-box proteins that were not found using the HMM profile because of their diverged F-box domains. A second scan for the F-box domain was performed for PSI-BLAST hits.

In order to detect pseudogenes related to F-box genes, all of the F-box protein sequences retrieved from the hmmsearch and PSI-BLAST programs were used as queries to perform a TBLASTN search of the entire set of annotated pseudogenes of each species with an e-value cut-off of 1e-40. Finally, TBLASTN hits were translated and then scanned for presence of F-box domains.

The genomic distribution of identified F-box genes was evaluated by comparing the observed number of genes in each chromosome with its expected number under a Poisson distribution.

Phylogenetic analysis

Multiple sequence alignments of the F-box protein sequences, which are shown in File S1, were generated using MUSCLE [22] and then manually checked and trimmed with TRIMAL 1.2 (gt = 0.3) [23]. Subsequently, following the Akaike Information Criterion (AIC) computed with ProtTest 3.2 [24], the JTT+F model was chosen to construct a maximum likelihood (ML) tree with PhyML 3.0 [25]. Topological robustness of the phylogenetic tree was assessed by bootstrapping with 100 replicates.

Orthology assignments of F-box genes from the eight species were downloaded from the ENSEMBL database using Biomart. Orthologous groups (also known as orthogroups) were inferred based on phylogenetic relationships and confirmed by reciprocal BLAST. Subsequently, homology relationships of F-box genes in the same orthogroup were checked against the data obtained from ENSEMBL. We nominated human and mouse F-box genes

according to the nomenclature proposed by Jin et al. [26]. F-box genes in the other six species considered in this study had the same names as their orthologs in humans.

Inference of gene-gain and gene-loss events and their underlying mechanisms

Gene-gain and gene-loss events were inferred using the species/gene tree reconciliation approach with NOTUNG software [27]. For these analyses, the reference species tree used was reconstructed according to TIMETREE [28]. The reconciled tree was manually adjusted by applying information on orthologs (from outgroups *Danio rerio* and *Gallus gallus*), because the gene tree may not always be reliable for species/gene tree reconciliation.

We investigated several potential underlying mechanisms responsible for gene-gain events inferred from previous estimations. First, we explored whether retrotransposition may have contributed to such duplications. The nucleotide sequences of duplicated genes were inspected for signatures of retrosequences, such as lack of introns, stretches of poly (A) at the 3′ end, and short direct repeats at both ends. Second, if the duplicated gene was not found to have been generated by retrotransposition, we considered two types of segmental duplications (>90% identity and >1 kb in length) [29]: (*i*) tandem duplication, where the two genes are located within the same chromosomal region (i.e., fewer than 20 genes apart from each other) [30], or (*ii*) interspersed duplication.

Next, we performed a more exhaustive search for genes that are absent in certain lineages. The orthologs present in closely related species were used as queries for BLASTN searches against the genomic sequences in the NCBI database. When a high-identity match (identity >60%) was produced, we examined whether it was annotated as a gene by NCBI; if it wasn't, we manually annotated it using FGENESH+ (www.softberry.com). Next, the proteins annotated were scanned for known functional domains by InterProScan. If an F-box domain was found, the protein was designated as an F-box protein; otherwise, the protein was noted as having lost its F-box domain. If a high-identity match was not obtained, the gene was assumed to have been removed from the genome by deletion.

Selective pressure analyses

For each orthogroup, orthologous amino-acid sequences with similar length were aligned with MUSCLE [22] and then manually checked and trimmed with TRIMAL 1.2 [23]. TRIMAL removed poorly aligned columns and incomplete sequences, considering the remaining sequences in the MSA using three specified thresholds (- resoverlap 0.75, -seqoverlap 80, -gt 0.7). Subsequently, the sequences that did not pass the sequence overlap threshold were replaced by another transcript of the associated genome. Finally, only columns and sequences that passed the thresholds were retained in the final alignments. These retained protein alignments were used to guide the alignments of the corresponding CDSs using TRIMAL. The multiple coding sequence alignments thus generated are shown in File S3.

Average codon-based evolutionary divergence over all sequence pairs within each orthologous group was measured in terms of Ka (the number of non-synonymous substitutions per non-synonymous site) and Ks (the number of synonymous substitutions per synonymous site) using MEGA [31]. Next, the codon-based Z-test was used to evaluate the significance of the Ka/Ks substitution rate.

The distribution of selective pressure across the gene was investigated using the sliding window method in which Ka/Ks (ω) ratios were calculated by DnaSP with a window length of 30 bp and a step size of 6 bp [32]. The statistic ω was calculated in each window, and its value was assigned to the nucleotide at the midpoint of the window. Values of ω from each window were plotted against the nucleotide position, thus enabling the visualization of distinct selective pressure acting across the gene.

To assess variations in selective pressure among sites, the site-specific models were tested comparatively using M0 (one ratio) and M3 (discrete), both of which are implemented in the PAML software package [33]. To detect positive selection affecting a few sites along particular lineages, we applied branch-site model A for analysis.

The ω value for each codon was calculated using the MEC model that is implemented in the Selecton software package [34]. The MEC model is a combination of empirical and mechanistic model, which accounts for differing empirical amino acid mutation probabilities. Compared with the more conservative M8 model, a smaller proportion of sites (with ω values >1) may be sufficient in the MEC model to indicate positive selection despite a low global ω value for the protein [35]. Program Selecton was run with the MEC model and M8a (a null model that only allows purifying and neutral selection), and their second-order Akaike Information Criterion (AICc) score was compared. To map the detected positive selection sites onto the protein's three-dimensional (3D) structure, we used the homology modeling method to construct a 3D structure of F-box proteins in the SWISS-MODEL workspace [36]. The PyMOL (http://www.pymol.org) graphical interface was used to manipulate and display the F-box protein 3D structure.

Single-nucleotide polymorphism (SNP) analysis of F-box genes in the human population

Variant calls from the 1000 Genomes project [37] Phase 1 release v3 were obtained from the NCBI ftp server (ftp://ftp-trace.ncbi.nih.gov/1000genomes/ftp/release/20110521/). Variants were annotated using the SNPEff (v3.4) software (http://snpeff.sourceforge.net/). The annotated variants were filtered to retrieve SNP variants in F-box protein coding regions. Statistical analyses were performed using distribution-free non-parametric tests in the R program. The significance of differences in SNP density distributions in different regions of F-box genes was assessed using the Kruskal-Wallis test or the Mann-Whitney U test. The Fisher's exact test was used to compare ratios of non-synonymous to synonymous substitution SNPs between different regions of F-box genes.

Results

Identifying F-box proteins and their domain architectures

In order to comprehensively identify F-box genes, we used an integrated hmmsearch–BLAST–InterProScan approach. A total of 559 protein-coding F-box genes were identified in eight genomes (Table S1 A). The F-box gene number in all genomes was approximately 70 (except in the mouse genome), accounting for over 0.3% of the total protein-coding genes (Table S1 B). We also identified nine annotated pseudogenes duplicated from protein-coding F-box genes and 21 corresponding homologous DNA regions (Table S2).

The chromosomal distribution of F-box gene was uniform across all autosomes, except for several chromosomes (Table S3). In five primate genomes, F-box genes were significantly enriched in chromosomes 16 and 19, compared with other chromosomes. Chromosomal translocation may have contributed to the higher F-box gene density in chromosome 16. On the other hand, the high density of F-box genes in human chromosome 19 may be attributed simply to the fact that human chromosome 19 is the most gene-rich chromosome [38]. Interestingly, in all genomes analyzed, F-box genes were absent from chromosome X, and this

absence was not random (Table S3). The absence of F-box genes on chromosome X of all the eight species examined is consistent with the fact that mammalian X chromosomes are highly conserved across species [39]. No F-box gene was found on chromosome Y, but the departure from uniformity was statistically insignificant. The mechanisms underlying the absence of F-box genes from sex chromosomes remain unclear.

Twenty-seven types of domains (excluding the F-box domain) were identified within F-box protein sets, and all but nine were located at the C-terminus (Figure 1). Among them, the proportion of the LRR domain was the largest, followed by the WD40 domain. The remaining domains were present in small subsets of F-box proteins. Except the F-box domain, no other recognizable domain was found in 30% of the 559 F-box proteins identified in this study.

Clustering orthogroups

The 559 F-box genes identified may be clustered into 71 distinct orthogroups (Figure S1). The deep branches were poorly supported by bootstrapping, because F-box genes from different subgroups are of great divergence. However, the vast majority of the single orthogroups reached well-resolved topology with strong bootstrap support. Furthermore, the members in each identified orthogroup were the best blast hits of each other (File S2), and their orthology relationships were confirmed by data obtained from ENSEMBL. These orthogroups include stable and unstable clades based on evolutionary stability. Stable clades are those that contain at least one member of each species (but species-specific duplications could as well result in more copies). Unstable clades in turn do not contain genes of each species. Fifty-seven orthogroups were evolutionarily stable, and the remaining 14 orthogroups were unstable but varied only slightly. The gene orthology relationships among the eight organisms were one-to-one, one-to-many/one-to-one and one-to-zero/one-to-one for 54, 3, and 14 orthogroups, respectively. The one-to-many and one-to-zero orthology relationships are the results of lineage-specific gene duplications and losses, respectively. Although proteins in the majority of the orthogroups contained the same C-terminal domain, species-specific domain accretion or reduction also took place (Figures 1 and S1). For example, the primate FBXL22 lost the LRR domain, and the PRY domain was accreted in the FBXW10 of both gorilla and orangutan. Moreover, orthogroups with the same or similar domain architecture were not necessarily the closest neighbors. For instance, several LRR-domain-containing orthogroups were scattered across the phylogenetic tree.

All F-box genes were directly clustered into their corresponding orthogroups, with the exception of Fbxo6 and Fbxo44. Gene conversion may be a potential source of conflict between a gene tree and a species tree [40]. Hence, gene conversion tests were performed using Geneconv (http://.math.wustl.edu/~sawyer). Statistically significant evidence of a gene conversion event was found between *Fbxo6* and *Fbxo44* of human, chimpanzee, gorilla, orangutan, and macaque at nucleotide 1477 and nucleotide 1707 (site numbering refers to human *Fbxo44* with Ensembl Gene ID ENSG00000132879) (Figure S2). These results suggest that a gene conversion event occurred in the ancestor of these five primates (*Fbxo44* was not found in marmoset, so this species was excluded from our analysis). These gene conversion regions contained sequences of F-box functional domains and partial F-box associated (FBA) region.

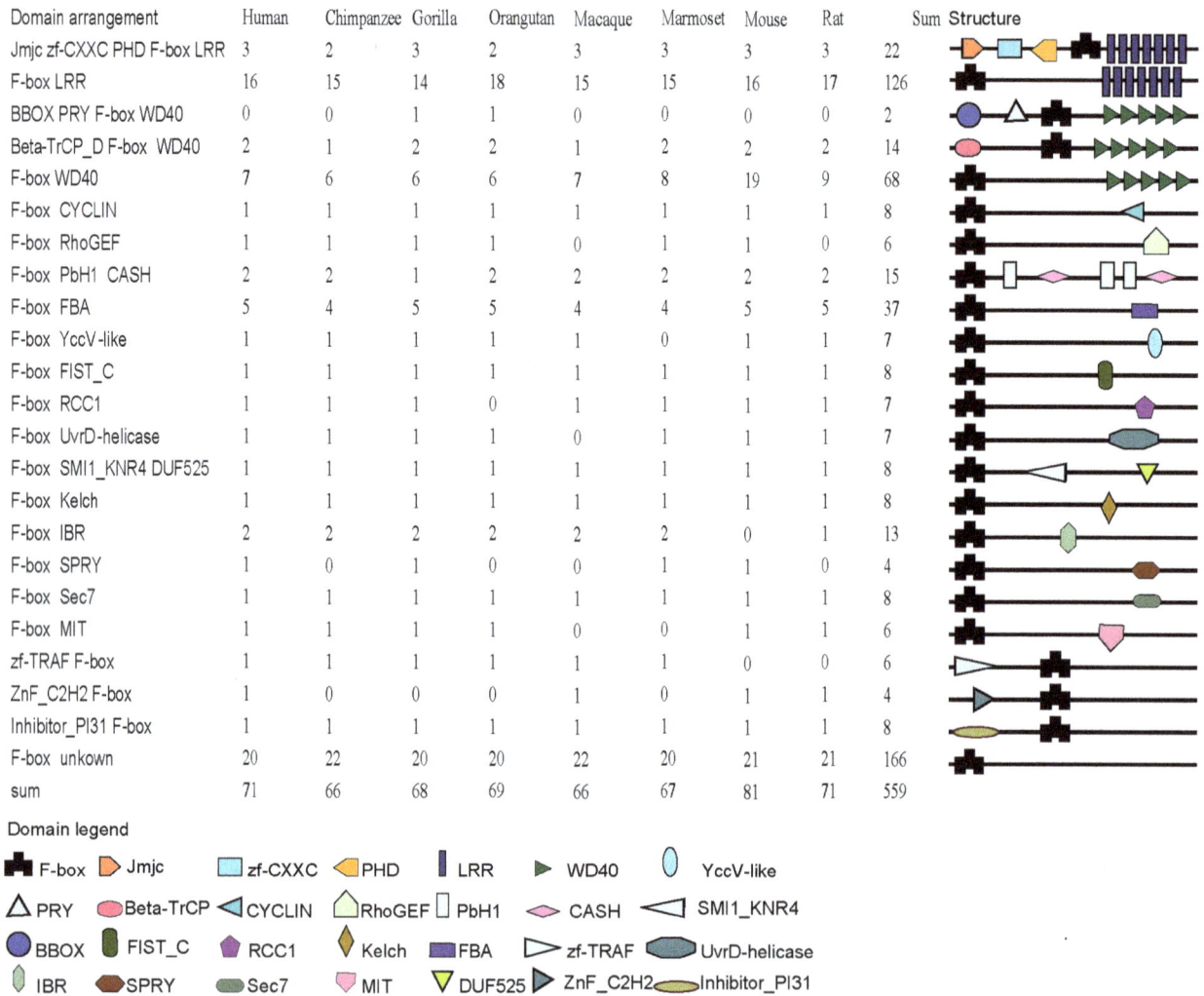

Domain arrangement	Human	Chimpanzee	Gorilla	Orangutan	Macaque	Marmoset	Mouse	Rat	Sum	Structure
Jmjc zf-CXXC PHD F-box LRR	3	2	3	2	3	3	3	3	22	
F-box LRR	16	15	14	18	15	15	16	17	126	
BBOX PRY F-box WD40	0	0	1	1	0	0	0	0	2	
Beta-TrCP_D F-box WD40	2	1	2	2	1	2	2	2	14	
F-box WD40	7	6	6	6	7	8	19	9	68	
F-box CYCLIN	1	1	1	1	1	1	1	1	8	
F-box RhoGEF	1	1	1	1	0	1	1	0	6	
F-box PbH1 CASH	2	2	1	2	2	2	2	2	15	
F-box FBA	5	4	5	5	4	4	5	5	37	
F-box YccV-like	1	1	1	1	1	0	1	1	7	
F-box FIST_C	1	1	1	1	1	1	1	1	8	
F-box RCC1	1	1	1	0	1	1	1	1	7	
F-box UvrD-helicase	1	1	1	1	0	1	1	1	7	
F-box SMI1_KNR4 DUF525	1	1	1	1	1	1	1	1	8	
F-box Kelch	1	1	1	1	1	1	1	1	8	
F-box IBR	2	2	2	2	2	2	0	1	13	
F-box SPRY	1	0	1	0	0	1	1	0	4	
F-box Sec7	1	1	1	1	1	1	1	1	8	
F-box MIT	1	1	1	1	0	0	1	1	6	
zf-TRAF F-box	1	1	1	1	1	1	0	0	6	
ZnF_C2H2 F-box	1	0	0	0	1	0	1	1	4	
Inhibitor_PI31 F-box	1	1	1	1	1	1	1	1	8	
F-box unkown	20	22	20	20	22	20	21	21	166	
sum	71	66	68	69	66	67	81	71	559	

Figure 1. Number and domain structure of F-box proteins from marmoset, gorilla, human, macaque, mouse, chimpanzee, orangutan, and rat.

Evolutionary change of F-box gene number and underlying mechanisms

As mentioned above, lineage-specific gene gain and loss events appeared to have given rise to one-to-many or one-to-zero orthologous relationships among the eight organisms. Therefore, we estimated the number of F-box genes in all ancestral organisms and their change at different stages of the evolution of *Euarchontoglire* animals. Evolutionary changes in the number of the F-box genes are shown in Figure 2. Although the number of F-box genes was found to be conserved, as a general trend, during the evolution of *Euarchontoglires*, lineage-specific gene gain and loss events still took place.

Multiple mechanisms contributed to F-box gene number variation among these organisms. It should be noticed that a closely linked gene cluster includes 12 neighboring genes in mouse chromosome 9 with similar gene structures and a high level of sequence identity, ranging from 63.7% to 97.9%, in their coding regions (Figure 3). Therefore, this gene cluster likely arose from a series of tandem duplication events. However, these genes appear to have diverged in terms of their intron length, which we identified to be caused by variations in internal intron sequence repeats, such as the 510 bp repeats in intron 6 of *Fbxw22* (Figure 3).

Both primates and mice contain a single *Fbxl18* gene, while rats contain two distinct *Fbxl18* genes. A schematic illustration of the gene structures of *Fbxl18I* and its paralogue *Fbxl18II* are shown in Figure 4. Evidences that *Fbxl18II* was formed by retrotransposition of *Fbxl18I* are as follows: (*i*) *Fbxl18I* and *Fbxl18II* are located on chromosomes 5 and 12, respectively; (*ii*) the two sequences are identical over 111 bp in their 5′ UTRs and their 3′ UTR sequences have over 80% similarity; (*iii*) *Fbxl18II* is flanked by a short sequence repeat of 'agaagaagggaga'; (*iv*) stretches of 12 adenine (A) occur in *Fbxl18II* as relics of poly (A) structure; and (*v*) *Fbxl18II* contains a 114-bp intron in its 5′ UTR. In addition, a 91-bp intron whose counterpart in *Fbxl18I* is flanked by an intron sequence with a GT-AG boundary is also observed (Figure 4). Based on these facts, rat *Fbxl18II* likely represents a semi-processed retrogene. Although functional experimental evidence for rat *Fbxl18II* is lacking, an expressed sequence tag (EST) (GenBank: CB765629.1) with sequence identity with *Fbxl18II*, rather than *Fbxl18I*, was found in the NCBI database. *Fbxl18II* sequence analysis showed that (*i*) the DNA sequence of *Fbxl18II* has a perfect open reading frame (ORF); (*ii*) using RepeatMasker (http://www.repeatmasker.org), a 937 bp sequence upstream of the start codon was predicted as a long interspersed nuclear element (LINE) that

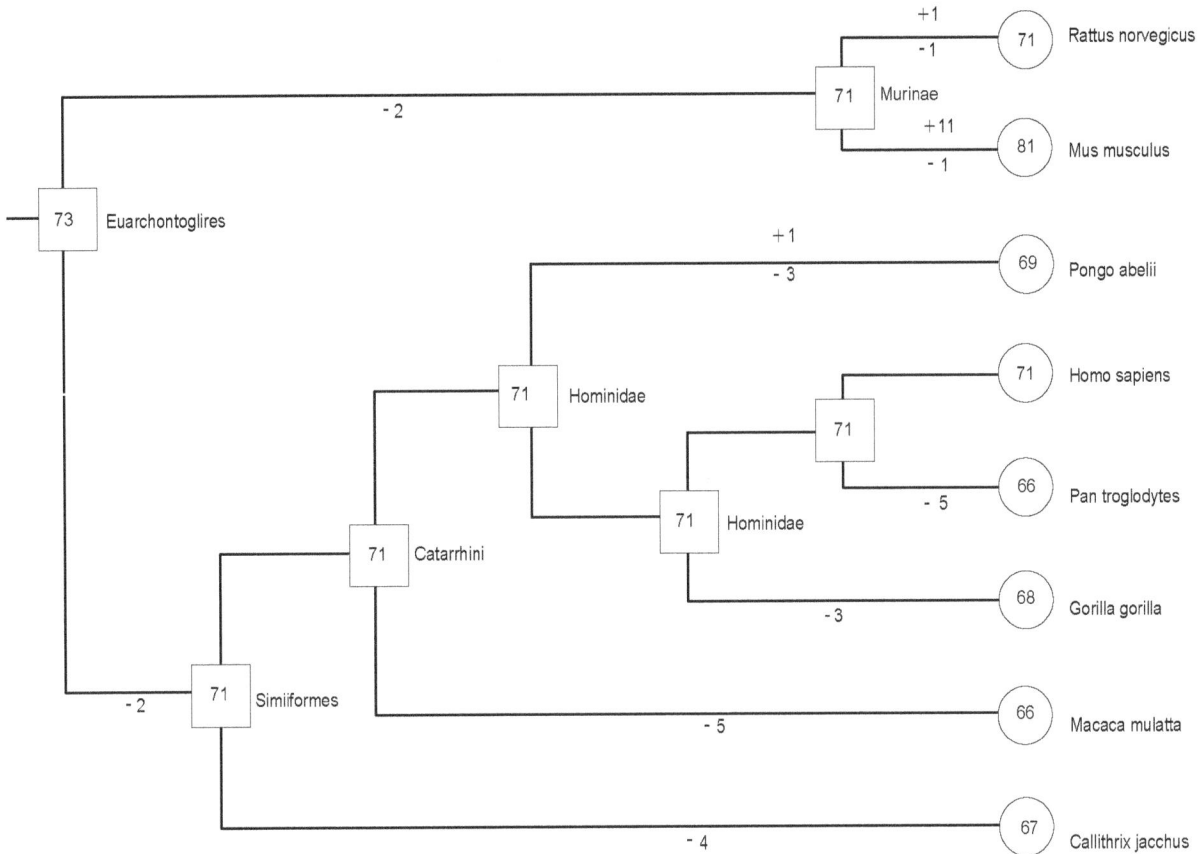

Figure 2. Estimated numbers of ancestral, gained, and lost genes during the evolution of *Euarchontoglire* animals. Names of extant and ancestral species are on the right-hand side of each external and internal node. Numbers within circles and boxes indicate the numbers of genes in each extant and ancestral species, respectively. Branches are not drawn in proportion to their lengths. The numbers above and below each branch are the numbers of genes gained and lost, respectively.

provides original evolutionary materials for promoter formation, and *(iii)* a highly likely transcription start site was predicted at about 200 bp upstream of the start codon using FirstEF [41] ($p = 0.422$) and Promoter 2.0 [42] (prediction score $= 1.224$) (Figure S3). These findings suggest that the retrogene *Fbxl18II* may indeed be a functional gene.

Absence of a gene in some genomes does not necessarily indicate removal by deletion. In fact, F-box gene sequences without typical F-box domains often remain in the genome, as supported by our findings that they exist as diverged orthologs of their counterparts with intact F-box domains across multiple species. For instance, the rat *Fbxo45* does not contain the typical F-box domain, while its orthologs in mouse and primate do.

Sequence divergence of orthologs and region- and lineage-specific positive selection

To explore the selection forces driving the evolution of the F-box gene family, we first calculated sequence divergence in the complete coding sequences (CDSs) for each orthogroup. The values of *Ka*, *Ks*, and ω showed extensive discrepancy among orthogroups (Figure S4). *Ks* was positively correlated with *Ka*, as assessed using the Kendall test (tau $= 0.419$, $p<0.001$). Codon-based Z-test indicated that the ω ratios of all orthogroups were significantly lower than 1 ($p<0.001$), except for *Lrc29* ($p = 0.107$).

To determine the presence of specific regions within F-box genes under positive selection, sliding window analyses were

performed. As expected, ω ratios greater than 1 were observed in regions such as WD40, LRR, and Cyclin_C domains (Figure 5), as well as other uncharacterized domains (Figure S5). Contrary to this pattern, the ω ratios of F-box domains were mostly lower than those of other regions within F-box genes. Such functional selective constraints on the F-box domain may be a consequence of co-evolution of the F-box domain and the SKP1 protein since SKP1 residues are highly conserved, particularly in the core portion of the SKP1–F-box protein interface [43]. For many orthogroups, all regions were under selective constraint with $\omega <1$ although the exact ω value varied among regions (Figure S5).

To verify the statistical significance of variation in selective pressure among sites, we applied a pair of models, namely M0 and M3, to perform likelihood ratio tests. Model M0 assigns a homogeneous ω among sites, whereas M3 assumes several ω site classes. The likelihood ratio test indicated that model M3 exhibited significantly better fit with the data than model M0 for all orthogroups, with the exception of *Fbxw7*, *Fbxl7*, *Fbxl14*, *Fbxl15*, *Fbxl18*, *Fbxl20*, *Kdm2A*, *Lrc29*, *Fbxo2*, *Fbxo4*, *Fbxo11* *Fbxo25*, *Fbxo33*, and *Fbxo44*, the p values of which were greater than 0.05 (Table S5).

Next, we investigated whether some sites within the F-box gene were under positive selection along specific lineages using branch-site models. The orthogroups that included eight orthologs from the eight species were selected for this analysis. Results showed that some sites were under positive selection along specific

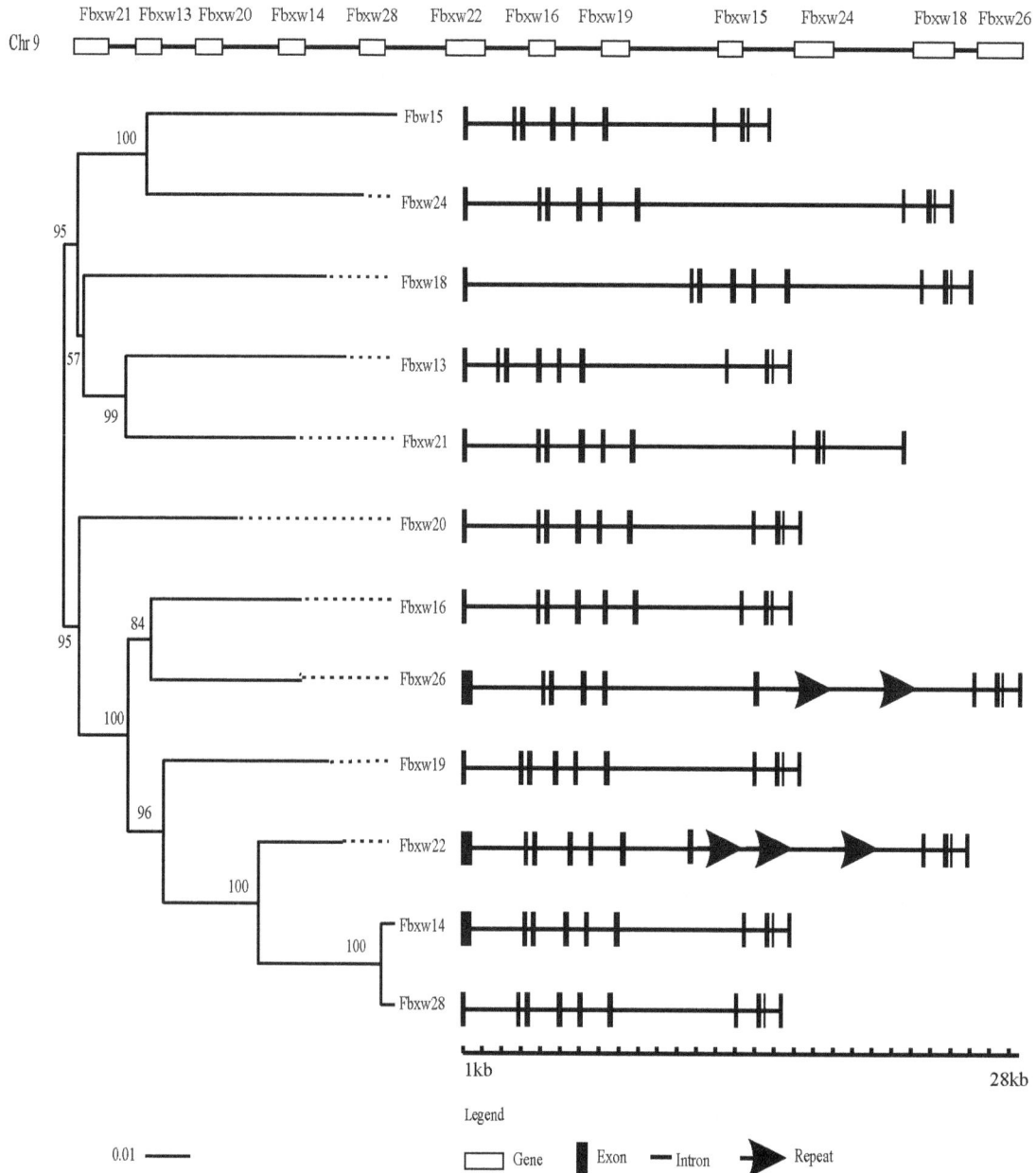

Figure 3. Chromosomal location, phylogenetic relationships and gene structure of the mouse *Fbxw12* gene cluster. These genes diverged at intron length, caused in part by internal sequence repeats.

lineages. A total of 22 out of 34 orthogroups exhibited site-specific positive selection along a specific branch (Table S6). Furthermore, certain F-box genes, such as *Fbxw12*, underwent adaptive evolution independently in different lineages.

Molecular adaptation at individual sites and their effects on function

Specific residues may be selected individually, regardless of the pattern of selection governing the global sequence. We calculated ω values at each codon position. Fourteen out of 71 orthogroups underwent adaptive evolution, as indicated by the presence of some positively selected residues in them (Akaike information criterion score < M8a, and posterior mean of ω >1.5) (Table 1). A large number of individual positively selected sites were observed

in *Fbxw12* and *Fbxl13*. Although positive selection on these sites were not statistically significant (the lower bound of the confidence interval ω was not >1), the results were in agreement with the interpretation that a large number of sites in the two genes had diverged in *Euarchontoglires* (File S3). Some of these positive selection residues were located in identified substrate-interacting domains.

The relationship between positively selected amino acid sites and their effects on function was preliminarily obtained by mapping the sites onto the 3D structure of the functional domain. A 3D structure was built for FBXW12 in automated mode (Figure 6). Currently known WD40 domain-peptide interaction sites were located on all three major surfaces: top, bottom and circumference [44]. The α-helix region was populated with

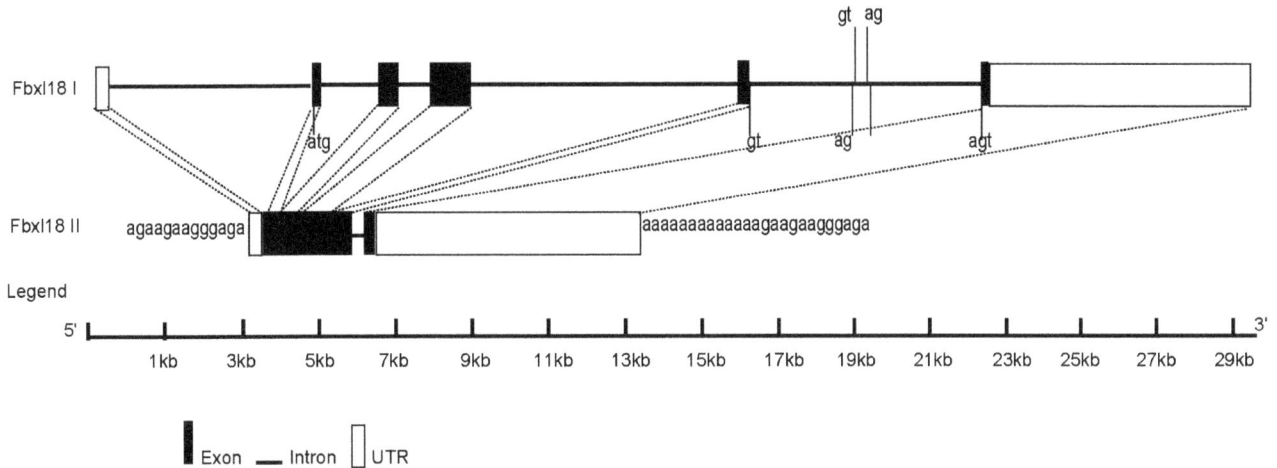

Figure 4. Gene structures of two paralogs *Fbxl18I* **(Ensembl Gene ID: ENSRNOG00000001117) and** *Fbxl18II* **(Ensembl Gene ID: ENSRNOG00000033326).** Dotted lines between the two paralogs indicate identical sequence fragments.

residues with low ω values, whereby the pivotal corrected 3D structure was safeguarded through purifying selection. By contrast, the majority of residues under positive selection of FBXW12 were located in the substrate-binding channel of the WD40 domain.

Stronger negative selection on domains than on non-domains of F-box genes in the human population

The 1000 Genomes data provided an opportunity to compare the mutational burden within different regions of the F-box genes. Here, we analyzed SNP density distribution and the difference of the ratios of non-synonymous to synonymous SNP numbers in the domains and non-domains of F-box genes. We mapped 1,254 SNPs genotyped in the 1000 Genomes project to coding regions of human F-box genes, of which 391 mapped to predicted domain regions (File S4). The average SNP density in the F-box genes was 1.080 per 100 bp, indicating that the F-box genes in humans had a high occurrence of single nucleotide substitutions. This result is similar to that found by Clark et al. in *Arabidopsis thaliana* [45]. In the current study, 15 SNPs were found in the coding regions of *Fbxo32* from 1,092 persons. However, only one SNP was found among the coding regions of the same gene from 1,313 cattle [46].

The average SNP density in the F-box domain, other domain and non-domain regions was 1.042, 1.040, and 1.100 SNPs per 100 bp, respectively (Figure 7). The Kruskal-Wallis test for differences in SNP density across regions of F-box genes indicated a significantly uneven distribution (Kruskal-Wallis test $p = 0.023$, Table S7A). Non-synonymous SNP density in domain regions was significantly lower than that in non-domain regions (Mann-Whitney U test $p = 0.038$) (Table S7B). Consistent with this, the ratio of non-synonymous to synonymous SNP numbers in the domain regions (0.769) was significantly lower than that in the non-domain regions (1.391) (Fisher's exact test $p = 1.403e-06$, Table S7C), reflecting stronger purifying selection on domain regions. Contrary to what might have been expected, the F-box domain region was not found to be more conservative than other domain regions based on both SNP density distribution and the ratio of non-synonymous to synonymous SNP number (Figure 7, Table S7C). However, F-box domain regions were the most conservative according to sequence diversity analyses conducted based on the orthogroups (Figures 5 and S5). The differences in results were likely caused by the following factors. First, after the species diverged from *Euarchontoglires*, certain lineages maintained

advantageous mutations in the C-terminal functional domains of F-box genes that facilitated adaptation to changing environments by altering their substrate specificities. Second, in humans, C-terminal functional domains of F-box genes have evolved under strong purifying selection to accurately recognize specific substrates.

Discussion

Evolutionary conservation and variation in F-box gene number

Conflicting reports exist in the literature regarding the number of F-box genes in humans [26,47]. We performed a comprehensive search of F-box genes using an integrated method. Approximately 70 F-box genes were identified in each *Euarchontoglire* species, which is much less than the hundreds of F-box genes known to exist in plants. This might be due to the sessile life-style of plants, which prevents them from escaping unfavorable environmental conditions, and necessitates more extensive molecular signaling machineries. Compared with the F-box genes predicted in humans by Jin et al. [26], only *Lmo7* was not detected in our study. A published report indicates that an isoform of LMO7 can bind to SKP1 [48]. The F-box domain of LMO7 contains significant changes compared with the consensus sequence. However, LMO7 can still bind to SKP1 because the mutations occur in non-key residues. In such a case, the F-box domain cannot be predicted by a profile search. The method of identifying functional F-box proteins solely based on the consensus sequence inevitably leads to limitations such as filtering out of diverged F-box proteins like *Fbox45* (Table S4). We also assigned the genes *Ect2l*, *Fbxo47*, and *Fbxo48*, all of which contain F-box domains, to the F-box gene family. In addition, Jin et al. [26] had noted that *Fbxw12*-related sequences are expanded over a cluster of six genes on chromosome 9 in mouse. Surprisingly, we found six other paralogs, namely *Fbxw20*, *Fbxw21*, *Fbxw22*, *Fbxw24*, *Fbxw26*, and *Fbxw28*, in this gene cluster (Figure 3). Unequal crossover at meiosis is one of the possible reasons for tandem gene duplication, which occurs more readily in the presence of tandem-repeating genes [49]. Therefore, we inferred that the *Fbxw12* gene cluster may have been formed by tandem duplication through a series of unequal crossover events. Genes included in a gene cluster often differentiate from each other with respect to expression patterns,

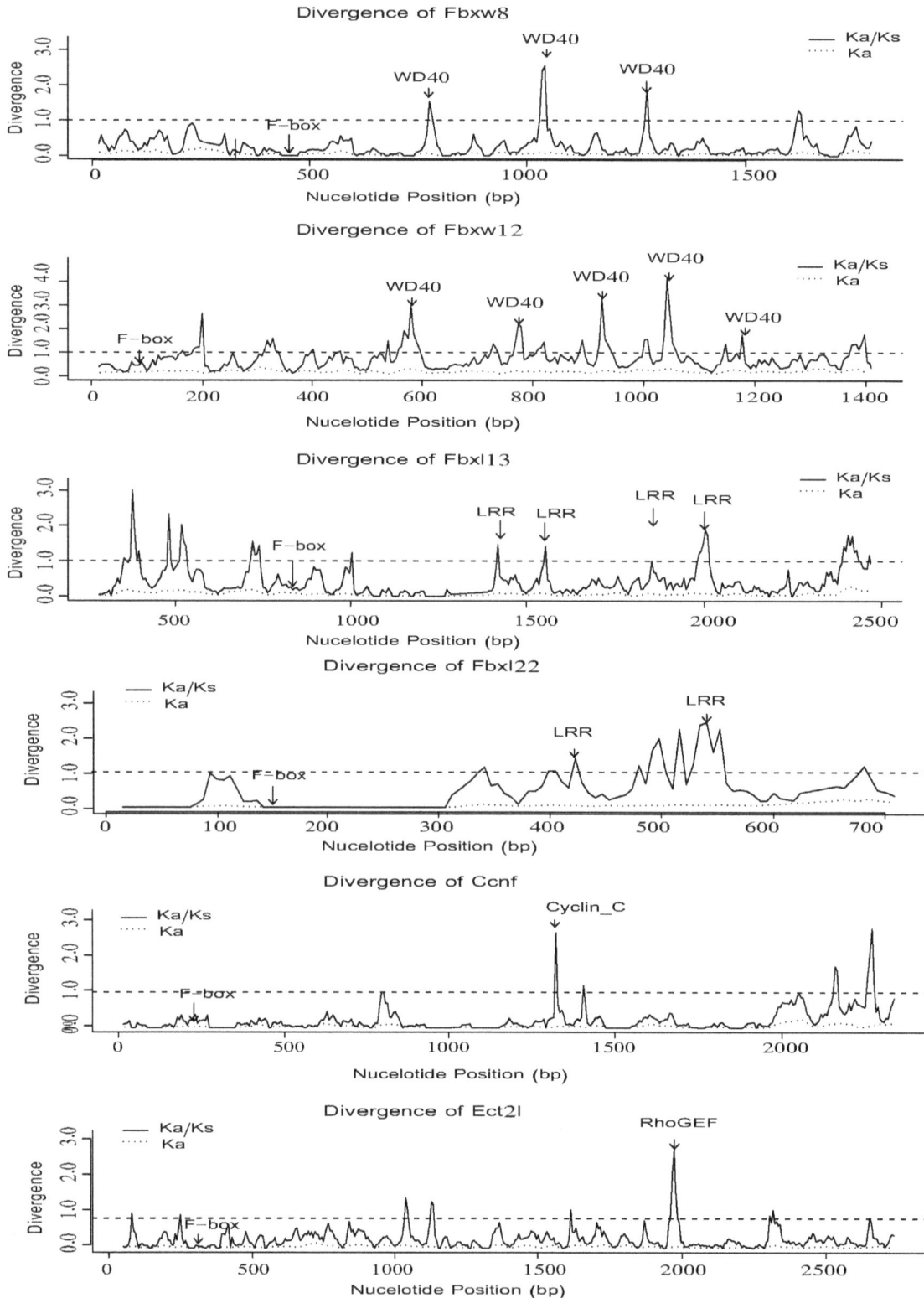

Figure 5. Sliding window analysis of sequence divergence across the protein-coding regions of six orthogroups using a window length of 30 bp and a step size of 6 bp.

such as the mammalian *Hox* gene cluster [50]. Therefore, future experimental investigations should provide insight into the significance of expansion of the *Fbxw12* gene cluster in mouse.

In general, F-box gene expansion or contraction events did not appear to occur massively during the course of *Euarchontoglires* evolution. Hence, F-box gene numbers are conserved among the

Table 1. Detection of positive selection at individual sites by Bayesian method with MEC model using Selecton.

Gene	AICc of M8a	AICc of MEC	Positive site
Fbxw9	7882	7866	13T **173V**
Fbxw10	16530	16488	352G 359T **772G**
Fbxw12	11143	11075	31H 35I 40Y 44S 45L 57N 85H 95I 97F 98E 99T 100E 101L **121S 127E 146E 147F**
			148H 150S 151N 166R 167K 187P 188Q 189P 192C 236L 253S 280P 282K 285A 303S
			304S 305T 306G 316L 330Y 331A 338A 339A 340H 343C 345I 375R 377E 381A 382A
			384N 391C 398E 412H 422E 426K 427D 430T 431D 447R 450K 451V 452S 453D 464T
Fbxl13	11932	11860	36V 66D 88T 97T 100H 130A 136F 139R 145F 150T 157L 187L 191L 192N 232L **308R**
			353M 381L 420F **427N 470K 480R 491R 495A 498M 524G 541D 563E 565Y 566R 570D 607N**
			623A 644L 651E 670N 673K 674K 701R 710D 712I 714S 717G 718A 727T 728Y 733Q 735A
Fbxl5	8092	8080	**562***
Ccnf	12375	12357	220T* **446A*** 691R*
Fbxo5	7538	7506	8C 18S 19A
Fbxo6	5619	5586	212T
Fbxo7	9792	9742	25H 27R 28S 32Q 140L 222L 286C 287K 421T
Fbxo15	9591	9553	65M 267L 269D 270S 275L 276H 288G 291Y 294G 301T 302K 376L 386Q 389N 395A
Fbxo18	14881	14780	190R* 238V* 568G* 1080N*
Fbxo36	2970	2960	100D 107S 132K
Fbxo47	7242	7207	20S 209Q 230R 231S
Fbxo48	2734	2709	2H 4N 9N 11L 15H 18A 28N 36E 40A* 42I 44F 53R 60L

Note: The residues written in bold letters are located in carboxyl-terminal functional domains. Asterisks represents ω confidence interval lower boundary of >1.

eight organisms investigated in the current study. Orthologous F-box proteins generally contain conserved domain architectures in these organisms. As such, they are likely to mediate essential biological pathways by interacting with similar substrates. In eukaryotes, although orthologous proteins typically have the same domain architectures and functions, significant exceptions and

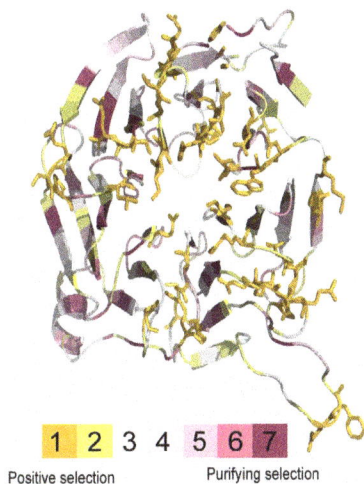

Figure 6. Projection of positive-selection sites onto WD40 domains. The Ka/Ks (ω) value for each residue is color coded on a seven-step scale from violet (<<1) to dark yellow (posterior mean of ω >1.5). Based on the template 3ow8, a 3D structure was constructed for a peptide fragment (residues 89—408) of FBXW12. Most of the residues under strong positive selection (orange-colored stick surface) were mapped to the substrate binding channel.

complications to this generalization may be observed [51]. This assumption can also be applied to F-box genes, since evidence of both lineage-specific domain accretion and reduction was found in this study. Differences in homologous protein domain architectures among species may play a role in the functional diversification of orthologs.

Generally speaking, gene duplication can eventually lead to an imbalance in gene quality, and most of the extra gene copies degenerate via accumulation of mutations and become pseudogenes [49]. For instance, several CYP2D genes and olfactory receptor gene cluster have been reported to be pseudogenes in humans [52,53]. In *Euarchontoglires*, nine F-box related pseudogenes and their corresponding orthologous DNA regions were found. Certain homologous F-box genes were absent in specific lineages, which may be due to bias caused by incomplete genome sequence, variations in genome assembly quality, or loss of the homologous gene from the genome. Although some homologous genes were still present, their F-box domains could hardly be detected because of long-term divergence. However, it should be noted that loss of the F-box domain in an F-box protein homologue does not necessarily imply loss of function of the ubiquitin ligase. For example, NIPA binds SKP1 and CUL1 despite not having a clear F-box domain, and appears to function as an F-box-like protein [54]. Taken together, our results indicate that although F-box gene gain and loss events do not occur as frequently in *Euarchontoglires* as they do in plants, the evolutionary pattern of the F-box gene family in these species is consistent with the birth-and-death evolution model [30,55].

C-terminal domains of F-box proteins have undergone adaptive evolution

C-terminal regions of F-box proteins may have evolved under different selective pressures, such as strong purifying, neutral or

A

B

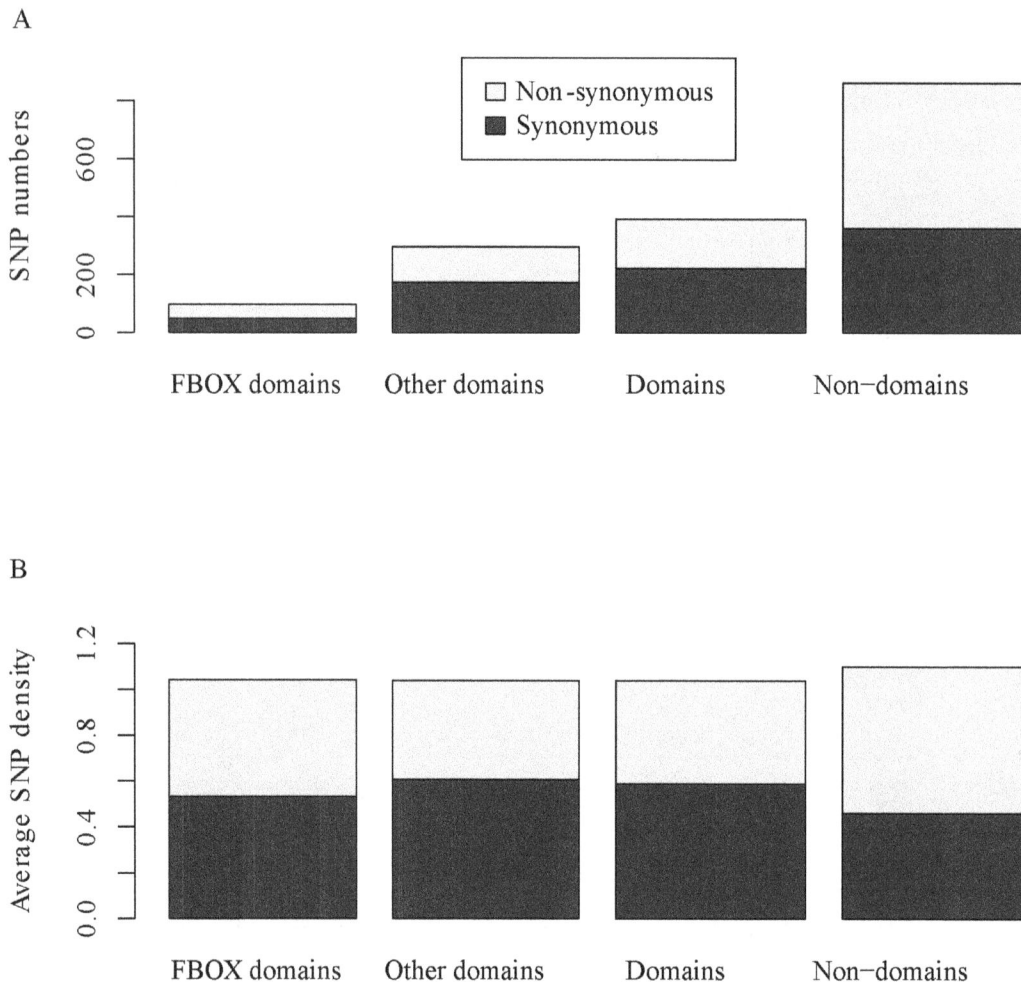

Figure 7. SNP distributions across different regions of F-box genes suggest stronger negative selection on domain regions compared with non-domain regions. A. The total number of synonymous and non-synonymous substitution SNPs in F-box domain, other domain (non F-box domain), domain (including F-box and other domains) and non-domain regions. B. The distributions of average SNP density (SNP numbers per 100 bp) across different regions of F-box genes.

positive selection. The protein structure may only be conserved in some parts of an active site responsible for catalysis, while the remaining peripheral regions may have changed considerably, causing change in substrate specificity [56]. Indeed, the analysis of selective pressure at individual sites showed that the vast majority of functional domains are rich in residues with low ω, whereas the residues in the core portion of the protein–protein interface underwent excess amino-acid fixation during the course of *Euarchontoglires* evolution. These results are consistent with previous findings in other organisms. In plants, while the F-box domain appears rather stable, some C-terminal protein-protein interaction domains such as Kelch and FBA show strong signatures of positive selection [12,13]. Our results are also in agreement with previous findings in nematodes [14]. Co-evolution of the substrate and F-box protein interface may explain the apparent fast evolution of the substrate-binding domain. Alternatively, mutation of residues that are solvent-exposed to a structural fold may be more tolerated than those located at the highly structured core. Therefore, further extensive research is required before the cause of this positive selection can be definitively determined.

Seventy-one F-box gene orthogroups were found to have been under significantly different selective pressure over the course of

Euarchontoglire evolutionary history. We propose that the orthogroups under strongly purifying selection pressure across the genes still recognize the same or similar targets in the eight organisms, whereas others under positive selection pressure conferring adaptive evolution have evolved to recognize different targets. These findings are in contradiction with previous reports that samples of F-box genes from several specific mammalian families show no evidence of positive selection. This discrepancy is likely caused by sample bias [14]. We suggest that mammalian members of the F-box gene family may be involved in both endogenous and exogenous protein degradation.

Comparison of F-box genes between animals and plants

F-box genes are small in number and quite conserved in *Euarchontoglires*. Much like in *Euarchontoglires*, no evidence of drastic changes in the total number of F-box genes (42–47) was found in the 12 extant *Drosophila* species considered in a previous study [57]. However, F-box gene family is one of the largest and fastest evolving gene families in plants [12,13]. For instance, the number of F-box Kelch genes (FBKs) varies dramatically among *Arabidopsis thaliana*, *Oryza sativa*, *Populus trichocarpa* and *Vitis vinifera*[12]. The large number of F-box proteins in plants might be required by their species-specific physiology, such as responses to various hormones [58,59], the circadian clock and photomorphogenesis [60,61], flower development [62], and defense responses [63].

For both plants and animals, F-box proteins often carry one of a variety of protein-protein interaction domains in the C-terminal regions in addition to the loosely conserved N-terminal F-box domain [7,64]. The differences in domain distributions among *Euarchontoglires*, nematodes and plants were striking. Among 27 identified C-terminal domains in *Euarchontoglires*, LRRs, one of the most abundant C-terminal domains in plants [30], was the most common. However, LRRs was rare in nematodes (unpublished data from our laboratory). In C. *elegans*, most F-box proteins contain either the FTH or FBA2 domain [14], both of which are absent in plants and *Euarchontoglires*. In addition to FTH and FBA2, many other distinct domains were also found to be lineage-specific. By contrast, only a single member of the F-box protein family contained the Kelch domain in each species of *Euarchontoglires* as well as in C. *elegans*. However, it has been reported that Kelch-containing F-box proteins expanded dramatically among terrestrial plants [12]. Thus, the very distinctive domain distribution of F-box proteins may reflect their divergent functional roles in plants and animals.

Conclusions

This study explored the evolutionary forces driving conservation and divergence of the F-box gene family in *Euarchontoglires*. Lineage-specific gene tandem duplication, mRNA-mediated retrotransposition, and gene loss contributed to F-box gene number variation in the eight organisms examined in this study. The evolutionary pattern of the F-box gene family in *Euarchontoglires* was in line with the birth-and-death evolution model, although some genes were found to be subject to concerted evolution. Certain F-box genes undergo adaptive evolution in specific lineages, although the majority of the orthogroups are under strong selective constraint. In addition, population genetic analyses indicated that the evolution of domain regions within F-box genes was shaped by stronger purifying selection compared to that of non-domain regions. Future studies employing proteomics and functional genomics approaches will be essential for the identification of human F-box protein targets and for determination of the cellular biological processes involved. The results of this work significantly improve our understanding of SCF biology. Given the roles of F-box proteins in many diseases, development of new therapies targeting F-box proteins may be expected in the future.

Supporting Information

Figure S1   A phylogenetic tree was created using F-box protein sequences of eight species (marmoset, gorilla, human, macaque, mouse, chimpanzee, orangutan, and rat) by the maximum likelihood (ML) method. Values above branches denote percent support for clades based on 100 bootstrap replicates. The interior colored strip corresponds to the distribution of species in each orthogroup. The outer colored strip represents the C-terminal domain contained in the protein from the corresponding interior species.

Figure S2   Gene conversion events between *Fbxo6* and *Fbxo44*. HSA, PTR, GGO, PPY, MMU, MUS, and RNO represent the species human, chimpanzee, gorilla, orangutan, macaque, mouse, and rat, respectively. Red box indicates gene conversion tracts.

Figure S3   Prediction of potential functional gene elements in rat *Fbxl18II*.

Figure S4   Average sequence divergence in protein-coding regions of orthologs from 71 orthogroups. *Ka*, *Ks*, and ω represent average non-synonymous substitution rate per site, synonymous substitution rate per site, and their ratios between orthologs, respectively.

Figure S5   Sliding window analysis of sequence divergence across protein-coding regions of 65 orthogroups using a window length of 30 bp and a step size of 6 bp.

File S1   Multiple sequence alignment of the F-box protein sequences.

File S2   An all-against-all BLAST (-e 1e-50) between every pair of F-box protein sequences from each of the eight organisms.

File S3   Multiple sequence alignments of coding region sequences for each orthologous group.

File S4   SNPs in F-box genes from the 1000 Genomes Project.

Table S1   F-box gene numbers and their accession numbers in the eight genomes.

Table S2   F-box gene-related pseudogenes in the eight genomes.

Table S3   Chromosomal distributions of F-box genes in the eight genomes.

Table S4   Events causing F-box gene number variation during the evolution of *Euarchontoglires*.

Table S5   Tests of variable ω among sites for 71 orthogroups using models M0-M3 comparison.

Table S6   Lineage-specific positive selection was identified using branch-site selection models.

Table S7   Statistical tests for differences in SNP distributions across different regions of F-box genes.

Acknowledgments

We would like to thank the members of the Bioinformatics Center of Northwest A&F University for their useful discussion. We are particularly grateful to Dr. Wenwu Wu for his valuable advice during the early stages of this work.

Author Contributions

Conceived and designed the experiments: ALW SHT. Performed the experiments: ALW MCF. Analyzed the data: ALW SHT. Contributed reagents/materials/analysis tools: ALW YHM XCL. Wrote the paper: ALW XQJ SHT.

References

1. Yen H-CS, Xu Q, Chou DM, Zhao Z, Elledge SJ (2008) Global protein stability profiling in mammalian cells. Science 322: 918–923.
2. Hershko A, Ciechanover A (1998) The ubiquitin system. Annu Rev Biochem 67: 425–479.
3. Feldman R, Correll CC, Kaplan KB, Deshaies RJ (1997) A complex of Cdc4p, Skp1p, and Cdc53p/cullin catalyzes ubiquitination of the phosphorylated CDK inhibitor Sic1p. Cell 91: 221–230.
4. Winston JT, Koepp DM, Zhu C, Elledge SJ, Harper JW (1999) A family of mammalian F-box proteins. Curr Biol 9: 1180–1183.
5. Wang C, Gale Jr M, Keller BC, Huang H, Brown MS, et al. (2005) Identification of FBL2 as a geranylgeranylated cellular protein required for hepatitis C virus RNA replication. Mol Cell 18: 425–434.
6. Moberg KH, Bell DW, Wahrer DC, Haber DA, Hariharan IK (2001) Archipelago regulates Cyclin E levels in Drosophila and is mutated in human cancer cell lines. Nature 413: 311–316.
7. Kipreos ET, Pagano M (2000) The F-box protein family. Genome Biol 1: 3002.
8. Yang X, Kalluri UC, Jawdy S, Gunter LE, Yin T, et al. (2008) The F-box gene family is expanded in herbaceous annual plants relative to woody perennial plants. Plant Physiol 148: 1189–1200.
9. Hua Z, Zou C, Shiu S-H, Vierstra RD (2011) Phylogenetic comparison of F-Box (FBX) gene superfamily within the plant kingdom reveals divergent evolutionary histories indicative of genomic drift. PLoS One 6: e16219.
10. Bellieny-Rabelo D, Oliveira AEA, Venancio TM (2013) Impact of Whole-Genome and Tandem Duplications in the Expansion and Functional Diversification of the F-Box Family in Legumes (Fabaceae). PloS One 8: e55127.
11. Gagne JM, Downes BP, Shiu S-H, Durski AM, Vierstra RD (2002) The F-box subunit of the SCF E3 complex is encoded by a diverse superfamily of genes in Arabidopsis. Proc Natl Acad Sci U S A 99: 11519–11524.
12. Schumann N, Navarro-Quezada A, Ullrich K, Kuhl C, Quint M (2011) Molecular evolution and selection patterns of plant F-box proteins with C-terminal kelch repeats. Plant Physiol 155: 835–850.
13. Navarro-Quezada A, Schumann N, Quint M (2013) Plant F-Box Protein Evolution Is Determined by Lineage-Specific Timing of Major Gene Family Expansion Waves. PLoS One 8: e68672.
14. Thomas JH (2006) Adaptive evolution in two large families of ubiquitin-ligase adapters in nematodes and plants. Genome Res 16: 1017–1030.
15. Graur D, Li W-H (2000) Fundamentals of molecular evolution. Sunderland: Sinauer Associates. 111 p.
16. Bateman A, Coin L, Durbin R, Finn RD, Hollich V, et al. (2004) The Pfam protein families database. Nucleic Acids Res 32: D138–D141.
17. Flicek P, Amode MR, Barrell D, Beal K, Brent S, et al. (2012) Ensembl 2012. Nucleic Acids Res 40: D84–D90.
18. Eddy SR (1998) Profile hidden Markov models. Bioinformatics 14: 755–763.
19. Quevillon E, Silventoinen V, Pillai S, Harte N, Mulder N, et al. (2005) InterProScan: protein domains identifier. Nucleic Acids Res 33: W116–W120.
20. Zdobnov EM, Apweiler R (2001) InterProScan–an integration platform for the signature-recognition methods in InterPro. Bioinformatics 17: 847–848.
21. Altschul SF, Madden TL, Schäffer AA, Zhang J, Zhang Z, et al. (1997) Gapped BLAST and PSI-BLAST: a new generation of protein database search programs. Nucleic Acids Res 25: 3389–3402.
22. Edgar RC (2004) MUSCLE: multiple sequence alignment with high accuracy and high throughput. Nucleic Acids Res 32: 1792–1797.
23. Capella-Gutiérrez S, Silla-Martínez JM, Gabaldón T (2009) trimAl: a tool for automated alignment trimming in large-scale phylogenetic analyses. Bioinformatics 25: 1972–1973.
24. Darriba D, Taboada GL, Doallo R, Posada D (2011) ProtTest 3: fast selection of best-fit models of protein evolution. Bioinformatics 27: 1164–1165.
25. Guindon S, Dufayard J-F, Lefort V, Anisimova M, Hordijk W, et al. (2010) New algorithms and methods to estimate maximum-likelihood phylogenies: assessing the performance of PhyML 3.0. Syst Biol 59: 307–321.
26. Jin J, Cardozo T, Lovering RC, Elledge SJ, Pagano M, et al. (2004) Systematic analysis and nomenclature of mammalian F-box proteins. Genes Dev 18: 2573–2580.
27. Chen K, Durand D, Farach-Colton M (2000) NOTUNG: a program for dating gene duplications and optimizing gene family trees. J Comput Biol 7: 429–447.
28. Hedges SB, Dudley J, Kumar S (2006) TimeTree: a public knowledge-base of divergence times among organisms. Bioinformatics 22: 2971–2972.
29. Bailey JA, Eichler EE (2006) Primate segmental duplications: crucibles of evolution, diversity and disease. Nat Rev Genet 7: 552–564.
30. Xu G, Ma H, Nei M, Kong H (2009) Evolution of F-box genes in plants: different modes of sequence divergence and their relationships with functional diversification. Proc Natl Acad Sci U S A 106: 835–840.
31. Tamura K, Peterson D, Peterson N, Stecher G, Nei M, et al. (2011) MEGA5: molecular evolutionary genetics analysis using maximum likelihood, evolutionary distance, and maximum parsimony methods. Mol Biol Evol 28: 2731–2739.
32. Librado P, Rozas J (2009) DnaSP v5: a software for comprehensive analysis of DNA polymorphism data. Bioinformatics 25: 1451–1452.
33. Yang Z (2007) PAML 4: phylogenetic analysis by maximum likelihood. Mol Biol Evol 24: 1586–1591.
34. Doron-Faigenboim A, Stern A, Mayrose I, Bacharach E, Pupko T (2005) Selecton: a server for detecting evolutionary forces at a single amino-acid site. Bioinformatics 21: 2101–2103.
35. Doron-Faigenboim A, Pupko T (2007) A combined empirical and mechanistic codon model. Mol Biol Evol 24: 388–397.
36. Arnold K, Bordoli L, Kopp J, Schwede T (2006) The SWISS-MODEL workspace: a web-based environment for protein structure homology modelling. Bioinformatics 22: 195–201.
37. Consortium TGP (2012) An integrated map of genetic variation from 1,092 human genomes. Nature 491: 56–113.
38. Mohrenweiser H, Johnson K (1996) Report of the third international workshop on human chromosome 19 mapping 1996. Cytogenet Genome Res 74: 161–186.
39. Ohno S (1967) Sex chromosomes and sex-linked genes. Berlin: Springer.
40. Ohta T (1990) How gene families evolve. Theor Popul Biol 37: 213–219.
41. Davuluri RV, Grosse I, Zhang MQ (2001) Computational identification of promoters and first exons in the human genome. Nat Genet 29: 412–417.
42. Knudsen S (1999) Promoter 2. 0: for the recognition of PolII promoter sequences. Bioinformatics 15: 356–361.
43. Schulman BA, Carrano AC, Jeffrey PD, Bowen Z, Kinnucan ER, et al. (2000) Insights into SCF ubiquitin ligases from the structure of the Skp1–Skp2 complex. Nature 408: 381–386.
44. Stirnimann CU, Petsalaki E, Russell RB, Müller CW (2010) WD40 proteins propel cellular networks. Trends Biochem Sci 35: 565–574.
45. Clark RM, Schweikert G, Toomajian C, Ossowski S, Zeller G, et al. (2007) Common sequence polymorphisms shaping genetic diversity in Arabidopsis thaliana. Science 317: 338–342.
46. Wang A, Zhang Y, Li M, Lan X, Wang J, et al. (2012) SNPs identification in FBXO32 gene and their associations with growth traits in cattle. Gene: 181–186.
47. Skaar JR, Pagan JK, Pagano M (2009) SnapShot: F box proteins I. Cell 137: 1160.
48. Cenciarelli C, Chiaur D, Guardavaccaro D, Parks W, Vidal M (1999) Identification of a family of human F-box proteins. Curr Biol 9: 1177–1179.
49. Ohta T (2008) Gene Families: Formation and Evolution. In Encyclopedia of the Human Genome. Macmillan Publishers Ltd., London.
50. Lufkin T (2005) Hox Genes: Embryonic Development. In Encyclopedia of the Human Genome. Macmillan Publishers Ltd., London.
51. Chervitz SA, Aravind L, Sherlock G, Ball CA, Koonin EV, et al. (1998) Comparison of the complete protein sets of worm and yeast: orthology and divergence. Science 282: 2022–2028.
52. Heim MH, Meyer UA (1992) Evolution of a highly polymorphic human cytochrome P450 gene cluster: CYP2D6. Genomics 14: 49–58.
53. Glusman G, Sosinsky A, Ben-Asher E, Avidan N, Sonkin D, et al. (2000) Sequence, structure, and evolution of a complete human olfactory receptor gene cluster. Genomics 63: 227–245.
54. Bassermann F, von Klitzing C, Münch S, Bai R-Y, Kawaguchi H, et al. (2005) NIPA defines an SCF-type mammalian E3 ligase that regulates mitotic entry. Cell 122: 45–57.
55. Nei M, Rooney AP (2005) Concerted and birth-and-death evolution of multigene families. Annu Rev Genet 39: 121–152.
56. Yeats CA, Orengo CA (2007) Evolution of Protein Domains. In Encyclopedia of the Human Genome. Macmillan Publishers Ltd., London.
57. Li A, Xu G, Kong H (2011) Mechanisms underlying copy number variation in F-box genes: evidence from comparison of 12 Drosophila species. Biodiversity Science 19: 3–16.
58. Binder BM, Walker JM, Gagne JM, Emborg TJ, Hemmann G, et al. (2007) The Arabidopsis EIN3 binding F-Box proteins EBF1 and EBF2 have distinct but overlapping roles in ethylene signaling. Plant Cell 19: 509–523.
59. Dill A, Thomas SG, Hu J, Steber CM, Sun T-p (2004) The Arabidopsis F-box protein SLEEPY1 targets gibberellin signaling repressors for gibberellin-induced degradation. Plant Cell 16: 1392–1405.
60. Han L, Mason M, Risseeuw EP, Crosby WL, Somers DE (2004) Formation of an SCFZTL complex is required for proper regulation of circadian timing. Plant J 40: 291–301.
61. Kim W-Y, Fujiwara S, Suh S-S, Kim J, Kim Y, et al. (2007) ZEITLUPE is a circadian photoreceptor stabilized by GIGANTEA in blue light. Nature 449: 356–360.
62. Chae E, Tan QK-G, Hill TA, Irish VF (2008) An Arabidopsis F-box protein acts as a transcriptional co-factor to regulate floral development. Development 135: 1235–1245.
63. Kim HS, Delaney TP (2002) Arabidopsis SON1 is an F-box protein that regulates a novel induced defense response independent of both salicylic acid and systemic acquired resistance. Sci Signal 14: 1469.
64. Lechner E, Achard P, Vansiri A, Potuschak T, Genschik P (2006) F-box proteins everywhere. Curr Opin Plant Biol 9: 631–638.

14

DNA as a Phosphate Storage Polymer and the Alternative Advantages of Polyploidy for Growth or Survival

Karolin Zerulla[1], Scott Chimileski[2], Daniela Näther[1], Uri Gophna[3], R. Thane Papke[2], Jörg Soppa[*]

1 Institute for Molecular Biosciences, Biocentre, Goethe-University, Frankfurt, Germany, 2 Department of Molecular and Cell Biology, University of Connecticut, Storrs, Connecticut, United States of America, 3 Department of Molecular Microbiology and Biotechnology, George S. Wise Faculty of Life Sciences, Tel Aviv University, Ramat Aviv, Tel Aviv, Israel

Abstract

Haloferax volcanii uses extracellular DNA as a source for carbon, nitrogen, and phosphorous. However, it can also grow to a limited extend in the absence of added phosphorous, indicating that it contains an intracellular phosphate storage molecule. As *Hfx. volcanii* is polyploid, it was investigated whether DNA might be used as storage polymer, in addition to its role as genetic material. It could be verified that during phosphate starvation cells multiply by distributing as well as by degrading their chromosomes. In contrast, the number of ribosomes stayed constant, revealing that ribosomes are distributed to descendant cells, but not degraded. These results suggest that the phosphate of phosphate-containing biomolecules (other than DNA and RNA) originates from that stored in DNA, not in rRNA. Adding phosphate to chromosome depleted cells rapidly restores polyploidy. Quantification of desiccation survival of cells with different ploidy levels showed that under phosphate starvation *Hfx. volcanii* diminishes genetic advantages of polyploidy in favor of cell multiplication. The consequences of the usage of genomic DNA as phosphate storage polymer are discussed as well as the hypothesis that DNA might have initially evolved in evolution as a storage polymer, and the various genetic benefits evolved later.

Editor: Shree Ram Singh, National Cancer Institute, United States of America

Funding: This project has been supported by the German Research Council (DFG grant So264/16), the National Science Foundation, USA (0919290 and 0830024), the US-Israel Binational Science Foundation (award No. 2007043), and the NASA Astrobiology Program (Grant NNX12AD70G). The funders had no role in study design, data collection and analysis, decision to publish, or preparation of the manuscript.

Competing Interests: The authors have declared that no competing interests exist.

* E-mail: soppa@bio.uni-frankfurt.de

Introduction

The advantages of polyploidy that led to its development in evolution has long been discussed in the framework of eukaryotes, because prokaryotes were long thought to be typically monoploid (a single copy of the chromosome before replication), which is often erroneously termed "haploid". Evolutionary explanations for organisms with homologous sets of chromosomes have long been linked to the invention of sexual reproduction [1], and have been developed from mathematical modeling using population genetics principles and assumptions. Those analyses indicate that ploidy levels ≥ 2 n could be selectively advantageous by preventing the expression of deleterious recessive alleles [2]. Additional hypotheses are interconnected with high recombination rates [2] or cell size and r vs. K selection [3]. However, in recent years polyploidy has been demonstrated to be widespread in bacteria and archaea as well [4–7], indicating that it is an ancient trait preceding eukaryotes, and that any explanation for the origin and maintenance of higher ploidy levels must address asexually reproducing prokaryotes.

A few polyploid prokaryotic species and their probable selective advantage of polyploidy have been well characterized. For example, cells from the unusually large bacterium *Epulospicium* type B, whose dimensions make it visible to the naked eye, are

estimated to contain 50,000–120,000 chromosome copies per cell, which are positively correlated with cytoplasmic volume [8]. Because of diffusion limitations, the extreme polyploidy of *Epulospicium* is thought to be necessary for efficient gene expression. Though interesting and biological relevant, this polyploidy system probably has evolved rather late in evolution because a giant cell size requires a cytoskeleton and advanced intracellular transport. Another example is the bacterium *Deinococcus radiodurans*, which survives high doses of ionizing radiation that generate hundreds of double strand breaks. Its survival strategy relies on polyploidy for performing interchromosomal recombination, which is necessary for repairing its fragmented DNA [9,10]. While X-ray irradiation is used to induce double strand breaks in the laboratory, the cause of double strand breaks and chromosome fragmentation in nature is desiccation. Polyploidy as a basis for the repair of scattered chromosomes probably evolved early, nevertheless, it requires the pre-existence of a sophisticated DNA repair system. In summary, nearly 10 putative evolutionary advantages that led to the development of polyploidy at different times in different prokaryotic lineages have been discussed [4–7], most of which require the pre-existence of homologous recombination. Here we add an additional evolutionary advantage of polyploidy that does not require the pre-existence of homologous recombination, namely the usage of genomic DNA as a storage polymer. The experiments revealing that a prokaryotic species uses DNA as a storage polymer were performed with *Haloferax volcanii*, a halophilic archaeon.

Halophilic Archaea of the family *Halobacteriaceae* are polyploids with phenotypic traits consistent with polyploidy. Both species *Hfx. volcanii* and *Halobacterium salinarum* are demonstrated to contain more than 20 chromosome copies during exponential phase and 10 during stationary phase [4]. *Hbt. salinarum* has been shown to be very resistant to gamma radiation [11], and *Halorubrum chaoviator* strain Halo-G survived the conditions of outer space for two weeks [12], which would be unlikely if these species were monoploids. Furthermore, halobacteria in general experience homologous recombination and gene transfer from distant species [13,14] and *Halorubrum* populations exist in genetic equilibrium [15]. Haloarchaea produce heterozygous cells after fusion of membranes and cell walls [16]. This is even true for different species thus displaying an unusually low species barrier to homologous recombination [17] and thus can account for their genetic exchange partner promiscuity. These phenotypic characteristics of haloarchaea show that they make intensive use of various genetic advantages of polyploidy. However, here we show that nutrient availability determines ploidy level and that extracellular and intracellular genomic DNA is used as a storage polymer. Notably, it is also shown that *Hfx. volcanii* diminishes genetic advantages of polyploidy under conditions of phosphate starvation.

Results

Intracellular storage capacities and growth on external genomic DNA

The first aim of this study was to clarify whether *Hfx. volcanii* can use external (environmental) genomic DNA as a source of carbon (C), nitrogen (N), and/or phosphorous (P). Control cultures supplemented with all three nutrients in the form of glucose, ammonium chloride and potassium phosphate were compared to cultures in which each one of the substances, respectively, was omitted. In each case three independent cultures were grown, and average growth curves and their standard deviations are shown in Fig. 1. In the absence of externally added genomic DNA no growth occurred when C was omitted, indicating that *Hfx. volcanii* has no intracellular carbon storage (Fig. 1, curve −C). In contrast, considerable growth occurred when P was omitted, showing that *Hfx. volcanii* contains an intracellular phosphate storage pool. The growth yield was about 40% of the control culture grown in the presence of all three nutrients. Also the omission of ammonium chloride resulted in considerable growth with a growth yield of about 80% of the control culture. However, in preparation of future genetic experiments *Hfx. volcanii* strain H26 was used, which is auxotrophic for uracil. Therefore, uracil had to be supplemented, which might have been used as nitrogen source, and thus the experiment is uninformative about the absence or presence of an internal nitrogen storage pool.

The addition of external genomic DNA to cultures lacking any of the three nutrients in all three cases enhanced the growth yield, revealing that genomic DNA can be a source for C, N, and P for *Hfx. volcanii* (filled symbols and dotted lines in Fig. 1). The addition of genomic DNA to cultures lacking phosphate or ammonium resulted even in faster growth compared to the control culture grown in the presence of all three nutrients. Only the culture with genomic DNA instead of glucose as a C-source had a substantially lower growth rate, showing that genomic DNA is metabolized more slowly than glucose as a carbon source.

These results revealed on the one hand that external (environmental) DNA can be used as a source for C, N, and P,

and on the other hand showed that *Hfx. volcanii* must have intracellular storage capacities for P, but not for C. In the following experiments we concentrated on the usage of external genomic DNA as a source of P and the identity of the intracellular P storage polymer.

To further confirm that high molecular weight genomic DNA was indeed the source of the phosphorous, and not potential impurities or contaminations, *Hfx. volcanii* was grown in the presence of DNA and the absence of any other supplemented P. As a control, non-inoculated cultures were incubated under identical conditions. Fig. 2A shows average OD_{600} values of three independent cultures and their standard deviations. Notably, the OD_{600} values of Fig. 2A cannot be compared to that of Fig. 1, because in this experiment cultures were grown in Erlenmeyer flasks and not in microtiter plates and thus path length, photometer, and +/−dilution prior to measurements differ in the two experiments. At the eight time points indicated in Fig. 2 aliquots were removed and the cells were pelleted by centrifugation. The DNA content of the supernatant was analyzed by analytical agarose gel electrophoresis (after dialysis to remove the high salt concentration of the medium). Fig. 2B shows one representative gel of the mock treated culture. It can be seen that the high molecular weight input genomic DNA is broken into small fragments, either by chemical hydrolysis or, more probable, by mechanical shearing forces due to the shaking with 250 rpm. The amount of DNA was quantified using the program ImageJ and the result is included in Fig. 2A (filled squares, dotted line). Within 21 hours the values dropped to 80% and then stayed constant throughout the remaining 120 hours of the experiment. Most probably the initial drop of 20% in integrated signal intensity is not due to a real loss of DNA, but to a broader distribution of the fragments in the gel compared to the full-size genomic DNA. Fig. 2C shows one representative gel of the DNA content in the supernatants of the inoculated culture. In contrast to the mock treated culture the amount of DNA steadily decreased and less than 10% of the input DNA was left after 142 hours. Taken together, these results clearly show that *Hfx. volcanii* can use external (environmental) genomic DNA as a source of phosphorous (Fig. 1 and 2) and also as a source of carbon and nitrogen (Fig. 1).

Genomic DNA is the intracellular storage polymer of phosphate

For a further characterization of the growth of *Hfx. volcanii* in the absence of any externally added P source cultures were grown in the presence of two different phosphate concentrations (1 mM, and 10 mM) and in the absence of added P. The results are shown in Fig. 3. Growth with 1 mM and with 10 mM phosphate was identical, indicating that phosphate is not the limiting nutrient under these conditions. Again, considerable growth was observed in the absence of added P, indicating that the liberation of phosphate from the intracellular phosphate storage polymer is growth rate-limiting. The OD_{600} at the start of the experiment was about 0.05. The sterile controls showed that the microtiter plates had and OD_{600} of about 0.03 and thus the inoculum had an OD_{600} of about 0.02. After 140 h growth in the absence of added P the cells had an OD_{600} of about 0.17 (measured OD_{600} of 0.2 minus the OD_{600} of the sterile control, 0.03). This is an 8.5-fold increase in OD_{600}, which would be equivalent to about three doublings in the absence of added phosphate if the light scatter of the cells would not change. Microscopic observation of the cells indicated that they had normal morphology and were of similar size.

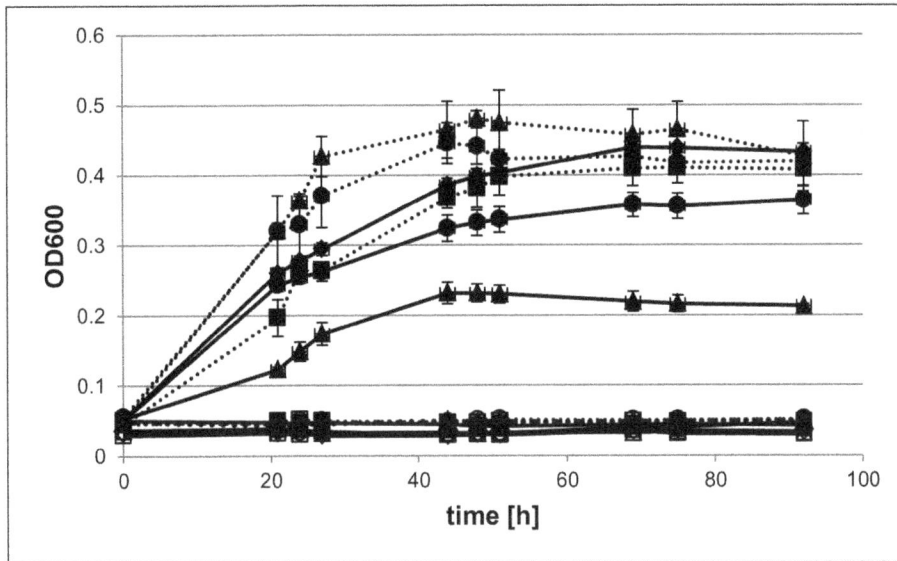

Figure 1. *Hfx. volcanii* uses external DNA as a nutrient source and contains internal P and N storages. *Hfx. volcanii* was grown in microtiter plates in synthetic medium with added carbon (C), nitrogen (N), and phosphate (P) as positive control (diaments). In additional cultures each one of the three nutrients was replaced with genomic DNA (dotted lines), i.e. C was replaced (squares), N was replaced (circles), and P was replaced (triangles). In further cultures each one of the respective nutrients was omitted without replacement (solid lines), i.e. C was omitted (squares), N was omitted (circle), and P was omitted (triangles). To verify that spill over did not occur, for each medium also non-inoculated controls (sterile controls) were performed (open symbols). In each case average values of three independent cultures and their standard deviations are shown.

These observations that *Hfx. volcanii* maintains an intracellular phosphorous storage and our previous results that *Hfx. volcanii* is highly polyploid and contains about 25–30 copies of the chromosome in exponential phase [4] led to the hypothesis that genomic DNA might be the intracellular phosphate storage polymer. To test our hypothesis we used *Hfx. volcanii* cells grown to exponential phase in complex medium as an inoculum for assessing growth in synthetic media supplemented with two different phosphate concentrations (1 mM, and 10 mM) and no added phosphate. Using quantitative PCR (qPCR), chromosome copy numbers were estimated for the inoculum (an estimate of the pre-growth condition) as well as cells grown to exponential phase and stationary phase without added phosphate (exponential phase: 9.4×10^7 cells/ml, stationary phase: 2.7×10^8 cells/ml) and with 1 mM and 10 mM phosphate supplementation (for both: exponential phase: 5.2×10^8 cells/ml, stationary phase: 1.3×10^9 cells/ml). During exponential growth, the phosphate concentration was found to influence the ploidy level, with 24 copies on average in cells grown with 10 mM phosphate, 19 copies in cells grown with 1 mM phosphate, and 14 in cells grown in the absence of an added source of phosphorous (Fig. 4). Stationary phase cells that were grown in the presence of added phosphate (10 and 1 mM) maintained approximately 13 chromosomal copies of their genome. However, in the absence of phosphate supplementation, stationary cells had on average reduced their genome copy number to two. This result showed that *Hfx. volcanii* indeed uses genomic DNA as a phosphate storage polymer and indicated that it diminishes the putative genetic and long-term advantages of polyploidy (e.g. DNA repair, desiccation resistance, long term survival) to enable short-term reproductive gains.

Polyploidy dependence upon nutrient availability was further substantiated when cells from P-starved stationary phase cultures that were depleted of extra chromosomes were amended with phosphate: within three hours the chromosome copy number more than tripled and within 24 hours they increased by greater than 10-fold to more than 40 copies per cell (Fig. 5). Thus phosphate-starved *Hfx. volcanii* cells take up phosphate very fast after re-addition and use it to re-establish the polyploid state, even with an overshoot phase with more than 40 chromosomal copies per cell.

The genome sequence of *Hfx. volcanii* contains five genes that are annotated to encode polyphosphate kinases (HVO_0074, HVO_0837, HVO_1650, HVO_2363, HVO_2598), opening the possibility that *Hfx. volcanii* might also use polyphosphate as a phosphate-storage polymer, in addition to genomic DNA. To investigate this possibility, *Hfx. volcanii* was grown in the presence of added phosphate and a culture aliquot was removed during exponential growth. The cells were fixed and stained with DAPI to simultaneously detect genomic DNA as well as polyphosphate based on the differential wavelengths of fluorescence emission for these biopolymers [18]. Chromosomal DNA was readily observed in cells using this approach, in contrast to polyphosphate, which was not detected (data not shown). Therefore, at least under the conditions of the experiments of this study, *Hfx. volcanii* appears to use only genomic DNA as polymer for the storage of phosphate and not polyphosphate. However, it should be noted that even the detection of polyphosphate would not have disproven our observation that *Hfx. volcanii* uses genomic DNA as a phosphate storage polymer.

During phosphate starvation other phosphate containing biomolecules are produced from genomic DNA, not from rRNA

The results showed that during growth under phosphate starvation *Hfx. volcanii* dramatically decreased its chromosome copy number from about 30 to only 2, suggesting that it uses genomic DNA as a phosphate storage polymer. Another possible source of phosphate might be ribosomal RNA. The numbers of ribosomes per cell are influenced by parameters like growth rate and it can vary widely, both in *E. coli* [19] and in *Hfx. volcanii* [20].

A

B

C

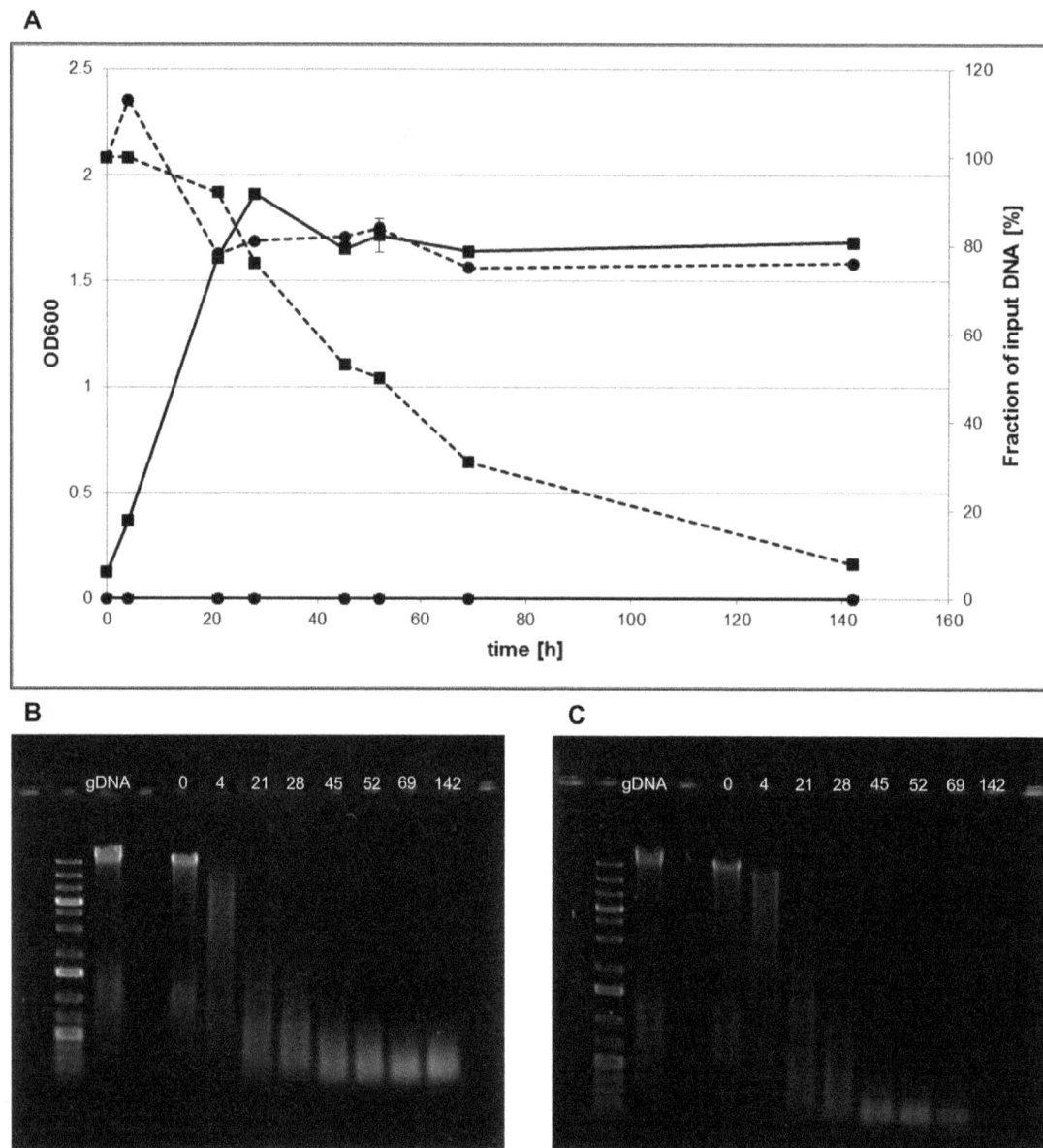

Figure 2. *Hfx. volcanii* **consumes high molecular weight chromosomal DNA.** Three *Hfx. volcanii* cultures were grown in synthetic medium with chromosomal DNA as sole source of phosphorous and a growth curve was recorded (solid line, squares). As negative controls three non-inoculated cultures were incubated under identical conditions (solid line, circles). At the indicated times the optical densities were recorded and aliquots were removed for the quantification of the DNA content. Average optical densities and their standard deviations are shown (solid lines). The cells were pelleted by centrifugation and the DNA content of the supernatants was analyzed by analytical agarose gel electrophoresis (compare B and C). The DNA concentration was quantified using ImageJ, and average values and their standard deviations are shown (dotted lines, circles for the mock-treated non-inoculated control, squares for the inoculated culture). **B.** The supernatants of the aliquots of non-inoculated negative control cultures were dialyzed to remove salts and analyzed by analytical agarose gel electrophoresis. One representative gel is shown. For comparison the input DNA (gDNA) and a size marker (1 kb plus) were included. **C.** The supernatants of the aliquots of cultures grown with genomic DNA as phosphate source were dialyzed to remove salts and analyzed by analytical agarose gel electrophoresis. One representative gel is shown. For comparison the input DNA (gDNA) and a size marker (1 kb plus) were included.

Therefore, for a better understanding of the phosphate balance of cells during phosphate starvation, also the number of ribosomes was quantified.

Hfx. volcanii cultures were again grown in the absence of added P. The cell density was quantified and increased from 3.22×10^7 cells ml^{-1} to 2.70×10^8 cells ml^{-1}. This is an 8.4-fold increase in cell number, which is in excellent agreement with the 8.5-fold increase in OD$_{600}$ observed in previous experiments

(compare Fig. 3). The number of ribosomes prior to and after phosphate starvation was quantified using a previously described approach [20]. Cells of the preculture grown in complex medium contained 29250 ribosomes (SD 1290, n = 3), a number similar to the number of 26000 ribosomes per cell determined earlier [20]. Stationary phase cells after phosphate starvation contained on average 3290 ribosomes (SD 97, n = 3). This is an 8.8-fold reduction of the number of ribosomes per cell during phosphate

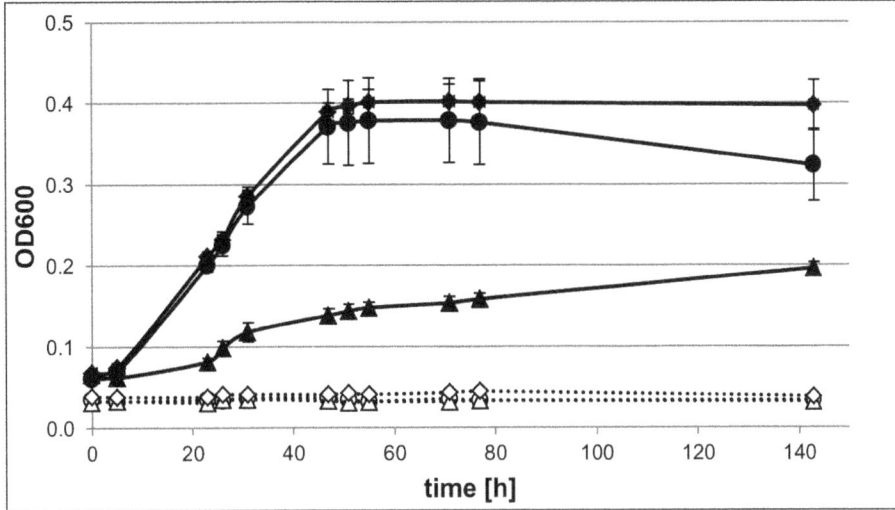

Figure 3. Comparison of growth with different phosphate concentrations and with DNA. *Hfx. volcanii* was grown in microtiter plates in synthetic medium in the absence of any added phosphate source (triangles) and in the presence of, respectively, 1 mM (standard concentration, solid circles) and 10 mM phosphate (diamonds). Non-inoculated sterile controls were also incubated (dotted lines). Growth was followed by measuring the optical density at 600 nm. Average values of three independent cultures and their standard deviations are shown.

starvation, a value that is very similar to the 8.4-fold increase in cell number during phosphate starvation. Together these results revealed that ribosomal RNA is neither source nor sink of phosphate during phosphate starvation, but that ribosomes are distributed among the daughter cells and that the phosphate content bound in rRNA is self-sufficient during phosphate starvation.

Hfx. volcanii harbors not only the major chromosome, but also three additional small chromosomes and a very small plasmid. To enable a comprehensive comparison of the total amount of phosphate bound in rRNA and in DNA, three independent cultures were again grown in the absence of added phosphate and

the numbers of four replicons were quantified prior to and after phosphate starvation (the replicon pHV2 is a very small plasmid that is not present in strain H26). The results are summarized in Table 1. As expected, the numbers of all replicons were severely reduced after phosphate starvation. It was revealed that a polyploid *Haloferax* cell growing exponentially in complex medium contains about 2.2×10^8 molecules of phosphate in its DNA. With approximately 4600 nucleotides per ribosome and 29250 ribosomes per cell, the estimated total amount of phosphate in rRNA is 1.2×10^8 molecules per cell. Thus in the polyploid *Hfx. volcanii* the amount of phosphate bound in DNA is about twice that

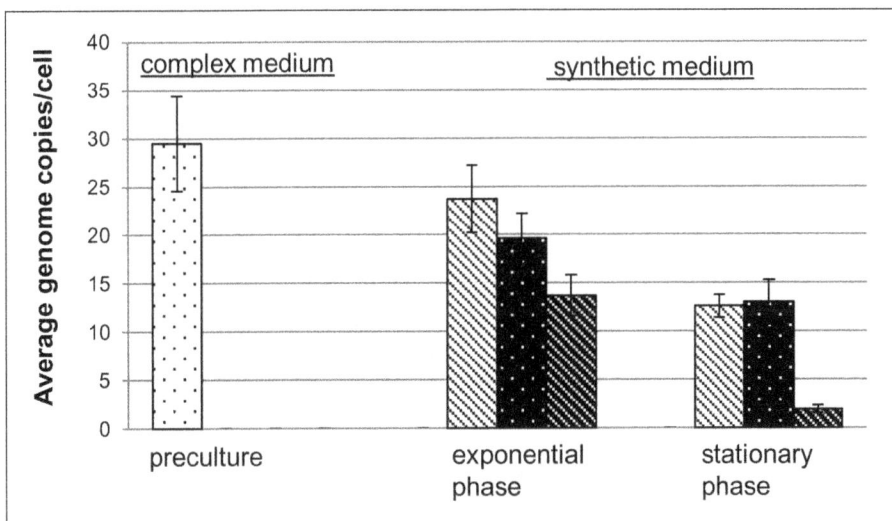

Figure 4. Chromosomal copy numbers during and after growth with and without added phosphate. *Hfx. volcanii* was grown in synthetic medium in the presence of 10 mM and 1 mM phosphate and in the absence of phosphate, respectively. Aliquots were removed during mid-exponential growth phase and at stationary phase (compare text). An aliquot from the pre-culture used for inoculation was also included. Cells were harvested by centrifugation and the chromosome copy number was quantified using Real Time PCR. Three biological replicates were performed and average values and standard deviations are shown, from left to right 10 mM phosphate, 1 mM phosphate, and no externally added phosphate.

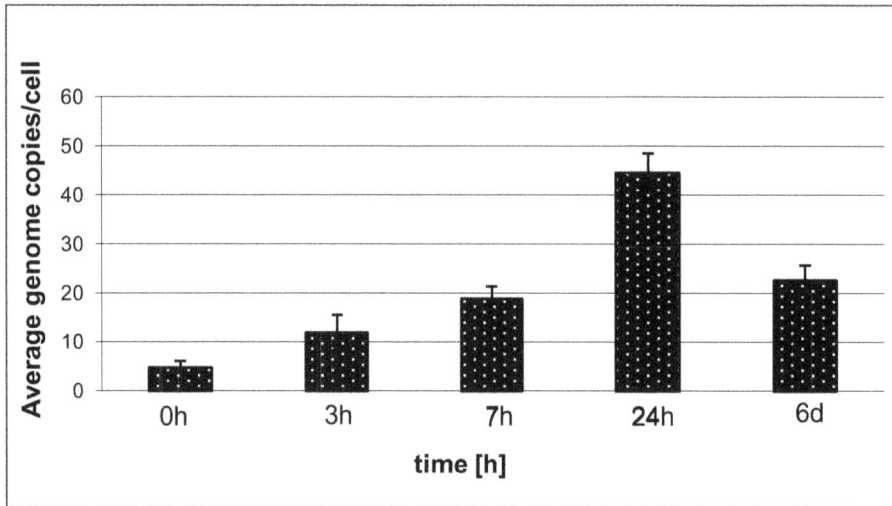

Figure 5. Chromosome copy numbers after re-addition of phosphate to starved cells. Stationary phase, phosphate-starved, chromosome-depleted cells were resuspended in medium containing 1 mM phosphate. At various times, as indicated, aliquots were removed and the chromosome copy number was determined using Real Time PCR. Three biological replicates were performed and average values and standard deviations are shown.

bound in ribosomes. This is contrast to monoploid species, which contain more phosphate in rRNA than in genomic DNA.

Fig. 6 summarizes the balancing of phosphate during growth of *Hfx. volcanii* under phosphate starvation, which is based on the quantification of the numbers of cells, chromosomes and ribosomes. Taken together, the results revealed that the numbers of rRNA-bound phosphate molecules are identical prior to and after phosphate starvation, and thus ribosomal RNA is neither source nor sink of phosphate. In contrast, only about 2/3 of the phosphate that was DNA-bound prior to phosphate starvation was still found in chromosomes after starvation. This indicates that 1/3 of the chromosomes had been degraded and suggests that they were the source of intracellular phosphate for the production of other phosphate-containing biomolecules, e.g. phospholipids, phosphoproteins, phosphosugars, ATP, $NADP^+$, etc. Thus it seems that the polyploid *Hfx. volcanii* uses chromosomal DNA as a phosphate storage polymer in two different ways: 1) cell division in the absence of replication is enabled by distribution of preexisting chromosomes to the daughter cells, and 2) chromosomal DNA is degraded to liberate phosphate needed for other biomolecules that do not have a storage pool in the cell.

Growth during phosphate starvation diminishes genetic advantages of polyploidy

Various genetic advantages of polyploidy for haloarchaea have been proposed [21]. The severe reduction of the ploidy level during growth under phosphate starvation indicates that *Hfx. volcanii* might diminish genetic advantages in favor of cell density increase. To experimentally test this hypothesis, the desiccation resistances of cultures with different chromosome copy numbers were quantified. The desiccation resistance of two types of cultures was analyzed, 1) cultures grown in synthetic medium with casamino acids and 1 mM phosphate to exponential phase, which contained 20 copies of the chromosome, and 2) cultures grown in the absence of phosphate, which contained 2 copies of the chromosome. Both types of cultures were exposed to a 12 day desiccation period. Prior to and after desiccation, the numbers of colony forming units were quantified, and the survival rates were calculated. Fig. 7 shows that 40% of the polyploid cells survived desiccation, while only 8% of the diploid cells survived desiccation. Thus the reduction of the ploidy level from 20 to 2 was accompanied by a fivefold reduction in desiccation resistance. These results show that during phosphate starvation *Hfx. volcanii* diminishes at least one genetic advantage of polyploidy and instead

Table 1. Phosphate content of the four *Hfx. volcanii* chromosomes before and after growth in the absence of phosphate.

Increase in cell No.		1×		8.4×	
Replicon	Size [bp]	Ploidy before.	P atoms	Ploidy after	P atoms
Chromosome	2.85×10^6	30	1.7×10^8	2	9.6×10^7
pHV1	8.51×10^4	21	3.5×10^6	2	2.9×10^6
pHV3	4.38×10^5	26	2.3×10^7	4	2.9×10^7
pHV4	6.36×10^5	17	2.1×10^7	2	2.1×10^7
sum [bp]	4.0×10^6	sum P atoms	2.2×10^8		1.5×10^8

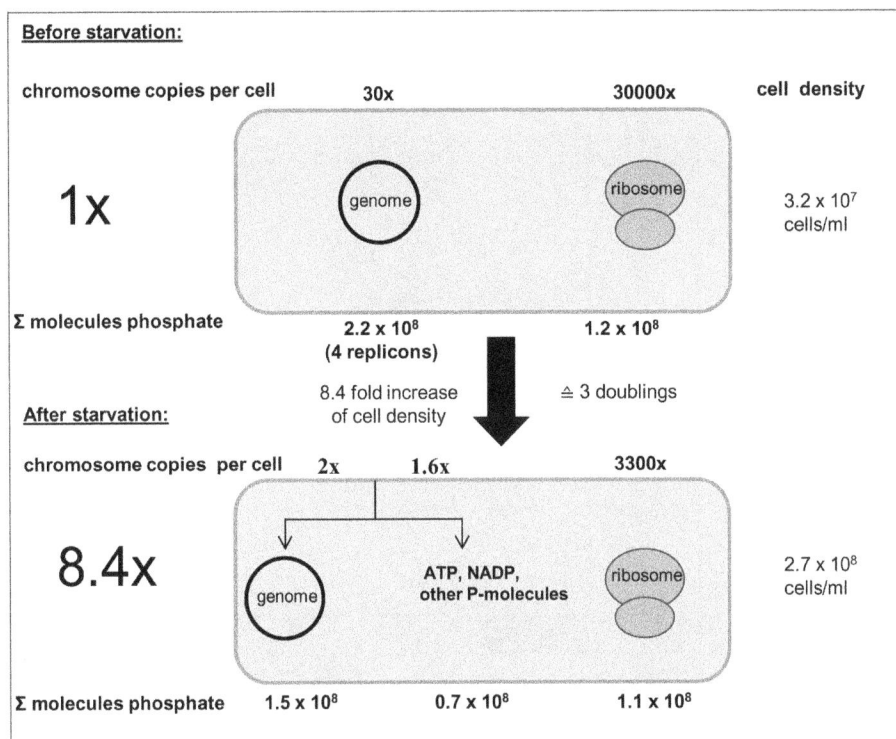

Figure 6. Phosphate balance in cells prior to and after growth in the absence of external phosphate. A preculture in complex medium was grown to mid-exponential phase. Aliquots were harvested, washed, and used to inoculate synthetic medium lacking any added phosphate source. Aliquots were removed at the beginning of the experiment and after growth in the absence of phosphate ceased. The cell densities were quantified using a counting chamber, the genome copy numbers were quantified by Real Time PCR, and the numbers of ribosomes were quantified after RNA isolation and two DNase treatments as described in the text. The figure gives a schematic overview of the phosphate balance prior to and after growth during phosphate starvation.

drastically reduces the copy number to enable sustained growth in the absence of external phosphate.

Discussion

Environmental genomic DNA as a nutrient source

DNA is an information storage polymer for all cells, yet for some species of prokaryotes this molecule also has a nutritional role. We could show that *Hfx. volcanii* can grow on external (environmental) genomic DNA as a sole source of carbon, nitrogen, or phosphate. This has also been observed for other species, e.g. the bacterium *Shewanella* [22] as well as for the cyanobacterium *Synechocystis* sp. strain PCC6803 (Zerulla and Soppa, unpublished data). Various additional bacteria can use DNA as a nutrient source [23]. Enrichment cultures with DNA as the sole source of C, N, P, and energy have been successful and resulted in the isolation of assemblages of diverse bacteria [24].

DNA is present in a variety of environments around the world with concentration from <1 µg/l to more than 70 µg/l [25–27]. Nitrogen and phosphorus are limiting nutrients in a variety of ecosystems and significant amounts of dissolved environmental phosphorus is often locked in DNA indicating that it is an excellent extracellular source of phosphorus [28].

Genomic DNA as an intracellular phosphate storage polymer

Many evolutionary advantages of polyploidy exist, and for haloarchaea alone nine different evolutionary advantages have recently been proposed, including low mutation rate, high

resistance to desiccation, gene redundancy, survival over a geological time scale and shifting gene regulation from stochastics to statistics [21]. Here we add an additional evolutionary advantage of polyploidy, i.e. the usage of genomic DNA as a storage polymer for phosphate. The results suggest that *Hfx. volcanii* uses genomic DNA as a phosphate storage polymer under phosphate starvation in two different ways, 1) the high copy number of chromosomes enables an eightfold increase in cell density, which is equivalent to three doublings, in the absence of replication, and 2) about 1/3 of the chromosomes are degraded to liberate phosphate needed for the synthesis of other phosphate containing biomolecules (see Fig. 6). It should be noted that the latter result is indirect and was calculated from the numbers of cells and chromosomes prior to and after phosphate starvation. If the preexisting chromosomes would only be distributed to the daughter cells and no DNA degradation would take place, the 8.4-fold increase in cell number would lead to an average copy number of 3.6 chromosomes per cell (Fig. 4). The experimentally determined real number of chromosomes per cell after starvation was 2.0 (very low standard deviation, see Fig. 4). To test the statistical significance of the difference between these two values, an unpaired t-test was performed. As three biological replicates were measured and for each replicate four different dilutions were quantified in duplicates, each value rests on 24 technical replicates. The P value was found to be 1.8×10^{-11} and thus the difference between these two values is highly significant, strongly suggesting that indeed about 1/3 of the chromosomes were degraded to liberate phosphate for other phosphate containing biomolecules.

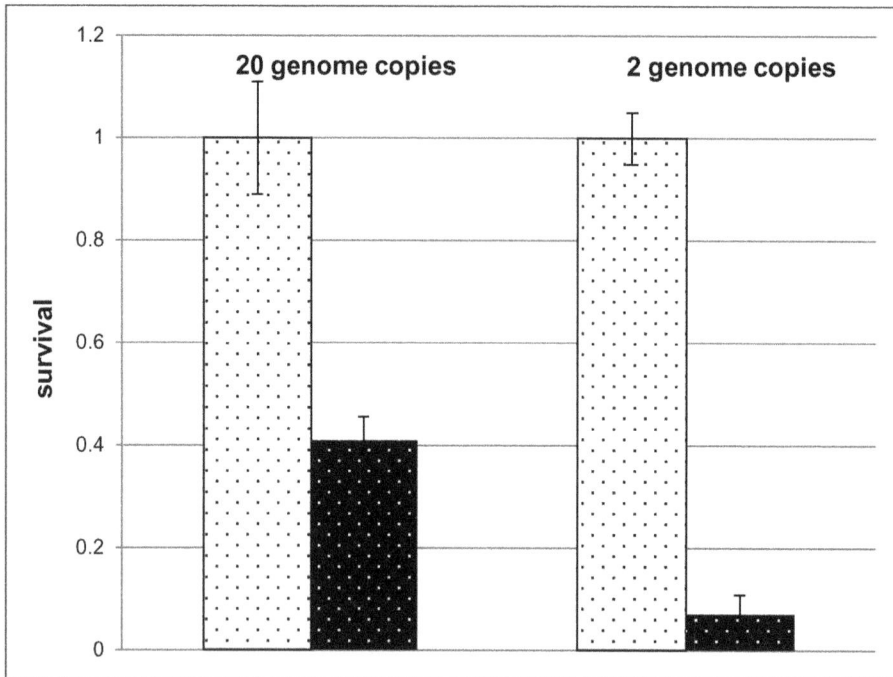

Figure 7. Desiccation resistances of cells of with different ploidy levels. Cultures were grown to mid-exponential phase in synthetic medium with casamino acids and 1 mM K_2HPO_4 to generate cells with 20 copies of the chromosome. For comparison, cultures were grown in synthetic medium in the absence of phosphate to generate cells with 2 copies of the chromosome. Both types of cells were exposed to a desiccation period of 12 days. Colony forming units (CFU) were quantified before and after desiccation, and the survival rates were calculated. Average results of three independent experiments and their standard deviations are shown. Left columns, prior to desiccation, right columns, after desiccation.

Notably, this was not the case for phosphate bound in ribosomes. The number of ribosomes per cell is severely decreased during growth under phosphate starvation (from 30000 to 3300). However, ribosomes are only distributed to the descendent cells and the sum of rRNA-bound phosphate molecules stays constant. Therefore, during growth under phosphate starvation ribosomes are neither a source nor a sink of phosphate, in contrast to genomic DNA. It should be noted that in polyploid species the fraction of phosphate bound in DNA is higher than that in RNA (see Fig. 6). In contrast, in monoploid species more phosphate is bound in RNA than in DNA. For example, *E. coli* cells growing with a generation time of 24 minutes contain 3.9×10^8 molecules of phosphate bound to RNA, but only 3.0×10^7 molecules of phosphate bound to DNA [19].

Chromosomes and ribosomes have high copy numbers that can be severely reduced during growth under phosphate starvation. This does not seem to be possible for other phosphate-containing biomolecules, e.g. phospho-lipids or ATP. The membrane of haloarchaea contain a high fraction of phospholipids [29,30], therefore, the 8.4-fold increase in cell number observed during phosphate starvation depends on extensive phospho-lipid synthesis. Similarly, it seems unlikely that the cells can grow with 8.4-fold reduced levels of ATP, $NADP^+$, etc. Therefore, the cells must have an internal phosphate storage pool to explain the observed growth in the absence of external phosphate. As phosphate was not liberated from ribosomes, polyphosphate was not detected, and no other phosphate-storage polymer is known in haloarchaea, this is another indication that genomic DNA was indeed used as a phosphate storage polymer that was partially degraded to liberate phosphate.

Recently it became clear that polyploid prokaryotes are no seldom exceptions, but that polyploidy is widespread in archaea and bacteria [4–9,31]. Therefore, it might well be that the usage of genomic DNA as intracellular nutrient storage polymer is not limited to *Hfx. volcanii*, but is present in many additional species. The analysis of selected additional polyploid species is currently under way.

That DNA has roles other than storing genetic information is exemplified by its stabilizing function in biofilms. Examples have been reported that DNA is essential for biofilm formation and that mutants unable to export DNA lose the ability to form biofilms [32]. Also haloarchaea including *Hbt. salinarum* and *Hfx. volcanii* have been shown to export DNA when they form biofilms [33]. Not only nature makes use of DNA that exceeds its role as genetic material, the many applications of "DNA origami" [34] demonstrate that DNA is an organic polymer that might have a bright future in bionanotechnology.

Growth during phosphate starvation diminishes genetic advantages of polyploidy

During phosphate starvation *Hfx. volcanii* dramatically decreases the number of chromosomes and in fact gives up polyploidy. This led to the prediction that the various genetic advantages of polyploidy should be lost at the end of growth in the absence of phosphate. This prediction was tested using one evolutionary advantage of polyploidy, i.e. the resistance to desiccation. It could indeed be shown that phosphate-starved cells with two chromosomes are five-fold more sensitive to desiccation than cells with the normal number of about 20 chromosomes (compare Fig. 7), and thus it could be experimentally verified that *Hfx. volcanii* diminishes at least one genetic advantage of polyploidy upon phosphate starvation. It will be interesting to investigate whether also other species prefer to grow under phosphate starvation and give up genetic advantages of polyploidy, or whether different strategies exist.

An alternative explanation for the evolutionary origin of DNA: a hypothesis

The distribution of polyploidy in archaea and bacteria indicates that polyploidy has evolved independently at different times for different reasons in various phylogenetic groups, and thus recent species would have developed polyploidy rather late in evolution. The universal conversation of RecA (RadA/Rad51) in all life forms indicates that homologous recombination is very ancient, and this might be also true for the initial development of all advantages of polyploidy that require homologous recombination. Nevertheless, polyploidy might even predate the invention of homologous recombination. Thus, we suggest that the evolutionary origin of polyploidy was initiated by a need to store phosphate intracellularly in a safe and readily utilizable form. In essence, we propose that the usage of DNA as a phosphate storage polymer might by far predate all other evolutionary advantages of polyploidy and that in fact the first "polyploid" cell had many "genome" copies without using DNA as genetic material.

Many controversial theories about the origin of life and the evolution to a free-living modern-type of cell exist, ranging from pyrite catalyzed early metabolism to a *bona fide* "RNA world". However, all of these theories agree that RNA by far predates DNA as the molecule encoding heritability. The current concept is that DNA evolved because its higher stability compared to RNA made it the material of choice to store genetic information, and DNA genomes replaced the previous RNA genomes. Given our evidence for DNA's dual nature as a molecule for information and phosphorus storage, we propose the parsimonious argument that the development of DNA might have stemmed from the need to store phosphate intracellularly, and then added the role of genetic storage once mechanisms for template-based replication and transcription were invented. The driving force for the early development of DNA as storage polymer would have been its much greater stability in comparison to an alternative phosphate storage polymer, e.g., polyphosphate. Hydrolysis of polyphosphate is exergonic, and its stability is highly influenced by pH and temperature [35]. While at neutral pH and room temperature polyphosphate is reasonably stable and complete hydrolysis requires several months, at acidic pH or elevated temperatures it occurs within hours [35]. Notably, even at neutral pH and room temperature DNA is much more stable than polyphosphate. Of course the proposal that the functionality as a storage polymer might have been the driving force in the development and acquisition of DNA during the early evolution of life is nothing else than a hypothesis. While it might not be true and cannot be proven, it opens an alternative view on possible developments in the pre-DNA world.

Methods

Haloarchaeal strain and media

Haloferax volcanii strain H26 was kindly provided by Thorsten Allers (University of Nottingham, UK). It was and grown in complex medium [36] or in synthetic medium [37] supplemented with 8 μM FeSO$_4$ (Roth, P015.1), 0,1% (v/v) SL-6 trace element solution [38] (all from Roth), 1 ml vitamin solution (Sigma Aldrich, B6891), 50 μg ml^{-1} uracil (Applichem, A0667) and 100 mM MOPS pH 7.2 (Sigma Aldrich, M3183). All components of the synthetic medium were of the grade "per analysis" and thus free of phosphate, e.g. K$_2$HPO$_4$ (Roth, 6878.2), NH$_4$Cl (Applichem, A0988), glucose (Merck, 1083441000), NaCl (Roth,

3957.5), MgCl$_2$ (Roth, 2189.1), MgSO$_4$ (Applichem, A1037), KCl (Roth, 6781,1), CaCl$_2$ (Applichem, A3587), and Tris (A1086). If not otherwise stated, the synthetic medium was also supplemented with 0.5% (w/v) glucose as a C source, 10 mM NH$_4$Cl as a N source, and 1 mM K$_2$HPO$_4$ as a P source. For growth experiments with DNA as a source of P K$_2$HPO$_4$ was omitted and genomic DNA was added to a final concentration of 250 μg/ml. Cultures were grown in Erlenmeyer flasks in a rotary shaker at 42 °C and 250 rpm or in microtiter plates as described below.

Growth in microtiter plates

For growth studies in microtiter plates a preculture was grown in an Erlenmeyer flask in complex medium to exponential phase (OD$_{600}$ 0.4), washed three times in basal salts (medium without a carbon, nitrogen and phosphate source), and resuspended in medium specific to the experiment. The OD$_{600}$ was determined and the cultures were diluted, so that the start OD$_{600}$ in the 150 μl culture volume in the microtiter plate was about 0.025 (which is equivalent to 0.5 using normal cuvettes with a pathlength of 1 cm). Microtiter plates were incubated in an orbital shaker (Heidolph, Schwalbach, Germany) at 42 °C with a shaking velocity of 1100 r.p.m. The OD$_{600}$ values of triplicate cultures was measured at the time points indicated in the respective figures using the microtiter plate photometer Spectramax 340 (Molecular Devices, Ismaning, Germany). Average values of three independent cultures and their standard deviations were calculated.

DNA extraction and purification

Chromosomal DNA isolated from *Hfx. volcanii* H26 cells by spooling and ethanol precipitation (www.haloarchaea.com/resources/halohandbook). DNA was dissolved in 10 mM Tris-Cl solution (pH 8, in DNA-grade water). Purified DNA was used fresh for growth experiments to avoid any subsequent hydrolysis. DNA concentrations were determined prior to supplementation spectrophotometrically using a Nanodrop ND-1000. DNA samples were visualized on agarose gels prior to supplementation and found to be of high molecular weight (data not shown).

Verification of the usage of high molecular weight chromosomal DNA

Hfx. volcanii was grown in Erlenmeyer flasks in synthetic medium with casamino acids (0.25% (w/v)) as carbon and energy source in the presence of 100 μg/ml genomic DNA as phosphate source. The culture was started with an OD$_{600}$ of 0.1 and grown into stationary phase. At various time points 1 ml aliquots were removed and cells were removed by centrifugation (8000 g, 5 min, room temperature). 30 μl aliquots of the supernatants were dialyzed on membrane filters (Millipore, 13 mm diameter, VSWP01300) against distilled water and analyzed by analytical agarose gel electrophoresis.

Quantification of ploidy levels using quantitative real-time PCR

Precultures were grown in complex medium to a cell density of about 2×10^8 cells/ml. Aliquots of 3×10^8 cells were removed and the cells were harvested by centrifugation (4000 rpm, 30 min) and washed in basal salt solution (medium without carbon and phosphate source). The cells were used to inoculate synthetic medium with glucose as carbon source and 10 mM NH$_4$Cl as nitrogen source. K$_2$HPO$_4$ was added as a phosphate source at concentrations described in the respective experiments, or phosphate was omitted. Samples for the quantification of the replicon copy number were collected during exponential phase (without added phosphate: 9.4×10^7 cells/ml; with 1 or 10 mM phosphate: 5.2×10^8 cells/ml) and at stationary phase (without added phosphate: 2.7×10^8 cells/ml; with 1 or 10 mM phosphate: 1.3×10^9 cells/ml). A RT-PCR approach was applied for the determination of chromosome copy numbers as described previously [4]. Standard fragments of about 1 kbp were amplified by PCR using total DNA of *Hfx. volcanii* as template (oligonucleotides see Table S1). Purification of the standard fragment and preparation of the cell extracts were essentially performed as described [38]. 3×10^8 cells were collected by centrifugation, resuspended in 100 µl basal salt solution, lysed by the addition of 900 µl water, and dialyzed on membrane filters (Millipore, 13 mm diameter, VSWP01300) against distilled water. RT-PCR conditions were 10 min 96°C, 40 cycles with 30 s 96°C, 30 s 62°C, 30 s 72°C followed by 5 min 72°C and a melt curve analysis from 62°C to 96°C in 1°C steps. In each case three independent cultures (biological replicates) were analyzed. For each replicate four different dilutions of the extracts were measured in duplicates, so that 24 technical replicates were used to calculate the average copy numbers and their standard deviations. The standard curve was comprised of serial tenfold dilutions of the standard fragment (10^3-fold to 10^8 fold) that were measured in duplicates. It was ensured that all PCR reactions were exponential, i.e. the C_T differences of tenfold dilutions were about 3.3. A negative control with the omission of template DNA was also performed.

Statistical analysis

The experimentally determined average number or genome copies after growth in the absence of phosphate was 2. The expected average value if the chromosomes present in the inoculum would have been distributed to the daughter cells and no degradation would have occurred was 3.6. To unravel whether these two values are significantly different an unpaired t-test was performed using Excel and the respective results of the technical replicates of the three biological replicates.

Analysis of potential polyphosphate formation

Hfx. volcanii was grown in synthetic medium with casamino acids as carbon and energy source in the presence of 1 mM phosphate to mid-exponential growth phase. Cells were harvested by centrifugation and fixed with formaldehyde as described previously [39]. They were stained with DAPI (500 µg/ml) for 10 minutes at room temperature as described by [18] and analyzed by fluorescence microscopy.

Quantification of the number of ribosomes

The number of ribosomes was quantified as described by [20]. In short, total RNA was isolated and residual DNA was removed by two consecutive treatments with RNase-free DNase according to the instructions of the supplier (Qiagen, 79254). Each time the DNase treatment was followed by an ethanol precipitation [40]. The RNA concentration was determined spectroscopically and the fraction of ribosomal RNA and the molecular weight of ribosomal RNA was used to calculate the number of ribosomes per cell [20].

Quantification of desiccation survival

Haloferax volcanii cultures were grown in synthetic medium with casamino acids and with 1 mM K_2HPO_4 to mid-exponential phase or without K_2HPO_4 to stationary phase. Subsequently the cells were washed in basal salt solution and were concentrated 20-fold in basal salt solution. Thereafter the cells were placed onto glass microscope cover slips (2×50 µl per cover slip as 4–5 dots) and allowed to dry completely. For desiccation at ambient pressure, cover slips were placed in an exsiccator with Drierite desiccant (97% $CaSO_4$ and 3% $CoCl_2$) at 37°C for 12 days. Cells were rehydrated by placing cover slips in 5 ml basal salt solution with moderate agitation for 30 minutes at room temperature. The resulting cell suspensions were concentrated by centrifugation with $5000 \times g$ at room temperature for 10 min, resuspended in 1 ml basal salt solution. Serial dilutions were prepared and triplicate aliquots were plated onto solid complex medium to quantify cell survival. Colony forming units were quantified after 4–7 days of incubation at 42°C. Control samples were processed in the same manner immediately after preparation and were not permitted to dry.

Acknowledgments

We thank Thorsten Allers from the University of Nottingham for the *Haloferax* strains and plasmids, Adit Naor from Tel Aviv University for discussion of the results and Olga Zhaxybayeva from Dartmouth University, Peter Gogarten of the University of Connecticut and Julie Maupin-Furlow of the University of Florida for comments on the manuscript.

Author Contributions

Conceived and designed the experiments: KZ SC UG RTP JS. Performed the experiments: KZ SC. Analyzed the data: KZ SC DN RTP JS. Wrote the paper: KZ SC RTP JS.

References

1. Crow JF, Kimura M (1965) Evolution in sexual and asexual populations. Am Nat. 99: 439–450.
2. Otto SP, Goldstein DB (1992) Recombination and the evolution of diploidy. Genetics 131: 745–751.
3. Cavalier-Smith T (1978) Nuclear volume control by nucleoskeletal DNA, selection for cell volume and cell growth rate, and the solution of the DNA C-value paradox. J Cell Sci 34: 247–278.
4. Breuert S, Allers T, Spohn G, Soppa J (2006) Regulated polyploidy in halophilic archaea. PLoS One 1: e92.
5. Griese M, Lange C, Soppa J (2011) Ploidy in cyanobacteria. FEMS Microbiol Lett 323: 124–131.
6. Pecoraro V, Zerulla K, Lange C, Soppa J (2011) Quantification of ploidy in proteobacteria revealed the existence of monoploid, (mero-)oligoploid and polyploid species. PLoS One 6: e16392.
7. Soppa J (2011) Ploidy and gene conversion in Archaea. Biochem Soc Trans 39: 150–154.
8. Mendell JE, Clements KD, Choat JH, Angert ER (2008) Extreme polyploidy in a large bacterium. Proc Natl Acad Sci U S A 105: 6730–6734.
9. Slade D, Lindner AB, Paul G, Radman M (2009) Recombination and replication in DNA repair of heavily irradiated *Deinococcus radiodurans*. Cell 136: 1044–1055.
10. Zahradka K, Slade D, Bailone A, Sommer S, Averbeck D, et al (2006) Reassembly of shattered chromosomes in *Deinococcus radiodurans*. Nature 443: 569–573.
11. Kottemann M, Kish A, Iloanusi C, Bjork S, DiRuggiero J (2005) Physiological responses of the halophilic archaeon *Halobacterium* sp. strain NRC1 to desiccation and gamma irradiation. Extremophiles 9: 219–227.
12. Mancinelli RL, White MR, Rothschild LJ (1998) Biopan-survival I: exposure of the osmophiles *Synechoccus* sp. (Nageli) and *Haloarcula* sp. to the space environment. Adv Space Res 22: 327–334.
13. Nelson-Sathi S, Dagan T, Landan G, Janssen A, Steel M, et al (2012) Acquisition of 1,000 eubacterial genes physiologically transformed a methanogen at the origin of Haloarchaea. Proc Natl Acad Sci U S A 109: 20537–20542.

14. Williams D, Gogarten JP, Papke RT (2012) Quantifying homologous replacement of loci between haloarchaeal species. Genome Biol Evol 4: 1223–1244.

15. Papke RT, Zhaxybayeva O, Feil EJ, Sommerfeld K, Muise D, et al (2007) Searching for species in haloarchaea. Proc Natl Acad Sci U S A 104: 14092–14097.

16. Rosenshine I, Tchelet R, Mevarech M (1989) The mechanism of DNA transfer in the mating system of an archaebacterium. Science 245: 1387–1389.

17. Naor A, Lapierre P, Mevarech M, Papke RT, Gophna U (2012) Low species barriers in halophilic archaea and the formation of recombinant hybrids. Curr Biol 22: 1444–1448.

18. Kulakova AN, Hobbs D, Smithen M, Pavlov E, Gilbert JA, et al (2011) Direct quantification of inorganic polyphosphate in microbiol cells usint 4'-6-dianino-2-phenylidone (DAPI). Enrivon Sci Technol 45: 7799–7803.

19. Bremer H, Dennis P (1996) Modulation of chemical composition and other parameters of the cell by growth rate. In: Escherichia coli and Salmonella. Neidhardt, F (edt.). ASM press Washington, 1553–1569.

20. Zaigler A, Schuster SC, Soppa J (2003) Construction and usage of a onefold-coverage shotgun DNA microarray to characterize the metabolism of the archaeon Haloferax volcanii. Mol Microbiol 48: 1089–1105.

21. Soppa J (2013) Evolutionary advantages of polyploidy in halophilic Archaea. Biochem Soc Trans 41: 339–343.

22. Pinchuk GE, Ammons C, Culley DE, Li SW, McLean JS, et al (2008) Utilization of DNA as source of phosphorous, carbon, and energy by Shewanella spp.: Ecological and physiological implications for dissimilatory metal reduction. Appl. Environ. Microbiol. 74: 1198–1208.

23. Finkel SE, Kolter R (2001) DNA as a nutrient: novel role for bacterial competence gene homologs. J Bacteriol 183: 6288–6293.

24. Lennon JT (2007) Diversity and metabolism of marine bacteria cultivated on dissolved DNA. Appl Environ Microbiol 73: 2799–2805.

25. Karl DM, Bailiff MD (1989) The measurement and distribution of dissolved nucleic acids in aquatic environments. Limnol Oceanogr 34: 543–558.

26. Paul JH, Jiang SC, Rose JB (1991) Concentrations of viruses and dissolved DNA from aquatic environments by Vortex flow filtration. Appl Environ Microbiol 57: 2197–2204.

27. Siuda W, Güde H (1996) Determination of dissolved deoxyribonucleic acid concentration in lake water. Aqu Microbiol Ecol 11: 193–202.

28. Dell'Anno A, Danovaro R (2005) Extracellular DNA plays a key role in deep-sea ecosystem functioning. Science 309: 2179.

29. Kates M (1993) Biology of halophilic bacteria, Part II. Membrane lipids of extreme halophiles: biosynthesis, function and evolutionary significance. Experientia 12:1027–36.

30. Koga Y, Morii H (2007) Biosynthesis of ether-type polar lipids in Archaea and Evolutionary considerations. Microbiol Mol Biol Rev. 71(1): 97–120.

31. Hildenbrand C, Stock T, Lange C, Rother M, Soppa J (2011) Genome copy numbers and gene conversion in methanogenic archaea. J Bacteriol 193: 734–743.

32. Whitchurch CB, Tolker-Nielsen T, Ragas PC, Mattick JS (2002) Extracellular DNA required for bacterial biofilm formation. Science 295: 1487.

33. Fröls S, Dyall-Smith M, Pfeifer F (2012) Biofilm formation by haloarchaea. Environ Microbiol 14: 3159–3174.

34. Kuzuya A, Sakai Y, Yamazaki T, Xu Y, Komiyama M (2011) Nanomechanical DNA origami 'single-molecule beacons' directly imaged by atomic force microscopy. Nat Commun 2: 449.

35. Rashchi F, Finch JA (2000) Polyphosphates: A review their chemistry and application with particular reference to mineral processing. Minerals Engineering 13: 1029–1035.

36. Allers T, Ngo HP, Mevarech M, Lloyd RG (2004) Development of additional selectable markers for the halophilic archaeon Haloferax volcanii based on the leuB and trpA genes. Appl Environ Microbiol 70: 943–953.

37. Nieuwlandt DT, Daniels CJ (1990) An expression vector for the archaebacterium Haloferax volcanii. J Bacteriol 172: 7104–7110.

38. Lange C, Zerulla K, Breuert S, Soppa J (2011) Gene conversion results in the equalization of genome copies in the polyploid haloarchaeon Haloferax volcanii. Mol Microbiol 80: 666–677.

39. Hermann U, Soppa J (2002) Cell cycle-dependent expression of an essential SMC-like protein and dynamic chromosome localization in the archaeon Halobacterium salinarum. Mol Microbiol 46: 395–409.

40. Sambrook J, Fritsch EF, Maniatis T (1989) Molecular cloning: a Laboratory manual. 2nd edition. Cold Spring Harbor, N.Y. Cold Spring Harbor Laboratory.

15

Joint Evolution of Kin Recognition and Cooperation in Spatially Structured Rhizobium Populations

Peter C. Zee***¤**, **James D. Bever**

Department of Biology, Indiana University, Bloomington, Indiana, United States of America

Abstract

In the face of costs, cooperative interactions maintained over evolutionary time present a central question in biology. What forces maintain this cooperation? Two potential ways to explain this problem are spatially structured environments (kin selection) and kin-recognition (directed benefits). In a two-locus population genetic model, we investigated the relative roles of spatial structure and kin recognition in the maintenance of cooperation among rhizobia within the rhizobia-legume mutualism. In the case where the cooperative and kin recognition loci are independently inherited, spatial structure alone maintains cooperation, while kin recognition decreases the equilibrium frequency of cooperators. In the case of co-inheritance, spatial structure remains a stronger force, but kin recognition can transiently increase the frequency of cooperators. Our results suggest that spatial structure can be a dominant force in maintaining cooperation in rhizobium populations, providing a mechanism for maintaining the mutualistic nodulation trait. Further, our model generates unique and testable predictions that could be evaluated empirically within the legume-rhizobium mutualism.

Editor: Angel Sánchez, Universidad Carlos III de Madrid, Spain

Funding: This research was supported by the Indiana University National Institutes of Health Genetics, Cellular, and Molecular Sciences Training Grant T32-GM007757 (PCZ), and NSF DEB-0919434, National Science Foundation DEB-1050237 and NIH-5 R01 GM092660 (JDB). The funders had no role in study design, data collection and analysis, decision to publish, or preparation of the manuscript.

Competing Interests: The authors have declared that no competing interests exist.

* E-mail: peterzee@stanford.edu

¤ Current address: Department of Biology, Stanford University, Stanford, United States of America

Introduction

The evolution and maintenance of cooperative traits in nature is a central question in evolutionary biology. Hamilton's rule [1] states that for an altruistic trait to evolve, the costs to the actor must be outweighed by the benefits to recipients. However, this benefit must be weighted by their genetic relatedness ($C<rB$). Two ways that Hamilton's rule can be satisfied are if the population is structured [2,3] or if the actors can recognize and direct beneficial behaviours to genetically similar individuals [4,5].

Population structure may play a major role in the evolution of cooperation. In a population structured by local dispersal (e.g., in a viscous environment), neighbouring individuals are more likely to share a common ancestry than in a fully mixed environment. This viscosity allows the kin of a cooperative individual to receive more benefit than unrelated individuals. Conversely, in an unstructured environment, the benefits of the cooperative behaviour are equally likely to affect the fitness of any genotype. Experimental studies have shown that cooperative traits are lost during extended evolution in unstructured environments [6] and favoured in a structured environment [7].

Kin recognition allows organisms to discriminate behaviour towards kin and may enhance the likelihood of cooperative traits evolving. If individuals can preferentially direct cooperative behaviours toward kin, unrelated individuals will not receive the benefit, resulting in lower fitness. The role of kin recognition in nature is unclear and debated in the literature [8–10]. These kin recognition mechanisms often entail a cost of expression [11].

Here, we present a model to investigate the joint effects of spatial structure and kin recognition on the evolution of cooperation, allowing us to disentangle the relative contribution of each. To ground our study in a biological system, we modelled intraspecific cooperation among rhizobia in the biological context of the interspecific plant-microbe mutualism. Within this mutualism, there is an important component of intraspecific cooperation within the bacterial population, as the rewards of nitrogen fixation are potentially available to many bacterial individuals [12–14]. Rhizobia are an ideal biological system for our study because: *(i)* the rhizosphere is a spatially structured environment [15]; *(ii)* rhizobia have a greenbeard-like recognition mechanism (rhizopines) [16]; *(iii)* the nodule environment locally increases the carrying capacity of rhizobium populations, assuaging concerns regarding the strength of local competition [17,18].

Biological scenario: *the rhizobium – legume mutualism*

Rhizobia are soil bacteria that engage in a mutualistic interaction with leguminous plants, for which they can fix nitrogen otherwise unavailable to the plant. In this resource exchange mutualism, the bacteria receive carbon from the host plant. Rhizobia cells infect plant root cells, where they differentiate into bacteroids inside a tumor-like growth on the root called a nodule. In these nodules, bacteroids fix atmospheric nitrogen in exchange for carbon, but also stimulate the plant to release nutrient rich resources into the surrounding rhizosphere [19,20].

Rhizobia carry a locus encoding the ability to nodulate plant roots on extrachromosomal, symbiotic plasmids [21]. Cells with functional alleles (*Nod+*) will infect plant roots, differentiate into bacteroids, and fix nitrogen. These are mutualists. Nodulation not only offers the plant benefit, but also offers an indirect intraspecific benefit to other rhizobia because it increases local resources which benefit both free-living cells in close proximity to the nodule and undifferentiated cells within the nodule (henceforth called '*adjacent*'). Hence, *Nod+* bacteria are also cooperators. We recognize two potential costs of functional *Nod+*genes: energetic costs of plasmid carriage and sterilization (as differentiation can be terminal) or growth rate inhibition due to differentiation into bacteroids. In this work, we focus on the energetic carriage cost of the plasmid (See Text S1 for a brief discussion on costs of nodulation). As *Nod−* bacteria receive the exuded benefits of nodulation, but do not pay the costs, they disproportionally benefit from nodulation (i.e., "cheaters").

We follow the approach of Bever and Simms [14], and assume a fixed *cost of plasmid carriage*, c_N. The basal fitness of the *Nod−* cells is set to unity, while the basal fitness of mutualistic cells is $(1-c_N)$ in all environments. The fitness of rhizobia cells also depends on their surrounding environment. Population densities of soil microbes are highest in the immediate vicinity of plant roots, where exuded nutrients are concentrated in the soil [19,20]. Nodulation stimulates a local increase in these exudates from roots, allowing for increased bacterial population densities around nodules [17,18]. This increase in resources available to cells adjacent to nodules is the *benefit of nodulation*, b_N, to all rhizobia. The fitness of undifferentiated cells adjacent to nodules is thus increased by a factor of $(1+b_N)$; around un-nodulated roots, fitness is 1 (i.e., the standing, background availability of resources in the soil). As local population density increases, concerns have been raised regarding the potential ability of kin competition to offset the benefits of spatial structure [22]. However, these concerns are misplaced given that carrying capacity locally increases within and around nodules [23,24]. This biological system presents a situation with elastic local carrying capacities, similar to that shown for opines in *Agrobacterium tumefaciens* [18].

In addition to the mutualistic *Nod+* locus, rhizobia can also carry loci for the production and catabolism of rhizopines (*Rhiz*) [25,26]. Rhizopines are carbon-rich compounds produced by the plant after stimulation by nodulating bacteroids [26,27]. Like the benefits of nodulation, rhizopines are available as nutrients to rhizobia within and in close proximity to the nodule. However, non-rhizopine individuals (i.e., *Rhiz−* individuals) are unable to catabolize rhizopines [16], rendering rhizopines a private resource for *Rhiz+* individuals [27]. Effectively, this constitutes a kin recognition system equivalent to a "greenbeard" trait [28–30], where the production of rhizopines is the "greenbeard", and the unique ability for rhizopine catabolism ensures directed benefits. Like *Nod*, the *Rhiz* loci are also carried on an extrachromosomal plasmid, and production is coupled with nodulation [26]. When a *Rhiz+* cell generates a nodule, we assume a *proportion, d, of general root exudates are diverted* to rhizopine production, reducing the resources available to *Rhiz−* cells. In addition, the synthesis and catabolism of rhizopines involve a carbon cost, c, that detracts from the total exudate available to all bacteria. We note that rhizopines have the characteristics of a "spiteful" trait [31] because it is costly for both the cells expressing the trait (c) and the non-rhizopine individuals (d). As with the *Nod* locus, we assume an energetic carriage cost of the *Rhiz+* allele, c_r. *Rhiz+* cells occur in the population at frequency equal to e, while *Rhiz−* have frequency equal to $f (= 1-e)$.

Access to increased exudates from nodulation will depend on the structure of the soil environment. The probability of a nodule forming on a plant root at any given time is dependent on the genotypic constitution of the rhizobium population present at the infection site. In an environment with no spatial structure (i.e., complete mixing), every genotype of reproductive rhizobia is equally likely to be adjacent (either within or in close proximity) to the nodule, and thereby equally likely to receive the benefits of nodulation. Conversely, in a spatially structured environment (i.e., limited mixing), reproductive cells receiving benefits of nodulation will be more likely to be of the same genotype as the nodule-founding bacteroid. We describe the level of environmental mixing with a *coefficient of relationship*, ϕ, between the bacteroid generating a nodule and the rhizobia adjacent to the nodule. This coefficient of relationship can take on values between zero and one, where $\phi = 0$ represents the situation where there is complete mixing (i.e., no spatial structure), and $\phi = 1$ represents a completely viscous environment (i.e., complete spatial structure). When $\phi = 1$, the benefits of nodulation go exclusively to *Nod+* cells; when $\phi = 0$ these benefits are randomly distributed with respect to genotype. Model parameters are summarized in Table 1. Figure 1 displays the distribution of exuded resources in the soil (a), and graphical annotations of costs and benefits (b).

Spatial structure (ϕ) has opposing influences on the evolution of nodulation and rhizopines in rhizobium populations. Bever and Simms [14] showed that in sufficiently structured environments, the legume-rhizobia interspecific mutualism can be maintained through the intraspecific cooperation. However, when mixing in the population became too high, *Nod−* cells are increasingly likely to receive the benefits of nodulation, and the magnitude of the relative benefit to *Nod+* does not outweigh costs of being a mutualist. Negative frequency-dependent dynamics at the *Nod* locus result in a stable internal equilibrium, and introducing spatial structure alters the location of this equilibrium, with increasing spatial structure shifting it towards fixation of *Nod+*. Conversely, Simms and Bever [32] found that the evolution of rhizopines is facilitated in well-mixed populations fixed for *Nod+* (i.e., in a mutualistic population). When the environment is well mixed, the advantage of kin recognition (i.e., rhizopines) is high because it allows for private sharing of resources among *Rhiz+* cells. However, when the environment is highly spatially structured, local groups of cells are likely to be related, thereby eroding the advantage of the directed benefits of rhizopines, and magnifying the cost. This is because increasing spatial structure raises the likelihood of a cell being adjacent to kin, which makes paying a cost to direct rhizopines to kin superfluous because benefits are likely to reach kin with such a mechanism. This translates to positive frequency-dependent dynamics at the *Rhiz* locus, with an unstable internal equilibrium. Increasing spatial structure decreases the equilibrium frequency of *Rhiz+*, thus widening the initial conditions that lead to loss of rhizopines.

Model

We analyze a population genetic model with two di-allelic loci (one for nodulation, *Nod*, and one for rhizopine, *Rhiz*) that are either inherited independently (no linkage; *Unlinked case*) or coinherited (complete linkage; *Linked case*). Prior research has identified spatial structure as being a key determinant of the dynamics at these loci. In this work, we focus on whether the kin recognition system of rhizopines qualitatively alters the evolutionary fate of the mutualism. We focus on the spatial structure term φ because we are primarily interested in how environmental structure influences the evolutionary dynamics in the rhizobium population, and its role in the legume-rhizobia mutualism.

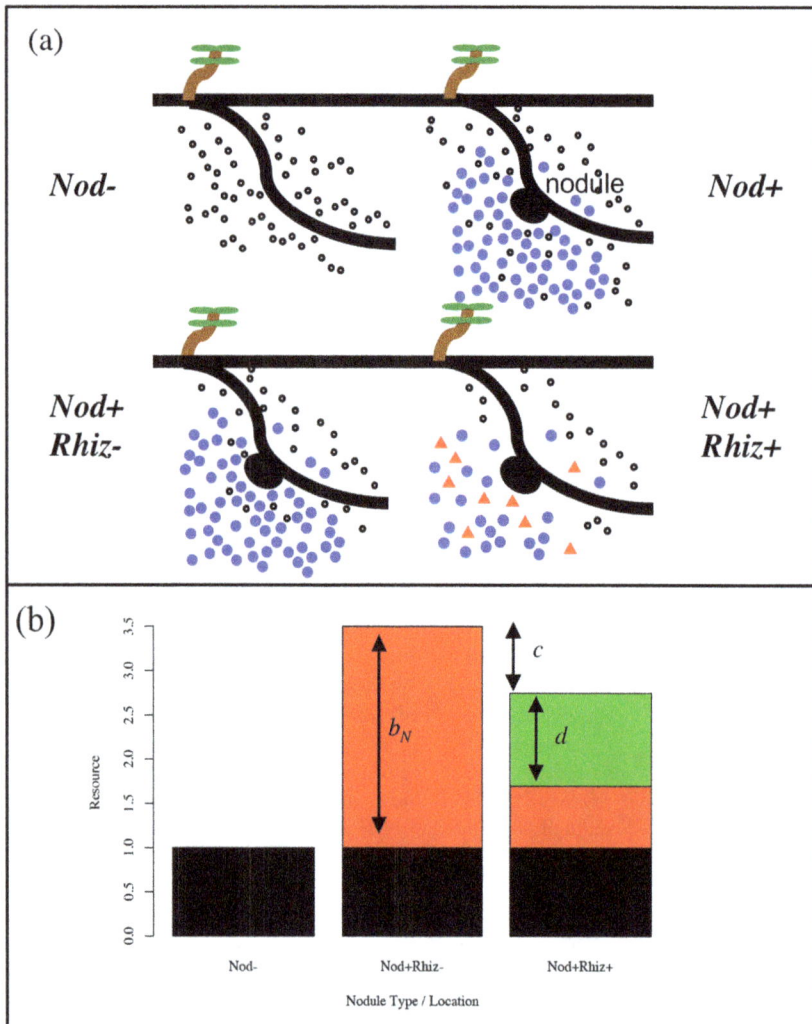

Figure 1. Root exudates and local resource environments. (a) Schematic of root exudates in the model. Small open circles are *general exudates* that are usable by any free-living cells. Blue circles are *nodulation induced* exudates (b_N), also available to all free-living cells. Red triangles are *rhizopines*, which are only available to *Rhiz+* cells. (b) Resources in local environments. Black portions of the bars represent the *general* exudates that are usable by all types. Red portions of bars show general use exudates induced by nodulation. Green portions of the bar represent the *rhizopines*. In the *Nod+Rhiz+* bar, the two costs of *Rhiz* (c and d) can be seen to decrease the induced benefits of nodulation.

Model I: Unlinked case

Independent inheritance of *Nod* and *Rhiz*

We analyze an unlinked model as a heuristic to understand the dynamics of the two loci independently. This approach assumes that dynamics of alleles of the two loci are unconstrained by linkage. While this assumption is biologically unreasonable because plasmids are not independently distributed among bacterial cells and physical linkage likely alters the transitory dynamics, analysis of the unlinked model allows application of analytical tools to capture the qualitative dynamics of the system around the equilibria. This gives us an analytical point of comparison for the more biologically realistic linked model discussed below, and in fact, the equilibria are unaffected by linkage.

We calculate fitness at the *Nod* and *Rhiz* loci as the product of two functions, G and E, that measure the constitutive growth and environment specific growth, respectively. G is a function of the constitutive costs and benefits (i.e., c_N, c_r) of a genotype. These

fitness effects are experienced across all environments. E is a function of genotype frequency, environment-specific costs and benefits (i.e., d, c, b_N), and the level of environmental mixing (φ). Together, these two functions measure the costs and benefits of being in each nodule environment, and the probability of being in each environment. Full exposition of the model can be found in Text S2.

$$Fitness = G(constitutive\ costs\ and\ benefits)$$

$$E(genotype\ freq.,\ env.\ costs\ and\ benefits,\ space)$$

Because the loci are segregating independently, the system of equations can be reduced to two equations by noting that allele frequencies at each locus sum to unity (e.g, $x = (1-y)$ (*Nod* locus) and $e = (1-f)$ (*Rhiz* locus)). This reduction allows us to visualize the dynamics on a standard phase plane, with a zero net growth isocline for each locus. Changes in allele frequencies at both loci can be derived from these fitness equations (Text S2),

Table 1. Description of model parameters.

Parameter	Biological meaning	Model
x	Nod+ frequency	unlinked
$y\ (=1-x)$	Nod− frequency	unlinked
e	Rhiz+ frequency	unlinked
$f\ (=1-e)$	Rhiz− frequency	unlinked
x	Nod+Rhiz+ frequency	linked
y	Nod+Rhiz− frequency	linked
q	Nod−Rhiz+ frequency	linked
z	Nod−Rhiz− frequency	linked
φ	Spatial structure; probability that bacteroids are identical to vegetative cells exterior of nodule	both
b_N	Benefit of nodulation to surrounding cells	both
c_N	Cost of carrying Nod+ allele	both
c_r	Cost of carrying Rhiz+ allele	both
c	Cost of rhizopine production/synthesis; decreased general exudate output	both
d	Amount of exudate produced not usable by Rhiz− (i.e., rhizopines)	both

allowing us to monitor the joint influence of spatial structure and recognition (i.e., rhizopines) on cooperation (i.e., nodulation).

Results I: Unlinked case

First, increasing spatial structure in the population greatly facilitates the evolution of mutualism (Figure 2). With elevated spatial structure undifferentiated Nod+ individuals will tend to be adjacent to nodules and able to receive the full benefits of the root exudates (Figure 2f). Alternatively, as the environment becomes increasingly mixed, nodulation is less likely to evolve (Figure 2a), because the carriage cost associated with nodulation (c_N) is not recovered through the indirect benefits of root exudates because Nod− individuals are increasingly likely to receive the benefit (Figure 2d). This retrieves the result of Bever and Simms [14] modelling the evolution of nodulation in a population without rhizopines.

Second, increasing spatial structure constrains the evolution of rhizopines (Figure 2b). In well-mixed environments, benefits of rhizopines are directed to other Rhiz+ individuals. However, as spatial structure increases and Rhiz+ individuals tend to be clustered, the relative benefits of kin recognition decrease because of a decreasing need for preferential allocation. Without non-rhizopine genotypes to compete with, rhizopines offer no advantage, and Rhiz+ cells suffer the cost of rhizopine production. Additionally, some level of mixing in the environment must occur for the maintenance of the kin recognition system. These results recover those of the Simms and Bever [32] modelling the evolution of rhizopines in a population fixed for Nod+.

Finally, increasing frequencies of rhizopines decrease the equilibrium frequency of Nod+ (illustrated by the negative slope of the Nod isocline in Figure 2b; Figure S1). The dependence of the evolution of cooperation on kin recognition can be derived as the partial derivative of the Nod+ frequency with respect to Rhiz+ frequency:

$$\frac{\partial x}{\partial e} = \frac{-c(1+b_N)}{(1-\phi)(b_N - ce^2(1+b_N))}$$

This expression is always negative as long as costs of producing rhizopines (c) are positive. (This expression is undefined when $\phi = 1$, a biologically unreal scenario). This indicates that, at equilibrium, rhizopines (i.e., kin recognition) are not beneficial to the maintenance of cooperation when not coinherited with the Nod locus. This result emerges because costs of rhizopine production act as an additional cost to the cooperative Nod trait, which leads to a decrease in cooperation at equilibrium.

Model II: Linked case

Coinheritance of Nod and Rhiz

We now consider a more realistic linked model, where the Nod and Rhiz loci are coinherited on the same plasmid; this is equivalent to complete linkage between the two loci. This is more biologically realistic than the unlinked case because these loci are often located on the same symbiotic plasmid in rhizobium species [33,26]. We now follow four distinct two-locus genotypes: Nod+Rhiz+, Nod+Rhiz−, Nod−Rhiz+, and Nod−Rhiz−. The fitnesses of these genotypes are calculated in the same way as in the unlinked case, as the product of the functions G and E, with x, y, q, and z denoting the frequencies of the four genotypes, respectively. A key distinction between the unlinked and linked models is that in the linked case, the costs and benefits of rhizopine production are coinherited with the costs and benefits of the nodulation. This makes it possible for rhizopines to more directly influence the evolution of nodulation, thus altering the transient genotype dynamics.

Unlike the unlinked, this linked model does not lend itself to analytical tractability. However, we use several approaches to understand the qualitative behaviour of the model in relation to the results from the unlinked model. First, we analyze the invasion conditions for each genotype. By assuming fixation of one genotype, we can determine which (if any) of the three remaining genotypes can invade by comparing fitnesses. While this approach allows us to qualitatively determine which genotypes are stable at fixation, it does not reveal any information regarding internal dynamics. We use numerical iterations of the full model to map genotype dynamics over time. Finally, we turn to a weak-selection approximation of the model that allows us to plot internal

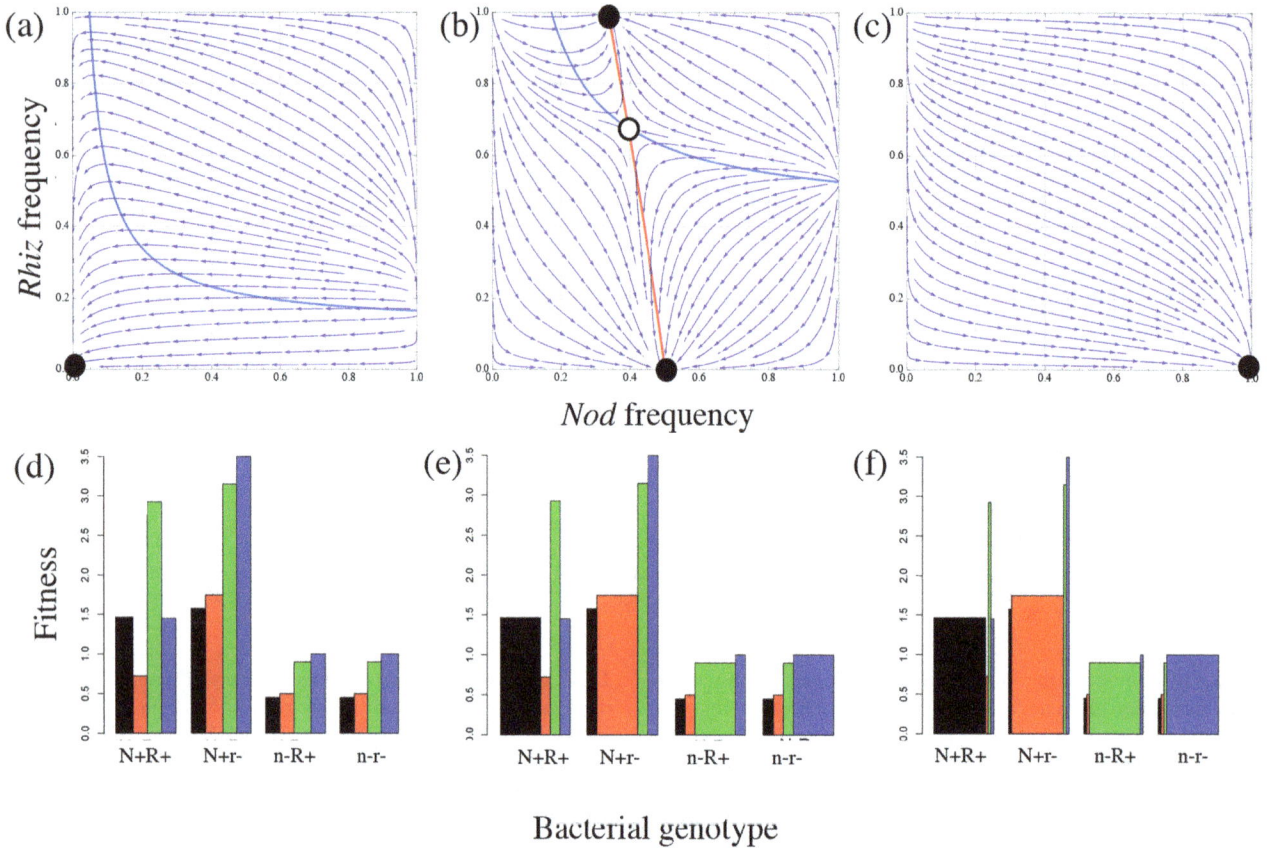

Figure 2. Dynamics and fitness of the unlinked model. (a–c) Isoclines and dynamics. Zero growth net growth isoclines for the unlinked model for three different levels of spatial structure ($\varphi = 0$, 0.5, 1). The blue, curved isocline represents the equilibrium for the *Rhiz* locus and is unstable. The linear isocline is the equilibrium for the *Nod* locus and is stable. Vectors on the phase plane represent the evolutionary dynamics towards the equilibria. (d–f) Fitness of genotypes in each nodule environment. These panels of display the fitness of each cell type in each environment (i.e., nodule adjacency). Width of bars is proportional the probability of being found in that environment, as altered by degree of spatial structure ($\varphi = 0$, 0.5, 1). Black, red, green and blue bars (left to right within each cluster of bars) represent *Nod+Rhiz+*, *Nod+Rhiz−*, *Nod−Rhiz+* and *Nod−Rhiz−*, respectively.

dynamics. From these three approaches, we are able to achieve a qualitative understanding of the linked model.

Results II: Linked case

The invasibility of genotypes in the linked model changes with spatial structure (Figure 3). At low spatial structure (Figure 3a), the non-interacting genotype (*Nod−Rhiz−*) is globally stable. All other genotypes are invaded by this genotype because it is able to reap benefits of general exudates, yet suffers none of the costs. As spatial structure increases (Figure 3b), we see that no genotype is globally stable; rather, the equilibrium is a polymorphic population. This matches the results of the unlinked model, where an increase in spatial structure leads to the evolution of nodulation. When the environment is highly structured (Figure 3c), the system moves towards the *Nod+Rhiz−* genotype, as in the unlinked model. These invasion results echo the qualitative dynamics of the unlinked case, and the quantitative conditions for the stable corner equilibria (*Nod+Rhiz−* and *Nod−Rhiz−*) are identical. Invasion criteria for each of the genotypes are presented in terms of ϕ in Text S3.

At intermediate values of ϕ, the invasion analysis cannot determine the stability of internal equilibria. There are potential internal equilbria (Figure 3c; gray circles) at three locations in the genotype-space: *(i)* between the *Nod+Rhiz+* and *Nod−Rhiz+*

genotypes; *(ii)* between the *Nod−Rhiz+* and *Nod+Rhiz−* genotypes; and *(iii)* between the *Nod+Rhiz−* and *Nod−Rhiz−* genotypes. To investigate the stability of these points, we turn towards numerical simulation and analytical approximations.

Numerical simulations show that increasing the spatial structure greatly facilitates the evolution of cooperation (Figure 4). In structured environments, *Nod+Rhiz−* quickly sweeps to fixation (Figure 4c,f). In less structured populations, the *Rhiz+* allele facilitates a transient increase in the frequency of cooperation. At this intermediate level of mixing, we see that stable equilibrium is a population polymorphic for the *Nod+Rhiz−* and *Nod−Rhiz−* genotypes (Figure 4b,e), as in the unlinked model. In unstructured populations, the non-interacting, saprophytic genotype (*Nod−Rhiz−*) invariably fixes (Figure 4a,d). In Text S4, we discuss a weak selection approximation that enables model simplification. With these functions, it is possible to visualize isoplanes for each of the genotypes (Figure S2).

The transient increase seen in the frequency of the *Nod+Rhiz+* at intermediate and low spatial structure is a striking result of the linked model (Figure 4b,e). The magnitude of the transient gain in nodulation – quantified as the area under the curve of the *Nod+Rhiz+* frequency dynamics that is greater than the equilibrium frequency reached by the *Nod+Rhiz−* genotype – measures the increase in frequency of mutualism that would not be realized in

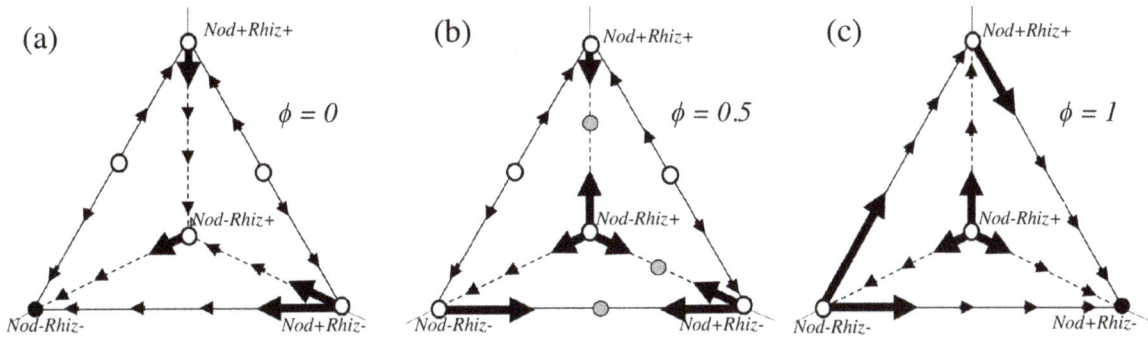

Figure 3. Invasibility at different levels of spatial structures. (a–c) Increasing levels of spatial structure ($\varphi = 0$, 0.5, 1). Black filled circles represent stable equilibria, grey filled circles represent unstable internal equilibria, while open circles are unstable. Arrows represent the movement of the population along the edges of this genotype space.

the absence of rhizopines. Though this transient gain in frequency of cooperation is not sustained over evolutionary time, the magnitude of increase is highest in mixed environments (where the evolution of cooperation is otherwise restricted), but disappears as spatial structure increases (Figure S3).

The *Nod+Rhiz+* genotype is not stable; it is invaded by the *Nod−Rhiz+* genotype (Figure 4a,b,d,e) as it approaches fixation, leading to the pattern of transience. The *Nod−Rhiz+* genotype is a non-cooperative 'cheating' genotype. However, this genotype is in turn unstable (Figure 3). Where the population moves in genotypic space after this invasion is determined by the level of spatial structure: in more structured environments, the population will move to a stable polymorphism of *Nod+Rhiz−* and *Nod−Rhiz−* (Figure 4b,e); in mixed environments, the non-interacting genotype (*Nod−Rhiz−*) will sweep to fixation (Figure 4a,d).

Discussion

The evolution and maintenance of cooperative traits in the face of countervailing forces is a longstanding question in population biology. Here, we have shown that spatial structure plays a dominant role relative to that of kin recognition in the evolution of cooperation. Spatial structure both promotes the evolution of cooperative traits, as well as maintains them over evolutionary time. When environments become increasingly mixed, non-cooperative individuals easily invade cooperative populations. We used the analytical results from the unlinked model as a basis for comparison to the more intractable linked model, which recovered many of the qualitative results.

We show that while kin recognition can favour the evolution of cooperation, it is a transient effect. As recognition-enabled cooperation becomes common, it becomes vulnerable to invasion

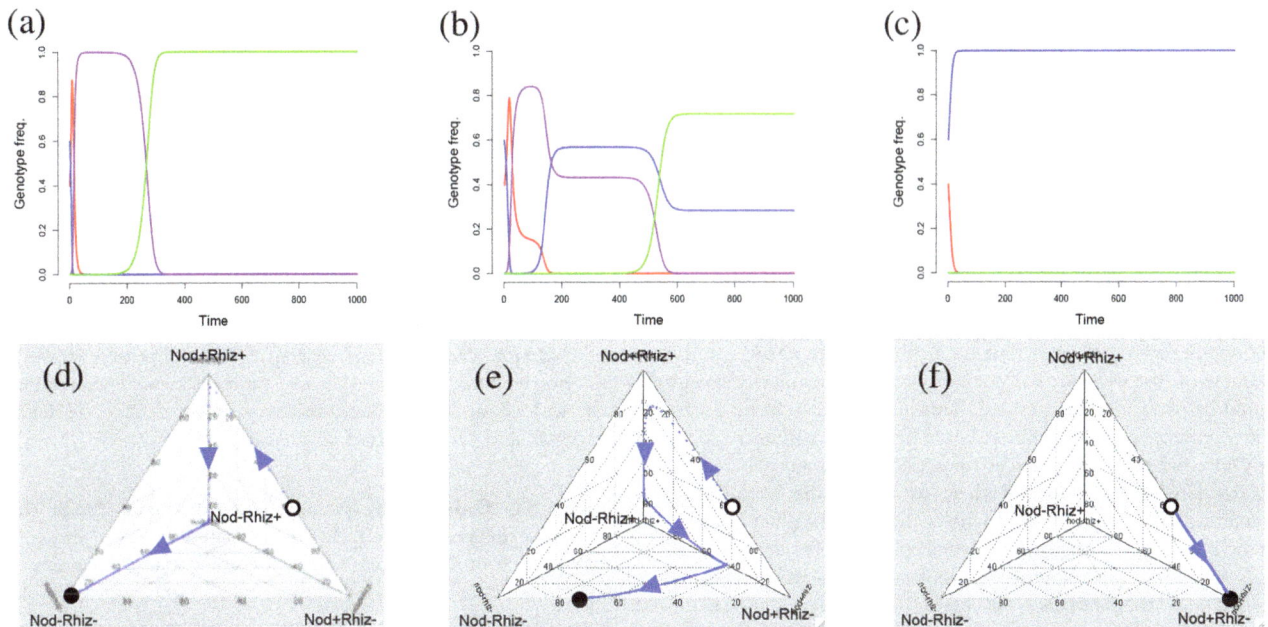

Figure 4. Linked genotype frequency dynamics. (a–c) Genotype frequency dynamics of the linked model for $\varphi = \{0, 0.5, 1\}$. *Nod+Rhiz+*, *Nod+Rhiz−*, *Nod−Rhiz+* and *Nod−Rhiz−* are represented by red, blue, purple, and green lines, respectively. At low spatial structure, there is a transient increase in *Nod+Rhiz+* frequency (red line). At higher spatial structures, this increase disappears, and *Nod+Rhiz−* goes to fixation. Genotype frequency is plotted on the y-axis, and time in generations is on the x-axis. (d–f) Evolutionary dynamics in the genotype space simplex. Blue arrows represents evolutionary trajectory, and black point represents the evolutionary endpoints. Open circles show initial condition.

by non-cooperative kin recognition 'cheaters'. Moreover, we show that in some cases, kin recognition – because of additional costs – actually constrains the evolution of cooperation. This instability suggests that cooperation founded solely upon kin recognition mechanisms is unlikely. The evolutionary stability of kin recognition in the population is reliant on the sustained association between kin recognition and the cooperative trait. If these can be uncoupled from each other, genotypes that do not suffer the cost of expression will be able to invade the system. Additional instability of kin recognition stems from the fact that effective kin recognition requires fidelity between two elements: the production of a signal and the ability to recognize that signal [29,30,34]. The potential disentanglement of these elements can further destabilize cooperation reliant on kin recognition mechanisms. In rhizobia, the different genes underlying rhizopine synthesis and catabolism could independently mutate, generating destabilizing genotypes (i.e. genotypes that can catabolize rhizopines but do not bear the costs of rhizopine synthesis genes). By contrast, cooperation based on spatial structure does not share the same vulnerabilities to cheating as kin recognition. The analogous destabilization of cooperation based on spatial structure requires evolution of increased dispersal in non-cooperators [35]. However, constraints on the evolution of genotype-specific dispersal phenotypes are more severe than the simple loss of function mutation required in kin recognition systems. Bacteria, for example, can swim through the soil via flagella or cooperative swarming [36,37], but even these local scale mobilities can be curtailed by drying of the soil [38]. As a result, spatial structure is a more resilient mechanism for increasing the frequency of contact between cooperators.

Though our study strengthens the view that population structure is the dominant factor in the evolution of cooperation, it does not exclude the possibility that kin recognition mechanisms are important in social evolution. In fact, the transient increase we find occurs in relatively unstructured environments, where cooperation is otherwise unlikely to evolve. Indeed, the presence of rhizopine genotypes in rhizobium populations [26,39,40] suggests an evolutionary force is maintaining rhizopines. One potential mechanism for this maintenance of polymorphism is broader meta-population dynamics, with migration among patches. The scale of a rhizobium-plant mutualism meta-population is not easy to identify. Due to overlap of root systems of individual plants in a local area, individual plants are unlikely to be the limit of a patch in a meta-population model. Rather, the root systems of a local cluster of plants may be more meaningful patches, and populations of bacterial cells could disperse among groups of plants. The transient increase in $Nod+Rhiz+$ cells in populations will increase opportunity for migration of $Nod+Rhiz+$ cells to other subpopulations, thereby maintaining rhizopines in nature. In this framework, the variation in local carrying capacities among groups could be explicitly incorporated. This meta-population model for the maintenance of rhizopines should be a focus of future study.

Our work does not consider an active role of the plant in the interaction with rhizobia. Rather, we model the evolutionary dynamics within rhizobia populations and treat the plant host as a static interactor through which cooperative benefits are delivered (i.e., exudates). This approach complements research seeking to understand the influence of plant 'sanctions' – the plants physiological ability to discontinue or restrict carbon allotment to nodules infected by relatively ineffective rhizobium strains – on the system dynamics [41]. These studies focus on the interspecific dynamics of this interaction, and specifically how the plant can alter the bacterial populations. If plants can "choose" among genotypes of rhizobia partners, this process can stabilize the mutualistic interaction [42,43]. Alternatively, if plants 'sanction' or

preferentially allocate resources among nodules, these interspecific forces can maintain the mutualism as well [41,44]. Both of these interspecific processes presume that plant resources are reliably delivered to kin of nodulating rhizobia. Rather than address these interspecific processes, we have focused on the mechanisms underlying this reliable delivery of resources. As a result, our modeling approach is complementary to research efforts to understand the importance of partner choice, sanctions and preferential allocation in maintaining interspecific mutualisms.

By focusing our attention on how intraspecific rhizobia dynamics can maintain the legume-rhizobia mutualism, our work offers a unique perspective on this ecologically important interaction. The qualitative results from our model generate predictions for the changes in rhizobia genotype frequencies over time. Experimental tests of these predictions would be valuable contributions to understanding the evolution of the legume-rhizobia mutualism. By empirically evaluating these within-rhizobia predictions, we can move towards a more complete view of the ecology and evolution of the mutualism.

Our model suggests that spatial structure can be a dominant contributor to the maintenance of mutualistic genotypes in rhizobium populations relative to the directed benefits of rhizopines. Rhizopines (kin recognition) are removed from the population because they are vulnerable to invasion by non-cooperating 'cheater' genotypes ($Nod-Rhiz+$), which in turn are unstable. That the non-interacting genotype invariably goes to fixation in unstructured populations is indicative of the necessity of spatial structure of cooperation, and thus the maintenance of the plant-rhizobium mutualism. This disintegration of the mutualism is analogous to the experimental work where cooperative traits are lost during evolution in unstructured environments (*interspecific*: [45,46]; *intraspecific*: [6]).

Supporting Information

Figure S1 Equilibrium frequency of nodulation is limited by rhizopines. The equilibrium $Nod+$ frequency is shown as function of spatial structure. The solid curve shows the equilibrium level of cooperation in the absence of rhizopines, while the dashed curve represents the equilibrium frequency of $Nod+$ when $Rhiz+$ is initially fixed in the population. In well-mixed environments and in structured environments, rhizopines have no influence of the evolution of cooperation. At intermediate levels of mixing, rhizopines substantially limit nodulation.

Figure S2 Isoplanes and evolutionary trajectory of approximate linked model. The blue, green, and yellow represent the zero-growth isoplanes of the $Nod+Rhiz+$, $Nod+Rhiz-$, and $Nod-Rhiz-$ genotypes, respectively. The first two isoplanes are overlapping. The red trace represents an evolutionary trajectory. Note the transient increase towards the $Nod+Rhiz+$ genotype, and eventual fixation at $Nod+Rhiz-$.

Figure S3 Gain in frequency of nodulation from the presence of rhizopines. In lower spatial structure environments, the transient increase in cooperation is more substantial that in highly structured environments. This figure represents the space between the red ($Nod+Rhiz+$) and blue ($Nod+Rhiz-$) curves in the Figure 4a–c. It is interpreted as the increase in frequency of mutualism that would not be realized in the absence of rhizopines.

Text S1 Costs of nodulation. A brief discussion of multiple potential costs of nodulation.

Text S2 Model. Full exposition of the unlinked and linked models.

Text S3 Invasion conditions in the linked model. Identification of the invasion conditions for mutant genotypes in the linked model.

Text S4 Weak-selection approximation. Used for visualizing linked model isoplanes.

Acknowledgments

We thank G. Velicer, M. Wade, C. Lively, L. Barrett, and members of the Bever lab, for helpful comments and discussion. We also thank Angel Sánchez, and two anonymous reviewers for constructive criticism that improved the manuscript.

Author Contributions

Conceived and designed the experiments: PCZ JDB. Performed the experiments: PCZ JDB. Analyzed the data: PCZ JDB. Contributed reagents/materials/analysis tools: PCZ JDB. Wrote the paper: PCZ JDB.

References

1. Hamilton WD (1964) Genetical evolution of social behaviour I. J Theor Biol 7: 1–16.
2. Wade MJ (1985) Soft selection, hard selection, kin selection, and group selection. Am Nat 125: 61–73.
3. Queller DC (1992) A general model for kin selection. Evolution 46: 376–380.
4. Fletcher DJC, Michener CD (eds) (1989) Kin Recognition in Animals. New York: Wiley.
5. Sherman PW, Reeve HK, Pfennig DW (1997) Recognition systems. In: Krebs, J, Davies, N, editors. Behavioural ecology. Oxford: Blackwell Scientific Publications.
6. Velicer GJ, Kroos L, Lenski RE (1998) Loss of social behaviors by *Myxococcus xanthus* during evolution in an unstructured habitat. Proc Natl Acad Sci U S A 95: 12376–12380.
7. Wade MJ (1980) An experimental study of kin selection. Evolution 34: 844–855.
8. Crozier RH (1987) Genetic aspects of kin recognition: concepts, models, and synthesis. In: Fletcher, DJC, Michener, CD. editors Kin recognition in animals. New York: Wiley.
9. Grafen A (1990) Do Animals Really Recognize Kin? Anim Behav 39: 42–54.
10. Rousset F, Roze D (2007). Constraints on the origin and maintenance of genetic kin recognition. Evolution 61: 2320–2330.
11. Platt TG, Bever JD, Fuqua C (2012) A cooperative virulence plasmid imposes a high fitness cost under conditions that induce pathogenesis. Proc Roy Soc Lond B 279: 1691–1699.
12. Jimenez J, Casadesus J (1989) An altruistic model of the rhizobium-legume association. J Hered 80: 335–337.
13. Olivieri I, Frank SA (1994) The evolution of nodulation in rhizobium - altruism in the rhizosphere. J Hered. 85: 46–47.
14. Bever JD, Simms EL (2000) Evolution of nitrogen fixation in spatially structured populations of rhizobium. Heredity 85: 366–372.
15. Parco SZ, Dilworth MJ, Glenn AR (1994) Motility and the distribution of introduced root-nodule bacteria on the root-system of legumes. Soil Biol Biochem 26,:297–300.
16. Rossbach S, Rasul G, Schneider M, Eardly B, Debruijn FJ (1995) Structural and functional conservation of the rhizopine catabolism (*moc*) locus is limited to selected *Rhizobium meliloti* strains and unrelated to their geographical origin. Mol Plant Microbe Interact 8: 549–559.
17. Denison RF (2000) Legume sanctions and the evolution of symbiotic cooperation by rhizobia. Am Nat 156: 567–576.
18. Platt TG, Fuqua C, Bever JD (2012) Resource and Competitive Dynamics Shape the Benefits of Public Goods Cooperation in a Plant Pathogen. Evolution 66: 1953–1965.
19. Boivin C, Barran LR, Malpica CA, Rosenberg C (1991) Genetic analysis of a region of the *Rhizobium meliloti* psym plasmid specifying catabolism of trigonelline, a secondary metabolite present in legumes. J Bact 173: 2809–2817.
20. Hartwig UA, Joseph M, Phillips DA (1991) Flavonoids Released Naturally From Alfalfa Seeds Enhance Growth-Rate Of *Rhizobium meliloti*. Plant Physiol 95: 797–803.
21. Long SR (1992) Genetic analysis of Rhizobium nodulation. In: Stacey, G, Burris, R H, Evans, HJ. editors. Biological Nitrogen Fixation. New York: Chapman & Hall.
22. Taylor PD (1992) Altruism in viscous populations - an inclusive fitness model. Evol Ecol 6: 352–356.
23. Platt TG, Bever JD (2009) Kin competition and the evolution of cooperation. Trends Ecol Evol 24: 370–377.
24. Van Dyken JD (2010) The Components of Kin Competition. Evolution 64: 2840–2854.
25. Murphy PJ, Heycke N, Banfalvi Z, Tate ME, Debruijn F, et al (1987) Genes for the catabolism and synthesis of an opine-like compound in *Rhizobium meliloti* are closely linked and on the *Sym* plasmid. Proc Natl Acad Sci U S A 84: 493–497.
26. Murphy PJ, Heycke N, Trenz SP, Ratet P, Debruijn FJ, et al (1988) Synthesis of an opine-like compound, a rhizopine, in alfalfa nodules is symbiotically regulated. Proc Natl Acad Sci U S A 85: 9133–9137.
27. Rao JP, Grzemski W, Murphy PJ (1995) *Rhizobium meliloti* lacking *mosA* synthesizes the rhizopine scyllo-inosamine in-place of 3-O-methyl-scyllo-inosamine. Microbiology 141: 1683–1690.
28. Dawkins R (1976) The Selfish Gene. New York: Oxford University Press.
29. Gardner A, West SA (2010) Greenbeards. Evolution 64: 25–38.
30. Jansen VAA, Van Baalen M (2006) Altruism through beard chromodynamics. Nature 440: 663–666.
31. Hamilton WD (1970) Selfish and Spiteful Behaviour in an Evolutionary Model. Nature 228: 1218–1220.
32. Simms EL (1998) Evolutionary dynamics of rhizopine within spatially structured rhizobium populations. Proc Roy Soc Lond B 265: 1713–1719.
33. Tempe J, Petit A, Bannerot H (1982) Presence of opine-like substances in alfalfa nodules. Comptes Rendus De L'Académie Des Sciences Serie III-Sciences De La Vie 295: 413–416.
34. Agrawal AF (2001) Kin recognition and the evolution of altruism. Proc Roy Soc Lond B 268: 1099–1104.
35. Mack KML (2012) Selective feedback between dispersal distance and the stability of mutualism. Oikos 121: 442–448.
36. Ames P, Bergman K (1981) Competitive advantage provided by bacterial motility in the formation of nodules by *Rhizobium meliloti*. J Bact, 148(2): 728–729.
37. Velicer GJ, Vos M (2009) Sociobiology of the myxobacteria. Annual review of microbiology 63: 599–623.
38. Dechesne A, Wang G, Gülez G, Or D, Smets BF (2010) Hydration-controlled bacterial motility and dispersal on surfaces. Proc Natl Acad Sci U S A 107(32). 14369–14372.
39. Murphy PJ, Wexler W, Grzemski W, Rao JP, Gordon D (1995) Rhizopines - their role in symbiosis and competition. Soil Biol Biochem 27: 525–529.
40. Gardener BBM, de Bruijn FJ (1998) Detection and isolation of novel rhizopine-catabolizing bacteria from the environment. Appl Environ Microbiol 64: 4944–4949.
41. Kiers ET, Rousseau RA, West SA, Denison RF (2003) Host sanctions and the legume-rhizobium mutualism. Nature 425: 78–81.
42. Heath KD, Tiffin P (2009) Stabilizing mechanisms in a legume-rhizobium mutualism. Evolution 63: 652–662.
43. Sachs JL, Russell JE, Lii YE, Black KC, Lopez G, et al (2010) Host control over infection and proliferation of a cheater symbiont. J Evol Biol 23: 1919–1927.
44. West SA, Kiers ET, Simms EL, Denison RF (2002) Sanctions and mutualism stability: why do rhizobia fix nitrogen? Proc Roy Soc Lond B 269: 685–694.
45. Bever JD, Richardson SC, Lawrence BM, Holmes J, Watson M (2009) Preferential allocation to beneficial symbiont with spatial structure maintains mycorrhizal mutualism. Ecol Lett 12: 13–21.
46. Harcombe W (2010) Novel cooperation experimentally evolved between species. Evolution 64: 2166–2172.

16

Comparative Genomics of the Bacterial Genus *Streptococcus* Illuminates Evolutionary Implications of Species Groups

Xiao-Yang Gao[1,5]*, Xiao-Yang Zhi[2], Hong-Wei Li[2,3], Hans-Peter Klenk[4], Wen-Jun Li[1,2]*

1 Key Laboratory of Biogeography and Bioresource in Arid Land, Xinjiang Institute of Ecology and Geography, Chinese Academy of Sciences, Urumqi, China, 2 Key Laboratory of Microbial Diversity in Southwest China, Ministry of Education and the Laboratory for Conservation and Utilization of Bio-Resources, Yunnan Institute of Microbiology, Yunnan University, Kunming, China, 3 The First Hospital of Qujing City, Qujing Affiliated Hospital of Kunming Medical University, Qujing, China, 4 Leibniz-Institute DSMZ-German Collection of Microorganisms and Cell Cultures, Braunschweig, Germany, 5 University of Chinese Academy of Sciences, Beijing, China

Abstract

Members of the genus *Streptococcus* within the phylum *Firmicutes* are among the most diverse and significant zoonotic pathogens. This genus has gone through considerable taxonomic revision due to increasing improvements of chemotaxonomic approaches, DNA hybridization and 16S rRNA gene sequencing. It is proposed to place the majority of streptococci into "species groups". However, the evolutionary implications of species groups are not clear presently. We use comparative genomic approaches to yield a better understanding of the evolution of *Streptococcus* through genome dynamics, population structure, phylogenies and virulence factor distribution of species groups. Genome dynamics analyses indicate that the pan-genome size increases with the addition of newly sequenced strains, while the core genome size decreases with sequential addition at the genus level and species group level. Population structure analysis reveals two distinct lineages, one including Pyogenic, Bovis, Mutans and Salivarius groups, and the other including Mitis, Anginosus and Unknown groups. Phylogenetic dendrograms show that species within the same species group cluster together, and infer two main clades in accordance with population structure analysis. Distribution of streptococcal virulence factors has no obvious patterns among the species groups; however, the evolution of some common virulence factors is congruous with the evolution of species groups, according to phylogenetic inference. We suggest that the proposed streptococcal species groups are reasonable from the viewpoints of comparative genomics; evolution of the genus is congruent with the individual evolutionary trajectories of different species groups.

Editor: Sean D. Reid, Wake Forest University School of Medicine, United States of America

Funding: This research was supported by grants from the National Basic Research Program of China (No. 2010CB833801) and Key Project of International Cooperation of Ministry of Science & Technology (MOST) (No. 2013DFA31980), China and Key Project of Yunnan Provincial Natural Science Foundation (2013FA004). W-J Li was also supported by 'Hundred Talents Program' of the Chinese Academy of Sciences. The funders had no role in study design, data collection and analysis, decision to publish, or preparation of the manuscript.

Competing Interests: The authors have declared that no competing interests exist.

* Email: wjli@ynu.edu.cn (W-JL); gaoxiaoyang99@gmail.com (X-YG)

Introduction

The genus *Streptococcus* comprises a wide variety of pathogenic and commensal gram-positive bacteria [1]. Pathogens and some commensals of *Streptococcus* show a surprising capacity for adaptation to new hosts and resistance to antibiotics and immune responses. As a result, they have caused the spread of infection and significantly increased morbidity and mortality rates all over the world, leading to huge health and economic loss [2–7]. A small group of commensals are opportunistic pathogens like *Streptococcus oralis*, while others are harmless saprophytes like *Streptococcus thermophilus* used as starter cultures in the food industry [8]. Due to the diversity and clinical importance of this genus, *Streptococcus* has attracted the attention of medical scientists and microbiologists and has undergone considerable taxonomic revision.

Previously, the taxonomy of the genus *Streptococcus* mainly focused on morphological, biochemical and serological characterization, but it is still not very clear with modern genomic data as yet not adequately considered [9]. Recent applications of chemotaxonomic approaches, genomic DNA-DNA hybridization and 16S rRNA sequencing techniques have not only provided significant insights into the natural relationships among streptococci, but have also influenced significantly their taxonomy and nomenclature [10–13]. These revisions form the basis of delineation and reveal the natural grouping of species into "species groups" [14]. The species groups have been named "Pyogenic", "Mitis", "Anginosus", "Bovis", "Mutans" and "Salivarius" respectively, and they encompass the majority of described species (several species remain ungrouped). Although these polyphasic taxonomy approaches are still widely used in many laboratories, limits of biochemical determination, and low efficiency operation of DNA hybridization [15] as well as possible phenotypic and ecological differentiation underlying identical 16S rRNA genes [16] all inevitably hamper the evolutionary and taxonomic investigations of streptococci. Moreover, understanding of the species groups relies on relevant biochemical features, so the reliability of species groups under a larger molecular data set needs to be determined. Hence, investigation of their phylogenetic relationships and evolutionary implications is necessary to enrich our knowledge of the evolution of the genus *Streptococcus*.

With increasing advances in sequencing and computational technologies, application of genomic tools has revolutionized microbial ecological studies and has drastically expanded our view on the previously underappreciated microbial world [17,18]. In this context, the number of available streptococcal genomes is growing exponentially. Whole-genome sequencing has gained new insights into microevolution of streptococci, and also helped researchers to decipher their host adaptation [19,20], determine virulence factors [21] and track pathogenesis mechanisms, laying the foundation for vaccine candidate development [22,23]. Comparative genomics is primarily used to investigate intraspecies variation [24,25], which is extended to the diversity studies of closely related *Streptococcus* species [26,27]. As mentioned above, comparative genomic analyses of streptococci along with other bacteria have revealed microbial genomes as dynamic entities shaped by multiple forces, including genome reduction, genome rearrangements, gene duplication, and acquisition of new genes through lateral gene transfer [26,28]. As a large number of bacterial genomes are sequenced, it has become increasingly evident that one strain's genome sequence is not entirely representative of other members of the same species. Information from more genomes is needed to understand the dynamic nature of genomes, and to comprehend the evolutionary process at higher taxonomic levels [1,29,30]. Thus, evolution of the genus *Streptococcus* underscores the need to implement comprehensive whole-genome analyses with more extensive genomic sampling.

This study uses genomic data to explore the evolution of the genus *Streptococcus* within the context of proposed species groups. Here, we employ comparative genomic analyses of the genus *Streptococcus* to define the pan-genome and core genome, assess population structure, infer phylogenetic relationships and determine virulence factor distribution of species groups. Specifically, the analyses enabled us to test (1) pan-genome size and core genome size of *Streptococcus* and species groups; (2) the phylogenetic relationships among those groups based on genomic data; (3) the reasonableness of species groups raised by associated biochemical features and 16S rRNA gene analysis; and (4) distribution of virulence factor among species groups, in order to explore their implications in evolution of the genus.

Materials and Methods

1 Materials

This study used 138 streptococcal genomes covering most species in the genus (Figure S1). Most of them were divided into 6 species groups according the previous studies [10,11,14]. Because *Streptococcus suis* has not been assigned to an existing species groups, we named it as the "Unknown" group. The genomic data was obtained from genome release in the public database NCBI (ftp://ftp.ncbi.nlm.nih.gov/genomes/Bacteria/) as of May, 2013, including all complete genomes as well as draft genomes of type strains or strains for which a complete genome was not available. Characteristics of *Streptococcus* species and strains were acquired from NCBI (http://www.ncbi.nlm.nih.gov/genome) and JGI (http://genome.jgi.doe.gov/) as well as related genome publications [9,31–33].

2 Methods

2.1 Identification and functional classification of homologous clusters. Homologous clusters used for subsequent analyses were determined by the program OrthoMCL version 2.0 [34]. In our analyses, all extracted protein sequences

were adjusted to a prescribed format and were grouped into homologous clusters using OrthoMCL based on sequence similarity. The BLAST reciprocal best hit algorithm [35] was employed with 50% match cutoff and 1e-5 e-value cutoff, and Markov Cluster Algorithms (MCL) [36] were applied with an inflation index of 1.5. As a result, a matrix describing the genome gene content for 138 strains was constructed. The total 274,822 protein-coding genes were grouped into 18,528 homologous clusters, including common genes represented by 369 core homologous clusters. The functional category of each core homologous cluster was determined by performing BLAST program against Cluster of Orthologous Groups (COGs) database (http://www.ncbi.nih.gov/COG/) with 50% identity cutoff and 1e-5 e-value.

2.2 Pan-genome, core genome and unique genes. In order to predict the possible dynamic changes of genome size at the genus and species group levels, the sizes of pan-genome, core genome and unique genes were simulated. 18,528 clusters, from OrthoMCL program, were parsed by Perl scripts. Then pan-genome (gene repertoire), core genome (common genes, mutually conserved) and unique genes (specific genes, only found in one genome) [37] were estimated as done in previous studies [38–41]. For pan-genome analysis starting from one single genome to 138 genomes, genomes were added 1000 times in a randomized order without replacement at each fixed number of genomes, and the gene repertoire was accumulated. The statistical analyses of core genome and unique genes followed the above procedures. Gene accumulation curves describing the dynamic changes of gene repertoire, common genes and new genes with the addition of new comparative genomes were implemented by SigmaPlot version 12.5. Furthermore, we employed best fitting functions to predict possible distributions of pan-genome, core genome as well as unique genes for streptococci, using the median values as determined by IBM SPSS Statistics version 19 [42].

2.3 Population structure. In order to investigate the population structure of the genus *Streptococcus* and its relationships with species groups, the Markov chain Monte Carlo (MCMC) based program Structure version 2.3.4 [43,44] was used to cluster individuals into populations. Initially, we treated orthologous genes as MLST sequence data from Extended FASTA Format into the Structure Format using xmfa2struct (available from http://www.xavierdidelot.xtreemhost.com/clonalframe.htm). The admixture ancestry model with assumption of correlated allele frequencies among populations was used. We ran the simulation 10 times under a burn-in period of 100,000 and a run length of 1,000,000 MCMC, without prior population information. K values from 1 to 7 were tested to allow us to identify the best K value, represented by the highest value of K and DK [45]. Results of the ten independent runs were averaged for each K value to determine the most likely model, i.e., the one with the highest likelihood, and they were subsequently plotted using Distruct version 1.1 [46]. The identification of the best K was evaluated following the DK-method through online program Structure Harvester (available at: http://taylor0.biology.ucla.edu/struct_harvest/) [47].

2.4 Phylogenetic Analysis. To determine the phylogenetic relationships among *Streptococcus* species and species groups based on genomic data, both supermatrix and gene content methods were applied to infer phylogenetic trees. For the supermatrix method, we selected a set of orthologous genes shared by all 138 streptococcal strains (278 genes present in a single copy in all strains) according to the identification of homologous clusters. For each orthologous cluster, protein sequences were aligned using ClustalW version 2.1 [48] and the resulting alignments of individual proteins were concatenated to infer the organismal phylogeny using Neighbor-Joining (NJ) in MEGA version 5.20 [49] and the maximum likelihood algorithm (ML) in RAxML version 7.3.0 [50]. For the gene content method, a gene content matrix was parsed using a phyletic pattern indicating the presence (1) or absence (0) of the respective genes of all streptococcal strains. Jaccard distance (one minus the Jaccard coefficient) between pairwise genomes was calculated based on the gene content matrix. Hierarchical clustering (unweighted pair group method with arithmetic mean, UPGMA) in package PHYLIP version 3.6 [51] was employed to reconstruct the gene content dendrogram, using paired Jaccard distances.

2.5 Virulence factor determination. To explore the distribution of species group-specific virulence factors, we collected all streptococcal virulence factors in the Virulence Factor Database (VFDB, http://www.mgc.ac.cn/VFs/). The relevant gene sequences of virulence factors were extracted from genomes, and all the protein-coding sequences of 138 *Streptococcus* strains analyzed were incorporated as the database. The virulence factor distribution for 138 streptococcal genomes was determined by BLAST with 50% match cutoff, 50% coverage cutoff and 1e-5 e-value cutoff. In cases of shared homologous genes related to virulence factors, phylogenetic trees were inferred by ML algorithms in RAxML version 7.3.0.

Results and Discussion

1 Genomic size and GC content of species groups

According to previous studies on 16S rRNA gene sequence analysis and associated biochemical features of the genus *Streptococcus* (Table S1 in File S1) [10–12], 138 *Streptococcus* strains were divided into seven species groups: "Pyogenic", "Mitis", "Anginosus", "Bovis", "Mutans", "Salivarius" and "Unknown" (Table S2 in File S1). The genome size varied from 1.64 Mb (*Streptococcus peroris* D1) to 2.43 Mb (*Streptococcus salivarius* D1) with the average value of 2.05 Mb. Within species groups, the genome size range and average showed no significant variations: Pyogenic (range 1.75–2.27 Mb, average 2.00 Mb), Mitis (range 1.64–2.39 Mb, average 2.07 Mb), Anginosus (range 1.82–2.29 Mb, average 1.96 Mb), Bovis (range 1.74–2.38 Mb, average 2.12 Mb), Mutans (range 1.92–2.42 Mb, average 2.09 Mb), Salivarius (range 1.8–2.43 Mb, average 2.01 Mb), and Unknown (range 1.98–2.23 Mb, average 2.10 Mb). Streptococcal genome size is relatively small when compared to other bacteria, and may indicate an adaptation for reproductive efficiency or competitiveness for a new host environment [52]. The genus *Streptococcus* is a low GC content taxon, and genomic GC content of its representatives range from 33.79% (*Streptococcus urinalis* D2) to 43.40% (*Streptococcus sanguinis* W1) with an average of 39.25%. Genomic GC content results from mutation and selection [53] involving multiple factors, including environment, symbiotic lifestyle, aerobiosis, nitrogen fixation ability, and the combination of polIIIa subunits [54].

2 Distribution and identification of homologous clusters

Homologous genes evolve through two fundamentally different ways, either through speciation events (producing orthologs) or by gene duplication events (producing paralogs) [55]. A clear distinction between orthologs and paralogs is critical for the construction of a robust evolutionary classification of genes and reliable functional annotation of newly sequenced genomes [56]. In this study, 274,822 protein coding sequences from 138 genomes of streptococci were grouped into 18,528 homologous clusters, including 8,203 clusters unique to one proteome. Of the 274,822

proteins, the majority had homologous counterparts; however, some proteins were unique and could not be matched to any homologs in the pan-genome of *Streptococcus* (Table S3 in File S1). The 18,528 homologous clusters included both orthologous clusters and paralogous clusters, and a histogram of the number of clusters vs. the number of genomes was bimodal, with maxima at those present in only one genome and those present in all 138 streptococcal genomes (Figure S2A). 274,822 proteins including orthologous proteins and paralogous proteins across 138 streptococcal genomes also provide the same result (Figure S2B). The broad orthologous/paralogous cluster here is composed of both absolute and relative parts, namely orthologs/paralogs and semi-orthologs/paralogs (accessory genes). The number of orthologs within each streptococcal genome is 278 and the percentage ranges from 11.24% (*Streptococcus ictaluri* D1) to 17.90% (*Streptococcus salivarius* D3). However, the number of paralogs within each streptococcal genome is not constant, ranging from 3.84% (95, *S. ictaluri* D1) to 6.25% (97, *S. salivarius* D3). Therefore, the percentage of core genes ranges from 15.08% (373, *S. ictaluri* D1) to 24.15% (375, *S. salivarius* D3) (Table S4 in File S1). The percentage of unique proteins shows obvious differences in each genome, therefore the view of stable genomes that function as unchanging information repositories has given way to a more dynamic view in which genomes frequently lose genes and incorporate foreign genes [57,58]. Notably, the accessory genes account for significant portions in streptococcal genomes, since strains from same species or strains from different yet closely related species will share common genes.

A logical speculation often made in studying pathogen evolution implies that most host-specific adaption is associated with bacterial species-specific genes [26]. Previous studies revealed that there have been significant amounts of positive selection pressure on core genome components and that this selection pressure has occurred disproportionately in certain lineages [26]. According to COG classification analysis of core clusters, possible functions of 369 core clusters were identified and subdivided into 20 subcategories (Figure S3). There are 3 subcategories in information storage and processing, 7 subcategories in cellular processes and signaling, 8 subcategories in metabolism, and 2 subcategories were poorly characterized. Information storage and processing category makes up 38.2% of clusters, whereas cellular processes and signaling as well as metabolism categories make up 19.0% and 26.1% of clusters, respectively. Most of these genes are related to the colonization, persistence, and propensity to cause disease in these organisms [26]. Moreover, the poorly characterized part accounted for 16.8% may be involved in specific adaptations that help streptococci survive in novel environments.

3 Pan-genome and core genome analyses

3.1 Pan-genome. Estimation of the *Streptococcus* pan-genome indicates that the gene repertoire steadily increased with sequential addition of each new genome, and tendency was opening until the last addition (Figure 1A). In this study, we predicted that the gene repertoire of the genus *Streptococcus* could hold at least 21,446 genes. There is a tremendous increase from the first addition to thirtieth addition and the growth gradually becomes gentle, with acquisition of only 51 genes after addition of the last genome. We performed a power law fitting with median values as described previously [38,59] to model the possible trend and display the changing process through the function. The trend of streptococcal pan-genome size revealed that the genus possesses an open pan-genome for which the size increases with the addition of new sequenced strains. This was in accordance with previous studies on pan-genome of *Streptococcus* [24,26,27,59], which indicated that the

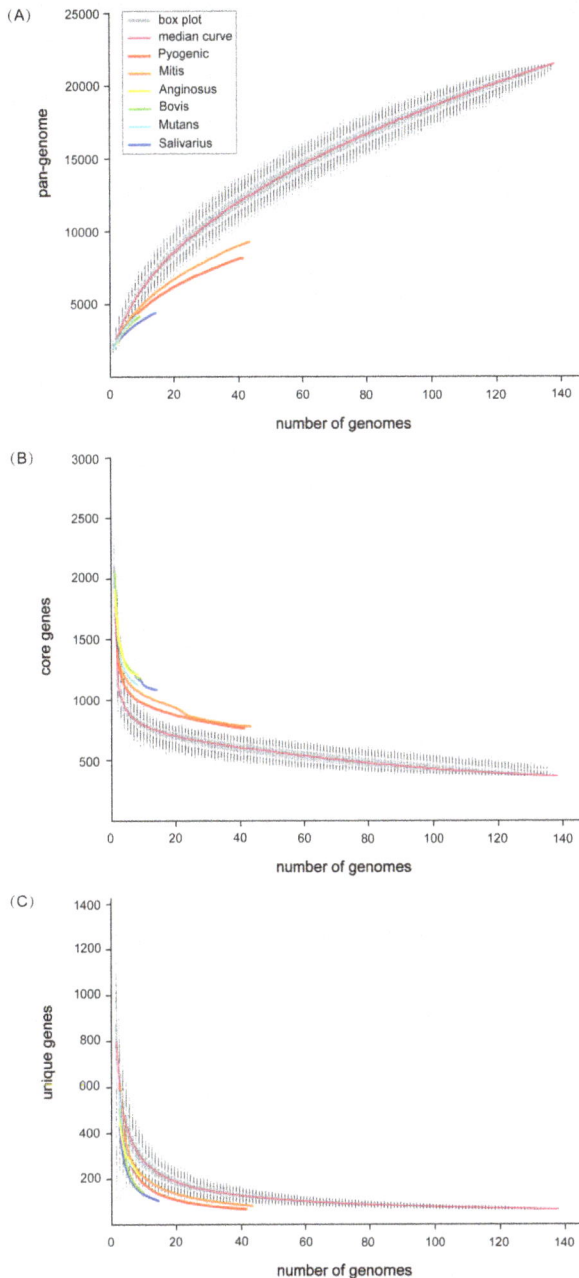

Figure 1. Size of pan-genome, core genome and unique genes for *Streptococcus*. (A) Total number of genes. The curve was fitted to the function $P(n) = Ap \cdot (n-1) \cdot n^{-\gamma} - Bp \cdot (n-1) + Cp$ and parameters $Ap = 1289 \pm 13.258$, $\gamma = 0.39$, $Bp = 49 \pm 2.42$, $Cp = 1809 \pm 21.73$ were determined under correlation $R^2 = 1$. (B) Number of genes in common. The curve was fitted to the function $C(n) = Ac^{-\alpha} + Bc$. The best fit was obtained with correlation $R^2 = 0.937$ for $Ac = 1560 \pm 38.49$, $\alpha = 0.34 \pm 0.02$, $Bc = 117 \pm 47.86$ (C) Number of unique genes. The curve was fitted to the function $U(n) = Au^{-\beta} + Bu$, the best fit was obtained with correlation $R^2 = 0.983$ for $Au = 1825 \pm 22.65$, $\beta = 0.994 \pm 0.02$, $Bu = 65 \pm 3.39$. The upper and lower edges of the boxes respectively indicate 25 and 75 percentiles, and the horizontal carmine lines indicate 50 percentile under 1,000 different random input orders of genomes. The central vertical lines extend from each box as far as the data extends to a distance of at most 1.5 interquartile ranges. Colors represent Pyogenic (red), Mitis (orange), Anginosus (yellow), Bovis (green), Mutans (cyan), and Salivarius (blue) species groups, respectively.

size of gene repertoire was underestimated and that the pan-genome size would continue to increase as more streptococcal genomes were sequenced.

In order to further verify that the pan-genome is open, the number of unique genes was calculated by incorporation of a new genome every time. In contrast to the pan-genome, the plot of new genes was fit well by a decaying function, and remarkably, the extrapolated curve reached an asymptotic value of 62, which meant that every newly sequenced genome could bring 62 new genes on average, even if many genomes were sequenced (Figure 1C). We therefore applied the exponential decay model to identify unique genes function using the median values. In light of the above analyses, we confirmed that the genus possesses an open pan-genome that increases in size with the addition of new sequenced strains. This was consistent with previous studies on the unique genes and pan-genome of *Streptococcus* [24,26,59,60].

3.2 Core genome. In contrast to pan-genome, estimation of the streptococcal core genome indicates that genes shared in all strains decreased with each addition, and it finally reached a plateau as the implication of keeping nearly constant over the last seven additions (Figure 1B). The decrease dropped from 1979 genes to 1179 genes at the first addition and kept stable at 369 genes since the next-to-last addition. As a result, the final constant of 369 shared genes was determined as the core genome size. The core gene number in each genome varied slightly because of involvement of duplicated genes and paralogs in shared clusters [59]. As observed for other bacterial species, the size of the *Streptococcus* core genome decreases as a function of genomes included, while the size of the pan-genome increases. The regression analysis of shared genes was extrapolated by fitting a decaying function, which was considered to provide the best fit to the dataset. Core and dispensable genes represent the essence and the diversity of the species, respectively [37]. As pointed out, this set of core genes does not correspond to the minimal set of genes necessary for an organism to survive and thrive in nature [5]. It is a backbone of essential components on which the rest of the genome is built [61].

The average gene content for *Streptococcus* genomes is $1,991 \pm 169$ genes and thus the core genome accounts for less than a fifth of the average gene content, and only 9.3% of the estimated pan-genome. In addition, this clear variability of gene content between species was also evident in comparison across strains of the same species. It once again implies the obvious genomic plasticity among streptococci living in different habits and possessing diverse lifestyles [1,62]. An open pan-genome is typical of those species that colonize multiple environments and have multiple ways of exchanging genetic material.

3.3 Genomes of Species groups. Estimations for genome sizes of six species groups were simultaneously carried out, except the "Unknown" group (Figure 1). The trends of the core genome, pan-genome and unique genes in these species groups were similar to those trends at the genus-level as described above. However, sizes of pan-genome and new genes of species groups were obviously smaller than the one at genus-level after each addition, and core genome sizes of species groups were larger than the one at genus-level after each addition. Moreover, there were subtle differences in genome sizes among species groups after each addition. This may be due to the fact that various species with diverse genome sizes were subsumed into different species groups. For example, Pyogenic and Mitis include more species and have more unique genes, and thus occupy a larger proportion of pan-genome than other species groups.

4 Population structure

The highest ΔK value (an ad hoc quantity related to the second order rate of change of the log probability of data with respect to the number of clusters) inferred from analysis using the program Structure [45] emerged when $K = 2$ (Figure 2B), indicating that streptococcal strains investigated here fall into two distinct populations (Figure 2A). The first of these populations included Pyogenic, Bovis, Mutans and Salivarius (orange color), and the second included Mitis, Anginosus and Unknown (blue color). Mutans appears to be a hybrid between the orange population and the blue population, with all individuals showing nearly 20% ancestry from the blue population composited of Mitis, Anginosus and Unknown. Mitis also shows fragmentary evidence for hybridization; genes from population including Pyogenic, Bovis, Mutans and Unknown were mixed into Mitis. The structures of two populations throw lights on evolutionary scenarios for streptococci and the relationships between populations and species groups.

5 Phylogenomic analyses

The inferred phylogeny of *Streptococcus* based on analysis of 138 genomes had a well-supported, consistent topology under Neighbor-joining both (NJ) and Maximum Likelihood (ML) algorithms (Figure 3). Strains within the same species clustered together, regardless of whether the data was derived from complete or draft genomes. Similarly, genomes from the same species group clustered together. Clearer phylogenetic relationships can be acquired through more extensive genomic sampling, particularly analyzing the whole set of conserved genes across a taxonomical level such as the genus level. Additionally, core genomes will shed light on evolutionary and functional relationships among the related species [63]. The existence of a core set of genes present in all bacteria is a testament to the conservative nature of evolution. Within several billion years of bacterial evolution, no successful replacement of the core genes evolved in any of the lineages leading to the studied genomes. The core set of genes is under high positive pressure for functions that prevent drastic changes.

The relationships among species and species groups were better understood from a gene content dendrogram (Figure 4), which used unweighted pair group method with arithmetic mean (UPGMA) algorithm [64]. Similar to the supermatrix tree, nearly all strains from the same species and most species from the same group cluster together. *Streptococcus infantis* D5 did not cluster with the other five strains in this species, which was likely caused by variation in gene composition as a result of gene annotation bias. The two dendrograms identify two main clades of species groups in accordance with the above Structure analysis (Figure 5A–B), one of which includes species from Mitis, Anginosus and Unknown, while the other one includes species from Salivarius, Mutant, Bovis and Pyogenic. To verify these relationships, we inferred the gene content dendrogram based on the core genome using UPGMA algorithm. The dendrogram topology based on the pan-genome most resembles that based on the core genome. Particularly, *Streptococcus infantis* D5 was incorporated into the Mitis species group, due to the fact that species-specific genes were removed and only shared genes were used for analysis (Figure S4A–B). The streptococci from Mutans are associated with dental plaque in human and animals. Here, Mutans group was divided into two subgroups, because this group overall is regarded as relatively loose with the member species having deep lines of descent [9]. Lateral gene transfer and recombination of genes have played a significant role in generating diversity in both Mitis and

Anginosus species groups [65–68]. Species from Mitis and Anginosus have a close relationship with one another, consistent with the suggestion that Mitis and Anginosus formed subgroups within a single "Oralis group" according to the classification of Schleifer and Kilpper-Balz [12]. Therefore, hybridization between populations of clusters identified in Structure analysis can effectively explain the polyphyly in the phylogenetic tree.

6 Distribution of virulence factors

Virulence factors of pathogenic bacteria, such as streptococci, play an important role in conquering various niches through infecting hosts and adapting to new environments. Particularly fascinating is the fact that some bacterial species can invade tissues and elicit different diseases by expression of different combinations of virulence factors. Therefore, we further compared the *Streptococcus* genomes with respect to virulence gene content to uncover additional insights into the biology and evolution of this genus. The determination of virulence factors in *Streptococcus* was investigated on the basis of VFDB, and virulence factors were mainly distributed in 135 representatives of the streptococci (Table S5 in File S1). Particularly, all of the streptococci have a number of genes associated with capsule production, which plays a significant role in immune evasion. Abundant production of capsular polysaccharide composed of hyaluronic acid results in mucoid strains of group A *Streptococcus* associated with outbreaks of acute rheumatic fever [69]. *S. pneumonia* strains with capsule quickly colonize and multiply because of their ability to evade phagocytosis, whereas *strains* lacking capsule suffer phagocyte killing [70].

The prevalent pathogens like *Streptococcus agalactiae*, *S. mutans*, *S. pneumonia*, *S. pyogenes* and *S. suis* possessed abundant genes related to virulence factors, and an obvious regular distribution of virulence factors among species groups was not discovered. Seven relatively prevalent virulence genes were selected to construct ML phylogenetic trees to reveal the evolution of virulence (Figure 6). The seven virulence genes used were *pavA* (fibronectin binding proteins), *srtA* (sortase A), *slrA* (streptococcal lipoprotein rotamase) and *plr/gapA* (streptococcal plasmin receptor/GAPDH) from adhesion, *eno* (streptococcal enolase) from exoenzyme, *htrA/degP* (Serine protease) and *tig/ropA* (trigger factor) from protease, respectively. Interestingly, the phylogenetic relationships from five genes (Figure 6A–C, F–G) share a similar topology in accordance with the phylogenomic analyses. This implies that evolution of adhesion genes (i.e., *pavA*, *srtA* and *slrA*) and protease genes (i.e., *htrA/degP* and *tig/ropA*) is in concordance with evolution of the genus, and these virulence genes are generally monophyletic within most species groups. In contrast, Anginosus and Mitis are not fully resolved, and sometimes are monophyletic with the Unknown group. Thus, the evolutionary relationships of virulence between Unknown group and other groups are needed to investigate in future studies.

The tree topologies of *eno* and *plr/gapA* (Figure 6D–E) are different from the topologies of the other five virulence genes, which indicate phylogenetic clusterings incongruent with the proposed species clusters. Enolase of prokaryotic pathogens represents a multifunctional protein involved in glycolytic and plasminogen binding and activation [71]. Also, it plays a crucial role in fibrinolysis, homeostasis and the degradation of extracellular matrix (ECM) [72–74], enabling infection of tissues and migration between organs. Enolases from *Ureaplasma* and *Mycoplasma* were found to be more similar to archaebacterial enolases than to their bacterial counterparts [75]. Besides, lateral transfer events between endosymbiont and apicomplexan account for evolution of cryptomonad and chlorarachniophyte algal enolases [76]. Genetic exchange of enolases between streptococci and hosts

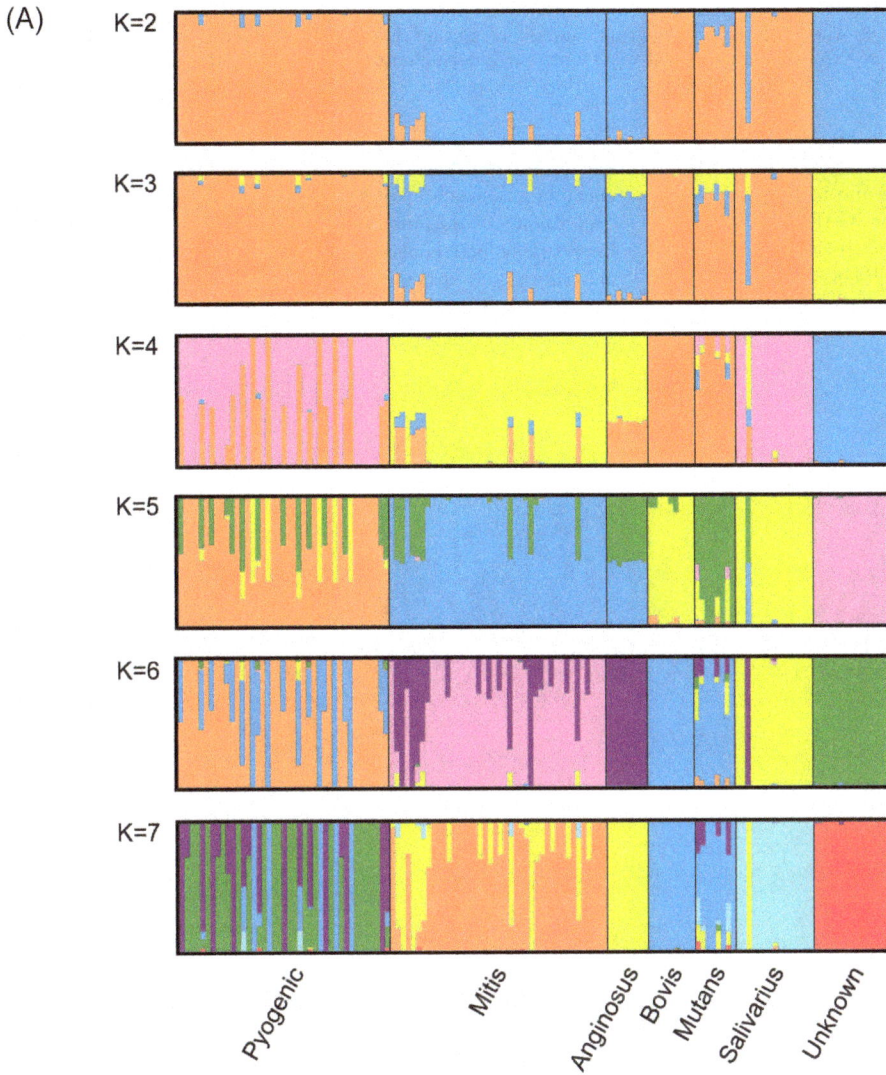

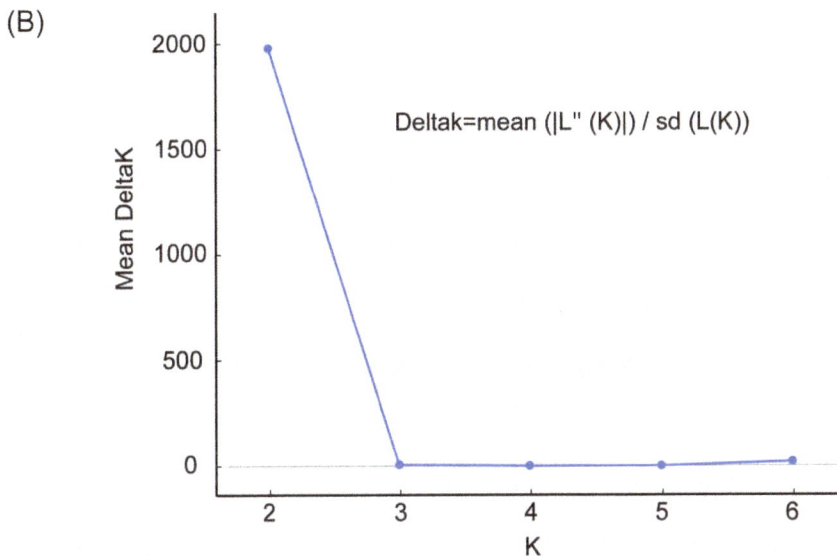

Figure 2. Population structure of streptococcal species groups. (A) The population memberships of the inspected species groups for a priori defined number of clusters K = 1–7 inferred by the Structure software. Each individual is represented by a thin vertical line divided into K colored segments that represent the individual's estimated membership fractions in K clusters. Black lines separate individuals of different populations.

Populations are labeled below the figure. (B) The detection of the true number of clusters inferred by the Structure software and set $\Delta K = mean(|L''(K)|)/sd(L(K))$ as a function of K. ΔK attains its highest value when K = 2, generated by Structure, according to Evanno et al.

could account for phylogram of enolase being incongruent with those of other markers. Streptococcal plasmin receptor, namely, glyceraldehyde-3-phosphate dehydrogenase (GAPDH) constitutes a protein family which displays diverse activities in different subcellular locations, in addition to its well-characterized role in glycolysis [77]. GAPDH of streptococci has been reported to bind fibronectin, lysozyme, the cytoskeletal proteins myosin and actin, affecting colonization of those bacteria [78]. LGT events have been frequently documented in the evolution of GAPDH [79,80]. Interestingly, both enolase and GAPDH are two main receptors of plasminogen in streptococci, and more efforts are required to enlighten their origin.

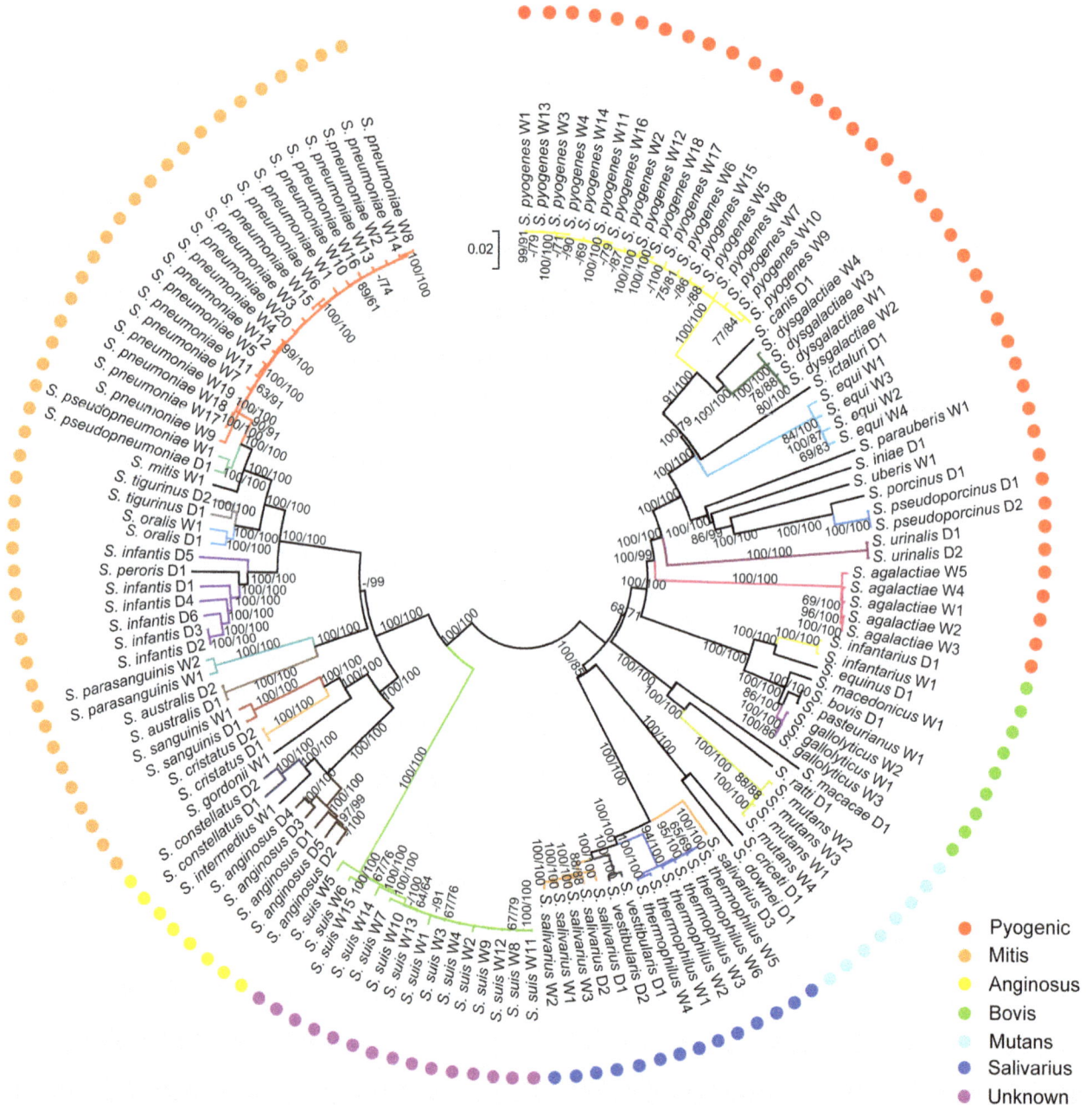

Figure 3. Phylogenomic tree of *Streptococcus.* The supermatrix tree was constructed based on maximum likelihood (ML, bootstrap value indicated as numerator) and neighbor-joining (NJ, bootstrap value indicated as denominator) algorithms, using a concatenated alignment of 278 orthologous proteins. All the 138 *Streptococcus* strains analyzed were assigned to the corresponding species groups and were marked with related colored circles. Different color-coded branches denoted different species.

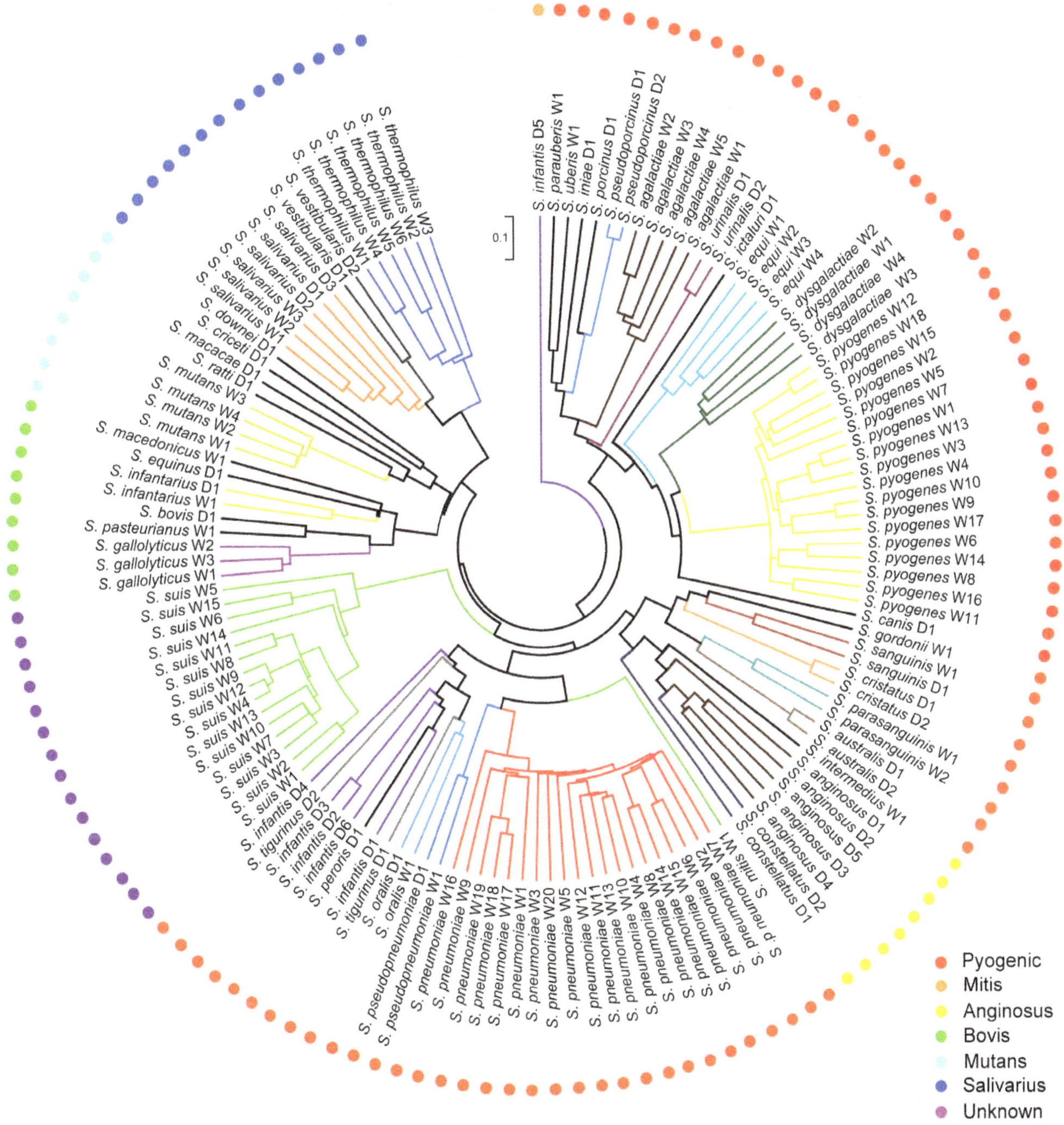

Figure 4. Gene content dendrogram of *Streptococcus*. The dendrogram was constructed by hierarchical clustering (UPGMA) based on the dissimilarities in gene content among 138 *Streptococcus* strains, using paired Jaccard distances which range from 0 to 1. Different color-coded branches denoted different species.

Conclusions

Applications of chemotaxonomic approaches, DNA hybridization and 16S rRNA gene sequencing have resulted in the proposal of "species groups" for streptococci with various lifestyles. Our study, using population structure, phylogenetic and phylogenomic analyses of 138 *Streptococcus* genomes, offers additional insights into the evolution of species and species groups within this genus. Population structure of streptococcal species groups indicated that all *Streptococcus* strains branched into two distinct populations, with Pyogenic, Bovis, Mutans and

Salivarius species groups forming one population, and Mitis, Anginosus and Unknown groups clustering into another population, suggesting that there are two major evolutionary lineages within this genus. Phylogenetic relationships based on core genome and pan-genome suggest that species from the same group are close to each other and indicate a pattern of different species groups accompanying the evolution of the genus *Streptococcus*, which is in accordance with the population structure analysis and provides supports for the proposed species groups based on comparative genomics approaches. Identifica-

(A)

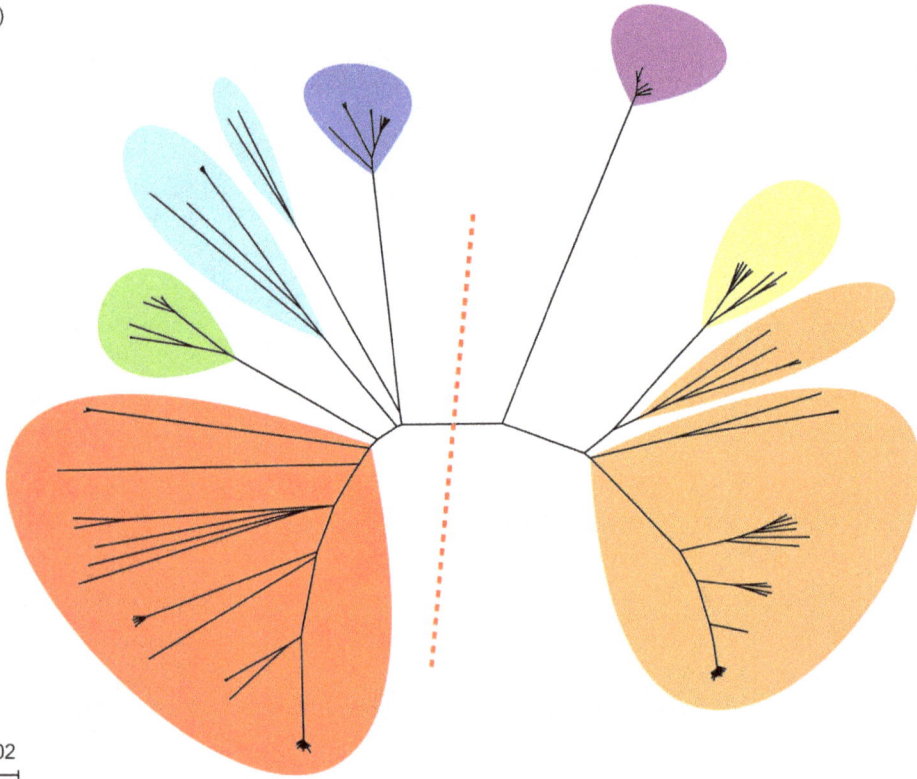

0.02

(B)

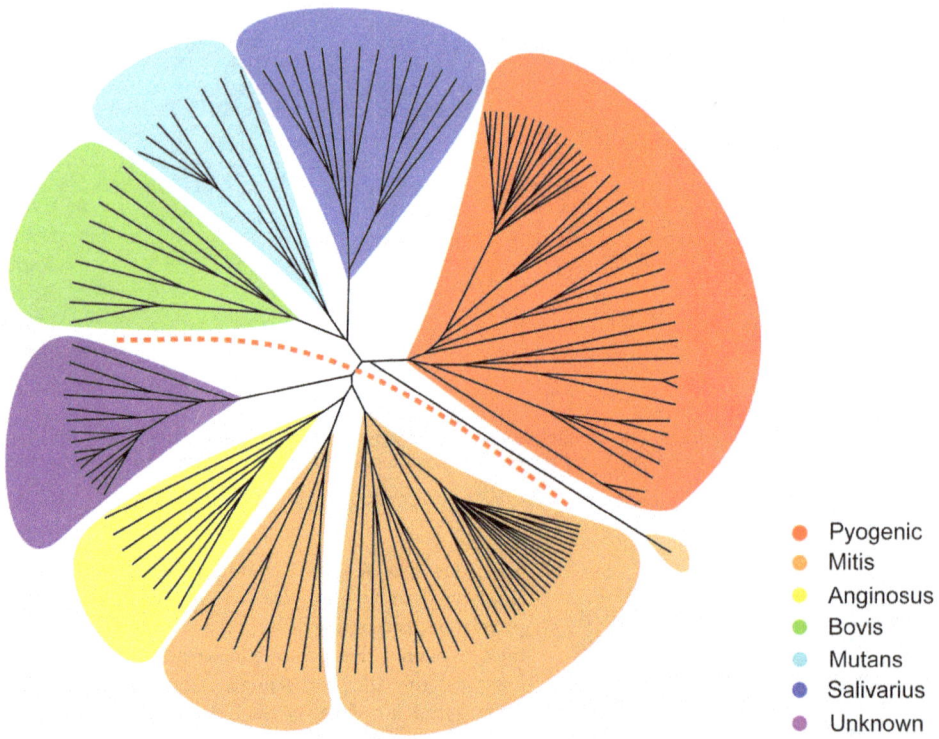

0.1

● Pyogenic
● Mitis
● Anginosus
● Bovis
● Mutans
● Salivarius
● Unknown

Figure 5. Phylogenomic relationships of streptococcal species groups. The clustering results of seven species groups were based on phylogenomic tree and gene content dendrogram. Each species group was painted with the assigned color as the above analysis.

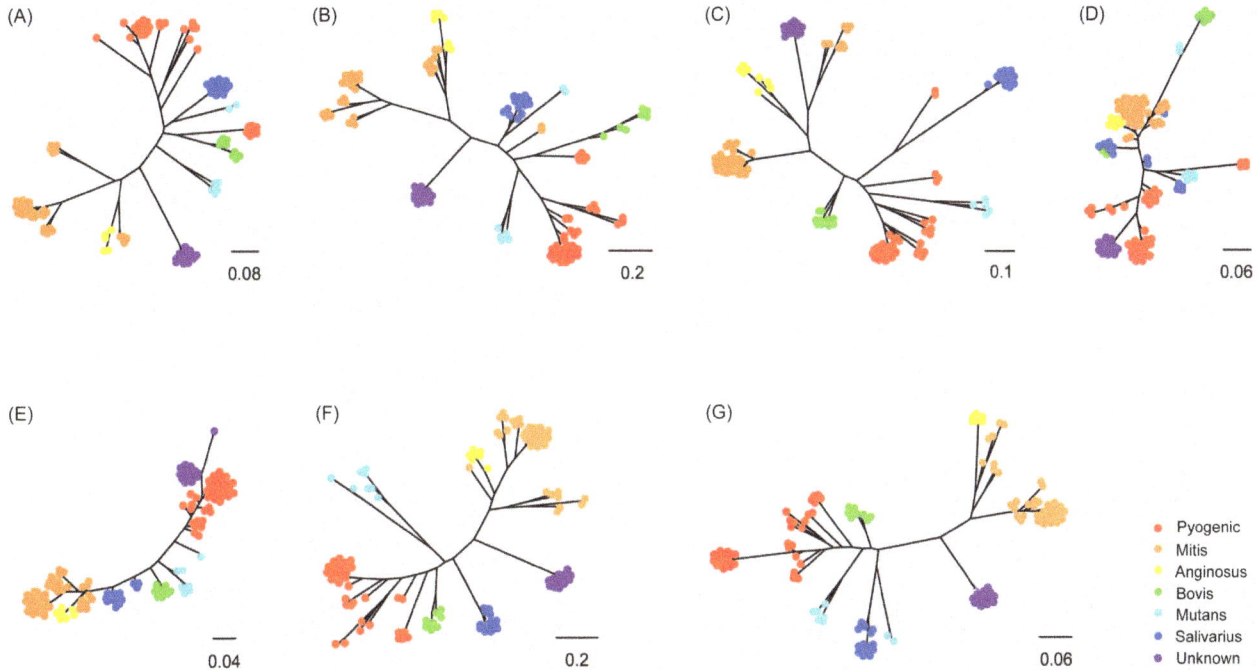

Figure 6. Phylogenetic dendrograms of seven conserved genes related to virulence factors. A, B, C and D represent trees of *pavA*, *srtA*, *slrA*, *plr/gapA* from adhesion factor, respectively; E represents tree of *eno* from exoenzyme factor; F and G represent trees of *htrA/degP* and *tig/ropA* from protease factor, respectively.

tion of virulence factors in streptococci revealed the toxin essence of highly pathogenic streptococci. Moreover, several virulence factors evolve in the same way as species groups according to phylogenies of their common virulence genes. All analyses indicate that the evolution of streptococci is congruent with the evolutionary pattern of species groups. The genus *Streptococcus* possesses an open pan-genome, thus the size of the pan-genome is yet underestimated and will increase as additional streptococcal strains are sequenced. Although the estimated genome size meshes with previous studies cited in the analysis, limitations in our abilities to accurately estimate genome size variation also limit the robustness of our inferences. These inferences should be accepted with caution and, as hypotheses, remain open for testing and refinement in future studies using dataset with more comprehensive sampling of streptococcal strains from a broader habitat range. Nonetheless, this study provides insights into streptococcal species differentiation and enriches our knowledge of evolution within the genus *Streptococcus*.

Supporting Information

Figure S1 Phylogenetic tree of the genus *Streptococcus* based on 16S rRNA gene sequences. The phylogenetic tree was constructed based on ML (bootstrap values on the left of slashes) and NJ (bootstrap values on the right of slashes) algorithms. Species with red fonts had genome data and were analyzed in this study. Species with asterisks possessed complete genome sequences.

Figure S2 Occurrence of homologous clusters and proteins within 138 *Streptococcus* proteomes ranged from 1 to 138. (A) At one extreme of the horizontal axis are the species-specific clusters (8344, 45.03%), while at the opposite end of the scale are clusters, which include genes from every proteome (369, 1.99%). (B) At one

extreme of the horizontal axis are the species-specific proteins present in a single proteome (8582, 3.12%), while at the opposite end of the scale are situated the genes found in all 138 proteomes (51318, 18.67%).

Figure S3 Histogram of core gene clusters assigned COG functional categories. COG categories are indicated to the right of the figure. The ordinate axis indicates the individual COG sub-categories for orthologous and paralogous clusters. The horizontal axis indicates the number of clusters assigned to each COG sub-category.

Figure S4 Comparison of phylogenetic relationships of seven species groups. The clustering results of seven species groups were obtained from gene content dendrograms using different dataset: (A) pan-genome and (B) core-genome.

File S1   Table S1, Classification of streptococcal species groups based on biochemical characteristics. **Table S2,** Genomic size and GC content of *Streptococcus* species and species groups. **Table S3,** Complete list of the 18,528 homologous clusters in 138 *Streptococcus* genomes. **Table S4,** Homologous genes proportion and distribution of *Streptococcus* species and species groups. **Table S5,** Determination of virulence factors in *Streptococcus*.

Acknowledgments

We would like to thank Dr. Jeremy A. Dodsworth (University of Nevada, Las Vegas) for English improvement on the manuscript and valuable suggestions and Dr. Liang-Liang Yue (Kunming Institute of Botany, CAS) for kind help with data analyses.

Author Contributions

Conceived and designed the experiments: XYG XYZ HWL WJL. Performed the experiments: XYG XYZ HWL. Analyzed the data: XYG XYZ HWL. Wrote the paper: XYG XYZ HPK HWL WJL.

References

1. Marri PR, Hao W, Golding GB (2006) Gene gain and gene loss in *Streptococcus*: is it driven by habitat? Mol Biol Evol 23: 2379–2391.
2. Gratten M, Morey F, Dixon J, Manning K, Torzillo P, et al. (1993) An outbreak of serotype 1 *Streptococcus pneumoniae* infection in central Australia. Med J Aust 158: 340–342.
3. Hoe NP, Nakashima K, Lukomski S, Grigsby D, Liu M, et al. (1999) Rapid selection of complement-inhibiting protein variants in group A *Streptococcus* epidemic waves. Nat Med 5: 924–929.
4. Guimbao Bescos J, Vergara Ugarriza A, Aspiroz Sancho C, Aldea Aldanondo MJ, Lazaro MA, et al. (2003) *Streptococcus pneumoniae* transmission in a nursing home: analysis of an epidemic outbreak. Med Clin 121: 48–52.
5. Evans JJ, Bohnsack JF, Klesius PH, Whiting AA, Garcia JC, et al. (2008) Phylogenetic relationships among *Streptococcus agalactiae* isolated from piscine, dolphin, bovine and human sources: a dolphin and piscine lineage associated with a fish epidemic in Kuwait is also associated with human neonatal infections in Japan. J Med Microbiol 57: 1369–1376.
6. Carroll RK, Beres SB, Sitkiewicz I, Peterson L, Matsunami RK, et al. (2011) Evolution of diversity in epidemics revealed by analysis of the human bacterial pathogen group A *Streptococcus*. Epidemics 3: 159–170.
7. Lee S, Kim SH, Park M, Bae S (2013) High prevalence of multiresistance in levofloxacin-nonsusceptible *Streptococcus pneumoniae* isolates in Korea. Diagn Microbiol Infect Dis 76: 227–231.
8. Law BA, Sharpe ME (1978) Formation of methanethiol by bacteria isolated from raw milk and Cheddar cheese. J Dairy Res 45: 267–275.
9. De Vos P, Garrity G, Jones D, Krieg NR, Ludwig W, Rainey FA, Schleifer K, Whitman WB (2009) *Bacillus*. Bergey's manual of systematic Bacteriology 3: 655–710.
10. Bentley RW, Leigh JA, Collins MD (1991) Intrageneric structure of *Streptococcus* based on comparative analysis of small-subunit rRNA sequences. Int J Syst Bacteriol 41: 487–494.
11. Kawamura Y, Hou X-G, Sultana F, Miura H, Ezaki T (1995) Determination of 16S rRNA sequences of *Streptococcus mitis* and *Streptococcus gordonii* and phylogenetic relationships among members of the genus *Streptococcus*. Int J Syst Bacteriol 45: 406–408.
12. Schleifer K, Kilpper-Bälz R (1987) Molecular and chemotaxonomic approaches to the classification of streptococci, enterococci and lactococci: a review. Syst Appl Microbiol 10: 1–19.
13. Drucker D (1974) Chemotaxonomic fatty-acid fingerprints of some streptococci with subsequent statistical analysis. Can J Microbiol 20: 1723–1728.
14. Stackebrandt E, Frederiksen W, Garrity GM, Grimont PA, Kämpfer P, et al. (2002) Report of the ad hoc committee for the re-evaluation of the species definition in bacteriology. Int J Syst Evol Microbiol 52: 1043–1047.
15. Li Z, Yang H, He N, Liang W, Ma C, et al. (2013) Solid-Phase hybridization efficiency improvement on the magnetic nanoparticle surface by using dextran as molecular arms. J Biomed Nanotechnol 9: 1945–1949.
16. Konstantinidis KT, Tiedje JM (2007) Prokaryotic taxonomy and phylogeny in the genomic era: advancements and challenges ahead. Curr Opin Microbiol 10: 504–509.
17. Xu J (2006) Invited review: microbial ecology in the age of genomics and metagenomics: concepts, tools, and recent advances. Mol Ecol 15: 1713–1731.
18. Hajibabaei M, Singer GA, Hebert PD, Hickey DA (2007) DNA barcoding: how it complements taxonomy, molecular phylogenetics and population genetics. Trends Genet 23: 167–172.
19. Bolotin A, Quinquis B, Renault P, Sorokin A, Ehrlich SD, et al. (2004) Complete sequence and comparative genome analysis of the dairy bacterium *Streptococcus thermophilus*. Nat Biotechnol 22: 1554–1558.
20. Rusniok C, Couvé E, Da Cunha V, El Gana R, Zidane N, et al. (2010) Genome sequence of *Streptococcus gallolyticus*: insights into its adaptation to the bovine rumen and its ability to cause endocarditis. J Bacteriol 192: 2266–2276.
21. Kreikemeyer B, McIver KS, Podbielski A (2003) Virulence factor regulation and regulatory networks in *Streptococcus pyogenes* and their impact on pathogen-host interactions. Trends Microbiol 11: 224–232.
22. Johri AK, Paoletti LC, Glaser P, Dua M, Sharma PK, et al. (2006) Group B *Streptococcus*: global incidence and vaccine development. Nat Rev Microbiol 4: 932–942.
23. Maione D, Margarit I, Rinaudo CD, Masignani V, Mora M, et al. (2005) Identification of a universal Group B *Streptococcus* vaccine by multiple genome screen. Science 309: 148–150.
24. Tettelin H, Masignani V, Cieslewicz MJ, Donati C, Medini D, et al. (2005) Genome analysis of multiple pathogenic isolates of *Streptococcus agalactiae*: implications for the microbial "pan-genome". Proc Natl Acad Sci U S A 102: 13950–13955.
25. Hiller NL, Janto B, Hogg JS, Boissy R, Yu S, et al. (2007) Comparative genomic analyses of seventeen *Streptococcus pneumoniae* strains: insights into the pneumococcal supragenome. J Bacteriol 189: 8186–8195.
26. Lefebure T, Stanhope MJ (2007) Evolution of the core and pan-genome of *Streptococcus*: positive selection, recombination, and genome composition. Genome Biol 8: R71.
27. Donati C, Hiller NL, Tettelin H, Muzzi A, Croucher NJ, et al. (2010) Structure and dynamics of the pan-genome of *Streptococcus pneumoniae* and closely related species. Genome Biol 11: R107.
28. Fraser-Liggett CM (2005) Insights on biology and evolution from microbial genome sequencing. Genome Res 15: 1603–1610.
29. Dobrindt U, Hacker J (2001) Whole genome plasticity in pathogenic bacteria. Curr Opin Microbiol 4: 550–557.
30. Barocchi MA, Censini S, Rappuoli R (2007) Vaccines in the era of genomics: the pneumococcal challenge. Vaccine 25: 2963–2973.
31. Murray PR, Drew WL, Kobayashi GS, Thompson J Jr (1990) Medical microbiology: Wolfe Medical Publications Ltd.
32. Facklam R (2002) What happened to the streptococci: overview of taxonomic and nomenclature changes. Clin Microbiol Rev 15: 613–630.
33. Köhler W (2007) The present state of species within the genera *Streptococcus* and *Enterococcus*. Int J Med Microbiol 297: 133–150.
34. Li L, Stoeckert CJ, Roos DS (2003) OrthoMCL: identification of ortholog groups for eukaryotic genomes. Genome Res 13: 2178–2189.
35. Moreno-Hagelsieb G, Latimer K (2008) Choosing BLAST options for better detection of orthologs as reciprocal best hits. Bioinformatics 24: 319–324.
36. Enright AJ, Van Dongen S, Ouzounis CA (2002) An efficient algorithm for large-scale detection of protein families. Nucleic Acids Res 30: 1575–1584.
37. Medini D, Donati C, Tettelin H, Masignani V, Rappuoli R (2005) The microbial pan-genome. Curr Opin Genet Dev 15: 589–594.
38. Tettelin H, Masignani V, Cieslewicz MJ, Donati C, Medini D, et al. (2005) Genome analysis of multiple pathogenic isolates of *Streptococcus agalactiae*: implications for the microbial "pan-genome". Proc Natl Acad Sci U S A 102: 13950–13955.
39. Touchon M, Hoede C, Tenaillon O, Barbe V, Baeriswyl S, et al. (2009) Organised genome dynamics in the *Escherichia coli* species results in highly diverse adaptive paths. PLoS Genet 5: e1000344.
40. Tenaillon O, Skurnik D, Picard B, Denamur E (2010) The population genetics of commensal *Escherichia coli*. Nat Rev Microbiol 8: 207–217.
41. Li HW, Zhi XY, Yao JC, Zhou Y, Tang SK, et al. (2013) Comparative genomic analysis of the genus *Nocardiopsis* provides new insights into its genetic mechanisms of environmental adaptability. PLoS One 8: e61528.
42. Gray CD, Kinnear PR (2012) IBM SPSS Statistics 19 made simple: Psychology Press.
43. Pritchard JK, Stephens M, Donnelly P (2000) Inference of population structure using multilocus genotype data. Genetics 155: 945–959.
44. Falush D, Stephens M, Pritchard JK (2007) Inference of population structure using multilocus genotype data: dominant markers and null alleles. Mol Ecol Notes 7: 574–578.
45. Evanno G, Regnaut S, Goudet J (2005) Detecting the number of clusters of individuals using the software STRUCTURE: a simulation study. Mol Ecol 14: 2611–2620.
46. Rosenberg NA (2004) DISTRUCT: a program for the graphical display of population structure. Mol Ecol Notes 4: 137–138.
47. Earl DA (2012) STRUCTURE HARVESTER: a website and program for visualizing STRUCTURE output and implementing the Evanno method. Conserv Genet Resour 4: 359–361.
48. Thompson JD, Gibson T, Higgins DG (2002) Multiple sequence alignment using ClustalW and ClustalX. Curr Protoc Bioinformatics: 2.3. 1–2.3. 22.
49. Tamura K, Peterson D, Peterson N, Stecher G, Nei M, et al. (2011) MEGA5: molecular evolutionary genetics analysis using maximum likelihood, evolutionary distance, and maximum parsimony methods. Mol Biol Evol 28: 2731–2739.
50. Stamatakis A (2006) RAxML-VI-HPC: maximum likelihood-based phylogenetic analyses with thousands of taxa and mixed models. Bioinformatics 22: 2688–2690.
51. Plotree D, Plotgram D (1989) PHYLIP-phylogeny inference package (version 3.2).
52. Burke GR, Moran NA (2011) Massive genomic decay in *Serratia symbiotica*, a recently evolved symbiont of aphids. Genome Biol Evol 3: 195.
53. Hildebrand F, Meyer A, Eyre-Walker A (2010) Evidence of selection upon genomic GC-content in bacteria. PLoS Genet 6: e1001107.
54. Wu H, Zhang Z, Hu S, Yu J (2012) On the molecular mechanism of GC content variation among eubacterial genomes. Biol Direct 7.
55. Jensen RA (2001) Orthologs and paralogs-we need to get it right. Genome Biol 2: 1002.1–1002.3.
56. Koonin EV (2005) Orthologs, paralogs, and evolutionary genomics. Ann Rev Genet 39: 309–338.
57. Snel B, Bork P, Huynen MA (2002) Genomes in flux: the evolution of archaeal and proteobacterial gene content. Genome Res 12: 17–25.

58. Koonin EV, Wolf YI (2008) Genomics of bacteria and archaea: the emerging dynamic view of the prokaryotic world. Nucleic Acids Res 36: 6688–6719.

59. Zhang A, Yang M, Hu P, Wu J, Chen B, et al. (2011) Comparative genomic analysis of *Streptococcus suis* reveals significant genomic diversity among different serotypes. BMC Genomics 12: 523.

60. Donati C, Hiller NL, Tettelin H, Muzzi A, Croucher N, et al. (2010) Structure and dynamics of the pan-genome of *Streptococcus pneumoniae* and closely related species. Genome Biol 11: R107.

61. Lapierre P, Gogarten JP (2009) Estimating the size of the bacterial pan-genome. Trends Genet 25: 107–110.

62. Hohwy J, Reinholdt J, Kilian M (2001) Population dynamics of *Streptococcus mitis* in its natural habitat. Infect Immu 69: 6055–6063.

63. Alcaraz L, Moreno-Hagelsieb G, Eguiarte L, Souza V, Herrera-Estrella L, et al. (2010) Understanding the evolutionary relationships and major traits of *Bacillus* through comparative genomics. BMC Genomics 11: 332.

64. Phylip LH, Richards AD, Kay J, Konvalinka J, Strop P, et al. (1990) Hydrolysis of synthetic chromogenic substrates by HIV-1 and HIV-2 proteinases. Biochem Biophysl Res Commun 171: 439–444.

65. Lunsford RD, London J (1996) Natural genetic transformation in *Streptococcus gordonii*: comX imparts spontaneous competence on strain wicky. J Bacterioly 178: 5831–5835.

66. Shanley TP, Schrier D, Kapur V, Kehoe M, Musser JM, et al. (1996) Streptococcal cysteine protease augments lung injury induced by products of group A streptococci. Infect Immu 64: 870–877.

67. Enright MC, Spratt BG, Kalia A, Cross JH, Bessen DE (2001) Multilocus sequence typing of *Streptococcus pyogenes* and the relationships between emm type and clone. Infect Immu 69: 2416–2427.

68. Dowson CG, Hutchison A, Brannigan JA, George RC, Hansman D, et al. (1989) Horizontal transfer of penicillin-binding protein genes in penicillin-resistant clinical isolates of *Streptococcus pneumoniae*. Proc Natl Acad Sci U S A 86: 8842–8846.

69. Wessels MR, Moses AE, Goldberg JB, DiCesare TJ (1991) Hyaluronic acid capsule is a virulence factor for mucoid group A streptococci. Proc Natl Acad Sci U S A 88: 8317–8321.

70. Hyams C, Camberlein E, Cohen JM, Bax K, Brown JS (2010) The *Streptococcus pneumoniae* capsule inhibits complement activity and neutrophil phagocytosis by multiple mechanisms. Infect Immu 78: 704–715.

71. Pancholi V (2001) Multifunctional α-enolase: its role in diseases. Cell Mol Life Sci 58: 902–920.

72. Collen D, Verstraete M (1975) Molecular biology of human plasminogen. II. Metabolism in physiological and some pathological conditions in man. Thromb Diath Haemorrh 34: 403.

73. Saksela O, Rifkin DB (1988) Cell-associated plasminogen activation: regulation and physiological functions. Ann Rev Cell Biol 4: 93–120.

74. Vassalli J-D, Sappino A-P, Belin D (1991) The plasminogen activator/plasmin system. J Clin Invest 88: 1067.

75. Piast M, Kustrzeba-Wójcicka I, Matusiewicz M, Banas T (2005) Molecular evolution of enolase. Acta Biochem Pol 52: 507.

76. Keeling PJ, Palmer JD (2001) Lateral transfer at the gene and subgenic levels in the evolution of eukaryotic enolase. Proc Natl Acad Sci U S A 98: 10745–10750.

77. Sirover MA (1999) New insights into an old protein: the functional diversity of mammalian glyceraldehyde-3-phosphate dehydrogenase. Biochim Biophys Acta 1432: 159–184.

78. Pancholi V, Fischetti VA (1992) A major surface protein on group A streptococci is a glyceraldehyde-3-phosphate-dehydrogenase with multiple binding activity. J Exp Med 176: 415–426.

79. Takishita K, Inagaki Y (2009) Eukaryotic origin of glyceraldehyde-3-phosphate dehydrogenase genes in *Clostridium thermocellum* and *Clostridium cellulolyticum* genomes and putative fates of the exogenous gene in the subsequent genome evolution. Gene 441: 22–27.

80. Baibai T, Oukhattar L, Mountassif D, Assobhei O, Serrano A, et al. (2010) Comparative molecular analysis of evolutionarily distant glyceraldehyde-3-phosphate dehydrogenase from *Sardina pilchardus* and *Octopus vulgaris*. Acta Biochim Biophys Sin 42: 863–872.

17

The Effects of Extra-Somatic Weapons on the Evolution of Human Cooperation towards Non-Kin

Tim Phillips[1]*, Jiawei Li[2], Graham Kendall[2,3]

1 Independent Researcher, Birmingham, United Kingdom, 2 Automated Scheduling, Optimisation and Planning (ASAP) research group, School of Computer Science, University of Nottingham, Nottingham, United Kingdom, 3 University of Nottingham Malaysia Campus, Broga, Malaysia

Abstract

Human cooperation and altruism towards non-kin is a major evolutionary puzzle, as is 'strong reciprocity' where no present or future rewards accrue to the co-operator/altruist. Here, we test the hypothesis that the development of extra-somatic weapons could have influenced the evolution of human cooperative behaviour, thus providing a new explanation for these two puzzles. Widespread weapons use could have made disputes within hominin groups far more lethal and also equalized power between individuals. In such a cultural niche non-cooperators might well have become involved in such lethal disputes at a higher frequency than cooperators, thereby increasing the relative fitness of genes associated with cooperative behaviour. We employ two versions of the evolutionary Iterated Prisoner's Dilemma (IPD) model – one where weapons use is simulated and one where it is not. We then measured the performance of 25 IPD strategies to evaluate the effects of weapons use on them. We found that cooperative strategies performed significantly better, and non-cooperative strategies significantly worse, under simulated weapons use. Importantly, the performance of an 'Always Cooperate' IPD strategy, equivalent to that of 'strong reciprocity', improved significantly more than that of all other cooperative strategies. We conclude that the development of extra-somatic weapons throws new light on the evolution of human altruistic and cooperative behaviour, and particularly 'strong reciprocity'. The notion that distinctively human altruism and cooperation could have been an adaptive trait in a past environment that is no longer evident in the modern world provides a novel addition to theory that seeks to account for this major evolutionary puzzle.

Editor: Alex Mesoudi, Durham University, United Kingdom

Funding: The authors have no support or funding to report.

Competing Interests: The authors have declared that no competing interests exist.

* E-mail: Ptjp2749@aol.com

Introduction

The puzzle of cooperation and strong reciprocity

Human cooperation and altruism towards non-kin poses two fundamental questions for biology and the behavioural sciences. Firstly, why do genes associated with these behaviours survive when evolutionary theory appears to predict that a 'cheating' strategy is fitter and will thus drive such genes to extinction [1], [2]? Secondly, 'strong reciprocity' is defined as a propensity to cooperate unconditionally even if this is costly and provides neither present nor future rewards to the co-operator/altruist [3], [4]. Why, therefore, does this behaviour persist as it incurs a cost and/or offers no present or future rewards [3], [4]?

Much current theory suggests that the first question can be resolved by punishment of 'cheats' who do not reciprocate altruistic acts i.e. reciprocal altruism [5], [6] and/or by cooperative individuals recognizing each other by reputation and gaining fitness by only associating with other cooperative individuals i.e. indirect reciprocity [7], [8], [9], [10], [11].

'Strong reciprocity', however, has been observed in one-off, anonymous encounters under experimental conditions [12] where the reciprocal and reputation effects required by reciprocal altruism or indirect reciprocity theory could not influence the behaviour of participants [4]. Instead, it has been explained by 'new' group selection linked to extinction-threatening events [3],

[13], to cultural evolution [4], [14], to gene-culture co-evolution [15] and to a combination of the last two factors [16], [17]. Others explain it as a response to conditions of uncertainty in reciprocal relationships [18] or as a maladaptive 'misfiring' of evolutionary mechanisms in modern, experimental settings [19].

Extra-somatic weapons in human evolution

Human cooperation towards non-kin has often been seen as unique in nature [8], [12], [14], [16], [20]. Here, we propose that this very uniqueness might be attributable to unique selection pressures likely to have been present in human evolution itself. The invention of tools and weapons is generally seen as being an important step in human evolution as it enabled hominins to consume a high protein diet through the hunting of game and/or scavenging of meat [21]. Such weapons should be distinguished from somatic weapons (e.g. prominent canine teeth, pronounced musculature), with only humans employing extra-somatic weapons on a widespread and systematic basis. Here, we explore whether their invention might have led to a cultural niche that could have resulted in the evolution of distinctively human cooperation. It has been claimed that cultural processes, particularly in humans, can lead to the creation of niches that change selection pressures to which individuals are exposed, thereby influencing their evolution [22], [23], [24].

The oldest complete hunting weapons yet discovered have been dated to approximately 400,000 years ago [25], although it is speculated that wooden weapons could have existed up to 1 million years before the present [26]. These wooden hunting spears resemble projectile weapons [25] although there is some question as to whether they would have been too heavy to have been used as true projectiles [27] and may have been used instead as thrusting weapons. Recent analysis has suggested that humans first developed long-range projectile weaponry in Africa 70–90,000 years ago [27]. Nearly every known human population over the past 50,000 years has used projectile weapons (e.g. spears and darts, bows and arrows) [28] and those few groups who lacked them (e.g. Tasmanian Aborigines) are known to be descended from populations that did possess them [28].

It has been suggested that the invention of extra-somatic weapons in human evolution could have had an important impact on relationships within hominin groups [5], [6], [29]. In many species, intraspecific conflict has been observed to be of the 'limited war' type in which ritualized tactics or inefficient somatic weapons are used to settle disputes without death or serious injury resulting [30]. However, in the case of early humans, the lethal effectiveness of extra-somatic weapons is likely to have selected against those behavioural adaptations that inhibit intraspecific violence observed in many other species. Specifically, the development of extra-somatic weapons could have resulted in:

- an increased frequency of agonistic encounters as dominant individuals, hitherto reliant on physical size, strength and intimidation alone to gain access to valued resources (e.g. mating with females, hunted meat), would have been more open to challenge [29];

- a greater likelihood that such encounters would have proved lethal for the protagonists due to the effectiveness of the weapons and the speed with which they could have resulted in serious injury or death [29] and;

- a consequently greater symmetry or equalization in power between individuals [5], [6], [29] as it could have proved as easy for a subordinate to kill a dominant as *vice versa*.

In such a cultural niche it is likely that natural selection would have discriminated strongly against individuals whose behaviour provoked an above average level of within-group aggression against them. 'Cheats' and non-cooperators might well have experienced involvement in such lethal disputes at a higher frequency than others and thus faced a correspondingly greater risk of injury or death. Providing that the costs of involvement in agonistic encounters, where weapons were employed, exceeded the benefits of 'cheating' and non-cooperation then such strategies would have proved maladaptive.

In contrast, cooperators inclined to reduce their fitness in order to help others within their group could as a result have been less likely to have become involved in lethal disputes and might consequently have experienced relatively greater fitness. Providing that the cost of this helping behaviour was less than the selective benefit of a reduced frequency of involvement in lethal fights then cooperation could have proved an adaptive strategy in this cultural niche.

To test this hypothesis, we employed two versions of the evolutionary Iterated Prisoner's Dilemma (IPD) model – one in which weapons use was absent and one in which it was present. We used computer simulation to measure the performance of a range of IPD strategies, both cooperative and non-cooperative, in order to quantify the effects of weapons use on them.

Methods

The traditional iterated prisoner's dilemma (IPD) model

The IPD model [31], [32] is a classic formulation of how mutual co-operation can evolve in a world of selfish individuals and its process is well known (see Supporting Information, File S1. The iterated prisoner's dilemma (IPD) model). Each player has a choice of whether to cooperate or defect on each move. If both cooperate each receives three points but if both defect each gets only one point. If one cooperates and the other defects then the former receives no points and the latter five points. On any single move it always pays to defect but cooperation has been found to emerge where the endpoint of the series of interactions between two players is unknown [32]. In modelling the IPD, this uncertainty is reflected in the duration of each interaction being determined by a certain probability or discount parameter (see File S1). Computer tournaments have been used to identify those strategies that perform best, with 'Tit for Tat' (TFT) (cooperate on the first move and then copy the opponent's last move) [31] and Pavlov [33] generally found to be most successful.

The evolutionary IPD model

The evolutionary IPD model [34] extends the principle of the traditional IPD model by reflecting the payoffs received by players in one generation in terms of copies of themselves represented in the next generation. Stochastic universal sampling is used to ensure that players produce offspring in proportion to payoffs received so that those with higher payoffs reproduce at a proportionately higher rate than those with lower payoffs. In so far as payoffs reflect fitness the evolutionary IPD model can be seen to mimic natural selection, although recombination and mutation are not simulated.

In our evolutionary IPD model a population of 40 players compete against each other on a round robin basis. We chose 25 widely recognized IPD strategies taken from the scientific literature [34], of which 14 have been classified as cooperative and 11 as non-cooperative [35] (see Supporting Information, File S2: IPD strategies employed).

In each round (see Supporting Information, File S3: Definitions of terms) eight strategies were chosen at random with five players initially adopting a particular strategy. To ensure that each strategy was simulated a sufficient number of times to remove the effects of chance we repeated each round 100 times per generation and averaged the payoffs. Strategies were chosen randomly so that, for example in the initial round, there was an $8/25 = 0.32$ probability of any one strategy being chosen. With 100 rounds in that game, each strategy was therefore run an average of $100*0.32 = 32$ times. As strategies were subsequently eliminated, this probability was adjusted throughout the competition. With stochastic universal sampling used to choose players for the next generation on the basis of payoffs received in the previous one, the game was repeated for 100 generations, showing how each of the 25 strategies increased, decreased or were eliminated over the duration of the competition.

The performance of each strategy was measured by the number of times each survived for the full 100 generations. As long as a single player adopting a particular strategy was present at the end then that strategy was deemed to have survived. Survival time is seen as providing a comprehensive index of the performance of IPD strategies [35] and in this context offers a measure of the relative fitness of individuals adopting each strategy.

The 'weapons use' IPD model

The 'weapons use' IPD model we designed for this study was based on the evolutionary IPD model but we also set out to simulate the effects of extra-somatic weapons on the fitness of players. Our model works on the basis that a defection is, in effect, a refusal to cooperate and an attempt to exploit the other player. It therefore assumes that the more defections there are between players, the greater will be the chance of a dispute occurring between them. It thus captures a key aspect of the real world – that, in relationships with others, 'cheats' and uncooperative individuals (i.e. those inclined to defect) are more likely to be involved in disputes than 'nice' individuals who are more inclined to cooperate.

When a player accumulated 200 defections as a result of moves both by the player and the player's opponents a dispute was deemed to occur. To simulate an environment of equalized power, each player had the same probability of being eliminated in each dispute (i.e. $p = 0.05$). If either or both were eliminated they took no further part in the competition. If they were not, the process continued until a further 200 defections were accumulated and another dispute was deemed to occur.

Thus, as well as stochastic universal sampling reflecting the payoffs of players who adopt a particular strategy, our model quite separately reflected the effects on fitness of elimination as a result of disputes. In the 'weapons use' IPD model, therefore, the performance of each strategy was a measure of both decisions on whether to cooperate or defect and also the effects of the number of defections leading to disputes.

Finally, in all simulations we contrasted the performance of strategies under the evolutionary IPD model (i.e. without weapons use) with that under the 'weapons use' IPD model to quantify the impact of weapons use. We were thus able to simulate the relative impact of weapons use on the fitness of individuals adopting each strategy in this environment.

Other points

Our population of 40 players lies within the typical size range of 15–50 observed in modern hunter/gatherer groups [36]. Nevertheless we are aware that there is mobility between hunter/gatherer groups [37] and that therefore a typical hunter/gatherer would interact with more than 40 individuals over a lifetime. However, we make the simplifying assumption that transfers in and out of groups would have had a broadly neutral effect on the frequency with which each player would encounter the 25 IPD strategies modelled. This point is reflected in the re-running of each round 100 times per generation with different combinations of strategies being encountered in each round.

The 'weapons use' IPD model has three key parameters (i.e. average number of interactions with other players in a generation, number of defections before a dispute occurs, probability of being eliminated as a result of a dispute) that, taken together, determine the fitness of each player. The values chosen, however, were not arbitrary but based on a population consisting entirely of players who always defect (AllD) and who are thus at the extreme end of the continuum of behaviour being examined. Based on the values selected, in an AllD population there is a probability of $p = 0.50$ of being eliminated in an average lifetime (see Supporting Information, File S4: The parameters and values used). These values therefore provide a clear benchmark against which the performance of the 25 IPD strategies was calibrated.

Results

The performance of the cooperative and non-cooperative strategies in our sample is illustrated in Figures 1 and 2 respectively. We found that all cooperative strategies survived, on average, an additional 5.6 generations under simulated weapons use (see Table 1), a significant variation (paired samples t-test: $t_{13} = 3.22$; $p = 0.003$ one-tailed). In contrast, we found that non-cooperative strategies survived, on average, 8.8 fewer generations under simulated weapons use (see Table 2), which was also a significant variation (paired samples t-test: $t_{10} = 7.47$; $p = 0.00001$ one-tailed).

We were surprised at the success of one strategy in particular - 'Always Cooperate' (AllC) - in our simulations. When we contrasted the performance of AllC with all other cooperative strategies we found it survived an additional 23.5 generations under simulated weapons use as opposed to an average of only an additional 4.2 generations for all other cooperative strategies (see Table 1). This demonstrated a very significant improvement in performance (one-sample t-test: $t_{11} = 16.03$; $p = 5.7*1^{-9}$ one-tailed) for AllC over all other cooperative strategies in this cultural niche.

Finally, throughout our simulations we have assumed an elimination rate of $p = 0.05$. To explore whether variation in the negative impact of weapons use on fitness might produce different trends we adjusted the elimination rate by stages from $p = 0.00$ (without weapons use) to $p = 0.25$ (or five times the rate used in our original simulation). To simplify illustration of the key patterns we averaged the performance of non-cooperative strategies and contrasted that of AllC with the average for other cooperative strategies (see Figure 3). We found that for non-cooperative strategies and for all cooperative strategies other than AllC the influence of weapons use was most marked between $p = 0.00$ and $p = 0.05$ but thereafter the effect tailed off. In contrast, the performance of AllC generally continued to improve above $p = 0.05$ rather than tailing off. In response to simulated weapons use, AllC improved from being the least successful cooperative strategy at $p = 0.00$ to being the most successful at $p = 0.15$ and $p = 0.25$ (see Table 3).

Discussion

Under simulated weapons use, the performance of cooperative strategies improved significantly compared with an environment where weapons use was absent. In contrast, the performance of non-cooperative strategies declined significantly under simulated weapons use compared with an environment where this effect was not modelled. Our hypothesis was therefore supported. The higher incidence of disputes encountered between non-cooperative players and the adverse effect this had on individual fitness appears to explain the patterns found, despite the usual payoffs from the IPD model.

We believe that these findings have important implications for understanding how distinctively human cooperation might have evolved. The traditional IPD model shows how cooperation can emerge in a world of selfish individuals but our model demonstrates that, in an environment of widespread weapons use, this tendency could have been boosted to a substantially greater degree. The invention of extra-somatic weapons has been rightly recognized as an important step in human evolution but this is the first time, as far as we are aware, that its likely indirect impact on human cooperation has been modelled. We therefore consider that these findings provide an important new perspective in helping us to better understand the evolution of human cooperation.

Average survival time

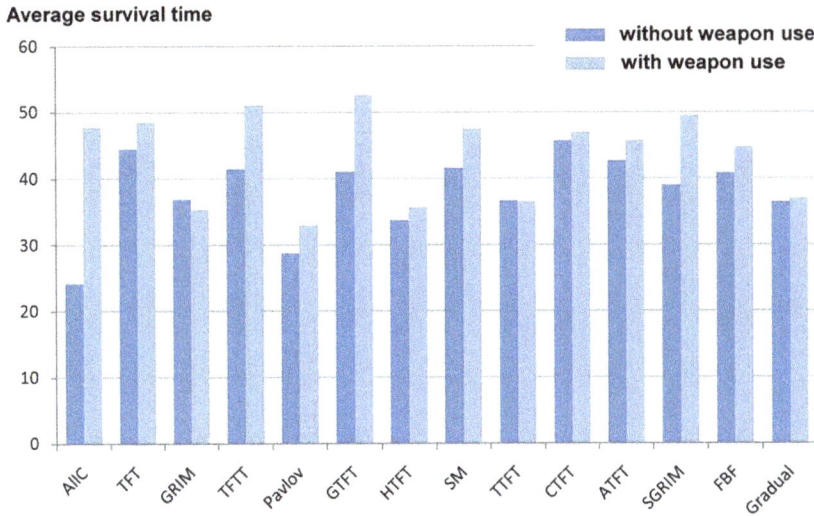

Figure 1. Cooperative strategies: average number of generations survived out of 100 generations (without and with weapons use).

As discussed above, the traditional IPD model and reciprocity theory have difficulty accounting for the evolutionary puzzle of 'strong reciprocity' and other forms of the costly punishment of 'cheats'. The unexpected success of the 'Always Cooperate' (AllC) strategy in our simulations, we believe, has important implications for resolving this puzzle. With the traditional IPD model, AllC is easily exploited and is thus relatively unsuccessful [32]. Under the evolutionary IPD model (i.e. without simulated weapons use), AllC proved the least successful of all cooperative strategies (see Table 1). However, despite the lower payoffs that it usually receives and the additional burden of fighting defectors with weapons, an AllC strategy flourished under the 'weapons use' IDP model. Its performance improved substantially more than that of all other cooperative strategies in an environment of simulated weapons use (see Table 1) and, from being the poorest performing cooperative strategy in the absence of weapons use, it became the best

performing when the negative effect of weapons use was increased to $p = 0.15$ and $p = 0.25$ (see Table 3).

We suggest that the unexpected success of an AllC strategy has important implications for understanding the evolution of 'strong reciprocity'. The two distinguishing features of 'strong reciprocity' are that (i) it is unconditionally cooperative and (ii) it is prepared to inflict costly punishment on 'cheats' [3], [4]. Under the 'weapons use' evolutionary IPD model, an AllC strategy (i) by definition, cooperates unconditionally and (ii) becomes involved in disputes with defecting strategies that involve an equal probability of elimination from the game. An AllC strategy is therefore equivalent to one of 'strong reciprocity' in this cultural niche. The fact that, among all the strategies simulated, AllC responded best to an environment where lethal weapons use was common (see Figure 3 and Table 3) thus provides a new explanation for the evolutionary puzzle of 'strong reciprocity'.

Average survival time

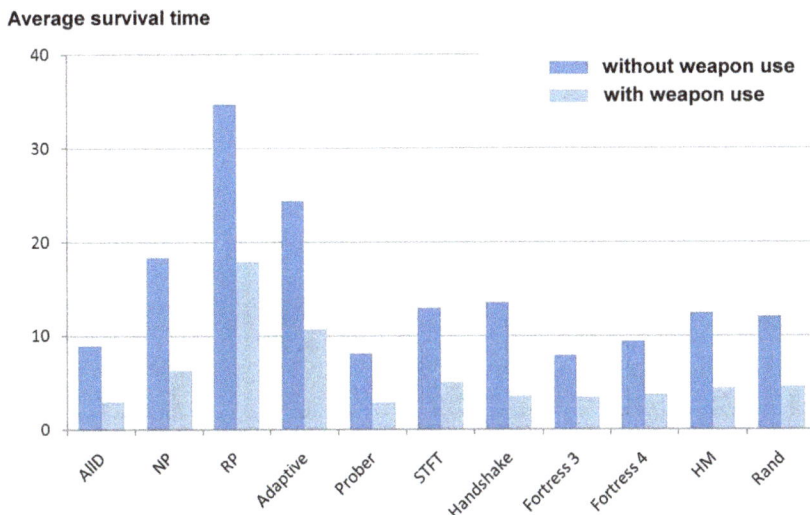

Figure 2. Non-cooperative strategies: average number of generations survived out of 100 generations (without and with weapons use).

Table 1. Cooperative strategies: average survival time in 100 evolutionary games (for details of strategies see S 2: IPD strategies employed).

Strategy	Without weapon use (generations) ($p=0.00$)	With weapon use (generations) ($p=0.05$)	Change (generations)
AllC	24.26	47.72	23.46
TFT	44.56	48.46	3.90
GRIM	36.84	35.29	−1.55
TFTT	41.44	51.05	9.61
Pavlov	28.75	32.96	4.21
GTFT	41.01	52.66	11.65
HTFT	33.76	35.61	1.85
SM	41.62	47.51	5.89
TTFT	36.62	36.54	−0.08
CTFT	45.65	47.05	1.40
ATFT	42.72	45.72	3.00
SGRIM	38.96	49.53	10.57
FBF	40.84	44.74	3.90
Gradual	36.36	36.94	0.58
Total	500.03	577.84	78.39

Note: Average change in performance for all strategies is 78.39/14 = 5.6 generations. Average change in performance of all strategies other than 'Always Cooperate' is 54.93/13 = 4.2 generations.

Our model does not allow players to avoid costly disputes and thus opt out of their adverse effects on individual fitness as this reflects a likely key feature of the development of extra-somatic weapons in human evolution. For, in the resulting environment or cultural niche, it would scarcely have been possible to 'disinvent' such weapons or opt out of their use. Attempted invasion of a population of weapons users by those who avoided the use of weapons, and the costs involved, is likely to have resulted in the death of such individuals at the hands of those in possession of superior weapons technology. It would not therefore have been possible to 'cheat' by adopting a less costly strategy that avoided

weapons use as doing so would have been likely to have resulted in extinction.

One argument that could be raised against our hypothesis is that the development of weapons would have led to variation in the skills needed for their use and would not therefore have resulted in the equalization of power within hominin social groups assumed by our model. We agree that it is likely that differences in skill in weapons use would have emerged in this cultural niche, as suggested by the variation in hunting skills observed in modern hunter/gatherer groups [38]. But this point need not invalidate our hypothesis. The very swiftness and effectiveness with which

Table 2. Non-cooperative strategies: average survival time in 100 evolutionary games (for details of strategies see S 2: IPD strategies employed).

Strategy	Without weapon use (generations) ($p=0.00$)	With weapon use (generations) ($p=0.05$)	Change (generations)
AllD	8.92	3.01	−5.91
NP	18.37	6.33	−12.04
RP	34.72	17.93	−16.79
Adaptive	24.41	10.67	−13.74
Prober	8.10	2.93	−5.17
STFT	12.97	5.02	−7.95
Handshake	13.55	3.58	−9.97
Fortress 3	7.90	3.48	−4.42
Fortress 4	9.41	3.75	−5.66
HM	12.43	4.42	−8.01
Rand	12.03	4.57	−7.46
Total	162.81	65.69	−97.12

Note: Average change in performance is −97.12/11 = −8.8.

Survival time (generations)

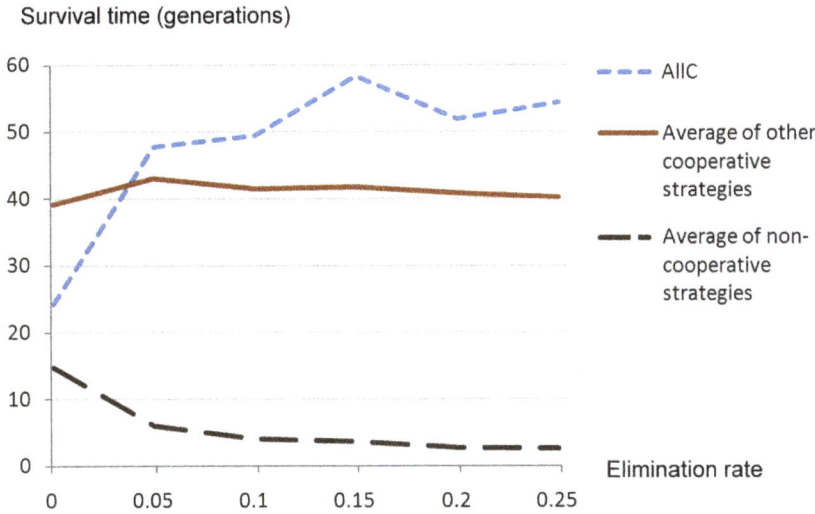

Figure 3. Effects of variation in the elimination rate on: 'Always Cooperate' (AllC), the average for other cooperative strategies and the average for all non-cooperative strategies.

such weapons could have been employed would have made it relatively easy for any individual to be killed [29]. No one, we suggest, however skilful in the use of weapons, would have been immune to the risk of serious injury or death if they defected too often in interactions with other group members.

Our view is supported by the egalitarian ethos observed in modern hunter/gatherer societies and the extraordinary extent to which such societies go to suppress potential causes of violence [29], [39], [40]. It is also supported by a study of the! Kung where 22 cases of homicide that occurred prior to the introduction of effective legal sanctions by outside civil authorities were examined [39]. Many of these homicides appeared to have had random causes. However, four of these cases of homicide were interpreted

as being sanctioned by the community against individuals with a propensity for disruption and violence [39] (i.e. those likely to have consistently defected against other group members). We therefore conclude that it is likely that selection in this cultural niche would have acted primarily against poor co-operators rather than those less skilled in weapons use.

The most important contribution that this study makes, however, is to suggest a novel alternative to explanations of altruism and cooperation based on reciprocal altruism [5] and indirect reciprocity [7], [8]. If the patterns discovered in these simulations were to have been reflected in human evolution, then genes associated with cooperation, altruism and 'strong reciprocity' are likely to have increased in frequency in ancestral

Table 3. Cooperative strategies: effects of variation in elimination rate on survival rate (for details of strategies see S2: IPD strategies employed).

Strategy	Without weapon use (p = 0.00)	With weapon use (varying elimination rates expressed by the value of p)				
		0.05	0.10	0.15	0.20	0.25
AllC	24.26	47.72	49.49	58.39	51.90	54.46
TFT	44.56	48.46	45.71	49.69	42.22	43.95
GRIM	36.84	35.29	28.93	31.92	30.68	29.40
TFTT	41.44	51.05	47.05	49.76	53.73	49.41
Pavlov	28.75	32.96	35.96	35.37	35.43	36.65
GTFT	41.01	52.66	53.15	52.25	51.05	51.36
HTFT	33.76	35.61	32.30	28.68	27.92	28.24
SM	41.62	47.51	41.67	41.94	39.20	40.71
TTFT	36.62	36.54	35.94	32.96	33.70	37.95
CTFT	45.65	47.05	41.37	45.41	44.54	40.14
ATFT	42.72	45.72	41.25	44.49	41.18	41.35
SGRIM	38.96	49.53	52.51	48.47	51.35	49.82
FBF	40.84	44.74	48.12	48.59	47.62	42.26
Gradual	36.36	36.94	34.79	33.35	32.66	31.69

populations. These selection pressures might even have favoured genes linked to closely allied behaviours and emotions such as prosociality, reciprocity, a sense of 'fairness', guilt and moralistic aggression. If so, then genes associated with distinctively human cooperation towards non-kin would come to be expressed in modern populations, like those of any other adaptive behavioural trait, given appropriate stimuli in contemporary environments.

What is being suggested here, therefore, is that distinctively human cooperative and altruistic behaviour evolved because it was an adaptive behavioural trait like any other - but in a cultural niche that no longer exists in the contemporary world (in this case, due to the development of civil societies and modern legal systems suppressing widespread weapons use). Another example of a past environment, no longer evident in the modern world, which might have favoured genes associated with human cooperation and altruism, has been examined elsewhere [41].

What could be called the 'altruism as an adaptive trait in past environments' hypothesis can be seen to offer a relatively parsimonious account of this major evolutionary puzzle. It avoids the complex and costly monitoring and punishment of the behaviour of others required by reciprocity theory [9], [42], [43], [44], [45], [46], [47], [48], [49] because the key determinant of altruistic behaviour would simply be the frequency of genes associated with this behaviour across the whole population over evolutionary time. It also avoids the need for repeated interactions [5] and reputation effects [7], [8] required by reciprocity theory. This, of course, has a special relevance to the evidence for 'strong reciprocity' described at the start of this article i.e. the experiment that demonstrated altruistic behaviour can persist in the form of altruistic punishment of cheats in one-off, anonymous encounters where reciprocity and reputation effects are not possible [12]. The 'altruism as an adaptive trait in past environments' hypothesis is therefore capable of resolving this puzzle by demonstrating that 'strong reciprocity' could have been a successful adaptation to a past cultural niche, thus requiring no other mechanism for it to be expressed in modern populations.

The hypothesis tested in this study is based on the intuitively simple notion that the invention of extra-somatic weapons would have made competition far more costly to individual fitness. Intuition, however, is not the same as rigorous testing of a hypothesis. What is much less clear or predictable is how the range of widely contrasting and often complex IPD strategies simulated in this study would perform in a cultural niche where such a condition is commonplace. This point is amply illustrated in the unexpected success of the AllC strategy and the implications which that success has for understanding the origins of human altruism and cooperative behaviour. We consider that the resulting patterns found in this study provide an important new insight into the two major evolutionary puzzles outlined in the Introduction. The challenge now is to test this novel hypothesis by using other methodologies (e.g. a population genetic model) to establish whether the same effects are replicated.

Acknowledgments

We thank Tom Reader, Francis Gilbert and two anonymous reviewers for advice and help in its preparation of this article.

Author Contributions

Conceived and designed the experiments: TP. Performed the experiments: JL GK. Analyzed the data: TP JL GK. Contributed reagents/materials/analysis tools: TP JL GK. Wrote the paper: TP.

References

1. Maynard Smith J (1964) Group selection and kin selection. Nature 201: 1145–1147.
2. Dawkins R (1976) The selfish gene. Oxford: Oxford University Press.
3. Gintis H (2000) Strong reciprocity and human sociality. J Theor Biol 206: 169–179.
4. Fehr E, Fischbacher U, Gachter S (2002) Strong reciprocity, human cooperation and the enforcement of social norms. Hum Nat 13: 1–25.
5. Trivers RL (1971) The evolution of reciprocal altruism. Q Rev Biol 46: 35–57.
6. Trivers RL (1985) Social evolution. Menlo Park, California: Benjamin Cummings Publishing Co Inc.
7. Alexander RD (1979) Darwinism and human affairs. London: Pitman Publishing Ltd.
8. Nowak MA, Sigmund K (1998) Evolution of indirect reciprocity by image scoring. Nature 393: 573–577.
9. Leimar O, Hammerstein P (2001) Evolution of cooperation through indirect reciprocity. Proc R Soc Lond B Biol Sci 268: 745–753.
10. Nowak MA, Sigmund K (2005) Evolution of indirect reciprocity. Nature 437: 1291–1298.
11. Nowak MA (2006) Five rules for the evolution of cooperation. Science 314: 1650–1653.
12. Fehr E, Gachter S (2002) Altruistic punishment in humans. Nature 415: 137–140.
13. Gintis H, Bowles S, Boyd R, Fehr E (2007) Explaining altruistic behavior in humans. In: Dunbar RIM, Barrett L editors. Oxford handbook of evolutionary psychology. Oxford: Oxford University Press 605–620.
14. Boyd R, Gintis H, Bowles S, Richerson PJ (2003) The evolution of altruistic punishment. Proc Natl Acad Sci USA 100: 3531–3535.
15. Bowles S, Gintis H (2003) The origins of human cooperation. In: Hammerstein P editor. Genetic and cultural evolution of cooperation. Cambridge, Mass: The MIT Press 429–443.
16. Fehr E, Fischbacher U (2003) The nature of human altruism. Nature 425: 785–791.
17. Fehr E, Henrich J (2003) Is strong reciprocity a maladaptation? On the evolutionary foundations of human altruism. In: Hammerstein P, editor. The genetic and cultural evolution of cooperation. Cambridge, Mass. The MIT Press.
18. Delton AW, Krasnow MM, Cosmides L, Tooby J (2011) Evolution of direct reciprocity under uncertainty can explain human generosity in one-shot encounters. Proc Natl Acad Sci USA 108: 13335–13340.
19. Johnson D, Stopka P, Knights S (2003) The puzzle of human cooperation. Nature 421: 911–912.
20. Gintis H (2003) Solving the puzzle of prosociality. Rationality and Society 15: 155–187.
21. Aiello L, Wheeler P (1995) The expensive-tissue hypothesis: The brain and the digestive system in human and primate evolution. Curr Anthrop 36: 199–221.
22. Lumsden CJ, Wilson EO (1981) Genes, mind and culture. Cambridge, Mass: Harvard University Press.
23. Laland KN, Odling-Smee FJ, Feldman MW (2001) Cultural niche construction and human evolution. J Evol Biol 14, 22–33.
24. Laland KN (2008) Exploring gene-culture interactions: insights from handedness, sexual selection and niche-construction case studies. Philos Trans R Soc Lond B Biol Sci 363: 3577–3589.
25. Thieme H (1997) Lower palaeolithic hunting spears from Germany. Nature 385: 807–810.
26. Kelly RC (2005) The evolution of lethal intergroup violence. Proc Natl Acad Sci USA, Series of Inaugural Articles by newly elected members.
27. Churchill SE, Rhodes JA (2009) The evolution of the human capacity for 'killing at a distance': The human fossil evidence for the evolution of projectile weaponry. In: Hublin J-J, Richards MP editors. The evolution of hominin diets: Integrating approaches to the study of palaeolithic subsistence. Springer 201–210.

28. Shea JJ (2009) The impact of projectile weaponry on late pleistocene hominin evolution. In: Hublin, J-J, Richards MP, editors. The evolution of hominin diets: Integrating approaches to the study of palaeolithic subsistence. Springer 189–199.

29. Boehm C (1999) Hierarchy in the forest: the evolution of egalitarian behavior. Harvard University Press. Cambridge, Mass. London.

30. Maynard Smith J, Price GR (1973) The logic of animal conflict. Nature 246: 15–18.

31. Axelrod R, Hamilton WD (1981) The evolution of cooperation. Science 211: 1390–96.

32. Axelrod R (1984) The evolution of cooperation. London: Penguin Books.

33. Nowak MA, Sigmund K (1993) A strategy of win-stay, lose-shift that outperforms tit-for-tat in the prisoner's dilemma games. Nature 364: 56–58.

34. Chong S Y, Humble J, Kendall G, Li J, Yao X (2007) Iterated prisoner's dilemma and evolutionary game theory. In: The iterated prisoners' dilemma: 20 years on. Singapore: World Scientific, Advances in Natural Computation 4, 23–62.

35. Li J, Kendall G (2009) A strategy with novel evolutionary features for the iterated prisoner's dilemma. Evol Comput 17: 257–274.

36. Schmitt DP (2005) Fundamentals of human strategies. In: Buss DM editor. The handbook of evolutionary psychology Hoboken, New Jersey: John Wiley and Sons Inc. pp. 258–291.

37. Hill KR, Walker RS, Božičević M, Eder J, Headland T, et al. (2011) Co-residence patterns in hunter-gatherer societies show unique human social structure. Science 331: 1286–89.

38. Hawkes K (1993) Why hunter-gatherers work. Curr Anthrop 34: 341–361.

39. Lee RB (1979) The! kung san: men, women, and work in a foraging society. Cambridge, UK: Cambridge University Press.

40. Knauft BM (1991) Violence and sociality in human evolution. Curr Anthrop 32: 391–429.

41. Phillips T, Barnard C, Ferguson E, Reader T (2008) Do humans prefer altruistic mates? Testing a link between sexual selection and altruism towards non-relatives. Br J Psychol 99: 555–572.

42. Boyd R, Richerson P (1988) The evolution of reciprocity in sizeable groups. J Theor Biol 132: 337–356.

43. Boyd R, Richerson PJ (1992) Punishment allows the evolution of cooperation (or anything else) in sizable groups. Ethol Sociobiol 13: 171–195.

44. Sigmund K (1993) Games of life: explorations in ecology, evolution and behaviour London: Penguin Books.

45. Stevens JR, Hauser MD (2004) Why be nice? Psychological constraints in the evolution of cooperation. Trends Cogn Sci 8: 60–65.

46. Stevens JR, Stephens DW (2004) The economic basis of cooperation: tradeoffs between selfishness and generosity. Behav Ecol 15: 255–261.

47. Suzuki S, Akiyama E (2005) Reputation and the evolution of cooperation in sizable groups. Proc R Soc Lond B Biol Sci 272: 1373–1377.

48. Dreber A, Rand DG, Fudenberg D, Nowak MA (2008) Winners don't punish. Nature 452: 348–351.

49. Egas M, Riedl A (2008) The economics of altruistic punishment and the maintenance of cooperation. Proc R Soc Lond B Biol Sci 275: 871–878.

18

Why Do Female *Callosobruchus maculatus* Kick Their Mates?

Emile van Lieshout*, Kathryn B. McNamara, Leigh W. Simmons

Centre for Evolutionary Biology, School of Animal Biology (M092), University of Western Australia, Crawley, Australia

Abstract

Sexual conflict is now recognised as an important driver of sexual trait evolution. However, due to their variable outcomes and effects on other fitness components, the detection of sexual conflicts on individual traits can be complicated. This difficulty is exemplified in the beetle *Callosobruchus maculatus*, where longer matings increase the size of nutritious ejaculates but simultaneously reduce female future receptivity. While previous studies show that females gain direct benefits from extended mating duration, females show conspicuous copulatory kicking behaviour, apparently to dislodge mating males prematurely. We explore the potential for sexual conflict by comparing several fitness components and remating propensity in pairs of full sibling females where each female mated with a male from an unrelated pair of full sibling males. For one female, matings were terminated at the onset of kicking, whereas the other's matings remained uninterrupted. While fecundity (number of eggs) was similar between treatments, uninterrupted matings enhanced adult offspring numbers and fractionally also longevity. However, females whose matings were interrupted at the onset of kicking exhibited an increased propensity to remate. Since polyandry can benefit female fitness in this species, we argue that kicking, rather than being maladaptive, may indicate that females prefer remating over increased ejaculate size. It may thus be difficult to assess the presence of sexual conflict over contested traits such as mating duration when females face a trade off between direct benefits gained from one mating and indirect benefits from additional matings.

Editor: Therésa M. Jones, University of Melbourne, Australia

Funding: EVL was funded by a Research Development Award from the University of Western Australia (http://www.research.uwa.edu.au/staff/funding/awards). KBM was funded by an ARC Australian Postdoctoral Fellowship (DP110101163)(http://arc.gov.au/ncgp/). LWS was supported by an ARC Australian Professorial Fellowship (DP110104594)(http://arc.gov.au/ncgp/). The funders had no role in study design, data collection and analysis, decision to publish, or preparation of the manuscript.

Competing Interests: The authors have declared that no competing interests exist.

* E-mail: emile.vanlieshout@uwa.edu.au

Introduction

Sexual conflict, the conflict between the evolutionary interests of individuals of the two sexes [1], is a fundamental driver of adaptation. Any trait that reduces genetic fitness in the other sex by definition imposes antagonistic selection on that sex for traits that counteract this cost [2]. Such conflicts can occur over traits like mating duration, where male and female adaptations and counteradaptations vie for sex-specific optima. Discrepancies between the costs of mating and resistance may 'resolve' conflicts in one sex's favour [3], although on a population level the linkage of male and female average fitness in populations at equal sex ratios should cause similar fitness declines in both sexes [4]. In other cases, ongoing coevolutionary processes can ultimately leave contested traits largely unchanged [5], although the suboptimal trait state for both sexes and the costs of engaging in antagonistic interactions similarly reduce fitness [2]. Assessment of sexual conflict may be further complicated when the consequences of sexual interactions manifest indirectly. This can occur when the direct costs of manipulation to female fecundity are outweighed by sexy-son-type benefits [3] or conversely when competitive males sire offspring of low fitness [6]. As a consequence, sexual conflicts may be difficult to detect, whichever way they are resolved.

A prime illustration of the complexity involved in demonstrating sexual conflict can be found in the seed beetle *Callosobruchus maculatus*. In this species, mating duration increases the degree of damage to the female reproductive tract made by male genital spines [7]. Longer matings also result in the transfer of larger ejaculates [8,9], which confer direct benefits [10] as well as costs [11] and may contain products that suppress remating [12]. Given these findings, the conspicuous kicking behaviour that females display in the last third of the mating [13] is generally interpreted as an attempt to dislodge the male and limit mating duration [14]. Females commence kicking earlier when mating with large males, which transfer ejaculate at a higher rate, suggesting the onset of kicking marks the receipt of a threshold quantity of ejaculate [9]. In turn, the spiny genitalia of males have been suggested to be a counteradaptation to resist dislodgement by females [14], although recent findings suggest that their length is not associated with variation in mating duration [15] and they serve to promote coupling [16] and the transmission of seminal products through the wall of the female reproductive tract [17].

Several studies have tested the idea that *C. maculatus* exhibits sexual conflict over mating duration. While theory suggests that male mating effort should covary positively with female fecundity [18], larger (and therefore more fecund) females are able to reduce mating duration and thus the size of the ejaculate received [9]. However, females evolved in high-sexual-conflict environments do not exhibit reduced mating duration against standard partners, and variation in the onset of kicking does not affect mating duration [9]. When female kicking legs are ablated, matings last longer and result in more damage from the genital spines [7] and females indeed suffer reduced fitness, although apparently without any benefit to males [14]. Edvardsson and Canal [8] examined conflict over mating duration by either terminating matings at the mean onset of kicking (assumed to be the female optimum), preventing females from kicking (the male optimum), or leaving matings unmanipulated ('contested'). While the onset of kicking could conceivably occur earlier than the female's optimal mating duration because kicking does not dislodge males immediately, it should approach female optima. Edvardsson and Canal [8] found that females experience increased fecundity after male-optimal matings, and detected no fitness difference between female-optimal and contested durations, suggesting that no sexual conflict exists. In this previous experiment, however, treatments were set to rough estimates of the typical onset of kicking and the durations of unmanipulated and male-controlled matings, ignoring considerable between-individual variation.

Here, we examine the existence of sexual conflict over mating duration by manipulating mating duration and looking at its effect on several fitness components and remating propensity in single- and twice-mated female *C. maculatus*. Following Edvardsson and Canal [8], we assume that the onset of kicking indicates that mating duration is approaching the female optimum. We improve on previous designs by terminating copulations at the exact onset of kicking in some females while allowing copulations to end naturally in their siblings. Given that ejaculates have properties that induce a refractory period [12], we expect remating propensity to increase when matings are interrupted at the onset of kicking. However, the fitness consequences of female mating duration may depend on mating frequency.

Materials and Methods

Experimental animals were sourced from a large outbred population that originated from a stock culture held by the Stored Grain Research Laboratory of CSIRO (Canberra, Australia). Under Australian guidelines, animal ethics approval for research on this species is not required. Experimental and stock individuals were maintained under constant conditions at the University of Western Australia for approximately 4 years at 30°C under a light cycle of 12 h light:12 h dark. Individuals were reared from eggs until adult on black-eyed beans (*Vigna unguiculata*) in several large populations of approximately 300 individuals to ensure a large effective population size.

To create individuals with known relationships, 40 virgin stock males were mated monandrously to 40 virgin stock females. Females were placed in 60 ml vials with approximately 40 black-eyed beans and allowed to oviposit. Males were discarded. Beans that contained larvae were isolated in microtubes just prior to adult emergence. As adults emerged they were isolated in separate microtubes and their emergence date, sex and weight were recorded. For each of the 40 parental pairs, 4 male and 4 female offspring were retained. All experimental animals were between one and three days old when assayed.

To examine the effect of female control over mating duration and subsequent reproductive success in once-mated females, two randomly assigned female siblings were mated to two male siblings from another family. The time taken for copulation to commence, the time taken for females to commence kicking, and the time for females to eject the aedeagus were recorded. For one of the sibling females, the male's abdomen was severed immediately following the commencement of kicking. This causes the male's aedeagus to deflate, allowing the female to eject the male's aedeagus more

quickly (Kruskal-Wallis $\chi^2_1 = 53.67$, P<0.0001). Male genital spines are located on the inflatable sac at the apex of the aedeagus and are exposed only upon inflation [19]. Hence, deflation of the sac also prevents further damage. The other sibling female's matings were not interrupted to reflect contested mating durations.

To examine the effect of female control over mating on remating propensity and reproductive success in twice-mated females, we used the remaining two siblings. Females were mated as above and then isolated in microtubes. Here, after 24 h, the females were placed with a newly emerged virgin stock male and allowed the opportunity to remate within ten minutes. If remating did occur, females received the same mating treatment as on their first day: for one, the male's abdomen was severed at the commencement of kicking, whereas for her sister, the copulation was uninterrupted. Again, the time taken until copulation, until the commencement of kicking, and until the aedeagus was ejected from the female were recorded.

To assess fitness consequences, females were placed into individual 60 ml vials containing 9 g of black-eyed beans immediately following the once- and twice-mated treatments and allowed to oviposit until death. The number of eggs visible on the exterior of each bean (fecundity), the number of adult offspring that subsequently emerged from these beans (fertilisation success) and the female's longevity were recorded.

Statistics

Data were analysed using generalised linear mixed models, with family as a random factor to account for the sibling design. The significance of the random variable (family) was tested using log-likelihood tests. Data were transformed to normality for analysis, where appropriate. Three trials were excluded from the twice-mated treatment (in one trial, the couple refused to mate, and in the two remaining trials, experimenter error resulted in a failed trial). Dependent variables in linear models were power transformed to maximise normality of residuals. Interactions between mating treatment and frequency were non-significant in all analyses, and were omitted [20].

Results

Remating Propensity

For 75 females that were assigned to mate twice, only 11 failed to re-mate (6 from interrupted and 5 from uninterrupted mating treatments; $\chi^2_1 = 0.14$, p = 0.71). To examine the mating treatment on the propensity to remate in the remaining females, we analysed copulation latency using mixed-effects modelling with female and family identity as random factors. Females whose first matings were interrupted at the onset of kicking commenced copulation considerably earlier than females that had received uninterrupted matings (exponent 0.32, mating×treatment, $\chi^2_1 = 4.34$, P = 0.037; Fig. 1), but was not affected by the female's age ($\chi^2_1 = 0.10$, P = 0.75), or her weight ($\chi^2_1 = 0.13$, P = 0.72). Exclusion of female or family identity did not affect the fit of the model ($\chi^2_1 = 0.27$, P = 0.60, $\chi^2_1 = 0.00$, P = 1.00 respectively). Furthermore, the median latency to copulation was considerably longer for second matings (32s vs. 72s; $\chi^2_1 = 27.56$, P<0.0001).

Latency to Kicking

The effect of mating treatment on the latency to kicking in the second mating was, like copulation latency, analysed using mixed-effects modelling to account for individual female variation. Kicking latency was not affected by whether the previous copulation duration was interrupted (exponent 0.88, mating× treatment, $\chi^2_1 = 0.26$, P = 0.61), nor the female's age ($\chi^2_1 = 0.02$,

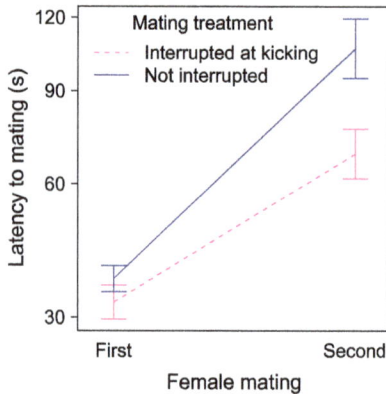

Figure 1. The latency to mating (mean±standard error) increased in the second mating depending on the mating treatment: when a female's first mating was interrupted at the onset of kicking, the latency to the second mating was reduced (increased remating propensity). Note the exponential scale on the y-axis.

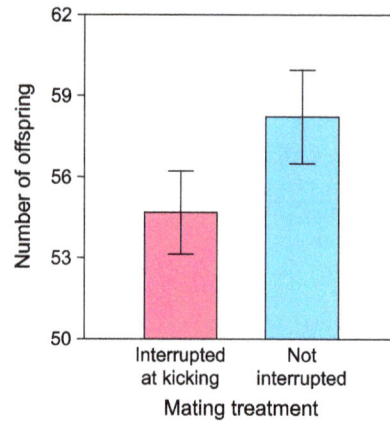

Figure 2. The number of hatched offspring produced by females (mean±standard error) was greater for females when their first mating was uninterrupted rather than terminated at the onset of kicking (interrupted).

$P = 0.88$), or weight ($\chi^2_1 = 1.85$, $P = 0.17$). Exclusion of female or family identity did not affect the fit of the model ($\chi^2_1 = 0.00$, $P = 0.96$, $\chi^2_1 = 0.00$, $P = 1.00$ respectively). Furthermore, the latency to kicking was significantly longer for second matings ($\chi^2_1 = 27.56$, $P < 0.0001$).

Female Reproductive Output

For the 153 females that successfully mated, two single-mated and one double-mated female failed to oviposit. These females were excluded from further analysis. For females that did lay eggs, fecundity increased with female mating frequency (single vs. double medians: 77 vs. 83 eggs; exponent 2.93, $\chi^2_1 = 8.57$, $P = 0.003$), female weight ($\chi^2_1 = 46.14$, $P < 0.0001$) and female age at first mating ($\chi^2_1 = 5.45$, $P = 0.02$). Fecundity, however, was not affected by whether the matings were interrupted ($\chi^2_1 = 1.81$, $P = 0.18$). Exclusion of family identity improved the fit of the model ($\chi^2_1 = 5.91$, $P = 0.02$).

One female laid only unviable eggs and was excluded from analysis of offspring production. The number of emerged offspring produced by a female increased when copulation duration was uninterrupted (exponent 2.36, $\chi^2_1 = 7.07$, $P = 0.008$; Fig. 2), and with female weight ($\chi^2_1 = 19.13$, $P < 0.0001$). Median offspring numbers in uninterrupted matings were 9% greater than in interrupted matings (61 vs. 56 offspring, respectively). Offspring numbers, however, were not affected by female mating frequency ($\chi^2_1 = 0.53$, $P = 0.47$) or the age of the female at her first mating ($\chi^2_1 = 1.42$, $P = 0.23$). Exclusion of family identity did not affect the fit of the model ($\chi^2_1 = 0.02$, $P = 0.90$).

Female Longevity

For those females that laid eggs, adult longevity was analysed. Longevity increased subtly but significantly when the matings were uninterrupted (exponent 0.005, $\chi^2_1 = 5.10$, $P = 0.02$; back-transformed means using model exponent: interrupted vs. uninterrupted = 6.98 vs. 7.05 days). Female longevity also increased with the age of the female at her first mating ($\chi^2_1 = 76.72$, $P < 0.0001$), and with female weight ($\chi^2_1 = 12.69$, $P < 0.0001$). Female longevity, however, was not affected by mating frequency ($\chi^2_1 = 0.88$, $P = 0.35$). Family identity improved the fit of the model ($\chi^2_1 = 4.94$, $P = 0.03$).

Discussion

In this study we examined whether sexual conflict over mating duration exists in *C. maculatus* by exploring the fitness consequences of interrupting copulations at the onset of female kicking, assuming this indicates that females are approaching their optimal mating duration. We show that some female fitness components clearly benefit from longer copulations. While lifetime fecundity was unaffected by mating duration, both in singly and doubly mated females, uninterrupted copulations slightly increased longevity and resulted in 9% greater offspring numbers. This increase in offspring production is unlikely to be a consequence of sperm limitation in matings interrupted at the onset of kicking: sperm transfer occurs from the start of copulation, and sperm numbers far exceed the requirements for fertilisation [21]. This effect thus appears driven by ejaculate properties associated with mating duration, potentially nutritional content. When mating was terminated at the onset of kicking, however, females had a greater propensity to remate.

Previous work suggests that no conflict over mating duration exists in *C. maculatus* [8]. Indeed, our results agree that receiving longer matings, and the larger ejaculates that accompany these [9], benefit female fitness when compared to interrupted matings at the same mating frequency. Yet, this would suggest that female copulatory kicking behaviour, widespread in seed beetles, is maladaptive yet evolutionarily persistent.

Strictly speaking, the fact that long copulations enhance female fitness does not represent evidence for the absence of conflict over mating duration. For males, the benefits of extending mating duration, increased fecundity, fertility, and paternity share, are obvious [this study; 14]. Yet, for females, both larger ejaculates and multiple mating, especially polyandrously, benefit fitness [22–25], in part because both tactics provide direct benefits via ejaculatory water [10]. Our results indicate that, in *C. maculatus*, females clearly show adaptive plasticity in accepting male courtship: when first matings approached female duration optima, females showed a reduced resistance to remating. Although we could not detect an effect on the acceptance of second matings, the cumulative effect of female-controlled mating durations is likely to affect lifetime remating rates. This notion is supported by both by the strong link between mating duration and ejaculate size [8,9], and the fact that the receptivity-suppressing properties of ejaculates are dose dependent [21,24].

Relationships between ejaculate size and remating propensity have been reported in other species. In the almond moth *Cadra cautella*, for example, males transfer large spermatophores that inhibit female remating [26]. Females possess chitinous teeth in their reproductive tract, thought to be counteradaptations intended to break down spermatophores and promote female remating [27,28]. As in the Lepidoptera, females in *C. maculatus* possess teeth in their bursa [29], which may serve a similar function. Inhibition of female remating appears to result from products of the male seminal vesicle [12], although this does not rule out physical effects of ejaculate size. Additionally, there is evidence that female reproductive tract scarring increases with mating duration [7]. Consequently, the apparently premature copulatory kicking of females may indicate that they prefer to gain fitness benefits through increased remating rather than increased ejaculate size. Since remating is unlikely to occur with the same male [22], remating may provide indirect, genetic benefits in addition to direct benefits. Previous studies have shown benefits of polyandry over monandrous multiple mating [22], and fertilization success is in part mediated by the compatibility of male and female genotypes [30]. Yet, females that received uninterrupted matings in our experiment had higher fitness than those whose matings were interrupted. This is perhaps unsurprising, because our maximum mating rate (2) is less than the typical female lifetime mating rate in this population [8], which itself is based on uninterrupted matings.

Male postcopulatory adaptations to reduce sperm competition often directly target female mating duration and frequency [31], to which females may develop a range of behavioural counter-adaptations. For example, a significant proportion of female hide beetles (*Dermestes maculatus*) are able to dislodge mating and mate-guarding males to gain benefits associated with polyandry [32].

Despite the indications of female preferences for shorter matings and increased remating in *C. maculatus*, it is unclear to what extent these are realised in a natural context. Although multiple studies show that ablation of females' kicking legs results in longer matings [7,8], van Lieshout, McNamara and Simmons [9] recently showed that the onset of kicking has no effect on the eventual mating duration. However, the evolutionary maintenance of kicking suggests that uninterrupted matings do not fully conform to male optima, and that this conflict is resolved at a suboptimal state for both sexes [2]. Consistent with this, Brown *et al* [33] showed both male and female genetic effects on virgin mating duration.

In conclusion, consistent with other studies, we show that some female fitness components benefit from longer, uninterrupted mating durations. However, we argue that this result alone cannot reveal whether sexual conflict over mating duration exists. By enforcing mating durations to be closer to the female optima, we find indications that the direct fitness benefits of mating duration may trade off with (genetic) benefits gained through additional polyandrous mating. Omission of alternative routes to fitness, such as polyandrous remating, from consideration when testing for sexual conflict could lead to underestimates of its pervasiveness.

Acknowledgments

We are very grateful to Joe Tomkins for advice and providing the initial stock population and thank Stephen Robinson for assistance in the lab.

Author Contributions

Conceived and designed the experiments: EVL KBM LWS. Performed the experiments: EVL KBM. Analyzed the data: EVL KBM LWS. Contributed reagents/materials/analysis tools: EVL KBM. Wrote the paper: EVL KBM LWS.

References

1. Parker GA (1979) Sexual selection and sexual conflict. In: M. S Blum and N. A Blum, editors. Sexual selection and reproductive competition in insects. New York: Academic Press. 123–166.

2. Lessells CM (2006) The evolutionary outcome of sexual conflict. Philos Trans R Soc Lond B Biol Sci 361: 301–317. doi:10.1098/rstb.2005.1795.

3. Kokko H, Brooks R, Jennions MD, Morley J (2003) The evolution of mate choice and mating biases. Proc R Soc Lond B Biol Sci 270: 653–664. doi:10.1098/rspb.2002.2235.

4. Arnqvist G, Rowe L (2005) Sexual conflict. Princeton: Princeton University Press. 330 p.

5. Arnqvist G, Rowe L (2002) Antagonistic coevolution between the sexes in a group of insects. Nature 415: 787–789. doi:10.1038/415787a.

6. Bilde T, Foged A, Schilling N, Arnqvist G (2009) Postmating sexual selection favors males shat sire offspring with low fitness. Science 324: 1705–1706. doi:10.1126/science.1171675.

7. Crudgington HS, Siva-Jothy MT (2000) Genital damage, kicking and early death - the battle of the sexes takes a sinister turn in the bean weevil. Nature 407: 855–856. doi:10.1038/35038154.

8. Edvardsson M, Canal D (2006) The effects of copulation duration in the bruchid beetle *Callosobruchus maculatus*. Behav Ecol 17: 430–434. doi:10.1093/beheco/arj045.

9. van Lieshout E, McNamara KB, Simmons LW (2014) Rapid loss of behavioural plasticity and immunocompetence under intense sexual selection. Evolution accepted.

10. Edvardsson M (2007) Female *Callosobruchus maculatus* mate when they are thirsty: resource-rich ejaculates as mating effort in a beetle. Anim Behav 74: 183–188. doi:10.1016/j.anbehav.2006.07.018.

11. Eady PE, Hamilton L, Lyons RE (2007) Copulation, genital damage and early death in *Callosobruchus maculatus*. Proc R Soc Lond B Biol Sci 274: 247–252. doi:10.1098/rspb.2006.3710.

12. Yamane T, Miyatake T, Kimura Y (2008) Female mating receptivity after injection of male-derived extracts in *Callosobruchus maculatus*. J Insect Physiol 54: 1522–1527. doi:10.1016/j.jinsphys.2008.09.001.

13. Eady PE (1991) Sperm competition in *Callosobruchus maculatus* (Coleoptera: Bruchidae): a comparison of two methods used to estimate paternity. Ecol Entomol 16: 45–53. doi:10.1111/j.1365–2311.1991.tb00191.x.

14. Edvardsson M, Tregenza T (2005) Why do male *Callosobruchus maculatus* harm their mates? Behav Ecol 16: 788–793. doi:10.1093/beheco/ari055.

15. Rönn JL, Hotzy C (2012) Do longer genital spines in male seed beetles function as better anchors during mating? Anim Behav 83: 75–79. doi:10.1016/j.anbehav.2011.10.007.

16. Polak M, Rashed A (2010) Microscale laser surgery reveals adaptive function of male intromittent genitalia. Proc R Soc Lond B Biol Sci 277: 1371–1376. doi:10.1098/rspb.2009.1720.

17. Hotzy C, Polak M, Rönn JL, Arnqvist G (2012) Phenotypic engineering unveils the function of genital morphology. Curr Biol 22: 2258–2261. doi:10.1016/J.Cub.2012.10.009.

18. Parker GA, Pizzari T (2010) Sperm competition and ejaculate economics. Biological Reviews 85: 897–934. doi:10.1111/j.1469-185X.2010.00140.x.

19. Hotzy C, Arnqvist G (2009) Sperm competition favors harmful males in seed beetles. Curr Biol 19: 404–407. doi:10.1016/j.cub.2009.01.045.

20. Engqvist L (2005) The mistreatment of covariate interaction terms in linear model analyses of behavioural and evolutionary ecology studies. Anim Behav 70: 967–971. doi:10.1016/j.anbehav.2005.01.016.

21. Eady PE (1995) Why do male *Callosobruchus maculatus* beetles inseminate so many sperm? Behav Ecol Sociobiol 36: 25–32. doi:10.1007/BF00175725.

22. Eady PE, Wilson N, Jackson M (2000) Copulating with multiple mates enhances female fecundity but not egg-to-adult survival in the bruchid beetle *Callosobruchus maculatus*. Evolution 54: 2161–2165. doi:10.1111/j.0014–3820.2000.tb01259.x.

23. Wilson N, Tufton TJ, Eady PE (1999) The effect of single, double, and triple matings on the lifetime fecundity of *Callosobruchus analis* and *Callosobruchus maculatus* (Coleoptera: Bruchidae). J Insect Behav 12: 295–306. doi:10.1023/A:1020883220643.

24. Savalli UM, Fox CW (1999) The effect of male mating history on paternal investment, fecundity and female remating in the seed beetle *Callosobruchus maculatus*. Funct Ecol 13: 167–177. doi:10.1046/j.1365–2435.1999.00287.x.

25. Fox CW (1993) Multiple mating, lifetime fecundity and female mortality of the bruchid beetle, *Callosobruchus maculatus* (Coleoptera: Bruchidae). Funct Ecol 7: 203–208.

26. McNamara KB, Elgar MA, Jones TM (2008) Seminal compounds, female receptivity and fitness in the almond moth, *Cadra cautella*. Anim Behav 76: 771–777. doi:10.1016/j.anbehav.2008.04.018.

27. Cordero C (2005) The evolutionary origin of signa in female Lepidoptera: natural and sexual selection hypotheses. J Theor Biol 232: 443–449. doi:10.1016/j.jtbi.2004.08.031.

28. Galicia I, Sánchez V, Cordero C (2008) On the function of signa, a genital trait of female Lepidoptera. Ann Entomol Soc Am 101: 786–793. doi:10.1603/0013–8746(2008)101[786:OTFOSA]2.0.CO;2.

29. Cayetano L, Maklakov AA, Brooks RC, Bonduriansky R (2011) Evolution of male and female genitalia following release from sexual selection. Evolution 65: 2171–2183. doi:10.1111/j.1558–5646.2011.01309.x.

30. Wilson N, Tubman SC, Eady PE, Robertson GW (1997) Female genotype affects male success in sperm competition. Proc R Soc Lond B Biol Sci 264: 1491–1495. doi:10.1098/rspb.1997.0206.

31. Simmons LW (2001) Sperm competition and its evolutionary consequences in the insects. Princeton, New Jersey: Princeton University Press.

32. Archer MS, Elgar MA (1999) Female preference for multiple partners: sperm competition in the hide beetle, *Dermestes maculatus* (DeGeer). Anim Behav 58: 669–675. doi:10.1006/anbe.1999.1172.

33. Brown EA, Gay L, Vasudev R, Tregenza T, Eady PE, et al. (2009) Negative phenotypic and genetic associations between copulation duration and longevity in male seed beetles. Heredity 103: 340–345. doi:10.1038/hdy.2009.80.

19

Functional Characterization of Duplicated *SUPPRESSOR OF OVEREXPRESSION OF CONSTANS 1*-Like Genes in Petunia

Jill C. Preston*, Stacy A. Jorgensen, Suryatapa G. Jha

Department of Plant Biology, The University of Vermont, Burlington, Vermont, United States of America

Abstract

Flowering time is strictly controlled by a combination of internal and external signals that match seed set with favorable environmental conditions. In the model plant species *Arabidopsis thaliana* (Brassicaceae), many of the genes underlying development and evolution of flowering have been discovered. However, much remains unknown about how conserved the flowering gene networks are in plants with different growth habits, gene duplication histories, and distributions. Here we functionally characterize three homologs of the flowering gene *SUPPRESSOR OF OVEREXPRESSION OF CONSTANS 1 (SOC1)* in the short-lived perennial *Petunia hybrida* (petunia, Solanaceae). Similar to *A. thaliana soc1* mutants, co-silencing of duplicated petunia *SOC1*-like genes results in late flowering. This phenotype is most severe when all three *SOC1*-like genes are silenced. Furthermore, expression levels of the *SOC1*-like genes *UNSHAVEN* (*UNS*) and *FLORAL BINDING PROTEIN 21* (*FBP21*), but not *FBP28*, are positively correlated with developmental age. In contrast to *A. thaliana*, petunia *SOC1*-like gene expression did not increase with longer photoperiods, and *FBP28* transcripts were actually more abundant under short days. Despite evidence of functional redundancy, differential spatio-temporal expression data suggest that *SOC1*-like genes might fine-tune petunia flowering in response to photoperiod and developmental stage. This likely resulted from modification of *SOC1*-like gene regulatory elements following recent duplication, and is a possible mechanism to ensure flowering under both inductive and non-inductive photoperiods.

Editor: Hector Candela, Universidad Miguel Hernández de Elche, Spain

Funding: This work was supported by the College of Agriculture and Life Sciences at the University of Vermont. The funders had no role in study design, data collection and analysis, decision to publish, or preparation of the manuscript.

Competing Interests: The authors have declared that no competing interests exist.

* E-mail: Jill.Preston@uvm.edu

Introduction

Flowering time is a complex trait shaped by internal and external signals that interplay to determine reproductive success [1,2]. On the one hand selection acts against mutations that cause flowering to occur during sub-optimal times of the year, such as in drought, freezing temperatures, or when pollinator abundances are low. On the other hand, plants that rarely experience optimal conditions must chance flowering eventually, or find alternative reproductive strategies (e.g. clonal reproduction and apomixis). Thus, flowering time pathways have evolved to both respond to, and buffer against, environmental variation. These seemingly opposing forces can be achieved either by integrating signals from parallel genetic pathways or by differentially utilizing related genes within the same pathway.

In the annual rosid species *Arabidopsis thaliana* (Brassicaceae) long day induced flowering is controlled by CONSTANS (CO)-mediated upregulation of the integrator protein FLOWERING LOCUS T (FT) in leaves [3–5]. FT protein is translocated through the phloem to the shoot apical meristem (SAM) where it binds with FLOWERING LOCUS D (FD) to induce the expression of floral promoters such as *SUPPRESSOR OF OVEREXPRESSION OF CONSTANS 1 (SOC1), APETALA1 (AP1), FRUITFULL (FUL)*, and *LEAFY (LFY)* [6–8]. Although largely functionally conserved across both long and short day-induced flowering plants [9–11]

alterations of *SOC1*-like gene function have been implicated in flowering time and plant habit evolution [11]. For example, in perennial short day strawberry (*Fragaria vesca*, Rosaceae) SOC1 represses flowering through a novel regulatory interaction with the flowering repressor *TERMINAL FLOWER1 (TFL1)* [12]. Thus, fine-scale characterization of *SOC1*-like genes in diverse species has great potential to illuminate our understanding of flowering phenology and its evolution.

Petunia (*Petunia hybrida*, Solanaceae) is a short-lived perennial in the asterid lineage of core eudicots that is induced to flower by long day photoperiods [13]. The petunia genome contains three *SOC1* homologs – *UNSHAVEN (UNS)/FLORAL BINDING PROTEIN 20 (FBP20), FBP21*, and *FBP28* – derived from two duplication events within the Solanaceae (Fig. 1). All three genes are strongly expressed in leaves [14] and at least *UNS* is expressed in vegetative apices, with expression becoming reduced following the transition to flowering [15]. Null mutations in *UNS* and *FBP28* cause no obvious mutant phenotype [15,16]. However, constitutive expression of *UNS* in petunia and *FBP21* in tobacco (*Solanum tabacum*) causes early flowering under short days, bract-like petals, and hairy ovaries [15], and accelerated flowering under long days, reduced plant height and leaf size, and increased flowering branches [17], respectively. The late flowering phenotype of both species suggests that *UNS* and *FBP21* promote the transition to

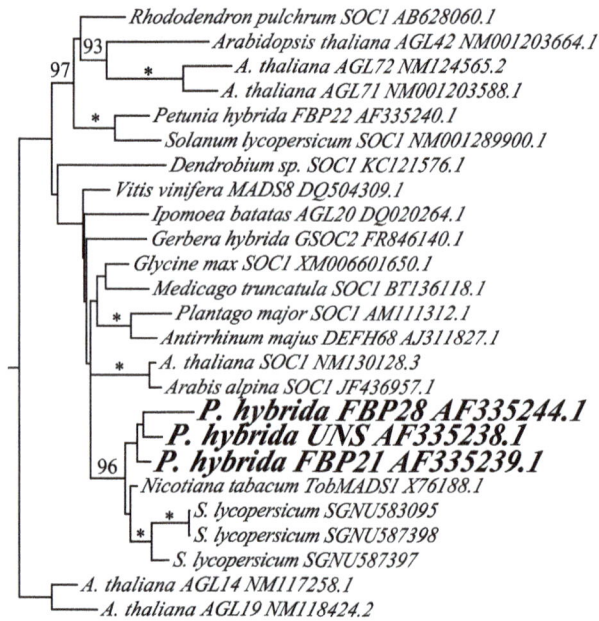

Figure 1. Maximum likelihood phylogeny of *SOC1*-like genes in petunia (bold) and other angiosperms. The tree is rooted on TM3 MADS-box genes in the *AGL14/19* clade sister to *SOC1/AGL42/AGL71/AGL72* genes [30]. Numbers indicate maximum likelihood bootstrap values, with 100% being represented by an asterisk. Sol Genomics Network (SGN) and Genbank accession numbers are shown after each gene name.

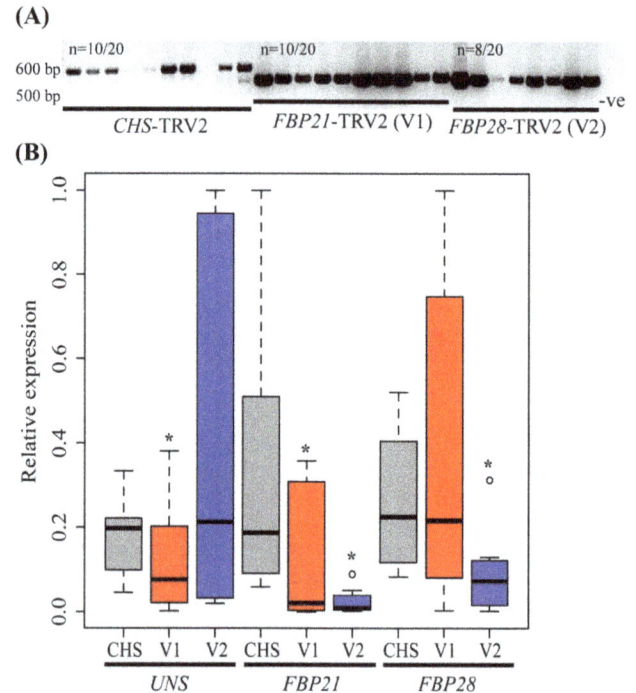

Figure 2. VIGS efficiency in petunia. (A) Typical agarose gel showing ethidium bromide stained amplicons using the TRV2-specific primers. Each lane represents a different individual in the VIGS experiment. −ve, negative control lacking cDNA. Of the 20 plants screened per treatment results for only 8 (*FBP28*-TRV2) or 10 (*CHS*-TRV2 and *FBP21*-TRV2) are shown as examples of efficiency. (B) Boxplots showing relative qPCR cT values for *UNS*, *FBP21*, and *FBP28* amplification. Each gene was amplified in ten plants infected with *CHS*-TRV2 (gray), *FBP21*-TRV2 (V1) (red), or *FBP28*-TRV2 (V2) (blue). Asterisks indicate at least a 2-fold average difference in gene expression between *FBP21*- or *FBP28*-TRV2 and control plants. Circles denote outliers.

flowering similar to *Arabidopsis thaliana SOC1* [18]. However, the relative importance of *UNS*, *FBP21*, and possibly *FBP28*, in flowering time regulation and plant architecture has yet to be rigorously tested.

Here we use virus-induced gene (co)-silencing (VIGS) and gene expression analyses to functionally characterize the petunia *SOC1*-like genes. Our data suggest a conserved role for all three genes in the promotion of flowering. However, the contribution of each gene to flowering likely varies under different photoperiods and in an age-dependent manner.

Materials and Methods

Plant Materials and Growth Conditions

Petunia 'Fantasy Blue' (2PET131) seed from Seedman.com were grown and selfed for at least two generations under standard greenhouse conditions. For the photoperiod experiment, plants were either grown under long day (16 h light: 8 dark) or short day (8 h light: 16 h dark) conditions at 22°C in controlled growth chambers. For VIGS experiments, plants were grown from seed in two batches at the University of Vermont greenhouse from September 2012 and August 2013 with supplemental light to mimic long days (16 h light: 8 dark). Diurnal temperatures ranged from 20 to 24°C. Significant differences in days to flowering and leaf number at flowering were determined using the analysis of variance (aov) and posthoc Tukey test functions in R.

Phylogenetic Analysis

SOC1-like gene sequences of representative angiosperms were downloaded from Genbank after a BLAST search with *SOC1* from *A. thaliana*. The closest *A. thaliana SOC1* paralogs *AGL14*, *AGL19*, *AGL42*, *AGL72*, and *AGL71* were also downloaded for reference and to root the tree. Amino acid sequences were aligned manually

in Mesquite version 2.75 [19] (File S1). Nucleotide sequences were then subjected to maximum likelihood phylogenetic analysis in GARLI under a GTR+I+Γ model of evolution with 200 bootstrap replicates [20].

Vector Construction and Plant Transformation

Previous results suggest functional redundancy of petunia *SOC1*-like genes under standard greenhouse conditions [15–17]. Thus, in order to target more than one petunia *SOC1*-like gene for silencing, but without affecting the expression of more distantly related paralogs, two VIGS constructs were designed based on approximately 200 bp sequences spanning part of the coiled-coil keratin-like and C-terminal domains of *FBP21* and *FBP28*. The *FBP21* region was 85, 68, and 40% identical to *UNS*, *FBP28*, and *FBP22* (next closest paralog), respectively; and the *FBP28* region was 83, 82, and 60% identical to *UNS*, *FBP21*, and *FBP22*, respectively. Gene fragments were amplified from petunia floral cDNA with gene specific primers containing restriction fragment ends (Table S1), sequence verified, and cloned into the tobacco rattle virus 2 (TRV2) vector. Each VIGS vector was transformed into *Agrobacterium tumefaciens* strain EHA105. A 194 bp fragment of the petal pigment gene *CHALCONE SYNTHASE* (*CHS*) was also amplified, cloned, and ligated into TRV2 for use as an experimental control as previously described [21]. Bacterial growth and plant infiltration methods followed Drea *et al.* (2007)

Figure 3. VIGS phenotypes. (A) Flowering *CHS*-TRV2 positive plants showing lack of anthocyanin pigment in normally purple petals. (B) Late flowering *FBP21*-TRV2 plant. (C) Later flowering *FBP28*-TRV2 plant. (D) Boxplots showing number of days to flowering for plants positive for *CHS*-TRV2 (gray), *FBP21*-TRV2 (red), *FBP28*-TRV2 (blue), or *FBP21/FBP28*-TRV2 (green) constructs. Vertical lines separate batches of plants that were grown at different times of the year. Circles denote outlier values. Three asterisks indicate p values <0.001 relative to control plants. A single asterisk indicates p values <0.05 relative to control plants.

and Hileman *et al.* (2005) [22,23]. Twenty plants at the four-leaf pair stage for each silencing target were infiltrated in half their leaves with a 1:1 ratio of TRV1 and TRV2 using a needleless syringe for a total of 60 plants (*CHS*-, *FBP21*-, and *FBP28*-TRV2). To target all three *SOC1*-like genes for silencing, a batch of 20 plants was also infiltrated with TRV1, *FBP21*-TRV2, and *FBP28*-TRV2 in a 2:1:1 ratio.

Gene Expression and Phenotyping

RNA was extracted from leaves, SAMs, and nodes on the main stem of wild type plants at different developmental stages three hours post-zeitgeber, and in leaves at different times post-zeitgeber. Total RNA was extracted using TriReagent (Life Technologies) according to the manufacturer's instructions, DNA was degraded using TURBO DNase (Life Technologies), and cDNA was synthesized using 1 μg RNA and iScript reverse transcriptase (BioRad). Primers for quantitative PCR (qPCR) of the housekeeping genes *EF1alpha* and *UBQ5*, *UNS*, *FBP21*, and *FBP28* were designed in Primer3 [24] and tested for efficiency using Fast SYBR Green Master Mix (Life Technologies) as previously described [25]. Primer pairs that gave high PCR efficiencies and single melt curves were selected for gene expression analyses (Table S1). After correcting for transcriptional activity, cycle threshold (C_T) values were normalized against the geomean of the two housekeeping genes for three technical replicates and two to three biological replicates. To verify infection

with TRV2 constructs and gene silencing, RNA was extracted from the tips of non-flowering branches (SAMs plus young leaves) in plants with the first flower at anthesis, and cDNA was synthesized. VIGS plants were screened for TRV2 using the primers pYL156F and pYL156R as previously described [23]. qPCR expression of *UNS*, *FBP21*, and *FBP28* was compared between ten plants positive for *FBP21*-TRV2 or *FBP28*-TRV2, and *CHS*-TRV2. Each VIGS positive plant was phenotyped for number of days from germination to anthesis, leaf number at anthesis, and petal whitening. Significant differences in gene expression were determined using the analysis of variance (aov) and posthoc Tukey test functions in R.

Results and Discussion

Petunia *SOC1*-like Genes are Partially Redundant in the Control of Flowering Time

Infiltration of petunia seedlings with *Agrobacterium* resulted in a total of 24, 21, 37, and 4 plants positive for the *CHS*-, *FBP21*-, *FBP28*-, and *FBP21/28*-TRV2 constructs, respectively (Fig. 2A). Relative expression analyses on vegetative SAMs of flowering individuals revealed significant silencing of *UNS* and *FBP21*, but not *FBP28*, by the *FBP21*-TRV2 construct (Fig. 2B). Although the *FBP28*-TRV2 construct had high sequence similarity to *UNS*, *FBP21*, and *FBP28*, only *FBP21* and *FBP28* were silenced using this construct. This disparity might be explained by the number of

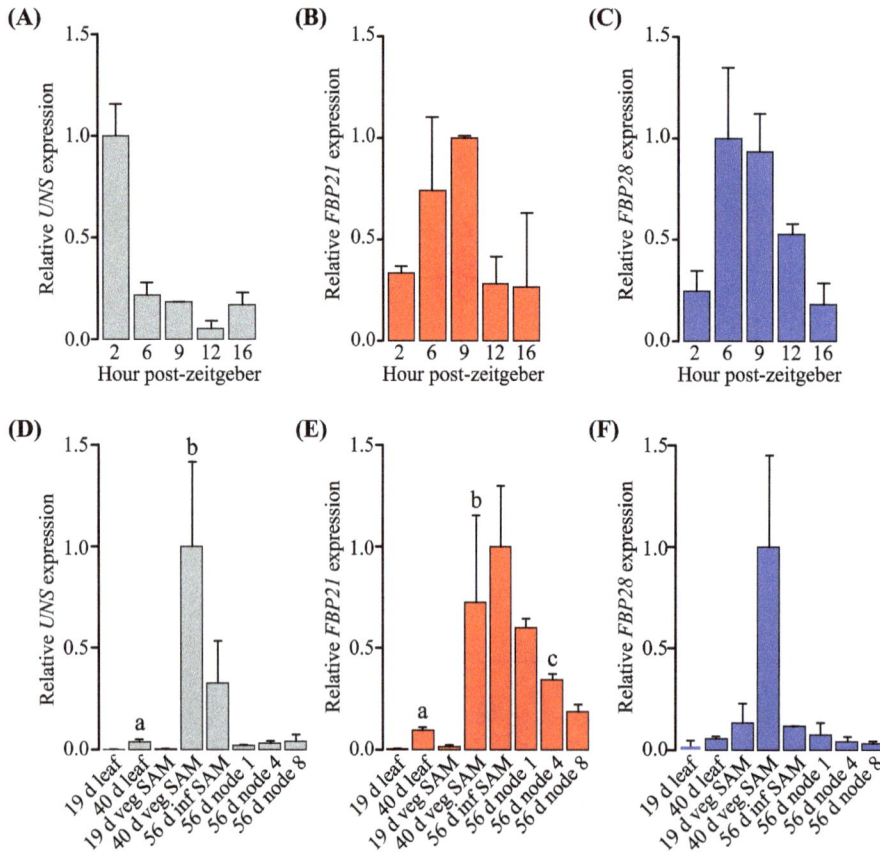

Figure 4. Relative expression of petunia *UNS* (gray), *FBP21* (red), and *FBP28* (blue). (A–C) Transcript levels of *SOC1*-like genes vary in the youngest (upper) leaves throughout 16 h long days relative to the zeitgeber (dawn), but these trends are not significant. (D–F) *SOC1*-like gene transcripts are most abundant in SAMs relative to leaves and nodes. However, the peak of expression in each tissue type varies between genes. Bars are averages for two to three biological replicates with standard deviations. Significant differences at the 0.05 level are denoted by letters: a, leaves; b, SAMs; and c, nodes.

contiguous 21 bp matches between the three target *SOC1*-like genes, which is the minimum recognition length for RNA interference in plants. Thus, all three genes were targeted for silencing when plants were positive for *FBP21*- and *FBP28*-TRV2, and different combinations of two genes were silenced when positive for one construct alone.

In order to test the hypothesis that *SOC1*-like genes positively regulate flowering in petunia, days to flowering were calculated for each infected group of plants (Fig. 3). Data were compared between *CHS*-infected control plants, which all showed loss of anthocyanin accumulation in petals (Fig. 3A), and *SOC1*-like infected VIGS plants (Fig. 3B–E). The flowering response varied across plants as expected for incomplete and variable silencing, and between treatment blocks as predicted based on differences in ambient greenhouse light. Despite this, *SOC1*-like silenced plants in every category were later flowering on average than *CHS*-TRV2 control plants grown at the same time (Fig. 3D). The most extreme late flowering phenotypes were observed for the triple silenced plants positive for both *FBP21*- and *FBP28*-TRV2 (p<0.001). Moreover, plants silenced for *UNS* and *FBP21* (*FBP21*-TRV2 vector) were later flowering (p<0.001) than plants silenced for *FBP21* and *FBP28* (*FBP28*-TRV2 vector) (p = 0.015) when grown under the same conditions (Fig. 3D).

Petunia *SOC1*-like Genes are Differentially Regulated by Photoperiod and Age

Heterogeneous protein function is often correlated with differences in the spatiotemporal pattern of underlying gene expression. In order to determine if petunia *SOC1*-like genes are differentially regulated diurnally, during development, and in response to different photoperiods, qPCR was conducted on wild type tissues. No significant differences in *SOC1*-like gene expression levels were found across 16 hours of daylight. This result is in contrast to *A. thaliana SOC1* expression, which peaks 12 hours post-zeitgeber in long day photoperiods [26]. Thus, petunia *SOC1*-like genes could potentially promote flowering in light under both long and short day conditions.

Transcript levels of *UNS* and *FBP21* significantly increased in leaves (p<0.05 and <0.01, respectively) and SAMs (p<0.05 and <0.05, respectively) during vegetative development (Fig. 4D and 4E). However, expression was not significantly different for *FBP28*, or any of the *SOC1*-like genes between vegetative and inflorescence apices (Fig. 4D–F). With the exception of *FBP28*, these data are consistent with reports of *A. thaliana SOC1* expression [18,27]. Following the transition to flowering at day 56, *FBP21* decreased significantly (p<0.001) in nodes from the plant base to the SAM (Fig. 4E). In contrast, no significant differences in nodes expression were detected for *UNS* and *FBP28* (Fig. 4D and 4F).

Although petunia is a long day plant, flowering earlier with increasing daylight hours, *FBP28* was significantly (p<0.05) more

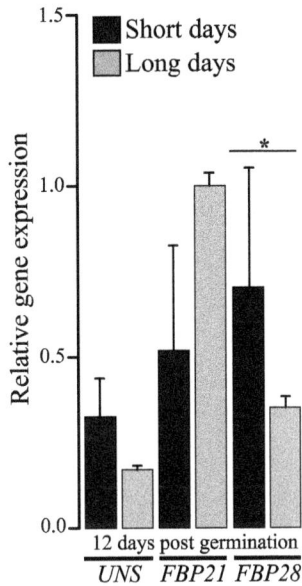

Figure 5. Relative leaf expression of *SOC1*-like genes in response to short (8 h light: 16 h dark) (black bars) versus long (16 h light: 8 h dark) (gray bars) days. Tissue was collected three hours post-zeitgeber under both photoperiod conditions. *FBP28*, but not *UNS* and *FBP21*, is more strongly expressed after 12 days grown under short versus long day conditions (p<0.05 denoted by an asterisk). Bars are averages for two to three biological replicates with standard deviations.

strongly expressed in the fully expanded top leaf 12 days post-germination when grown under short versus long days (Fig. 5). In contrast, there was no significant photoperiod effect on the expression of *UNS* and *FBP21* (Fig. 5). Taken together, data from this experiment suggest subtle differences in the spatiotemporal regulation of petunia *SOC1*-like genes during vegetative develop-

ment and by photoperiod, with *FBP28* expression being the most divergent compared to *A. thaliana SOC1* [18,27]. We hypothesize that *FBP28* has been recruited to ensure flowering under short days in petunia, and that the regulatory elements of all three *SOC1*-like genes have evolved following their recent duplication.

Conclusions

Our data support the hypothesis that petunia *SOC1*-like genes have retained their function as flowering time pathway integrators following duplication [15,17]. This is in contrast to the closely related *SOC1*-like gene *GhSOC1* in *Gerbera hybrida* (Asteraceae) that appears only to function in late flower development [28]. However, differential expression of *UNS*, *FBP21*, and *FBP28* in response to photoperiod and age suggests subtly different developmental roles to both ensure and fine-tune flowering under variable environmental conditions. A similar mechanism has been assigned to duplicated *FLOWERING LOCUS C (FLC)/MADS AFFECTING FLOWERING (MAF)* genes in *A. thaliana* that response to different low and high temperature cues [29]. Future studies determining the effect of co-silencing *SOC1*-like genes under variable photoperiod and temperature conditions will be needed to further test this hypothesis.

Author Contributions

Conceived and designed the experiments: JCP. Performed the experiments: JCP SAJ SGJ. Analyzed the data: JCP. Contributed reagents/materials/analysis tools: JCP. Wrote the paper: JCP SAJ SGJ.

References

1. Samis KE, Murren CJ, Bossdorf O, Donohue K, Fenster CB, et al. (2013) Longitudinal trends in climate drive flowering time clines in North American *Arabidopsis thaliana*. Ecol. Evol. 2: 1162–1180.
2. Colautti RI, Barrett SC (2013) Rapid adaptation to climate facilitates range expansion of an invasive plant. Science 342: 364–366.
3. Kardailsky I, Shukla VK, Ahn JH, Dagenais N, Christensen SK, et al. (1999) Activation tagging of the floral inducer FT. Science 286: 1962–1965.
4. Suárez-López P, Wheatley K, Robson F, Onouchi H, Valverde F, et al. (2001) CONSTANS mediates between the circadian clock and the control of flowering in *Arabidopsis*. Nature 410: 1116–1120.
5. Valverde F, Mouradov A, Soppe W, Ravenscroft D, Samach A, et al. (2004) Photoreceptor regulation of CONSTANS protein in photoperiod flowering. Science 303: 1003–1006.
6. An H, Roussot C, Suárez-López P, Corbesier L, Vincent C, et al. (2004) CONSTANS acts in the phloem to regulate a systemic signal that induces photoperiodic flowering of *Arabidopsis*. Development 131: 3615–3626.
7. Huang T, Bohlenius H, Eriksson S, Parcy F, Nilsson O (2005) The mRNA of the Arabidopsis gene FT moves from leaf to shoot apex and induced flowering. Science 309: 1694–1696.
8. Andres F, Coupland G (2012) The genetic basis of flowering responses to seasonal cues. *Nat. Rev.* Genet. 13: 627–639.
9. Lee S, Kim J, Han JJ, Han MJ, An G (2004) Functional analyses of the flowering time gene *OsMADS50*, the putative *SUPPRESSOR OF OVEREXPRESSION OF CO 1/AGAMOUS-LIKE 20 (SOC1/AGL20)* ortholog in rice. Plant J. 38: 754–764.
10. Ding L, Wang Y, Yu H (2013) Overexpression of *DOSOC1*, an ortholog of *Arabidopsis SOC1*, promotes flowering in the orchid *Dendrobium* Chao Parya Smile. Plant Cell Physiol. 54: 595–608.
11. Zhou C-M, Zhang T-Q, Wang X, Yu S, Lian H, et al. (2013) Molecular basis of age-dependent vernalization in *Cardamine flexuosa*. Science 340: 1097–1100.
12. Mouhu K, Kurokura T, Koskela EA, Albert VA, Elomaa P, et al. (2013) The *Fragaria vesca* homolog of *SUPPRESSOR OF OVEREXPRESSION OF CON-*

STANS1 represses flowering and promotes vegetative growth. Plant Cell 25: 3296–3310.
13. Snowden KC, Napoli CA (2003) A quantitative study of lateral branching in petunia. *Funct. Plant Biol.* 30: 987–994.
14. Immink RGH, Ferrario S, Busscher-Lange J, Kooiker M, Busscher M, et al. (2003) Analysis of the petunia MADS-box transcription factor family. *Mol. Genet.* Genom. 268: 598–606.
15. Ferrario S, Busscher J, Franken J, Gerats T, Vandenbussche M, et al. (2004) Ectopic expression of the petunia MADS box gene *UNSHAVEN* accelerates flowering and confers leaf-like characteristics to floral organs in a dominant-negative manner. Plant Cell 16: 1490–1505.
16. Rijpkema A, Gerats T, Vandenbussche M (2006) Genetics of floral development in *Petunia*. Adv. Bot. Res. 44: 238–278.
17. Ma G, Ning G, Zhang W, Zhan J, Lv H, et al. (2011) Overexpression of Petunia *SOC1*-like gene *FBP21* in tobacco promotes flowering without decreasing flower or fruit quality. *Plant Mol. Biol. Rep.* 29: 573–581.
18. Samach A, Onouchi H, Gold SE, Ditta GS, Schwarz-Sommer Z, et al. (2000) Distinct roles of CONSTANS target genes in reproductive development of Arabidopsis. Science 288: 1613–1616.
19. Maddison WP, Maddison DR (2011) Mesquite: a modular system for evolutionary analysis. Version 2.75. http://mesquiteproject.org.
20. Zwickl D (2006) Genetic algorithm approaches for the phylogenetic analysis of large biological sequence datasets under the maximum likelihood criterion. PhD thesis. TX University of Texas at Austin, Texas.
21. Chen J-C, Jiang C-Z, Gookin TE, Hunter DA, Clark DG, et al. (2004) Chalcone synthase as a reporter in virus-induced gene silencing studies of flower senescence. Plant Mol. Biol. 55: 521–530.
22. Drea S, Hileman LC, de Martino G, Irish VF (2007) Functional analyses of genetic pathways controlling petal specification in poppy. Development 134: 4157–4166.

23. Hileman LC, Drea S, Martino G, Litt A, Irish VF (2005) Virus-induced gene silencing is an effective tool for assaying gene function in the basal eudicot *Papaver somniferum* (opium poppy). Plant J. 44: 334–341.

24. Rozen S, Skaletsky H (2000) Primer3 on the WWW for general users and for biological programmers. In *Bioinformatics Methods and Protocols: Methods in Molecular Biology*. Edited by Krawetz S and Misener S. New Jersey: Humana Press; 365–386.

25. Preston JC, Hileman LC (2010) SQUAMOSA-PROMOTER BINDING PROTEIN 1 initiates flowering in *Antirrhinum majus* through the activation of meristem identity genes. Plant J. 62: 704–712.

26. Blázquez MA, Trénor M, Weigel D (2002) Independent control of gibberellin biosynthesis and flowering time by the circadian clock in Arabidopsis. Plant Physiol. 130: 1770–1775.

27. Winter D, Vinegar B, Nahal H, Ammar R, Wilson GV, et al. (2007) An "electronic fluorescent pictograph" browser for exploring and analyzing large-scale biological data sets. PLoS One 2: 718.

28. Ruokolainen S, Ng YP, Albert VA, Elomaa P, Teeri TH (2011) Over-expression of the *Gerbera hybrida At-SOC1-like1* gene *Gh-SOC1* leads to floral organ identity deterioration. Annals Bot. 107: 1491–1499.

29. Posé D, Verhage L, Ott F, Mathieu J, Angenenet GC, et al. (2013) Temperature-dependent regulation of flowering by antagonistic FLM variants. Nature 503: 414–417.

30. Becker A, Theissen G (2003) The major clades of MADS-box genes and their role in the development and evolution of flowering plants. Mol. Phyl. Evol. 29: 464–489.

20

Laboratory Selection Quickly Erases Historical Differentiation

Inês Fragata[1][*][9], Pedro Simões[1][*][9], Miguel Lopes-Cunha[1], Margarida Lima[1], Bárbara Kellen[1¤], Margarida Bárbaro[1], Josiane Santos[1], Michael R. Rose[2], Mauro Santos[3], Margarida Matos[1]

[1] Centro de Biologia Ambiental e Departamento de Biologia Animal, Faculdade de Ciências, Universidade de Lisboa, Lisboa, Portugal, [2] Department of Ecology and Evolutionary Biology, University of California Irvine, Irvine, California, United States of America, [3] Departament de Genètica i de Microbiologia, Universitat Autònoma de Barcelona, Bellaterra, Barcelona, Spain

Abstract

The roles of history, chance and selection have long been debated in evolutionary biology. Though uniform selection is expected to lead to convergent evolution between populations, contrasting histories and chance events might prevent them from attaining the same adaptive state, rendering evolution somewhat unpredictable. The predictability of evolution has been supported by several studies documenting repeatable adaptive radiations and convergence in both nature and laboratory. However, other studies suggest divergence among populations adapting to the same environment. Despite the relevance of this issue, empirical data is lacking for real-time adaptation of sexual populations with deeply divergent histories and ample standing genetic variation across fitness-related traits. Here we analyse the real-time evolutionary dynamics of *Drosophila subobscura* populations, previously differentiated along the European cline, when colonizing a new common environment. By analysing several life-history, physiological and morphological traits, we show that populations quickly converge to the same adaptive state through different evolutionary paths. In contrast with other studies, all analysed traits fully converged regardless of their association with fitness. Selection was able to erase the signature of history in highly differentiated populations after just a short number of generations, leading to consistent patterns of convergent evolution.

Editor: Thomas Flatt, University of Lausanne, Switzerland

Funding: This study was partially financed by Portuguese National Funds through "Fundação para a Ciência e a Tecnologia" (FCT, http://www.fct.pt/) within the projects ref. PTDC/BIA-BDE/65733/2006 and ref. PTDC/BIA-BEC/098213/2008. BK and ML had BTI grants from FCT within the project ref. PTDC/BIA-BDE/65733/2006. PS had a Post-Doc grant and ML-C had a BI grant within the project ref. PTDC/BIA-BEC/098213/2008. IF had a PhD grant (SFRH/BD/60734/2009) and JS had a PhD grant (SFRH/BD/28498/2006) from FCT. MS is supported by grant CGL2010-15395 from the Ministerio de Ciencia e Innovación (Spain), and by the ICREA Acadèmia program. The funders had no role in study design, data collection and analysis, decision to publish, or preparation of the manuscript.

Competing Interests: The authors have declared that no competing interests exist.

* E-mail: irfragata@fc.ul.pt (IF); pmsimoes@fc.ul.pt (PS)

¤ Current address: Instituto Gulbenkian da Ciência, Oeiras, Portugal

[9] These authors contributed equally to this work.

Introduction

The roles of history, chance, and selection in shaping the evolutionary processes of populations adapting to new environments is a long-standing topic of debate in evolutionary biology [1–7]. Although repeated adaptive radiations and convergent evolution support the view that evolution is generally predictable [7–9], most classic case studies involve morphological traits ([10–13], but see [14]). However, the relationships between morphological characters and Darwinian fitness are sometimes ambiguous, suggesting the value of testing for convergence among traits that are more straightforwardly related to fitness. Life-history traits are obvious candidates to study in this respect, because their known association with fitness gives a greater likelihood of interplay between history, chance, and selection that strongly depends on underlying genetic variation [15,16]. Furthermore, pleiotropy and epistasis are common among life-history traits [17], which should foster the dependence of selection outcome on genetic background [18,19]. Thus experimental studies of the evolution of life-history characters should allow better tests of

whether adaptive convergence occurs when the "tape of evolution" is replayed [20,21].

Real-time examples of convergent evolution using replicated laboratory populations have been found in several species (e.g. [5,22–28]). But there are also examples of divergence in experimental evolution [29,30]. However, in all instances where strong convergence was found with selection erasing historical signatures, the lines were recently derived from the same ancestral population and might thus not differ much in their genetic background. Whether historical effects derived from highly and long differentiated ancestral populations do or do not constrain evolution is still an open question (vid. [5]).

A clarification is in order here. The terms convergent evolution and parallel evolution have been used sometimes with the same meaning and sometimes with different meanings (e.g. [31,32]). A common distinction involves the underlying genetic mechanisms, with the term "convergence" used when the genetic mechanisms involved are different (e.g. in distantly related species) and parallel evolution when the same genetic mechanisms are thought or found to be involved. Here we are interested in phenotypic convergent evolution, and how it is affected by previous history. We will apply the term *convergent* phenotypic evolution to cases where populations start from contrasting adaptive states and evolve such that these differences are erased during adaptation to a common environment. This applies to the real-time evolution studies in bacteria done by Travisano et al. [5] or Melnyk and Kassen [28], or to the reverse evolution studies done by Teotónio et al. [23,24] among others. By contrast, we refer to examples of *parallel* evolution as those cases which involve populations that are not differentiated to start with or that maintain those differences throughout evolution (e.g. [33–35]; see also [36,37]).

The southern peninsulas of Europe acted as refugia for many species at the height of the last Weichselian ice age (20 kya), species which rapidly expanded northward as the climate warmed. It has become apparent over the last 30 years that many European species are genetically differentiated across at least five major geographic regions as a result of this postglacial expansion [38]. Their life-histories evolved differently due to contrasting environmental conditions, and in some cases differentiated populations have met and produced narrow hybrid zones [38]. This glacial-interglacial climatic reversal provides usefully differentiated wild populations with which to study convergent evolution in replicated laboratory lines derived from such wild populations. We have used the native Palearctic fly *Drosophila subobscura* for this purpose, because it is amenable to laboratory experimentation and exhibits wild genetic differentiation which reflects its postglacial expansion [39]. Moreover, *D. subobscura* exhibits latitudinal clines for body size and chromosomal inversions [40,41], with recent studies showing that northern populations are becoming more similar to southern populations in response to global warming [42]. Both local adaptation and gene flow may be involved [41,43].

Our team has been studying the repeatability of adaptive evolution of *D. subobscura* populations to the laboratory environment over repeated samplings from nearby Portuguese locations [26,44,45]. For each experimental population, we characterized the evolutionary trajectories of a set of life-history traits - particularly age of first reproduction, early and peak fecundity - as well as a "physiological trait" - starvation resistance (related to fat storage [46]), throughout numerous generations since laboratory introduction. We have shown that, although the laboratory evolution of these populations is repeatable in some respects, evolutionary contingencies often hamper quick convergence across populations [26,44]. This appears to be due to genetic drift effects during the first generations of laboratory adaptation, since all these populations were derived from the same geographical area, sharing a recent history [45].

However, whether prior history constrains subsequent evolution among geographically differentiated populations is unknown. Taking advantage of the clinal variation of European *D. subobscura* populations, we here report the experimental evolution during the first 22 generations of adaptation to the laboratory of populations derived from wild-caught samples in turn obtained from three contrasting latitudes: Adraga (Portugal), Montpellier (France) and Groningen (Netherlands).

Materials and Methods

Founding and Maintenance of populations

D. subobscura individuals were collected in August 2010 from three locations in Europe: Adraga (Portugal), Montpellier (France) and Groningen (Netherlands); these were used to start three foundations in the laboratory. The number of founding females was 234 for Adraga (Ad), 171 for Montpellier (Mo) and 160 for Groningen (Gro). F1 eggs and individuals were treated with tetracycline (25 mg/l) and F2 with ceftriaxone and spectinomycin (50 mg/l) due to the presence of pathogenic (not identified) bacteria that caused high larval mortality. No *Wolbachia* was present in founder or control individuals untreated with antibiotics. Females from these first two generations were maintained in separate vials, to equalize their contribution to the next generation. To avoid inbreeding, females were crossed with males from different vials (1st laboratory generation) or derived from a random sample from all vials (2nd generation). At the 3rd generation an equal number of offspring of each female were randomly mixed, giving rise to the outbred populations. At the 4th generation, replicate populations were formed by dividing the overall egg collection of each outbred population in three equal parts (e.g. originating Ad_1, Ad_2 and Ad_3 from the Adraga foundation). Three long established populations (TA – formerly "TW" populations - [26]), derived from a collection in Adraga in 2001, were used as controls and assayed in synchrony with the experimental populations. These populations were in the 115th generation at the time of founding of the newly introduced populations, and were also treated with antibiotics (which led to the new labelling TA) in the same period as the new populations to avoid differences arising from contrasting treatments. In a preliminary assay, no interaction was found between the antibiotic treatment and the different foundations (See Additional Methods S1).

All populations were maintained under the same conditions with synchronous discrete generations of 28 days, reproduction close to peak fecundity, photoperiod of 12 hours of light: 12 hours of darkness at 18°C, with census sizes between 500 and 1200 individuals (with an average census size during the first 22 generations of 782.3 for Ad_{1-3}, 607.0 for Mo_{1-3}, and 708.6 for Gro_{1-3}). Flies were kept in vials with controlled density both for eggs (around 70 eggs per vial) and adults (50 adults per vial). At each generation, flies of a given population emerging from the several vials were thoroughly randomized four to five days after emergence, using CO_2 anaesthesia. Egg collection for the next generation was done one week later, with flies having between 8 and 12 days of age after emergence (see also [26,44,45]).

Genetic Differentiation

Global genetic differentiation among the three foundations (Adraga, Montpellier and Groningen) was measured at generation 2 using haploid data from chromosomal inversion polymorphisms. Differentiation through F_{ST} was obtained with Arlequin v3.5 [47].

Phenotypic assays

Assays were performed at generations 6, 11, 14, 18 and 22 after introduction to the laboratory. Sample sizes per population and assay varied between 15 and 24 mated pairs of flies. Assayed flies were transferred daily and the number of eggs laid per female was counted during the first 12 days. At the 12th day, the flies were transferred to agar medium and starvation resistance was assayed. With this design we estimated three fecundity-related traits: age of first reproduction (number of days between emergence and the first egg laying – 'A1R'), early fecundity (total number of eggs laid during the first week of life – 'F1–7'), and peak fecundity (total number of eggs laid between days 8 and 12 – 'F8–12'). Female starvation resistance was estimated as the number of hours until death (registered every 6 h after transfer to agar – 'RF'). The latter is a trait strongly related to lipid content, and somewhat correlated with adult survival [46,48]. Considering the generation time in our populations and the maintenance regime, this trait is not expected to be strongly related to fitness. It may nevertheless have an indirect impact on acquisition and allocation of resources [49]. We also estimated body size ('BS') for all assayed females, a morphological trait expected to have effects on life-history traits [50]. As a proxy metric for this trait we used wing size estimated by geometric morphometric analysis (see [51], details in Additional Methods S1).

Statistical methods

Evolutionary trajectories of the several traits. Nested Analyses of Variance (ANOVA) were performed at each generation to test for differences among foundations in all traits. The linear model used was:

$$Y = \mu + Found + Pop\{Found\} + \varepsilon,$$

where Y refers to the trait analysed, *Found* refers to the fixed factor Foundations (with three categories, Adraga, Montpellier and Groningen), and *Pop{Found}* refers to the random factor Populations nested in each Foundation (*i.e*, the three replicate populations). Analyses including TA controls were also performed. A type III sum of squares was used in all tests, with the error being the population term. Pairwise comparisons between foundations were performed, with adjustment for multiple testing following the false discovery rate (FDR) procedure of Benjamini and Yekutieli (2001, theorem 1.3 [52]).

The evolutionary trajectory of each trait and foundation was estimated using in each generation the three average values of the replicate populations (as differences from to the average control values); the best linear model being calculated by Type I least-squares linear regression. To test for significance of the overall response across foundations as well as differences in evolutionary rate between them ANCOVA analyses were performed using the linear model:

$$Y = \mu + Found + Gen + Pop\{Found\} + Found * Gen$$
$$+ Pop\{Found\} * Gen + \varepsilon,$$

where *Y* refers to the different traits analysed, *Gen* to the generations assayed (as covariate) and the other factors as mentioned above. Analyses using body size as a covariate were also done to account for its effects on other traits.

All data analyses described above were performed using STATISTICA 10 and EXCEL.

Multivariate analysis of evolutionary dynamics. A Principal Component Analysis (PCA), using the correlation matrix, was performed with the mean differences between experimental replicate populations and the controls for all studied traits and generations. The multivariate phenotypic trajectories were analysed using the method described in Adams and Collyer [53]. This led to the estimation of differences between pairs of foundations in the following parameters: *magnitude* (differences between the first and the last generations), *direction* (standardized differences between the angles of the first axis of the PCA), and *shape* (deviations of corresponding generations between two scaled and aligned trajectories). To estimate the statistical significance of these differences, 1000 residual-randomization permutations were made. To estimate the distance at generations 6 and 22 between each pair of foundations, we calculated the Euclidean distance using the average scores per foundation for each principal component. In order to calculate the significance of the Euclidean distances, a null distribution was created using 9999 permutations of replicates. Confidence intervals were estimated by 9999 bootstraps at the replicate level within each foundation using the

average trait values per replicate. Multivariate analyses were performed using R [54] with the rgl package [55].

Estimates of causal components of variation. To analyse the contributions of history, chance, and selection throughout the study, we estimated several variance components for early fecundity and starvation resistance in each generation. We used the nested ANOVA model to estimate the variance components of History – as the differences among foundations - and Chance – as the differences among populations within foundations. To calculate the cumulative effect of Selection for each foundation and generation, we applied a mixed bi-factorial ANOVA. This effect was estimated as the variation between the earliest generation assayed and each one of the later generations. Confidence intervals of variance components were estimated by bootstrap at the level of the error term. For further details, see Additional Methods S1.

Results

Early Differentiation

As a measure of the early genetic differentiation between our foundations, we estimated F_{ST} using chromosomal inversion frequencies. We found highly significant differentiation between foundations at the 2nd generation ($F_{ST} = 0.204$, $P < 0.0001$).

All foundations were clearly differentiated in phenotypic traits at the 6th generation, with Groningen females having better performance for all life-history traits as well as female starvation resistance (Fig. 1, Table S1). Additionally, Groningen females had a significantly higher body size compared to other foundations.

Evolutionary trajectories of single traits

The initially low fecundity (and high age of first reproduction) of the three foundations quickly improved and they phenotypically converged through time, such that they were no longer significantly different by the 14th generation (Fig. 1A–C, Table S1a and b). By the 22nd generation, fecundity traits did not differ significantly between the foundations and the control baseline, with the exception of Montpellier-derived populations for early fecundity (Fig. 1B, Table S1c). By contrast, starvation resistance was initially about equal or higher (Groningen) than the control baseline, and quick convergence was observed because of a drop in starvation resistance among Groningen-derived flies (Fig. 1D). In fact, by the 11th generation the foundations were no longer differentiated between them or relative to the controls (Table S1). Convergence of the three foundations by means of different evolutionary rates can also be seen from the significant foundation*generation interaction term in the global ANCOVA and pairwise tests (see Table S2).

Convergence in body size was also observed among the three foundations, with the relative values of Montpellier and Adraga (measured as differences from the controls) increasing towards the values of Groningen, which was stable through time (Fig. 1E). This is seen in the contrasting smaller sizes of Adraga and Montpellier populations at the 6th generation, while by the 22nd generation all foundations had very similar sizes. Nevertheless, Adraga did not show a significant linear trend, or differences in evolutionary rate relative to Groningen, and differences between Montpellier and Groningen were only marginally significant (Fig 1E, Table S2b). Interestingly, foundations did not converge in body size to the control values (Fig. 1E, Table S1c). In absolute terms both Groningen and the controls decreased in body size with time (from G6 to G22 Groningen declined 2.9% and TA declined 2.4%), while both Adraga and Montpellier populations remained fairly stable (increases of 0.40% and 0.27%, respectively). We are

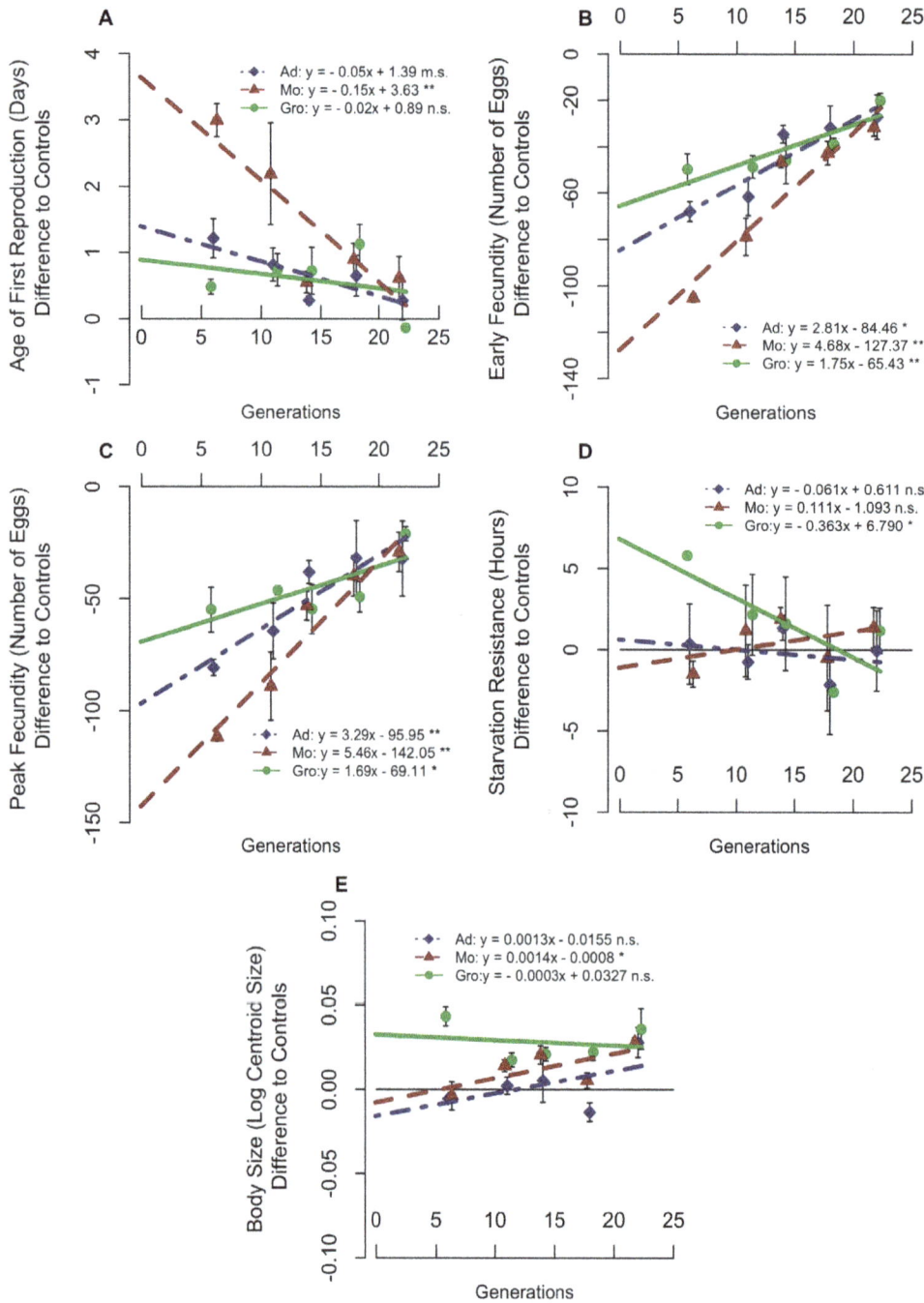

Figure 1. Evolutionary trajectories for the several traits analysed. Average differences from the controls for Age of First Reproduction (A), Early Fecundity (B), Peak Fecundity (C), Starvation Resistance (D) and Body Size (E) are presented for each foundation, as well as the corresponding linear regression models. Error bars correspond to variation between replicate populations of each foundation. Significance levels: $P>0.1$ n.s.; $0.1>P>0.05$ m.s.; $0.05>P>0.01$*; $0.01>P>0.001$**; $P<0.001$***.

assuming, as is common in experimental evolutionary designs that the long-established, TA populations, are close to stable genetic equilibrium, serving as controls [26,56]. Thus the temporal changes presented by these populations are likely environmental in origin and common to all foundations. Therefore differences from TA populations will give the evolutionary patterns of our experimental populations. Size differences did not account for the various evolutionary patterns in either fecundity or starvation

resistance, as analyses defining body size as covariate led to the same conclusions (Tables S2 and S3).

Consistently for all traits, we found that the populations showing a larger early differentiation also exhibited a higher evolutionary rate (Fig. 1).

Multivariate evolutionary trajectories

We integrated all phenotypic traits using Principal Component Analysis (PCA) to plot the multivariate evolutionary trajectories of

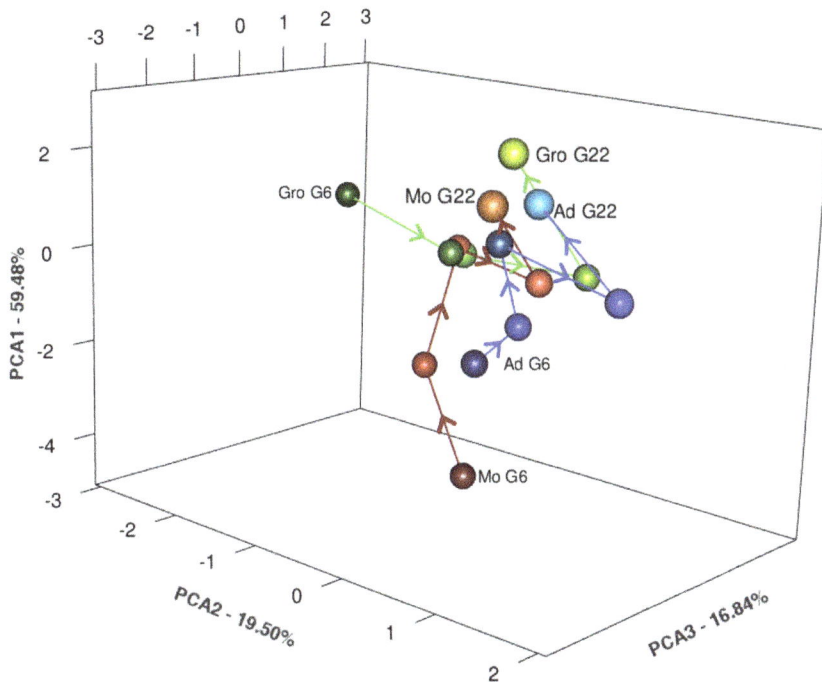

Figure 2. Multivariate evolutionary trajectories using Principal Component Analysis. All traits, generations and foundations were included.

the several populations (see Fig. 2 and Methods). The first axis of the PCA refers to changes involving life-history traits. The second and third axes show changes in starvation resistance and body size, respectively, possibly as correlated responses to selection (Table S4a).

We also estimated three parameters to compare the multivariate trajectories across populations: magnitude (rate of response), direction (convergence *vs* divergence), and shape (evolutionary path). Populations showed significant differences in magnitude and shape but not in direction, suggesting a clear convergence to the same adaptive state, though with contrasting routes and rates (Fig. 2, Table 1 and S4). Convergence was also confirmed by estimating the Euclidean distances between populations, which gave significant values in the initial generation, while by generation 22 they were no longer significantly higher than would be expected by chance alone (Fig. 2, Table S4b). Euclidean distance between all foundations decreased significantly between generation 6 and generation 22 (Table S4b).

The effect of History, Chance and Selection

Knowing that these populations converged to the same phenotypic state using different routes and rates, we measured the effects of history, chance and selection through time (Fig. 3). We focused this analysis on early fecundity and starvation resistance because of their contrasting associations with fitness (see Fig. 1B and 1D and [26,44]). Strong historical differentiation among foundations for early fecundity quickly faded as selection produced convergence after only 14 generations (Fig. 3, Tables S5 and S6). The initial historical variation for early fecundity was considerably higher than the variation due to sampling effects alone that was found for previous foundations from nearby locations. In fact, variance components estimated from our previous studies in 2001 and 2005 as a measure of sampling

effects [26,45] were at least eight times lower than the variance estimated among foundations in the present study (see Table S7).

A similar evolutionary pattern was observed for starvation resistance, starting from less differentiation (Fig. 3). However, when comparing the effect of chance to historical differentiation, there was a clear contrast between traits, particularly at generation 6 (ratio of history/chance = 25.3 for early fecundity and 3.6 for starvation resistance). Furthermore, whereas for early fecundity the role of chance was relatively small, for starvation resistance chance had similar effects to those of initial history. This suggests a bigger role for chance events during the evolution of starvation resistance (see Fig. 3A and 3C, Table S5).

The contrasting rates of convergence among foundations can be seen in the temporal changes of cumulative selective response for early fecundity (Fig. 3B). Higher temporal heterogeneity was observed for starvation resistance, with only the Groningen-derived populations changing markedly through time (see Table S6, Fig. 3D).

Discussion

In order to disentangle the relative impact of historical factors and selection during adaptation, we performed a real-time evolution study of highly differentiated *Drosophila subobscura* populations in nature during adaptation to a new environment under controlled conditions. Here we directly quantified the relative contribution of history and selection, an approach that has been used mostly in asexual organisms ([5,28,57,58], but see [25]). Combining this approach and the analysis of evolutionary multivariate trajectories we found a clear pattern of convergence despite the high level of differentiation in the initial foundations, suggesting that historical contingencies did not play a major role in adaptation.

Previous studies of experimental evolution, both in sexual and asexual populations, have also found convergence patterns,

Table 1. Pairwise differences and significance levels using Multivariate trajectory analysis.

Parameter	Found	Ad	Gro
Magnitude	Ad	—	—
	Gro	16.235 n.s.	—
	Mo	74.245 m.s.	90.480 *
Orientation	Ad	—	—
	Gro	2.532 n.s.	—
	Mo	1.708 n.s.	3.818 n.s.
Shape	Ad	—	—
	Gro	0.739 **	—
	Mo	0.178 n.s.	0.716 **

Note: significance levels: $P>0.1$ n.s.; $0.1>P>0.05$ m.s.; $0.05>P>0.01$*; $0.01>P>0.001$**.
Magnitude refers to the amount of evolutionary response, Orientation to the direction of the evolutionary path and Shape to the route of this path.

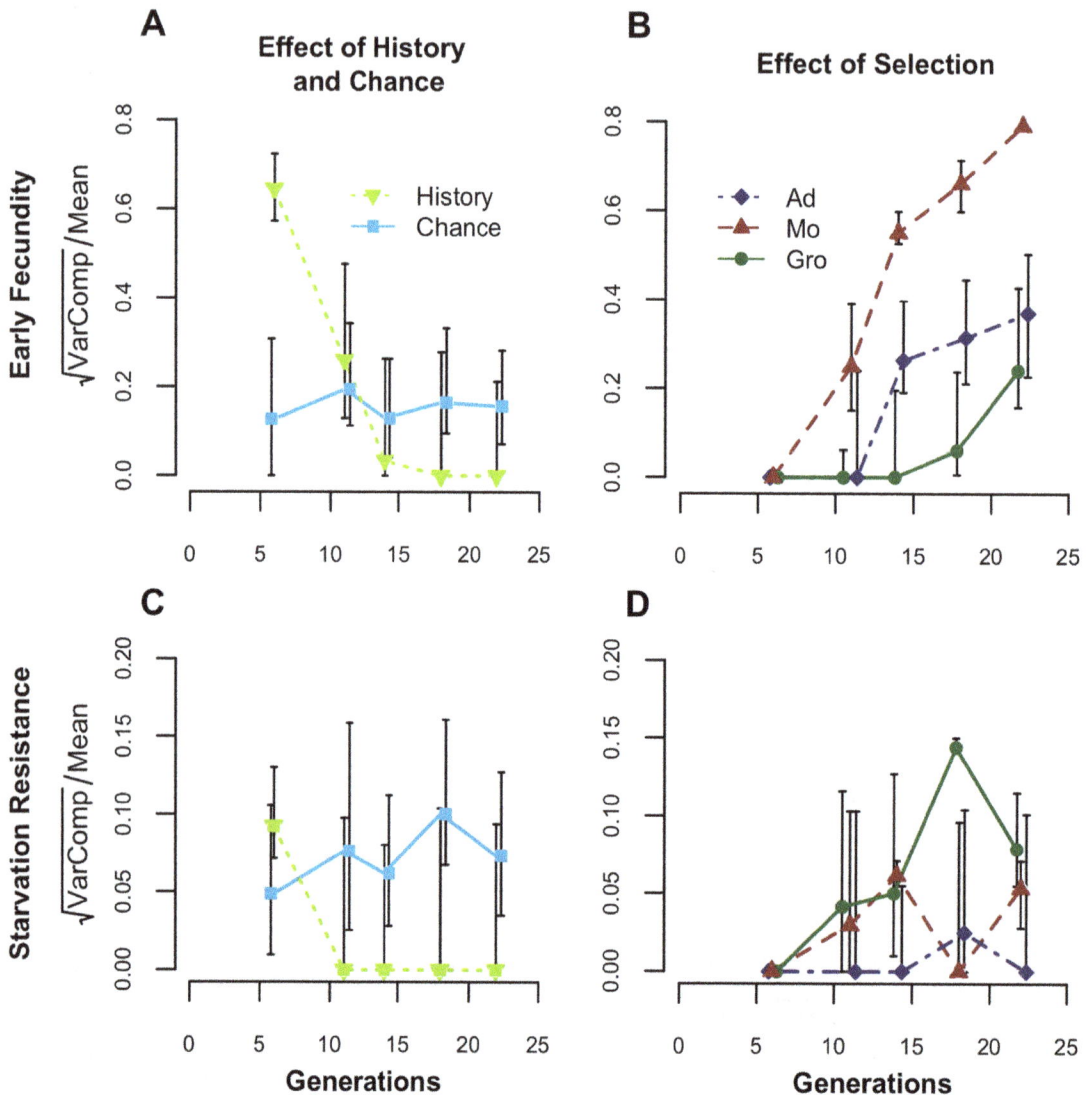

Figure 3. Variance components through time for history, chance and selection. Values presented are for Early Fecundity (A, B) and Starvation Resistance (C, D). Bars represent 95% confidence limits (see Material and Methods, Additional Methods S1 and Tables S5 and S6 for details).

although these patterns were less pervasive across traits [5,22,24–26,44]. Our results contrast with several studies in asexual populations, where history played an important role in traits weakly related to fitness [5,57] or even in traits strongly selected for [58]. The incomplete convergence observed by Spor et al [58] in *Saccharomyces cerevisiae* (initially highly differentiated) may have been due to the fact that their populations were still adapting at the end of the study. In our present study, the strong signature of history was quickly erased across all traits regardless of their association with fitness. It is possible that history has a greater impact in asexual populations, perhaps due to their relative lack of standing genetic variation and negative epistasis [23]. Blount et al [59] showed that historical contingencies have an important effect during adaptation in *Escherichia coli*, allowing the species to explore new ecological opportunities (see also [60] for an example in viruses). Furthermore, chance genetic associations during the initial stage of adaptation may even lead to divergence between asexual populations in traits not relevant to fitness [61]. Some experimental studies in sexual populations have also shown historical contingencies that prevented convergence (e.g. [29,30,62–65]), which illustrates the dangers of generalization.

We have previously shown that stochastic events during the early stages of laboratory colonization have some impact in the adaptive dynamics of *D. subobscura* populations sampled from nature in near-by or even the same location [26,44,45]. This might suggest an overall role of evolutionary contingencies hampering convergence. Nevertheless, the present study shows that strong initial differentiation among populations does not prevent convergent evolution. Our results differ from those abovementioned studies where populations of *Drosophila* derived from contrasting latitudes showed parallel [65] or even divergent [29,30] evolution under uniform selection. While the lack of convergence in those studies might be due to multiple solutions to the same problem [18], it is possible that the use of a limited number of isofemale lines in those studies contributed to their divergent outcomes.

The clear pattern of convergence across all traits in our study also suggests that the populations evolved to the same adaptive equilibrium. It is an open question whether this observed pattern of convergent adaptation might not have occurred if we had used a different laboratory environment. In fact, Melnyk and Kassen [28] showed contrasting evolutionary patterns in the same *Pseudomonas fluorescens* populations adapting to two different laboratory environments. These patterns point to the possibility of different underlying adaptive landscapes and different contributions of history *vs* selection, among selection regimes. Whether this applies to sexual populations with contrasting histories across fitness-related traits and ample standing genetic variation such as ours remains to be seen. Additionally, it is worth noting that our populations evolved in a benign homogeneous environment, and it is an open question whether other evolutionary outcomes may emerge as a function of the complexity and harshness of the environment [33,66].

In a study of reverse evolution in *Drosophila melanogaster*, Teotónio et al. [22,24] found that convergence to the ancestral state was not universal across traits, but instead depended somewhat on history as a result of previous adaptation to different selective regimes. It may be the case that adaptation to a novel environment, such as ours, is fostered by positive additive variance/covariance matrices across traits [49,67]. In the case of Teotónio et al's. study, the hypothesis that genetic variation affecting adaptation to the lab environment may have been exhausted is not likely, given the genome-wide absence of selective

sweeps leading to low heterozygosity found by Burke et al. [68] in their study of ten of those populations. Antagonistic pleiotropy seems to have played a role in our Groningen populations, since starvation resistance quickly declined during adaptation of those populations. This possible trade-off between starvation resistance and other traits, found only in Groningen, suggests that different mechanisms/associations among traits may be involved, in spite of the general convergence across foundations. Differential mechanisms of acquisition versus allocation of resources might have had different impacts between foundations or can change during the different phases of adaptation [56]. In this regard it is tempting to suggest that the Groningen populations were already better suited to the laboratory environment. Alternatively relaxed selection might have also contributed, though it is unlikely that it could by itself lead to the quick evolutionary pattern observed. Future studies on the effect of specific environments on G matrices would be interesting.

The fact that we report here full phenotypic convergence in only 22 generations of adaptation calls for a word of caution, in the sense that these populations might diverge if they continue to evolve at a steady rate in the future. This is not a likely scenario, considering the evidence for a slowing down of evolutionary rates of adaptation in our longer-term studies of laboratory adaptation in populations like these (e.g. [69,70]). Of course, close study of these populations after more generations of laboratory adaptation would settle this issue. More generations may also help to clarify the evolutionary dynamics of body size. Though convergence occurred among foundations, they did not evolve towards the values of the long established, control populations. This may be due to complex trade-offs that might manifest at a later evolutionary phase [56]. Alternatively, founder events or genetic drift may maintain these differences.

It is important to note that convergence at the phenotypic level does not necessarily imply that the underlying genetic basis of convergence is uniform. Although some studies show convergent (or parallel) genotypic evolution [9,36,71,72], phenotypically convergent (or parallel) evolution through different genetic mechanisms has also been found in both sexual and asexual organisms [27,31,32,73–75]. A plethora of factors could be responsible for these varying genetic pathways: different levels of standing genetic variation, mutational input, epistasis and pleiotropic effects [23,27,32,73,74,76]. It will be interesting to analyse to what extent the fast phenotypic convergence of our populations is matched at the genomic level.

The results of our study highlight the potential for error in characterizing geographical patterns from comparisons of populations even after relatively few generations of laboratory adaptation. Here we show that starvation resistance started with higher values for the Groningen foundations but quickly converged to values similar to those of the other populations. The absence of a latitudinal European cline for this trait in the study by Gilchrist et al. [77] might thus be due to laboratory convergent evolution during the multiple generations of laboratory culture prior to their measurement of starvation resistance.

To sum up, we found that populations with clear initial differentiation quickly converged phenotypically during adaptation to a new, common laboratory environment. Thus, selection was able to quickly overcome the effects of history even in laboratory populations founded from populations highly differentiated in nature. In this sense, phenotypic evolution was generally predictable, even across a set of complex traits, and was not significantly dependent on chance events or historical constraints.

Supporting Information

Table S1 Analyses of differences in life-history traits at each generation among and within foundations.

Table S2 ANCOVA models for each trait with Generation as covariate.

Table S3 ANCOVA models for each trait with Generation and Body Size as covariates.

Table S4 Principal Component Analysis including all traits and generations.

Table S5 Variance components of History and Chance for Early Fecundity and Starvation Resistance.

Table S6 Variance components of Selection for Early Fecundity and Starvation Resistance.

Table S7 Comparisons of the initial differences between the three foundations, relative to differences in previous studies.

Additional Methods S1

Acknowledgments

We thank Larry Mueller and Sara Magalhães for useful discussion and suggestions.

Author Contributions

Conceived and designed the experiments: IF PS MS MM. Performed the experiments: IF PS ML-C MB ML BK JS. Analyzed the data: IF PS MM. Contributed reagents/materials/analysis tools: MM. Wrote the paper: IF PS MM MS MRR.

References

1. Fisher RA (1930) The Genetical Theory of Natural Selection. Oxford: The Clarendon Press.
2. Wright S (1931) Evolution in Mendelian Populations. Genetics.
3. Kimura M (1968) Evolutionary rate at the molecular level. Nature 217: 624–626.
4. Gould SJ, Lewontin RC (1979) The Spandrels of San Marco and the Panglossian Paradigm: A Critique of the Adaptationist Programme. Proc R Soc B Biol Sci 205: 581–598.
5. Travisano M, Mongold JA, Bennett AF, Lenski RE (1995) Experimental Tests of the Roles of Adaptation, Chance, and History in Evolution. Science 267: 87–90.
6. Lenormand T, Roze D, Rousset F (2009) Stochasticity in evolution. Trends Ecol Evol 24: 157–165.
7. Losos JB (2011) Convergence, adaptation, and constraint. Evolution 65: 1827–1840.
8. Schluter D (2000) The Ecology of Adaptive Radiation. Oxford Uni. Oxford.
9. Stern DL (2013) The genetic causes of convergent evolution. Nat Rev Genet 14: 751–764.
10. Losos J, Jackman T, Larson A, Queiroz K, Rodriguez-Schettino L (1998) Contingency and determinism in replicated adaptive radiations of island lizards. Science 279: 2115–2118.
11. McKinnon JS, Rundle HD (2002) Speciation in nature: the threespine stickleback model systems. Trends Ecol Evol 17: 480–488.
12. Young KA, Snoeks J, Seehausen O (2009) Morphological diversity and the roles of contingency, chance and determinism in african cichlid radiations. PLoS One 4: e4740.
13. Manceau M, Domingues VS, Linnen CR, Rosenblum EB, Hoekstra HE (2010) Convergence in pigmentation at multiple levels: mutations, genes and function. Philos Trans R Soc Lond B Biol Sci 365: 2439–2450.
14. Johnson MA, Revell LJ, Losos JB (2010) Behavioral convergence and adaptive radiation: effects of habitat use on territorial behavior in Anolis lizards. Evolution 64: 1151–1159.
15. Lande R, Arnold S (1983) The measurement of selection on correlated characters. Evolution 37: 1210–1226.
16. Flatt T, Heyland A (2011) Mechanisms of Life History Evolution. The Genetics and Physiology of Life History Traits and Trade-Offs. Oxford Uni. Oxford.
17. Roff DA, Emerson K (2006) Epistasis and Dominance: Evidence for differential effects in life-history versus morphological traits. Evolution 60: 1981–1990.
18. Whitlock MC, Phillips PC, Moore FB-G, Tonsor SJ (1995) Multiple Fitness Peaks and Epistasis. Annu Rev Ecol Syst 601–629.
19. Gavrilets S (2010) High-dimensional fitness landscapes and speciation. In: Pigliucci M, Muller G, editors. Evolution: the extended synthesis. Cambridge, MA: MIT Press. pp. 45–80.
20. Gould SJ (1989) Wonderful Life: The Burgess Shale and the Nature of History. W. W. Norton & Company.
21. Lobkovsky AE, Koonin EV (2012) Replaying the tape of life: quantification of the predictability of evolution. Front Genet 3: 246.
22. Teotónio H, Rose MR (2000) Variation in the reversibility of evolution. Nature 408: 463–466. doi:10.1038/35044070.
23. Teotónio H, Rose MR (2001) Perspective: Reverse Evolution. Evolution 55: 653–660.
24. Teotónio H, Matos M, Rose MR (2002) Reverse evolution of fitness in Drosophila melanogaster. J Evol Biol 15: 608–617.
25. Joshi A, Castillo RB, Mueller LD (2003) The contribution of ancestry, chance, and past and ongoing selection to adaptive evolution. J Genet 82: 147–162.
26. Simões P, Santos J, Fragata I, Mueller LD, Rose MR, et al. (2008) How repeatable is adaptive evolution? The role of geographical origin and founder effects in laboratory adaptation. Evolution 62: 1817–1829.
27. Fox CW, Wagner JD, Cline S, Thomas FA, Messina FJ (2011) Rapid Evolution of Lifespan in a Novel Environment: Sex-Specific Responses and Underlying Genetic Architecture. Evol Biol 38: 182–196.
28. Melnyk AH, Kassen R (2011) Adaptive landscapes in evolving populations of Pseudomonas fluorescens. Evolution 65: 3048–3059.
29. Cohan FM, Hoffmann AA (1986) Genetic divergence under uniform selection. II Different responses to selection for knockdown resistance to ethanol among Drosophila melanogaster populations and their replicate lines. Genetics 114: 145–163.
30. Cohan FM, Hoffmann AA (1989) Uniform Selection as a Diversifying Force in Evolution: Evidence from Drosophila. Am Nat 134: 613–637.
31. Arendt J, Reznick D (2007) Convergence and parallelism reconsidered: what have we learned about the genetics of adaptation? Trends Ecol Evol 23: 26–32.
32. Elmer KR, Meyer A (2011) Adaptation in the age of ecological genomics: insights from parallelism and convergence. Trends Ecol Evol 26: 298–306.
33. Colegrave N, Buckling A (2005) Microbial experiments on adaptive landscapes. BioEssays 27: 1167–1173.
34. Collins S, Sültemeyer D, Bell G (2006) Rewinding the tape: selection of algae adapted to high CO_2 at current and pleistocene levels of CO_2. Evolution 60: 1392–1401.
35. Lee CE, Kiergaard M, Gelembiuk GW, Eads BD, Posavi M (2011) Pumping ions: rapid parallel evolution of ionic regulation following habitat invasions. Evolution 65: 2229–2244.
36. Wood TE, Burke JM, Rieseberg LH (2005) Parallel genotypic adaptation: when evolution repeats itself. Genetica 123: 157–170.
37. Saxer G, Doebeli M, Travisano M (2010) The repeatability of adaptive radiation during long-term experimental evolution of Escherichia coli in a multiple nutrient environment. PLoS One 5: e14184.
38. Hewitt GM (2000) The genetic legacy of the Quaternary ice ages. Nature 405: 907–913.
39. Krimbas CB, Loukas M (1980) The inversion polymorphism of Drosophila subobscura. Evol Biol 12: 163–234.
40. Huey RB, Gilchrist GW, Carlson ML, Berrigan D, Serra L (2000) Rapid evolution of a geographic cline in size in an introduced fly. Science 287: 308–309.
41. Rezende EL, Balanyà J, Rodríguez-Trelles F, Rego C, Fragata I, et al. (2010) Climate change and chromosomal inversions in Drosophila subobscura. Clim Res 43: 103–114.
42. Balanyà J, Oller JM, Huey RB, Gilchrist GW, Serra L (2006) Global genetic change tracks global climate warming in Drosophila subobscura. Science 313: 1773–1775.
43. Santos M (2007) Evolution of total net fitness in thermal lines: Drosophila subobscura likes it "warm." J Evol Biol 20: 2361–2370.

44. Simões P, Rose MR, Duarte A, Gonçalves R, Matos M (2007) Evolutionary domestication in *Drosophila subobscura*. J Evol Biol 20: 758–766.
45. Santos J, Pascual M, Simões P, Fragata I, Lima M, et al. (2012) From nature to the laboratory: the impact of founder effects on adaptation. J Evol Biol 25: 2607–2622.
46. Service PM (1987) Physiological Mechanisms of Increased Stress Resistance in *Drosophila melanogaster* Selected for Postponed Senescence. Physiol Zool 60: 321–326.
47. Excoffier L, Lischer H (2010) Arlequin suite ver 3.5: A new series of programs to perform population genetics analyses under Linux and Windows. Mol Ecol Resour 10: 564–567.
48. Rose MR, Passananti H, Matos M, editors (2004) Methuselah Flies: A Case Study in the Evolution of Aging. Singapure: World Scientific Publishing.
49. Jong G de, Noordwijk AJ van (1992) Acquisition and allocation of resources: genetic (co) variances, selection, and life histories. Am Nat 139: 749–770.
50. Monclus M, Prevosti A (1971) The relationship between mating speed and wing length in *Drosophila subobscura*. Evolution 25: 214–217.
51. Santos J, Iriarte PF, Céspedes W (2005) Genetics and geometry of canalization and developmental stability in *Drosophila subobscura*. BMC Evol Biol 5: 7.
52. Benjamini Y, Yekutieli D (2001) The control of the false discovery rate in multiple testing under dependency. Ann Stat 29: 1165–1188.
53. Adams DC, Collyer ML (2009) A general framework for the analysis of phenotypic trajectories in evolutionary studies. Evolution 63: 1143–1154.
54. R CDT (2008) R: a language and environment for statistical computing. Version 2.70. R Found Stat Comput.
55. Adler D, D.Murdoch (2012) rgl: 3D visualization device system (OpenGL). R Package version 092894.
56. Chippindale AK (2006) Experimental Evolution. In: Fox C, Wolf J, editors. Evolutionary genetics: concepts and case studies. London. pp. 482–501.
57. Flores-Moya A, Costas E, López-Rodas V (2008) Roles of adaptation, chance and history in the evolution of the dinoflagellate *Prorocentrum triestinum*. Naturwissenschaften 95: 697–703.
58. Spor A, Kvitek DJ, Nidelet T, Martin J, Legrand J, et al. (2013) Phenotypic and Genotypic Convergences Are Influenced By Historical Contingency and Environment in Yeast. Evolution 1–19.
59. Blount ZD, Borland CZ, Lenski RE (2008) Historical contingency and the evolution of a key innovation in an experimental population of *Escherichia coli*. Proc Natl Acad Sci U S A 105: 7899–7906.
60. Meyer JR, Dobias DT, Weitz JS, Barrick JE, Ryan T, et al. (2012) Repeatability and Contingency in the Evolution of a Key Innovation in Phage Lambda. Science 335: 428–432.
61. Bell G (2013) The incidental response to uniform natural selection. Biol Lett 9: 2013–2015.
62. Bieri J, Kawecki TJ (2003) Genetic architecture of differences between populations of cowpea weevil (*Callosobruchus maculatus*) evolved in the same environment. Evolution 57: 274–287.
63. Kawecki TJ, Mery F (2003) Evolutionary conservatism of geographic variation in host preference in *Callosobruchus maculatus*. Ecol Entomol 28: 449–456.
64. Kawecki TJ, Mery F (2006) Genetically idiosyncratic responses of *Drosophila melanogaster* populations to selection for improved learning ability. J Evol Biol 19: 1265–1274.
65. Griffiths J, Schiffer M, Hoffmann A (2005) Clinal variation and laboratory adaptation in the rainforest species *Drosophila birchii* for stress resistance, wing size, wing shape and development time. J Evol Biol 18: 213–222.
66. Colegrave N, Collins S (2008) Experimental evolution: experimental evolution and evolvability. Heredity (Edinb) 100: 464–470.
67. Service PM, Rose MR (1985) Genetic Covariation Among Life-History Components: The Effect of Novel Environments. Evolution 39: 943.
68. Burke MK, Dunham JP, Shahrestani P, Thornton KR, Rose MR, et al. (2010) Genome-wide analysis of a long-term evolution experiment with *Drosophila*. Nature 467: 587–590.
69. Matos M, Simões P, Duarte A, Rego C, Avelar T, et al. (2004) Convergence to a Novel Environment: Comparative Method versus Experimental Evolution. Evolution 58: 1503–1510.
70. Santos M, Fragata I, Santos J, Simões P, Marques A, et al. (2010) Playing Darwin. Part B. 20 years of domestication in *Drosophila subobscura*. Theory Biosci 129: 97–102.
71. Conte GL, Arnegard ME, Peichel CL, Schluter D (2012) The probability of genetic parallelism and convergence in natural populations. Proc Biol Sci 279: 5039–5047.
72. Martin A, Orgogozo V (2013) The Loci of repeated evolution: a catalog of genetic hotspots of phenotypic variation. Evolution 67: 1235–1250.
73. Tenaillon O, Rodríguez-Verdugo A, Gaut RL, McDonald P, Bennett AF, et al. (2012) The molecular diversity of adaptive convergence. Science 335: 457–461.
74. Bedhomme S, Lafforgue G, Elena SF (2013) Genotypic but not phenotypic historical contingency revealed by viral experimental evolution. BMC Evol Biol 13: 46.
75. Woods R, Schneider D, Winkworth CL, Riley MA, Lenski RE (2006) Tests of parallel molecular evolution in a long-term experiment with *Escherichia coli*. Proc Natl Acad Sci U S A 103: 9107–9112.
76. Teotónio H, Chelo IM, Bradić M, Rose MR, Long AD (2009) Experimental evolution reveals natural selection on standing genetic variation. Nat Genet 41: 251–257.
77. Gilchrist GW, Jeffers LM, West B, Folk D, Suess J, et al. (2008) Clinal patterns of desiccation and starvation resistance in ancestral and invading populations of *Drosophila subobscura*. Evol Appl 1: 513–523.

21

Specific Duplication and Dorsoventrally Asymmetric Expression Patterns of *Cycloidea*-Like Genes in Zygomorphic Species of Ranunculaceae

Florian Jabbour[1,2,3,4], Guillaume Cossard[2,5], Martine Le Guilloux[6], Julie Sannier[2], Sophie Nadot[2], Catherine Damerval[6]*

1 Université Paris-Sud, UMR 0320/UMR 8120, Génétique Végétale, Gif-sur-Yvette, France, 2 Université Paris-Sud, Laboratoire Ecologie, Systématique, Evolution, CNRS UMR 8079, AgroParisTech, Orsay, France, 3 Systematic Botany and Mycology, University of Munich (LMU), Munich, Germany, 4 Muséum National d'Histoire Naturelle, Institut de Systématique, Evolution, Biodiversité, UMR 7205 ISYEB MNHN-CNRS-UPMC-EPHE, Paris, France, 5 Department of Ecology and Evolution, Biophore, University of Lausanne, Lausanne, Switzerland, 6 CNRS, UMR 0320/UMR 8120, Génétique Végétale, Gif-sur-Yvette, France

Abstract

Floral bilateral symmetry (zygomorphy) has evolved several times independently in angiosperms from radially symmetrical (actinomorphic) ancestral states. Homologs of the *Antirrhinum majus* CYCLOIDEA gene (*Cyc*) have been shown to control floral symmetry in diverse groups in core eudicots. In the basal eudicot family Ranunculaceae, there is a single evolutionary transition from actinomorphy to zygomorphy in the stem lineage of the tribe Delphinieae. We characterized *Cyc* homologs in 18 genera of Ranunculaceae, including the four genera of Delphinieae, in a sampling that represents the floral morphological diversity of this tribe, and reconstructed the evolutionary history of this gene family in Ranunculaceae. Within each of the two *RanaCyL* (Ranunculaceae CYCLOIDEA-like) lineages previously identified, an additional duplication possibly predating the emergence of the Delphinieae was found, resulting in up to four gene copies in zygomorphic species. Expression analyses indicate that the *RanaCyL* paralogs are expressed early in floral buds and that the duration of their expression varies between species and paralog class. At most one *RanaCyL* paralog was expressed during the late stages of floral development in the actinomorphic species studied whereas all paralogs from the zygomorphic species were expressed, composing a species-specific identity code for perianth organs. The contrasted asymmetric patterns of expression observed in the two zygomorphic species is discussed in relation to their distinct perianth architecture.

Editor: Hirokazu Tsukaya, The University of Tokyo, Japan

Funding: FJ was supported by a PhD fellowship from the Ministère de l'Enseignement Supérieur et de la Recherche, France. FJ's postdoc lab work in Munich, Germany, was supported by the German Science Foundation (DFG) grant No. RE 603/12-1. This work was supported in 2007 by a grant from IFR 87 'La Plante et son Environnement', and the Agence Nationale de la Recherche through the ANR-07-BLAN-0112 grant. The funders had no role in study design, data collection and analysis, decision to publish, or preparation of the manuscript.

Competing Interests: The authors have declared that no competing interests exist.

* E-mail: damerval@moulon.inra.fr

Introduction

Bilaterally symmetric flowers are characterized by a single plane of symmetry that separates the flower into two mirror images. The zygomorphic phenotype can be more or less elaborate, involving only organ displacement around the receptacle, or various degrees of morphological differentiation [1,2]. It is considered an adaptive trait, promoting cross-pollination through optimized interaction between flower shape and pollinator type and behaviour. Comparison of species richness in sister clades that have or have not evolved zygomorphy revealed that bilateral symmetry might have promoted species diversification [3]. Reconstruction of the evolution of floral symmetry states on an angiosperm-wide phylogeny points to more than 70 transitions from an actinomorphic (radially symmetric) ancestral state to zygomorphy [4].

Among angiosperms, the eudicot clade is the most species-rich (about 200,000 species) and is also the one with the highest number of transitions from actinomorphy to zygomorphy [4]. At least 42 of the estimated 46 eudicot-specific transitions have taken place in the derived core eudicot clade. Among basal eudicots,

three independent transitions occurred in Ranunculales: one in Ranunculaceae, one in Papaveraceae, and one in Menispermaceae [5]. Ranunculaceae, with more than 2,500 species, is the largest family in the Ranunculales order. This family is especially remarkable by the diversity of its perianth form and architecture, exhibiting variable merism (2- to 5-merism), organ number (determinate or variable), phyllotaxis (spiral or whorled), and perianth differentiation (unipartite or bipartite with inner organs (petals) morphologically differentiated from outer ones (sepals)). The single evolutionary transition from actinomorphy (or pseudo-actinomorphy in the case of spirally inserted organs) to zygomorphy (or pseudo-zygomorphy) took place in the stem lineage of the species-rich Delphinieae tribe (accounting for 26% of all Ranunculaceae species), and has involved the asymmetric formation of spurs in both sepals and petals [5]. The five sepals are spirally inserted, with two ventral, two lateral, and a spurred or hooded dorsal one [6]. The corolla is reduced to four, two, or a single petal, all inserted in the dorsal half of the flower [7]. The other petal primordia stop developing shortly after organogenesis or develop into slender petaloid staminodes [6,8,9]. Petal spurs are nectariferous, and their size and shape may play a role in selecting pollinators [10]. Within Ranunculaceae, nectariferous petal spurs also exist in actinomorphic genera, such as *Aquilegia* and *Myosurus*.

Since the pioneering studies demonstrating the key role of CYCLOIDEA (CYC) and its paralog DICHOTOMA (DICH) in the establishment of bilateral symmetry in *Antirrhinum majus* [11,12], a number of studies have investigated the evolutionary history of CYC-like genes and their possible role in the evolution of floral symmetry in diverse plant groups. CYC encodes a transcription factor of the plant-specific TCP gene family [13,14] and is characterized by two typical domains, the TCP and R domains, and an additional 'ECE' motif in the inter-domain region [15]. The characterization of homologs of CYC in both monocots and eudicots reveals a complex history of gene duplications [15–17]. In the core eudicots, three paralogous lineages have been found: CYC1, CYC2, and CYC3 [15]. All the genes demonstrated or suspected to play a role in bilateral symmetry belong to the CYC2 clade but are not all orthologs due to the independent duplications in various families [18–22]. Studies in Poaceae, Zingiberales and Commelinales also suggest a possible role of *Tb1*-like genes, the closest CYC paralog in monocots, in the independent evolution of floral bilateral symmetry [16,23,24]. In parallel with the situation observed in core eudicots and monocots, independent duplications of CYC-like genes have been observed in all major lineages of basal eudicots, including the Ranunculales [25]. Studies in Fumarioideae (Papaveraceae) species showed that late expression of the two CYC-like genes was correlated with symmetry type [26,27]. In the Ranunculaceae, two paralogous lineages have also been found and seem to originate from a duplication event independent from the one that has occurred in the Papaveraceae [25].

In this study, we reconstructed the evolutionary history of CYC-like genes in Ranunculaceae with specific focus on Delphinieae, and compared their expression pattern in actinomorphic and zygomorphic species. We found that the zygomorphic species have generally more gene copies than actinomorphic species. The expression patterns observed in developing flowers varied from species to species, suggesting that the regulation of expression evolved not only after duplication but also after species divergence. Different asymmetric expression patterns were found in the two zygomorphic species studied, which may be related to their different perianth architecture.

Material and Methods

Species sampling and plant material

Forty-nine species (21 actinomorphic and 28 zygomorphic species) representing all subfamilies of Ranunculaceae, except the monotypic Hydrastidoideae (comprising a single genus, *Hydrastis*, which is actinomorphic), were selected for CYC-like gene characterization (Table S1). In particular, all genera, subgenera, and subgroups of Delphinieae, representing the tribe's floral morphological diversity (Table S1, Figure 1), were sampled: the genus *Aconitum* (subgenera *Aconitum* and *Lycoctonum*), the genus *Delphinium* (subgenera *Delphinium*, *Delphinastrum*, *Oligophyllon*, and subgroups *Consolida* and *Aconitella*), the genus *Staphisagria* (recently resurrected by [28]), and the monotypic genus *Gymnaconitum* (recently circumscribed by [29]). The taxonomic divisions within the tribe Delphinieae followed in this paper (except for *Staphisagria* and *Gymnaconitum*) were described in [30].

Material was collected from living plants and herbarium specimens (Table S1). Plants from ten species were grown from seed or were obtained from botanic gardens (the names of the authorities that issued the permit for each location are provided in Table S1) or garden centers, and grown outdoors or in a

greenhouse at the UMR de Génétique Végétale (Gif-sur-Yvette. France). Sampling the plants in the wild in France did not concern any endangered or protected species, and did not require any specific permission. Fresh leaf or floral material was harvested and immediately frozen in liquid nitrogen for DNA or RNA extraction, or stored in RNAlater™ (Qiagen) for subsequent RNA extraction.

Isolation of CYC-like homologs

Genomic DNA was extracted using the NucleoSpin plant kit (Macherey-Nagel, Düren, Germany) for herbarium samples, and following the procedure described in [26] for fresh tissue samples. Characterization of CYC homologs was a two-step process. First, a set of eleven species (six zygomorphic (tribe Delphinieae) and five actinomorphic (subfamilies Ranunculoideae, Thalictroideae, and Glaucidioideae), asterisks in Table S1) was chosen for PCR and cloning effort, and sequence elongation. Nested PCRs were performed on genomic DNA using degenerate oligonucleotides designed in the TCP and R domains (combination III and IV in [26], and additional primers in Table S2). PCR conditions followed [26]. Bands of the expected size (110 to 130 bp for TCP domain amplification, or 300 to 500 bp for the TCP to R amplification) were excised from agarose gel and purified using the QIAquick gel extraction kit (Qiagen). The fragment was ligated into the pGEM-T Easy plasmid (Promega) and transformed into competent subcloning efficiency DH5α *Escherichia coli* cells (Gibco-BRL) following the manufacturer's instructions. Twenty-three to 106 clones were sequenced depending on the species and primer combination. Sequencing was carried out by Genoscreen (Campus de l'Institut Pasteur de Lille, BP245, 59019 Lille, France). Clone sequence alignment distinguished one to three different sequence types (defined on the basis of shared similarities among clones) per species. These sequences were then elongated using inverse PCR. Genomic DNA (~200 ng) was digested by restriction enzymes without a cutting site within the known sequence in a final volume of 25 μL. Digested DNA (15 μL) was ligated using T4 DNA ligase (1 unit in a final volume of 25 μL). Ligations were heated for 5 minutes at 70°C to stop the reaction, and were purified using the Mini-elute purification Qiagen kit in a final volume of 10 μL. The purified product was used as a matrix for PCR with primers defined in the known sequences. Nested PCRs were performed to increase specificity. PCRs and nested PCRs were performed in 25 μL mix, with 2 μL of purified ligation and 1 μL of first PCR, respectively. Fragments obtained were either directly sequenced or cloned before sequencing, as above.

In a second step, the alignment of the 24 elongated sequences was used to define two sets of three degenerate primers in the TCP and R domains (Cossard *et al.* unpublished and Table S2). These primers were used to perform semi-nested PCRs on genomic DNA from 38 additional species (22 zygomorphic and 16 actinomorphic). PCR products were either sequenced directly or cloned (6 to 19 clones sequenced per species), as described above. Sequences were deposited in GenBank, with accession numbers KJ401946-KJ402052.

Phylogenetic analyses

Five CYC-like sequences from three species from other Ranunculales families were chosen to root the trees (*Akebia quinata* (Lardizabalaceae): GenBank accession numbers HQ599289-HQ599290, *Circaeaster agrestis* (Circaeasteraceae): HQ599293-HQ599294, and *Nandina domestica* (Berberidaceae) HQ599291). All CYC-like sequences were aligned using MAFFT v. 7.053 [31]; the alignment was visually refined on the basis of amino-acid translation. Highly divergent regions were deleted from the

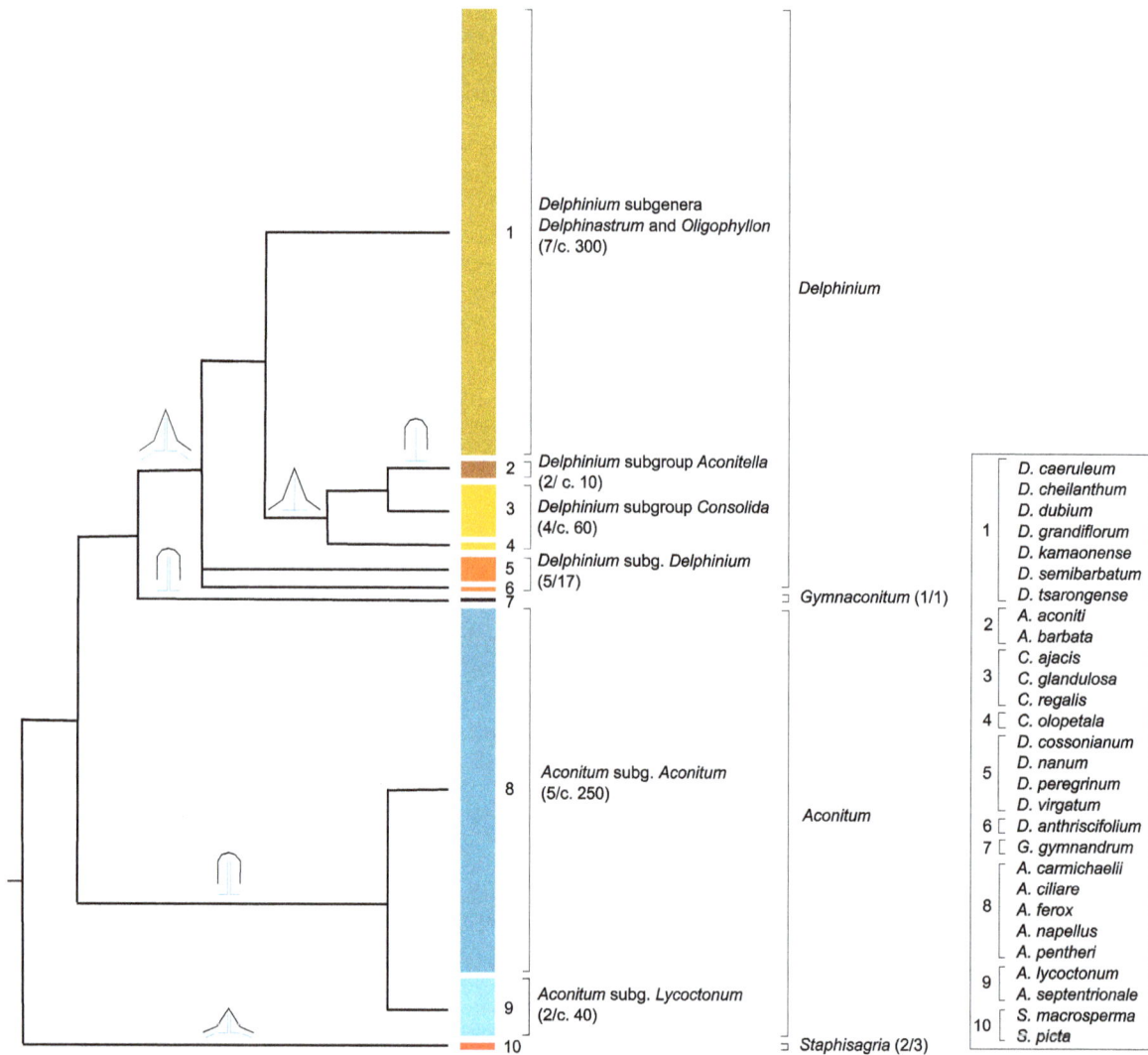

Figure 1. Summary of phylogenetic relationships in the tribe Delphinieae (cf. [7]). The sampling ratio for each of the genera, subgenera, and subgroups from this tribe is shown (the size of each colored rectangle representing the approximate number of species in each). At the bottom right hand corner, the names of the species used for the characterization of *Cyc*-like genes are specified for each of the ten groups. Schematic shapes for the dorsal sepal and corolla are shown above the branches of the tree: the dorsal sepal can be spurred (sharp black shape) or hooded (round black shape); the corolla (in blue) can be composed of four petals (two spurred and two flat), two petals (both spurred), or a single spurred petal.

alignment before phylogenetic analyses, resulting in a 381-nucleotide character matrix for global analysis (109 sequences from 48 species). Phylogenetic trees were inferred from these nucleotide alignments using Bayesian inference and maximum likelihood reconstruction methods. Maximum likelihood analyses were performed using raxmlGUI v. 1.3 [32,33]. Substitution model parameters were estimated using the general time-reversible model with a proportion of invariable sites and a γ-distribution for site-specific rates partitioned by codons (GTR+Γ+I) model. Statistical support for nodes was assessed by bootstrapping the data under the same model (1,000 replicates). Bayesian phylogenetic analyses were carried out using MrBayes v. 3.2.1 [34] using a GTR+Γ+I model. For all three analyses, four chains were run twice for 3,000,000 generations, with a burn-in of 7,500 samples (sample frequency = 100). Convergence was assessed using the potential scale reduction factor (PSRF = 1.0) and the average standard deviation of split frequencies (<0.01). A majority rule consensus tree with posterior probabilities of nodes was built.

Molecular evolution studies

We looked for particular selection regimes on the branches subtending the *RanaCyL1* and *RanaCyL2* clades of Delphinieae (see Figure 2 and Figure S1). For each paralog, we used sequences with a minimum of missing data in the TCP and R domains (respectively a 381-position matrix of 44 sequences for RanaCyL1, and a 357-position matrix of 49 sequences for *RanaCyL2*) to reconstruct phylogenetic trees. Maximum likelihood analyses were performed using the online version of PhyML v 3.0 [35], with a GTR+Γ+I evolutionary model; tree improvement was carried out by Subtree Pruning and Regrafting and Nearest-Neighbor Interchange, and branch support evaluated by approximate likelihood-ratio test (SH-like). We then performed tests on ω, calculated as the ratio of the non-synonymous (dN) over the synonymous (dS) substitution rates, using PAML v. 4.4 [36]. We first determined the most parsimonious model of codon frequency that fits the data using the Codonfreq option implemented in the codeml package under a unique ω model. To do so, we compared nested models (equal, F1×4, F3×4, F61) using likelihood ratio tests (LRTs), and retained the "equal" model for all subsequent analyses. Then, using the codeml package, we estimated ω in the branch of interest (foreground branch) and other branches in the tree (background branches). We tested this model against a null hypothesis in which a single ω value was estimated for all branches (model M0), using a hierarchical LRT with a single degree of freedom (df). We also used the Branch-Site model MA with four classes of sites (two classes allowing ω>1 specifically in the foreground branch), that we tested against MA0 (the two foreground specific classes are considered to evolve neutrally), and against model M1a that considers two classes of sites in all branches (one with 0<ω<1, the other with ω1 = 1). Testing model MA against MA0 allows detection of positive selection at some sites in the foreground branch using an LRT with 1 df. Comparing M1a and MA0 reveals relaxed constraints on the foreground branch (LRT with 1 df); the comparison between M1a and MA is a test for MA significance, but it confounds positive selection and relaxed constraints (LRT with 2 df [37]).

Expression studies

Expression of the *RanaCyL* genes was investigated by semi-quantitative RT-PCR in the leaves, floral buds, and floral dissections of four species: *Aquilegia alpina*, *Nigella damascena* (both actinomorphic), *Aconitum carmichaelii*, and *Consolida regalis* (both zygomorphic). In *A. alpina* and *N. damascena*, three bud sizes were sampled: <1 mm, 2 mm and 3–5 mm. In *A. carmichaelii* and *C. regalis*, buds were harvested at only two stages (<2 mm and 3–5 mm) because of limited material availability. Floral buds at advanced developmental stages were dissected in the four species in order to separate floral organs according to their dorsal/ventral position relatively to the inflorescence axis. For *A. alpina*, *A. carmichaelii*, and *C. regalis*, we could only dissect floral buds from 6–7 mm length. In *A. carmichaelii*, it was not possible to obtain sufficient material for the two dorsal petals that are tiny at this stage. Both 3–5 and 6–7 mm diameter buds were dissected in *N. damascena*. In all four species, calyx phyllotaxis is spiral, and sepal aestivation is quincuncial (imbricate) with two sepals exterior, two interior, and one with one margin interior and the other exterior [38]. The dorsal sepal is the second one in the initiation sequence [6,39]; it was identified as the second most developed sepal, i.e. the youngest of the two exterior sepals. *Aquilegia alpina* has five deeply spurred petals alternating with the sepals, and *N. damascena* has five to ten (most generally eight) saccate petals. In the dissections of both actinomorphic species, we separated the perianth organs into the dorsal sepal/petal versus the other sepals/petals. The two zygomorphic species have a reduced corolla, with either only two petals (*A. carmichaelii*) or a single petal (*C. regalis*), each of these petals being spurred. This reduced corolla is located in the dorsal part of the flower, and is nested within the dorsal sepal (hooded in *A. carmichaelii*, spurred in *C. regalis*). The perianth organs were separated according to their dorsal, lateral, or ventral position.

Total RNA was extracted using the RNeasy Plant Mini Kit (Qiagen) and processed as described in [26]. PCRs for *actin* were carried out for all samples without RT (and additionally for the *RanaCyL* genes for samples where sufficient material was available) in order to check for absence of DNA contamination. *Actin* was used as a loading control between samples, and amplified under the following conditions: 3 min at 95°C, then 25 cycles of 95°C for 30 s, 52°C for 30 s, 72°C for 2 min, and final elongation at 72°C for 8 min [26]. *RanaCyL* genes were amplified using specific oligonucleotides (Table S2) under the same conditions as *actin* except for the annealing temperature (52°C for *AaCyL1*, 54°C for *NdCyL2*, 56°C for *CrCyL2a*, 58°C for the other genes) and number

of cycles (30 for *AaCyL1*, *AaCyL2*, and *AcCyL1a*, 33 cycles for the other genes). Two to three biological replicates and at least two technical replicates were done for each species to validate the reproducibility of the results.

Results

Evolutionary history of *Cyc*-like genes in Ranunculaceae

One hundred and nine sequences homologous to *Cyc* were found in the 49 species analyzed; no sequences homologous to *Cyc* were obtained from *Aconitum ciliare*. Most sequences contained the conserved ECE motif between the TCP and R domains. The number of copies was generally higher in Delphinieae than in the other species, with up to four sequences in *Delphinium caeruleum*, *Aconitella aconiti* and *Aconitum septentrionale*.

Maximum likelihood analysis and Bayesian inference were used to reconstruct the evolutionary history of *Cyc*-like genes in the family Ranunculaceae. Both methods produced similar topologies, revealing two well-supported large clades each including one or more non-Ranunculaceae sequences, suggesting a duplication event predating the divergence of Lardizabalaceae and Ranunculaceae (Figure 2). Within both clades (hereafter *RanaCyL1* and *RanaCyL2*), sequence relationship was broadly congruent with species phylogeny at the tribal level, however deeper nodes poorly supported. Separate analyses were carried out for *RanaCyL1* (48 sequences, 39 species, 369 characters) and *RanaCyL2* (61 sequences, 45 species, 363 characters) without improving the resolution (results not shown). In both *RanaCyL1* and *RanaCyL2*, the Delphinieae sequences formed two well-supported clades (respectively *RanaCyL1a* and b, and *RanaCyL2a* and b) whose relationship with each other and with other sequences was not resolved. Within Delphinieae, the relationship among the *RanaCyL* sequences was congruent with species phylogeny (Figures 1 and 2): *Staphisagria Cyc*-like sequences were the earliest-diverging ones (except in the *RanaCyL1a* lineage where their position within the clade is not well supported), *G. gymnandrum Cyc*-like sequences and the rest of *Aconitum Cyc*-like sequences formed two independent lineages, and *Consolida* and *Aconitella Cyc*-like sequences were nested within the *Delphinium* sequences.

Patterns of molecular evolution in Delphinieae *RanaCyL* sequences

The higher number of *RanaCyL* paralogs found in Delphinieae compared to other Ranunculaceae could suggest that some of these gene lineages are subject to particular evolutionary forces. To test this hypothesis, we investigated patterns of selection at the molecular level in each *RanaCyL* paralog independently, focusing on the branches subtending the Delphinieae *RanaCyL* clades (see Figure 2). For *RanaCyL1*, the branch model suggested higher levels of constraint on the *RanaCyL1a* branch $\omega_0 = 0.305$, $\omega_1 = 0.088$, p = 0.0046). None of the Branch-Site models were significant, indicating neither positive selection nor relaxed constraints at specific sites in either branch subtending *RanaCyL1a* or 1b. For *RanaCyL2*, none of the Branch or Branch-Site models were significant, indicating no specific evolutionary pressures on the *RanaCyL2a* or 2b branches compared to other branches in the phylogeny (Table 1).

RanaCyL expression during flower development in actinomorphic and zygomorphic species

Both *RanaCyL1* and *RanaCyL2* genes were found to be expressed in floral buds. Expression was also detected in leaves for three of the four species (Figure 3B, C and D). Their expression level in flowers decreases over time in the actinomorphic species (*AaCyL1*,

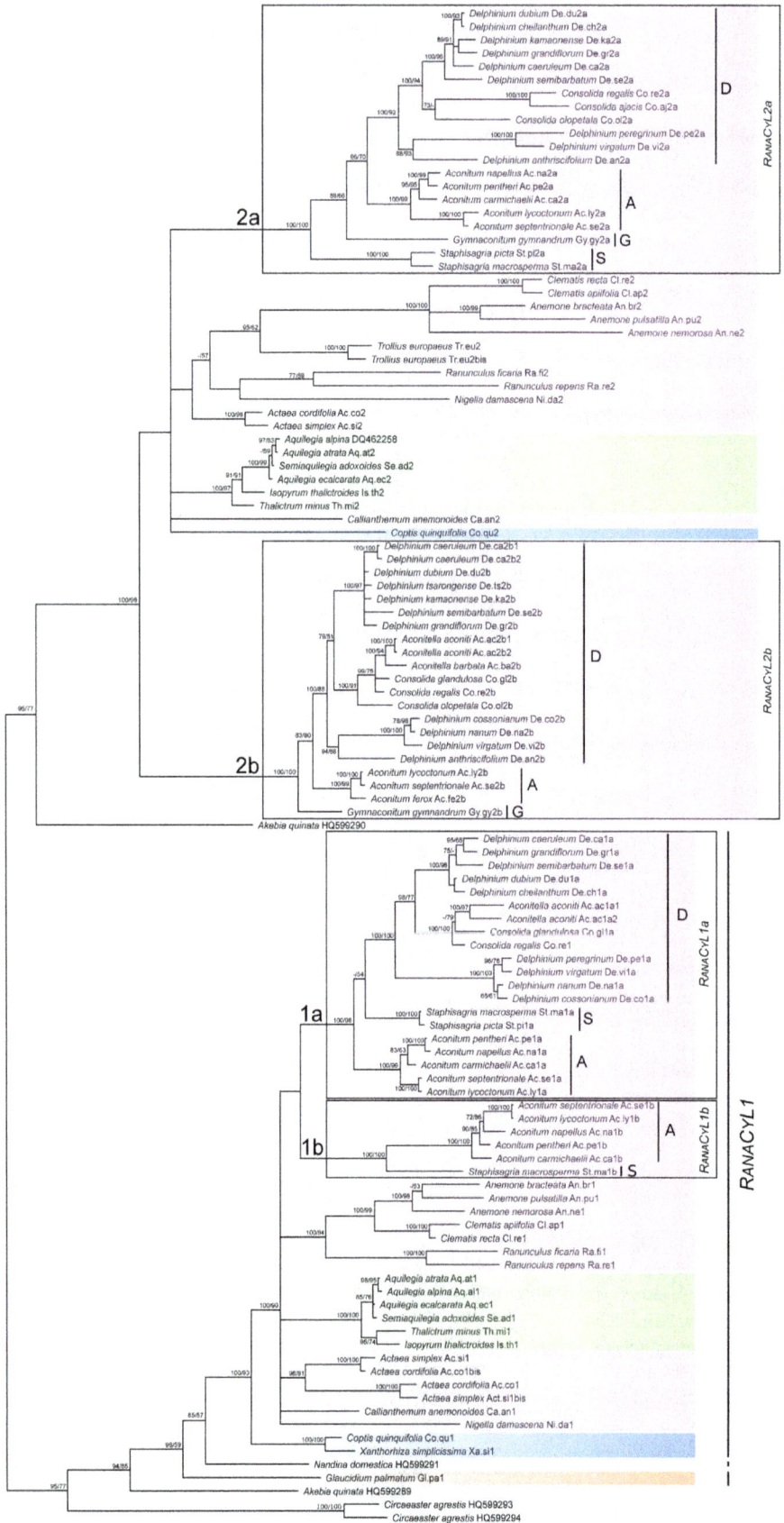

Figure 2. Phylogeny of *Crc*-like sequences in Ranunculaceae. The phylogeny is inferred using Bayesian and maximum likelihood analysis, rooted in order to group together all the *RanaCyL1* sequences in one clade, and all the *RanaCyL2* sequences in another clade. The tree topology obtained by Bayesian inference is shown. Bayesian posterior probability ≥70 and bootstrap support from maximum likelihood analysis ≥50 are given (hyphen if inferior to those thresholds). The 381-position alignment was generated from 109 *RanaCyL* (from 48 species including 27 Delphinieae) and five other Ranunculales *Crc*-like sequences (from three species). Portions of the tree are highlighted with colors according to the subfamilies of Ranunculaceae. The species sequenced for *RanaCyL* belong to: pink: Ranunculoideae, green: Thalictroideae, purple: Coptidoideae, and orange: Glaucidioideae. A: *Aconitum*, D: *Delphinium*, G: *Gymnaconitum*, S: *Staphisagria*. The *RanaCyL* branches tested in the molecular evolution analyses are 1a and 1b in *RanaCyL1*, and 2a and 2b in *RanaCyL2*.

AaCyL2, *NdCyL1*, *NdCyL2*) but less so in the zygomorphic species (*CrCyL2*a, *AcCyL1*a). In *Aquilegia alpina*, *RanaCyL1* is expressed at a higher level than *RanaCyL2* at the three bud stages whereas it is the opposite in *Nigella damascena* at the last two bud stages (Figure 3A and B). In the zygomorphic species, the signal for *CrCyL2*b appeared to be saturated, and *AcCyL2*a and *CrCyL1*a did not show any difference in the level of expression between the two bud stages investigated (Figure 3C, D). By contrast, duplicates within a same lineage show a differentiated expression pattern: *RanaCyL1*b is very faintly expressed compared to *RanaCyL1*a in *A. carmichaelii*, and *RanaCyL2*b is expressed at higher levels than *RanaCyL2*a in *C. regalis*. In floral dissections of late-stage floral buds, *RanaCyL1* but not *RanaCyL2* was found to be expressed in the perianth organs and stamens of 6–7 mm length buds in *A. alpina*; *AaCyL1* was less expressed in the dorsal sepal than in the set of other sepals (Figure 3A). In 3–5 mm diameter buds of *N. damascena*, *RanaCyL1* is faintly expressed in the dorsal sepal, and *RanaCyL2* expression is restricted to the stamens (Figure 3B); no expression was detected in any organ of 6–7 mm diameter buds (not shown). In *A. carmichaelii*, *RanaCyL1*a expression was detected in all sepals, with higher levels in the ventral ones; a similar expression pattern may exist for *RanaCyL1*b, but this is difficult to ascertain as the gene is very faintly expressed. *RanaCyL2*a expression was detected in all organs examined (Figure 3C). In *Consolida regalis*, *RanaCyL1*a is expressed equally in all sepals but not in petals or stamens. The two *RanaCyL2* paralogs are both expressed in stamens but exhibit a differential expression pattern in the perianth: *RanaCyL2* is more expressed in the dorsal sepal and petal, *RanaCyL2*a is more expressed in the ventral sepals but not in the petal (Figure 3D).

Table 1. Results of the tests of molecular evolution in each sublineage of *RanaCyL1* and *RanaCyL2*.

RanaCyL1		
Model	Loglikelihood	Significance of LRT and model parameters
Null model M0	−5392.85	ω = 0.296
Branch		
ω0 = ω1, ω2	−5392.84	ns
ω0 = ω2, ω1	−5388.83	0.0046 (ω0 = ω2 = 0.305, ω1 = 0.088)
ω0, ω1, ω2	−5388.82	0.0046 (ω0 = 0.306, ω2 = 0.287, ω1 = 0.088)
Site		
M1a	−5335.13	6×10^{-27}, ω = 0.187, p = 0.692
Branch-Site		
MA0 (foreground = branch 1a)	−5335.13	ns
MA selection (foreground = branch 1a)	−5335.13	ns
MA0 (foreground = branch 1b)	−5335.13	ns
MA selection (foreground = branch 1b)	−5335.13	ns
RanaCyL2		
Model	Loglikelihood	Significance of LRT and model parameters
Null Model M0	−5518.37	ω = 0.397
Branch		
ω0 = ω2, ω1	−5518.37	ns
ω0 = ω1, ω2	−5517.85	ns
ω0, ω1, ω2	−5517.84	ns
Site		
M1a	−5432.26	2×10^{-39}, ω = 0.196, p = 0.528
Branch-Site		
MA0 (foreground = branch 2a)	−5432.25	ns
MA selection (foreground = branch 2a)	−5432.18	ns
MA0 (foreground = branch 2b)	−5432.18	ns
MA selection (foreground = branch 2b)	−5431.68	ns

A. Aquilegia alpina

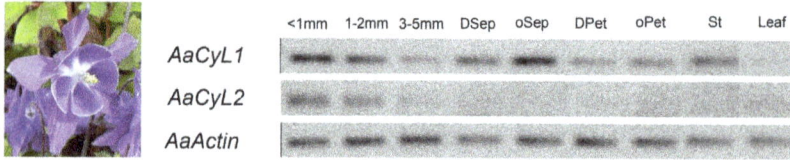

B. Nigella damascena

C. Aconitum carmichaelii

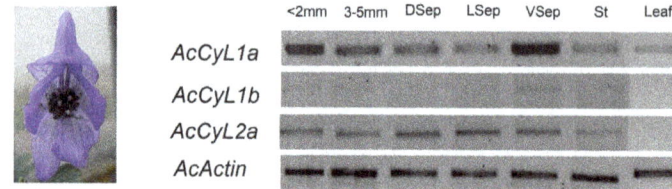

D. Consolida regalis

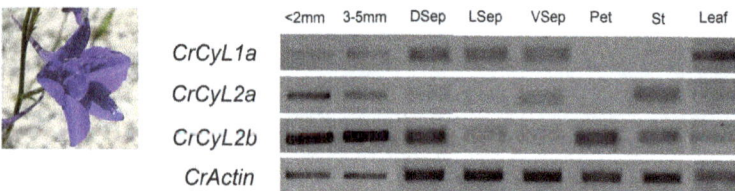

Figure 3. Semi-quantitative RT-PCR analysis of *RanaCyL* gene expression in floral buds, dissected floral organs, and leaves. (A) *Aquilegia alpina*. (B) *Nigella damascena*. (C) *Aconitum carmichaelii*. (D) *Consolida regalis*. For buds, diameter (*Aquilegia* and *Nigella*) or length (*Aconitum* and *Consolida*) are indicated. DSep, oSeP, LSep, VSep: dorsal, non-dorsal, lateral, and ventral sepal, respectively; DPet, oPet: dorsal, non-dorsal petals respectively (actinomorphic species); Pet: spurred petals (zygomorphic species); St: stamens.

Discussion

Lineage-specific duplications of *Cyc*-like genes in Delphinieae

The phylogenetic trees comprising 109 *Cyc*-like sequences from 48 Ranunculaceae species are consistent with the hypothesis of two paralogous lineages originating from a duplication predating the divergence of Lardizabalaceae and Ranunculaceae. These two lineages were previously identified from a small set of Ranunculales species by Citerne *et al.* [25] who suggested that they originated from an early duplication in the order and that the two paralogous lineages previously described in the Papaveraceae [26,40] were derived from a family-specific duplication in the *RanaCyL1* clade. Most actinomorphic Ranunculaceae species have one copy of each paralog, with the exception of the two *Actaea* species that underwent a specific duplication in the *RanaCyL1* lineage. The sequences obtained from species of the tribe Delphinieae, characterized by zygomorphic flowers, clustered in

two well-supported clades in each *RanaCyL* lineage. However, the lack of resolution at the deeper nodes in the phylogenetic trees leaves open the question of the origin of these duplicated lineages. They may have originated from duplication events having taken place in the stem lineage of the tribe, either independently in *RanaCyL1* and *RanaCyL2*, or through a genome wide duplication. However, no polyploidization event specific to Delphinieae has been reported so far. Alternatively, these duplications might have taken place earlier during the diversification of Ranunculaceae, followed by independent losses in the ancestors of the various tribes except Delphinieae. Differential gene loss probably also took place within each sublineage of Delphinieae, even though we cannot exclude that extensive sequence divergence in these groups resulted in amplification failure with our degenerate primers (perhaps the case for *A. ciliare* where we did not amplify any Cyc-like sequence). However, the absence of a *RanaCyL* paralog is often found at the level of entire sublineages, suggesting a non-artefactual result and a non-random genomic event. For instance, while copies from the four *RanaCyL* subclades were present in *Aconitum*, no copy of *RanaCyL1*b was found in *Delphinium*, suggesting gene loss in the ancestor of the genus. Similarly, no copy of *RanaCyL2*b was found in *Staphisagria*, no copy of *RanaCyL2*a was found in the *Delphinium* subgroup *Aconitella*, and no copies of *RanaCyL1* were found in *Gymnaconitum*.

In Dipsacales and Malpighiales (core eudicots), duplications in the CYC2 lineage have coincided with the evolution of zygomorphy [41,42]. Various studies suggest that in core eudicots, the CYC2 clade compared with the CYC3 and especially the CYC1 clade has undergone repeated duplication events and/or is prone to paralog retention [15,43,44]. Classical models predict that following gene duplication, duplicates experience functional divergence through neofunctionalization or subfunctionalization, or accumulate deleterious mutations and degenerate [45,46]. These processes theoretically leave different evolutionary signatures in gene sequences, but it remains difficult in practice to tease apart the first two models [37,47]. In sunflower, five Cyc2 genes have been found. Positive selection at four sites scattered in the TCP and R domains may have promoted functional divergence among three of these duplicates, with one paralog exhibiting a ray flower-specific expression whereas the others have wider expression in the inflorescence [48]. In *Lupinus*, a shift in floral architecture has been found to coincide with positive selection operating at a few sites in one CYC2 gene (*LegCyc1B*), suggesting neofunctionalization [49]. Our analyses failed to reveal any signature of positive selection in the Delphinieae sublineages, either in *RanaCyL1* or *RanaCyl2*. By contrast, we observed that the *RanaCyL1*a sublineage experienced a high level of purifying selection compared to other *RanaCyL1* or even *RanaCyL2* genes, suggesting a higher level of functional constraint on this gene, which is consistent with its presence in almost all sublineages of Delphinieae (except in the monotypic genus *Gymnaconitum*). This is similar to what has been described in Antirrhineae, where the Cyc lineage seems to experience a higher level of constraint than the Dich lineage [50]. In addition to a divergence in protein sequence, paralogs that are maintained through evolution generally acquire divergent expression patterns. In the case of transcription factors susceptible to heterodimerize such as TCP proteins, the novel organ and/or developmental stage specific combination of proteins resulting from these regulatory changes may promote the emergence of novel phenotypes by the recruitment of new sets of target genes [44,51,52] even in the absence of selection for amino acid replacement.

Diversity of perianth architecture and expression patterns of *Cyc*-like genes

In Ranunculaceae, the ancestral state for floral symmetry is actinomorphy [5]. An interesting finding of our study is that the two actinomorphic species, each one belonging to the two most species-rich tribes of Ranunculaceae, exhibit a different pattern of *RanaCyL* gene expression in perianth organs. In *A. alpina*, *RanaCyL1* is expressed in both sepals and petals, whereas it is faintly expressed only in the dorsal sepal in *N. damascena*. In core eudicot families where radial symmetry is considered to be the ancestral state, no general trend in the pattern of *Cyc*-like gene expression emerges: in Adoxaceae and Elatinaceae *Cyc*-like genes are broadly expressed across the perianth [42,53], while expression is absent in Centroplacaceae [42], and transitory in *Arabidopsis* (Brassicaceae) [54]. The same absence of general pattern is observed in cases where actinomorphy is derived from an ancestral zygomorphic state [55–61]. This suggests that a variety of processes, depending on *Cyc*-like genes or not, can produce a radially symmetric flower. Our results suggest that in Ranunculaceae, the expression of *Cyc*-like genes is dispensable as related to the elaboration of actinomorphy, or that the *Cyc*-dependent processes controlling actinomorphy have diverged since speciation. In *A. alpina*, the asymmetric expression observed in sepals that are spirally initiated is not observed in petals that are initiated in whorls, raising the hypothesis of a link between *Cyc*-like expression and phyllotaxis.

Consistent with the single evolutionary origin of zygomorphy in Delphinieae, the establishment of bilateral symmetry in this tribe results from similar ontogenetic processes that involve dorsoventralization through asymmetric corolla and spur formation during late developmental stages [6]. These features are superimposed on diverse perianth types that can be categorized according to the number and shape of the organs, and on the identity of the organs responsible for concealing nectar and for covering the sexual organs, thereby constraining the pollinator's movements to maximize pollen load (Figure 4, [7]). The *Aconitum* perianth type characterizes the flowers of the *Aconitum* and *Gymnaconitum* genera. The corolla is reduced to two dorsal petals, and the calyx is built upon a 3+2 pattern: three large sepals (a dorsal hooded sepal concealing the pair of nectariferous spurred petals, and two lateral rounded sepals protecting the sexual organs, see [62]) and two attenuate ventral sepals (Figure 4, and see Figure 1 in [7]). The *Consolida* perianth type is characterized by a pentamerous calyx with sepal limbs similar in shape and size, a spurred dorsal sepal, and a corolla reduced to a single spurred petal (resulting from the fusion of the two dorsal primordia [7]). In contrast with the *Aconitum* perianth type, the functions of nectar and sexual organ protection are achieved by the dorsal petal spur and by the lateral lobes of the petal limb, respectively. The *Delphinium* type is found in the *Delphinium* and *Staphisagria* genera (Figure 1). The corolla is reduced to four petals (two dorsal spurred petals and two lateral flat petals), and the calyx is built upon a 1+4 pattern: a dorsal spurred sepal and four other sepals (two lateral, two ventral). The petal spurs protect the nectar, and the lateral petals cover the sexual organs. The *Aconitella* type is similar to the *Consolida* type, except that the two lateral and two ventral sepals are grouped in a ventral position. This particular organ positioning, together with the nightcap shape of the dorsal spur, makes this flower resemble an *Aconitum* flower (Figure 1). In the *Aconitella* perianth type, the petal spur protects the nectar, and the two lowermost lobes of the petal limb (comprising a total of five lobes) cover the sexual organs (Figure 4).

Dorsoventral differentiation is the most frequent cause of floral zygomorphy in core eudicots. Since the first studies demonstrating the dorsal expression of *Cyc* and *Dich* in *Antirrhinum majus* [11,12], similar asymmetric expression patterns of *Cyc*-like genes have been reported in zygomorphic flowers from distantly related core eudicots species [4,53,63–65]. In the two Delphinieae species *A. carmichaelii* and *C. regalis*, the combined expression of the three *RanaCyL* paralogs composes a sepal identity code (Figure 4) that can be related with the calyx dorsoventral differentiation typical of respectively the *Aconitum* and *Consolida* perianth types. In *A. carmichaelii*, the lateral and dorsal sepals have a similar *RanaCyL* code, paralleling the morphological similarities of these organs typical of the *Aconitum* perianth type, and contrasting with the code of the narrower and flatter ventral sepals. By contrast, in *C. regalis* morphological similarities between lateral and ventral sepals is not echoed in the *RanaCyL* code, since each sepal type exhibits its own code. Comparing *A. carmichaelii* and *C. regalis*, the different expression patterns found for a given paralog result in different codes for homologous sepals, possibly related with their distinct perianth architectures. In particular, the dorsal sepal code differs between *C. regalis* and *A. carmichaelii*, suggesting that different

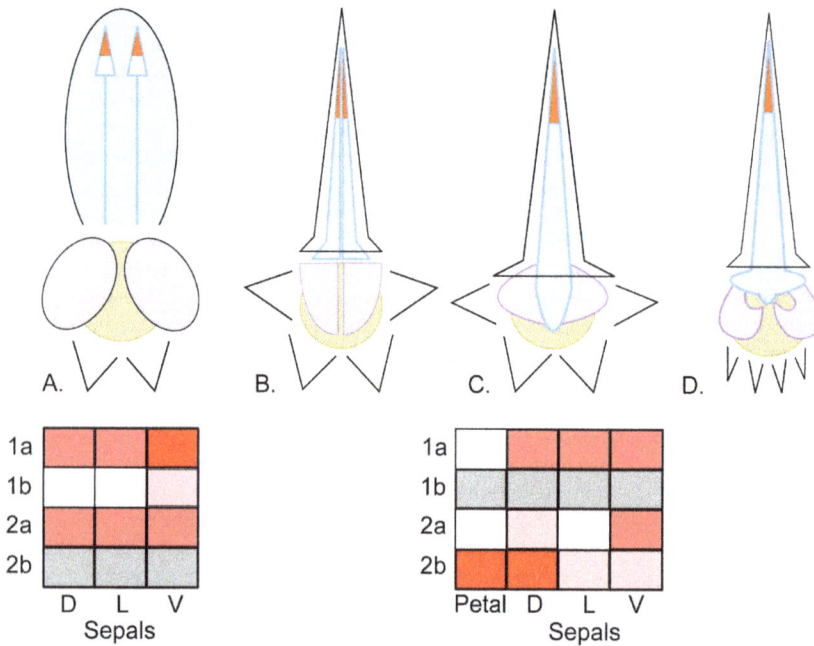

Figure 4. Main types of perianth organization in the tribe Delphinieae. The four sketches are not at the same scale. (A) *Aconitum* type (applicable to groups 7, 8, and 9 from Figure 1). (B) *Delphinium* type (applicable to groups 1, 5, 6, and 10 from Figure 1). (C) *Consolida* type (applicable to groups 3 and 4 from Figure 1). (D) *Aconitella* type (applicable to group 2 from Figure 1). Sepals, dorsal petals, and lateral petals (or lateral lobes of the single petal in *Consolida* and *Aconitella*) are delineated by black, blue, and purple lines, respectively. The organs responsible for covering-protecting the nectar and the sexual organs are colored in light blue and purple, respectively. Nectar is shown in orange; it is secreted and concealed in the spur of the dorsal petal(s). Androecium and gynoecium are presented as a beige disk. The limb of the single petal has three lobes (two lateral lobes, in purple, and one upper lobe, in blue) in *Consolida*, and five (two lower lobes, in purple, and three lateral and upper lobes, in blue) in *Aconitella*. The expression level of each of the four *RanaCyL* paralogs is shown in the grids for *Aconitum carmichaelii* and *Consolida regalis*, in the following perianth compartments: petals (results were available only for *C. regalis*), dorsal (D), lateral (L), and ventral (V) sepals. Increasing expression level is shown with increasing levels of red. White: no expression detected. Grey: the paralog could not be amplified. The expression profiles are schematized based on Figure 3.

morphogenetic processes may control their elaboration. In each species, the paralog without a counterpart (*RanaCyL2* in *A. carmichaelii* and *RanaCyL1* in *C. regalis*) exhibits an overall homogeneous expression amongst perianth organs, while differential expression patterns were observed for the other paralogs. In *C. regalis*, the two *RanaCyL2* paralogs have undergone complementary subfunctionalization, with a predominantly dorsal expression pattern for *RanaCyL2*b and a predominantly ventral expression pattern for *RanaCyL2*a. By contrast, in *A. carmichaelii* both *RanaCyL1* paralogs are overexpressed in the ventral sepals. Up to now, a primarily ventral expression of *Cyc*-like genes has been reported only in the zygomorphic ray flowers of Asteraceae [19], and in monocots in the ventral tepals of *Commelina* species [23] and in the ventral sepal, petals, and staminodial labellum of *Costus* [16]. Our results suggest that changes in the regulation of *RanaCyL* paralogs took place independently in the *Aconitum* and *Consolida* lineages. In the Papaveraceae, both *PapCyL* paralogs exhibit a late asymmetric expression along the transverse symmetry plane in zygomorphic Fumarioideae [27]. These results suggest a large flexibility and evolvability in the regulation of *Cyc*-like genes among the Ranunculales.

Conclusion

Our study suggests that within Ranunculaceae, additional *Cyc*-like gene duplications coincided with the divergence of Delphinieae and the evolution of zygomorphy in Ranunculaceae. The set of paralogs differed between genera and species in this tribe. In addition, expression patterns of *RanaCyL1* genes in floral organs varied between the two species of Delphinieae examined here, with differences correlating with their specific perianth architecture. Both observations raise the question of the role of *Cyc*-like genes in the control of floral architecture in Delphinieae, and the possibility that different evolutionary forces act on the paralogous lineages at the generic, subgeneric, or even subgroup level. This may account for the lack of a detectable selection signature on either *RanaCyL1*a/b or *RanaCyL2*a/b stem branches. To test the correlation between expression patterns and perianth type, a larger species sampling is required, with a more precise monitoring of expression levels over developmental stages. Additionally, our full understanding of the evolution of perianth architecture and *RanaCyL* expression in Delphinieae will also depend on our capacity to resolve the phylogenetic relationships within the Ranunculaceae, and especially to identify the sister group of Delphinieae.

Supporting Information

Figure S1 Unrooted phylogenetic trees of *RanaCyL* sequences used for the tests of molecular evolution. (A) *RanaCyL1* (44 sequences and 381 positions without gaps). (B) *RanaCyL2* (49 sequences and 357 positions without gaps). The backbones were obtained using PhyML v 3.0 and nodes with support below 0.60 (SH test) were collapsed. The branches tested are 1a and 1b in *RanaCyL1*, and 2a and 2b in *RanaCyL2*.

Table S1   List of species, voucher, and GenBank accessions for the sequences generated for the present study. Group: tribe or subfamily within the family Ranunculaceae. Subgroup: genus or subgenus (or subgroup within the genus). Asterisks point to the set of species used for extensive characterization of *Cyc*-like genes. G: plant grown by the authors, H: herbarium specimen, B: plant grown in a botanical garden, W: plant collected in the wild. M and MSB: herbaria codes, Munich, Germany. The De.gr2b sequence was too short to be submitted to GenBank (177 bp). Its nucleotide sequence is: AGACTCTCACTCGAGATCGCTCGTAAGTTCTTTAATCTTCAAGATATGCTTGGGTACGATAAGGCGAGTAAGACGGTCGAGTGGTTGCTGAGGAAGTCAAAGGATGCAATAAATGAGCTCAGCAAAGGGTCCTGTGGTGAGAATAAGAGTGCATCTTCTATTACTGACTGTGATGTG.

Table S2   List of primers. a = degenerate primers used to characterize *Cyc*-like genes, b = specific primers used for semi-quantitative PCR analysis. F: forward, R: reverse.

Acknowledgments

We thank E. Kramer for sharing the EST database relative to the 4X genome coverage of *Aquilegia formosa x pubescens*, and H. Citerne for fruitful comments on an earlier version of this manuscript. We also thank two anonymous reviewers for their suggestions and comments.

Author Contributions

Conceived and designed the experiments: FJ CD. Performed the experiments: FJ GC MG JS. Analyzed the data: FJ CD. Contributed reagents/materials/analysis tools: FJ SN CD. Wrote the paper: FJ CD.

References

1. Endress PK (1999) Symmetry in flowers: Diversity and evolution. Int J Plant Sci 160: S3–S23.
2. Rudall PJ, Bateman RM (2004) Evolution of zygomorphy in monocot flowers: iterative patterns and developmental constraints. New Phytol 162: 25–44.
3. Sargent RD (2004) Floral symmetry affects speciation rates in angiosperms. Proc R Soc London, B 271: 603–608.
4. Citerne H, Jabbour F, Nadot S, Damerval C (2010) The evolution of floral symmetry. In: Kader JC, Delseny M, editors. Advances in Botanical Research.Vol. 54 pp. 85–137.
5. Damerval C, Nadot S (2007) Evolution of perianth and stamen characteristics with respect to floral symmetry in Ranunculales. Ann Bot London 100: 631–640.
6. Jabbour F, Ronse De Craene LP, Nadot S, Damerval C (2009) Establishment of zygomorphy on an ontogenic spiral and evolution of perianth in the tribe Delphinieae (Ranunculaceae). Ann Bot London 104: 809–822.
7. Jabbour F, Renner SS (2012) Spurs in a spur: Perianth evolution in the Delphinieae (Ranunculaceae). Int J Plant Sci 173: 1036–1054.
8. Kosuge K, Tamura K (1989) Ontogenetic studies on petals of the Ranunculaceae. J Jpn Bot 64: 65–74.
9. Erbar C, Kusma S, Leins P (1998) Development and interpretation of nectary organs in Ranunculaceae. Flora 194: 317–332.
10. Whittall JB, Hodges SA (2007) Pollinator shifts drive increasingly long nectar spurs in columbine flowers. Nature (London) 447: 706–709.
11. Luo D, Carpenter R, Vincent C, Copsey L, Coen E (1996) Origin of floral asymmetry in *Antirrhinum*. Nature (London) 383: 794–799.
12. Luo D, Carpenter R, Copsey L, Vincent C, Clark J, et al. (1999) Control of organ asymmetry in flowers of *Antirrhinum*. Cell (Cambridge) 99: 367–376.
13. Cubas P, Lauter N, Doebley J, Coen E (1999) The TCP domain: a motif found in proteins regulating plant growth and development. Plant J 18: 215–222.
14. Navaud O, Dabos P, Carnus E, Tremousaygue D, Hervé C (2007) TCP transcription factors predate the emergence of land plants. J Mol Evol 65: 23–33.
15. Howarth DG, Donoghue MJ (2006) Phylogenetic analysis of the "ECE" (CYC/TB1) clade reveals duplications predating the core eudicots. Proc Natl Acad Sci USA 103: 9101–9106.
16. Bartlett ME, Specht CD (2011) Changes in expression pattern of the *TEOSINTE BRANCHED1*-like genes in the Zingiberales provide a mechanism for evolutionary shifts in symmetry across the order. Am J Bot 98: 227–243.
17. Mondragon-Palomino M, Trontin C (2011) High time for a roll call: gene duplication and phylogenetic relationships of TCP-like genes in monocots. Ann Bot London 107: 1533–1544.
18. Kim M, Cui ML, Cubas P, Gillies A, Lee K, et al. (2008) Regulatory genes control a key morphological and ecological trait transferred between species. Science 322: 1116–1119.
19. Broholm SK, Tähtiharju S, Laitinen RA, Albert VA, Teeri TH, et al. (2008) A TCP domain transcription factor controls flower type specification along the radial axis of the *Gerbera* (Asteraceae) inflorescence. Proc Natl Acad Sci USA 105: 9117–9122.
20. Busch A, Zachgo S (2007) Control of corolla monosymmetry in the Brassicaceae *Iberis amara*. Proc Natl Acad Sci USA 104: 16714–16719.
21. Feng X, Zhao Z, Tian Z, Xu S, Luo Y, et al. (2006) Control of petal shape and floral zygomorphy in *Lotus japonicus*. Proc Natl Acad Sci USA 103: 4970–4975.
22. Wang Z, Luo Y, Li X, Wang L, Xu S, et al. (2008) Genetic control of floral zygomorphy in pea (*Pisum sativum* L.). Proc Natl Acad Sci USA 105: 10414–10419.
23. Preston JC, Hileman LC (2012) Parallel evolution of TCP and B-class genes in Commelinaceae flower bilateral symmetry. EvoDevo 3: 6.
24. Yuan Z, Gao S, Xue DW, Luo D, Li LT, et al. (2009) *RETARDED PALEA1* controls palea development and floral zygomorphy in rice. Plant Physiol 149: 235–244.
25. Citerne H, Le Guilloux M, Sannier J, Nadot S, Damerval C (2013) Combining phylogenetic and syntenic analyses for understanding the evolution of TCP ECE genes in eudicots. PLoS One 8: e74803.
26. Damerval C, Le Guilloux M, Jager M, Charon C (2007) Diversity and evolution of *CYCLOIDEA*-Like TCP genes in relation to flower development in Papaveraceae. Plant Physiol 143: 759–772.
27. Damerval C, Citerne H, Le Guilloux M, Domenichini S, Dutheil J, et al. (2013) Asymmetric morphogenetic cues along the transverse plane: shift from disymmetry to zygomorphy in the flower of Fumarioideae. Am J Bot 100: 391–402.
28. Jabbour F, Renner SS (2011) Resurrection of the genus *Staphisagria* J. Hill, sister to all the other Delphinieae (Ranunculaceae). Phytokeys 7: 21–26.
29. Wang W, Liu Y, Yu SX, Gao TG, Chen ZD (2013) *Gymnaconitum*, a new genus of Ranunculaceae endemic to the Qinghai-Tibetan Plateau. Taxon 64: 713–722.
30. Jabbour F, Renner SS (2012) A phylogeny of Delphinieae (Ranunculaceae) shows that *Aconitum* is nested within *Delphinium* and that Late Miocene transitions to long life cycles in the Himalayas and Southwest China coincide with bursts in diversification. Mol Phyl Evol 62: 928–942.
31. Katoh K, Standley DM (2013) MAFFT multiple sequence alignment software version 7: Improvements in performance and usability. Mol Biol Evol 30: 772–780.
32. Stamatakis A (2006) RAxML-VI-HPC: maximum likelihood-based phylogenetic analyses with thousands of taxa and mixed models. Bioinformatics 22: 2688–2690.
33. Michalak S (2012) RaxmlGUI: a graphical front-end for RAxML. Org Divers Evol 12: 335–337.
34. Huelsenbeck JP, Ronquist F (2001) MrBAYES: Bayesian inference of phylogeny. Bioinformatics 17: 754–755.
35. Guindon S, Dufayard JF, Lefort V, Anisimova M, Hordijk W, et al. (2010) New algorithms and methods to estimate Maximum-Likelihood phylogenies: assessing the performance of PhyML 3.0 Syst Biol 59: 307–21.
36. Yang Z (2007) PAML 4: A program package for phylogenetic analysis by maximum likelihood. Mol Biol Evol 24: 1586–1591.
37. Zhang J, Nielsen R, Yang Z (2005) Evaluation of an improved Branch-Site Likelihood method for detecting positive selection at the molecular level. Mol Biol Evol 22: 2472–2479.
38. Beentje H (2010) The Kew Plant Glossary: An Illustrated Dictionary of Plant Terms (with illustrations by Juliet Williamson). Royal Botanic Gardens Kew, UK. 160 p.
39. Tucker SC, Hodges SA (2005) Floral ontogeny of *Aquilegia*, *Semiaquilegia*, and *Enemion* (Ranunculaceae). Int J Plant Sci 166: 557–574.
40. Kölsch A, Gleissberg S (2006) Diversification of *CYCLOIDEA*-like TCP genes in the basal eudicot families Fumariaceae and Papaveraceae s. str. Plant Biol 8: 680–687.
41. Howarth DG, Donoghue MJ (2005) Duplications in *CYC*-like genes from Dipsacales correlate with floral form. Int J Plant Sci 166: 357–370.
42. Zhang WH, Kramer EM, Davis CC (2010) Floral symmetry genes and the origin and maintenance of zygomorphy in a plant-pollinator mutualism. Proc Natl Acad Sci USA 107: 6388–6393.
43. Carlson SE, Howarth DG, Donoghue MJ (2011) Diversification of *CYCLOIDEA*-like genes in Dipsacaceae (Dipsacales): implications for the evolution of capitulum inflorescences. BMC Evol Biol 11: 325.
44. Tähtiharju S, Rijpkema AS, Vetterli A, Albert VA, Teeri TH, et al. (2012) Evolution and diversification of the *CYC/TB1* gene family in Asteraceae - a comparative study in *Gerbera* (Mutisieae) and Sunflower (Heliantheae). Mol Biol Evol 29: 1155–1166.
45. Force A, Lynch M, Pickett FB, Amores A, Yan YL, et al. (1999) Preservation of duplicate genes by complementary, degenerative mutations. Genetics 151: 1531–1545.

46. Ohno S (1970) Evolution by gene duplication. New-York: Springer-Verlag. 160 p.

47. Innan H, Kondrashov F (2010) The evolution of gene duplications: classifying and distinguishing between models. Nature Rev Genet 11: 97–108.

48. Chapman MA, Leebens-Mack JH, Burke JM (2008) Positive selection and expression divergence following gene duplication in the sunflower *CYCLOIDEA* gene family. Mol Biol Evol 25: 1260–1273.

49. Ree RH, Citerne HL, Lavin M, Cronk QCB (2004) Heterogeneous selection on *LEGCYC* paralogs in relation to flower morphology and the phylogeny of *Lupinus* (Leguminosae). Mol Biol Evol 21: 321–331.

50. Hileman LC, Baum DA (2003) Why do paralogs persist? Molecular evolution of *CYCLOIDEA* and related floral symmetry genes in Antirrhineae (Veronicaceae). Mol Biol Evol 20: 591–600.

51. Kosugi S, Ohashi Y (2002) DNA binding and dimerization specificity and potential targets for the TCP protein family. Plant J 30: 337–348.

52. Martin-Trillo M, Cubas P (2010) TCP genes: a family snapshot ten years later. Trends Plant Sci 15: 31–39.

53. Howarth DG, Martins T, Chimney E, Donoghue MJ (2011) Diversification of *CYCLOIDEA* expression in the evolution of bilateral flower symmetry in Caprifoliaceae and *Lonicera* (Dipsacales). Ann Bot London 107: 1521–1532.

54. Cubas P, Coen E, Zapater JMM (2001) Ancient asymmetries in the evolution of flowers. Curr Biol 11: 1050–1052.

55. Cubas P, Vincent C, Coen E (1999) An epigenetic mutation responsible for natural variation in floral symmetry. Nature (London) 401: 157–161.

56. Citerne HL, Pennington RT, Cronk QCB (2006) An apparent reversal in floral symmetry in the legume *Cadia* is a homeotic transformation. Proc Natl Acad Sci USA 103: 12017–12020.

57. Zhou XR, Wang YZ, Smith JF, Chen RJ (2008) Altered expression patterns of TCP and MYB genes relating to the floral developmental transition from initial zygomorphy to actinomorphy in *Bournea* (Gesneriaceae). New Phytol 178: 532–543.

58. Reardon W, Fitzpatrick DA, Fares MA, Nugent JM (2009) Evolution of flower shape in *Plantago lanceolata*. Plant Mol Biol 71: 241–250.

59. Pang HB, Sun QW, He SZ, Wang YZ (2010) Expression pattern of *CYC*-like genes relating to a dorsalized actinomorphic flower in *Tengia* (Gesneriaceae). J Syst Evol 48: 309–317.

60. Preston JC, Martinez CC, Hileman LC (2011) Gradual disintegration of the floral symmetry gene network is implicated in the evolution of a wind-pollination syndrome. Proc Natl Acad Sci USA 108: 2343–2348.

61. Zhang W, Steinmann VW, Nikolov L, Kramer EM, Davis CC (2013) Divergent genetic mechanisms underlie reversals to radial floral symmetry from diverse zygomorphic flowered ancestors. Front Plant Sci 4: 1–13.

62. Fukuda Y, Suzuki K, Murata J (2001) The function of each sepal in pollinator behavior and effective pollination in *Aconitum japonicum* var. *montanum*. Plant Spec Biol 16: 151–157.

63. Busch A, Horn S, Muhlhausen A, Mummenhoff K, Zachgo S (2012) Corolla monosymmetry: evolution of a morphological novelty in the Brassicaceae family. Mol Biol Evol 29: 1241–1254.

64. Zhang WH, Kramer EM, Davis CC (2012) Similar genetic mechanisms underlie the parallel evolution of floral phenotypes. PLoS One 7: e36033.

65. Hileman LC (2014) Bilateral flower symmetry – how, when and why? Curr Opin Plant Biol 17: 146–152.

Permissions

All chapters in this book were first published in PLOS ONE, by The Public Library of Science; hereby published with permission under the Creative Commons Attribution License or equivalent. Every chapter published in this book has been scrutinized by our experts. Their significance has been extensively debated. The topics covered herein carry significant findings which will fuel the growth of the discipline. They may even be implemented as practical applications or may be referred to as a beginning point for another development.

The contributors of this book come from diverse backgrounds, making this book a truly international effort. This book will bring forth new frontiers with its revolutionizing research information and detailed analysis of the nascent developments around the world.

We would like to thank all the contributing authors for lending their expertise to make the book truly unique. They have played a crucial role in the development of this book. Without their invaluable contributions this book wouldn't have been possible. They have made vital efforts to compile up to date information on the varied aspects of this subject to make this book a valuable addition to the collection of many professionals and students.

This book was conceptualized with the vision of imparting up-to-date information and advanced data in this field. To ensure the same, a matchless editorial board was set up. Every individual on the board went through rigorous rounds of assessment to prove their worth. After which they invested a large part of their time researching and compiling the most relevant data for our readers.

The editorial board has been involved in producing this book since its inception. They have spent rigorous hours researching and exploring the diverse topics which have resulted in the successful publishing of this book. They have passed on their knowledge of decades through this book. To expedite this challenging task, the publisher supported the team at every step. A small team of assistant editors was also appointed to further simplify the editing procedure and attain best results for the readers.

Apart from the editorial board, the designing team has also invested a significant amount of their time in understanding the subject and creating the most relevant covers. They scrutinized every image to scout for the most suitable representation of the subject and create an appropriate cover for the book.

The publishing team has been an ardent support to the editorial, designing and production team. Their endless efforts to recruit the best for this project, has resulted in the accomplishment of this book. They are a veteran in the field of academics and their pool of knowledge is as vast as their experience in printing. Their expertise and guidance has proved useful at every step. Their uncompromising quality standards have made this book an exceptional effort. Their encouragement from time to time has been an inspiration for everyone.

The publisher and the editorial board hope that this book will prove to be a valuable piece of knowledge for researchers, students, practitioners and scholars across the globe.

List of Contributors

David W. G. Stanton, Michael W. Bruford
School of Biosciences, Cardiff University, Cardiff, United Kingdom

John Hart
Lukuru Foundation, Projet Tshuapa-Lomami-Lualaba (TL2), Kinshasa, Democratic Republic of Congo

Peter Galbusera, Philippe Helsen, Jill Shephard
Centre for Research and Conservation, Royal Zoological Society of Antwerp, Antwerp, Belgium

Noëlle F. Kümpel
Conservation Programmes, Zoological Society of London, London, United Kingdom

Jinliang Wang, John G. Ewen
Institute of Zoology, Zoological Society of London, London, United Kingdom

Naoki Morimoto
Laboratory of Physical Anthropology, Graduate School of Science, Kyoto University, Kyoto, Japan

Marcia S. Ponce de León, Christoph P. E. Zollikofer
Anthropological Institute, University of Zurich, Zurich, Switzerland

Sofia Stathopoulos, Colleen O'Ryan
Department of Molecular and Cell Biology, University of Cape Town, Cape Town, Western Cape, South Africa

Jacqueline M. Bishop
Department of Biological Sciences, University of Cape Town, Cape Town, Western Cape, South Africa

Yaron Mosesson, Yoav Voichek, Naama Barkai
Department of Molecular Genetics, Weizmann Institute of Science, Rehovot, Israel

Scott T. Bickel
Department of Life Sciences, Scottsdale Community College, Scottsdale, Arizona, United States of America

Joseph D. Juliano
Department of Chemistry and Biochemistry, Arizona State University, Tempe, Arizona, United States of America

John D. Nagy
School of Mathematical and Statistical Sciences, Arizona State University, Tempe, Arizona, United States of America
Department of Life Sciences, Scottsdale Community College, Scottsdale, Arizona, United States of America

Adrien Perrard
UMR 7205 ISYEB, Muséum National d'Histoire Naturelle, Paris, France
Division of Invertebrate Zoology, American Museum of Natural History, New York, New York, United States of America

Mariangela Arca, Jean-François Silvain
Université Paris-Sud 11

Quentin Rome, Franck Muller, Michel Baylac, Claire Villemant
UMR 7205 ISYEB, Muséum National d'Histoire Naturelle, Paris, France

Jiangli Tan
School of Life Sciences, Northwest University, Xi'an, Shaanxi, China

Sanjaya Bista
Entomology Division, Nepal Agricultural Research Council (NARC), Lalitpur, Nepal

Hari Nugroho
Museum Zoologicum Bogoriense, Indonesian Institute of Sciences, Cibinong, Bogor, Indonesia

Raymond Baudoin
CBNBP, Muse´um National d'Histoire Naturelle, Paris, France

James M. Carpenter
Division of Invertebrate Zoology, American Museum of Natural History, New York, New York, United States of America

Teemu Smura
Department of Infectious Disease Surveillance and Control, National Institute for Health and Welfare (THL), Helsinki, Finland
Department of Virology, Haartman Institute, Faculty of Medicine, University of Helsinki, Helsinki, Finland

Soile Blomqvist, Haider Al-Hello, Carita Savolainen-Kopra, Tapani Hovi, Merja Roivainen
Department of Infectious Disease Surveillance and Control, National Institute for Health and Welfare (THL), Helsinki, Finland

Tytti Vuorinen
Department of Virology, University of Turku, Turku, Finland

Olga Ivanova
M.P. Chumakov Institute of Poliomyelitis and Viral Encephalitides, Russian Academy of Medical Sciences, Moscow, Russia

Elena Samoilovich
Republican Research and Practical Center for Epidemiology and Microbiology, Minsk, Republic of Belarus

Stefan Elfwing, Kenji Doya
Neural Computation Unit, Okinawa Institute of Science and Technology Graduate University, Okinawa, Japan

Wei Li, Wei Liu, Qiuling He, Jinhong Chen, Shuijin Zhu
Department of Agronomy, Zhejiang University, Hangzhou, Zhejiang, China

Hengling Wei
State Key Laboratory of Cotton Biology, Cotton Research Institute, Chinese Academy of Agricultural Sciences, Anyang, Henan, China

Baohong Zhang
Department of Biology, East Carolina University, Greenville, North Carolina, United States of America

Vasco Koch, Inga Nissen, Björn D. Schmitt, Martin Beye
Institute of Evolutionary Genetics, Heinrich Heine University Duesseldorf, Duesseldorf, Germany

Karolin Zerulla
Institute for Molecular Biosciences, Biocentre, Goethe-University, Frankfurt, Germany

Scott Chimileski
Department of Molecular and Cell Biology, University of Connecticut, Storrs, Connecticut, United States of America

Daniela Nather
Institute for Molecular Biosciences, Biocentre, Goethe-University, Frankfurt, Germany

Uri Gophna
Department of Molecular Microbiology and Biotechnology, George S. Wise Faculty of Life Sciences, Tel Aviv University, Ramat Aviv, Tel Aviv, Israel

R. Thane Papke
Institute for Molecular Biosciences, Biocentre, Goethe-University, Frankfurt, Germany

Peter C. Zee, James D. Bever
Department of Biology, Indiana University, Bloomington, Indiana, United States of America

Xiao-Yang Gao
Key Laboratory of Biogeography and Bioresource in Arid Land, Xinjiang Institute of Ecology and Geography, Chinese Academy of Sciences, Urumqi, China
University of Chinese Academy of Sciences, Beijing, China

Xiao-Yang Zhi
Key Laboratory of Microbial Diversity in Southwest China, Ministry of Education and the Laboratory for Conservation and Utilization of Bio-Resources, Yunnan Institute of Microbiology, Yunnan University, Kunming, China

Hong-Wei Li
Key Laboratory of Microbial Diversity in Southwest China, Ministry of Education and the Laboratory for Conservation and Utilization of Bio-Resources, Yunnan Institute of Microbiology, Yunnan University, Kunming, China
The First Hospital of Qujing City, Qujing Affiliated Hospital of Kunming Medical University, Qujing, China

Hans-Peter Klenk
Leibniz-Institute DSMZ-German Collection of Microorganisms and Cell Cultures, Braunschweig, Germany

Wen-Jun Li
Key Laboratory of Microbial Diversity in Southwest China, Ministry of Education and the Laboratory for Conservation and Utilization of Bio-Resources, Yunnan Institute of Microbiology, Yunnan University, Kunming, China

Tim Phillips
Independent Researcher, Birmingham, United Kingdom

Jiawei Li
Automated Scheduling, Optimisation and Planning (ASAP) research group, School of Computer Science, University of Nottingham, Nottingham, United Kingdom

Graham Kendall
University of Nottingham Malaysia Campus, Broga, Malaysia
Automated Scheduling, Optimisation and Planning (ASAP) research group, School of Computer Science, University of Nottingham, Nottingham, United Kingdom

Emile van Lieshout, Kathryn B. McNamara, Leigh W. Simmons
Centre for Evolutionary Biology, School of Animal Biology (M092), University of Western Australia, Crawley, Australia

Jill C. Preston, Stacy A. Jorgensen, Suryatapa G. Jha
Department of Plant Biology, The University of Vermont, Burlington, Vermont, United States of America

Florian Jabbour
Université Paris-Sud, UMR 0320/UMR 8120, Génétique Végétale, Gif-sur-Yvette, France
Université Paris-Sud, Laboratoire Ecologie, Systématique, Evolution, CNRS UMR 8079, AgroParisTech, Orsay, France
Systematic Botany and Mycology, University of Munich (LMU), Munich, Germany
Muséum National d'Histoire Naturelle, Institut de Systématique, Evolution, Biodiversité, UMR 7205 ISYEB MNHN-CNRS-UPMC-EPHE, Paris, France

Guillaume Cossard
Université Paris-Sud, Laboratoire Ecologie, Systématique, Evolution, CNRS UMR8079, AgroParisTech, Orsay, France
Department of Ecology and Evolution, Biophore, University of Lausanne, Lausanne, Switzerland
Martine Le Guilloux, Catherine Damerval
CNRS, UMR 0320/UMR 8120, Génétique Végétale, Gif-sur-Yvette, France

Julie Sannier, Sophie Nadot
Université Paris-Sud, Laboratoire Ecologie, Systématique, Evolution, CNRS UMR 8079, Agro Paris Tech, Orsay, France

Index

www.ingramcontent.com/pod-product-compliance
Lightning Source LLC
Chambersburg PA
CBHW061248190326
41458CB00011B/3616